VOLUME SEVENTY-FIVE

ADVANCES IN CARBOHYDRATE CHEMISTRY AND BIOCHEMISTRY

Sialic Acids, Part I: Historical Background and Development, and Chemical Synthesis

VOLUME SEVENTY-FIVE

ADVANCES IN CARBOHYDRATE CHEMISTRY AND BIOCHEMISTRY

Sialic Acids, Part I: Historical Background and Development, and Chemical Synthesis

Editor

DAVID C. BAKER
University of Tennessee,
Knoxville, TN

ELSEVIER

ACADEMIC PRESS
An imprint of Elsevier

Academic Press is an imprint of Elsevier
50 Hampshire Street, 5th Floor, Cambridge, MA 02139, United States
525 B Street, Suite 1650, San Diego, CA 92101, United States
The Boulevard, Langford Lane, Kidlington, Oxford OX5 1GB, United Kingdom
125 London Wall, London, EC2Y 5AS, United Kingdom

First edition 2018

ISBN: 978-0-12-815202-7
ISSN: 0065-2318

Publisher: Zoe Kruze
Acquisition Editor: Jason Mitchell
Editorial Project Manager: Peter Llewellyn
Production Project Manager: Vignesh Tamil
Cover Designer: Miles Hitchen

Typeset by SPi Global, India

CONTENTS

CONTRIBUTORS

Cristina De Meo
Department of Chemistry, Southern Illinois University—Edwardsville, Edwardsville, IL, United States

Bradley T. Jones
Department of Chemistry, Southern Illinois University—Edwardsville, Edwardsville, IL, United States

Johannis P. Kamerling
Bijvoet Center for Biomolecular Research, Utrecht University, Utrecht, The Netherlands

Roland Schauer
Biochemisches Institut, Christian-Albrechts-Universität zu Kiel, Kiel, Germany

PREFACE

This volume of *Advances in Carbohydrate Chemistry and Biochemistry* (*ACCB*) marks a departure from the usual content of the series in that it is focused on a single topic, "Sialic Acids." The idea for a volume on sialic acids sprang from an e-mail from Johannis P. (Hans) Kamerling, who proposed that he and Roland Schauer contribute a chapter that would historically trace the development of the field, especially the structural aspects of the sialic acids, which have undergone a most interesting metamorphosis over the decades. After some discussion, and encouraged by the continuing popularity of R. Schauer's *ACCB* chapter of some 36 years ago,[1] the editor asked for their assistance in formulating a broader coverage—a collection of chapters that would address, in depth, specific topics and applications within the field. Thus with their help in identifying topics and potential authors, a two-volume series has been assembled: Vol. 75, *Sialic Acids, Part I: Historical Background and Development, and Chemical Synthesis*, and Vol. 76, *Sialic Acids, Part II: Biological and Biomedical Aspects*.

Chapter 1, "Exploration of the Sialic Acid World," by Roland Schauer, of the Biochemisches Institut, Christian-Albrechts-Universität zu Kiel, Kiel, Germany, and Johannis P. (Hans) Kamerling, of the Bijvoet Center for Biomolecular Research, Utrecht University, Utrecht, the Netherlands, is a fascinating chapter that traces the sialic acid story from the earliest biological observations through the key developments in elucidating the structure (including stereochemical aspects, NMR spectroscopy, and X-ray studies) of these 9-carbon sugars. Topics extend to the biological/biochemical areas, including O- and N-glycans, glycolipids, glycoproteins, metabolism, and the role of sialic acids in various disease states. Written by two major players in the field, their first-hand observations make for interesting reading and provide the researcher with a comprehensive overview of the field. One eminent sialic acid researcher who reviewed the chapter stated, "… it is outstanding. This should be required reading for anyone who decides to study sialic acids!"

Chapter 2, "Chemical Synthesis of Glycosides of *N*-Acetylneuraminic Acid," by Cristina De Meo and Bradley T. Jones, of the Chemistry Department, Southern Illinois University—Edwardsville, Edwardsville, IL, USA, covers the development of modern chemical methods for the synthesis of an important class of sialic acid derivatives—sialic acid glycosides, with an emphasis on key strategies developed over the past decade. While most chemists concede that carbohydrate synthesis, in general, is challenging and difficult, those who have attempted these types of syntheses in the sialic acid field find the challenge beyond the pale, with the interplay of remote substituent groups and choice of C-2 leaving groups affecting the outcomes, sometimes in ways difficult to reconcile. Topics developed in the chapter include a discussion of substituent groups at key carbon atoms and their effects on the sialylation reaction outcomes, as well as the elegant mechanistic studies that have been carried out in an effort to put scientific rigor into predicting reactivities and products. While the authors concede that all answers to problems of sialylation have not been realized, this in-depth review of methodology certainly will serve to inspire those who choose to do synthetic chemistry in this field.

Volume 76, *Sialic Acids, Part II: Biological and Biomedical Aspects*, is scheduled to follow shortly after publication of the present volume.

David C. Baker
Knoxville, TN, USA
October 2018

REFERENCE

1. Schauer, R. Chemistry, Metabolism, and Biological Functions of Sialic Acids. *Adv. Carbohydr. Chem. Biochem.* **1982**, *40*, 131–234.

CHAPTER ONE

Exploration of the Sialic Acid World

Roland Schauer[*,1], **Johannis P. Kamerling**[†,1]
[*]Biochemisches Institut, Christian-Albrechts-Universität zu Kiel, Kiel, Germany
[†]Bijvoet Center for Biomolecular Research, Utrecht University, Utrecht, The Netherlands
[1]Corresponding authors: e-mail address: schauer@biochem.uni-kiel.de; j.p.kamerling@uu.nl

Dedicated to Prof. Dr. Hans-Dieter Klenk on the occasion of his 80th birthday.

Contents

Advances in Carbohydrate Chemistry and Biochemistry, Volume 75
ISSN 0065-2318
https://doi.org/10.1016/bs.accb.2018.09.001

ABBREVIATIONS

Aci	acinetaminic acid (for abbreviations of Aci variants, see Table 4)
ALL	acute lymphoblastic leukemia
Asn	asparagine
ATP	adenosine triphosphate
BHK	baby hamster kidney
CDP	cytidine diphosphate
CE	capillary electrophoresis
CIMS	chemical-ionization mass spectrometry
CMAH	CMP-Neu5Ac hydroxylase
CMP	cytidine monophosphate
CTP	cytidine triphosphate
DIFO	difluorinated cyclooctyne
DMB	1,2-diamino-4,5-methylenedioxybenzene

DTH	delayed-type hypersensitivity
EIMS	electron-impact mass spectrometry
EPR	electron paramagnetic resonance
ESIMS	electrospray ionization mass spectrometry
Fab	fragment antigen-binding (region on antibody)
FABMS	fast-atom bombardment mass spectrometry
Fc	fragment crystallizable (region on antibody)
FLAG	Asp-Tyr-Lys-Asp-Asp-Asp-Asp-Lys octapeptide
Fuc	fucose
Gal	galactose
GalNAc	*N*-acetylgalactosamine
Glc	glucose
GLC	gas–liquid chromatography
GlcNAc	*N*-acetylglucosamine
GlcNGc	*N*-glycolylglucosamine
GNE	UDP-GlcNAc 2-epimerase
HD	Hanganutziu–Deicher
HE	hemagglutinin-esterase
HEF	hemagglutinin-esterase-fusion
HEV	high endothelial venules
HFB	heptafluorobutyryl
HIBM	hereditary inclusion body myopathy
HPAEC–PAD	high-performance anion-exchange chromatography with pulsed amperometric detection
HPLC	high-performance liquid chromatography
HSEA	hard-sphere exo-anomeric
IR	infrared
ISAV	Infectious Salmon Anemia Virus
ISSD	Infantile Sialic Acid Storage Disease
Kdn	ketodeoxynononic acid
LAD II	leukocyte adhesion deficiency type II syndrome
Leg	legionaminic acid (for abbreviations of derivatives and variants of Leg, see Table 4)
LFA	*Limax flavus* agglutinin
MAA	*Maackia amurensis* agglutinin
MAG	myelin-associated glycoprotein
MALDI-TOFMS	matrix-assisted laser-desorption ionization time-of-flight mass spectrometry
Man	mannose
ManNAc	*N*-acetylmannosamine
ManNAz	*N*-azidoacetylmannosamine
ManNGc	*N*-glycolylmannosamine
MD	molecular dynamics
MERS-CoV	Middle East Respiratory Syndrome Coronavirus
MM	molecular mechanics
MNK	ManNAc kinase
MS	mass spectrometry

NADH/NAD$^+$	nicotinamide adenine dinucleotide (reduced and oxidized form)
NADPH	nicotinamide adenine dinucleotide phosphate
Neu	neuraminic acid [for abbreviations of derivatives (sialic acids), see Table 1]
NMR	nuclear magnetic resonance
NOE	nuclear Overhauser effect
NulO	non-2-ulosonic acid
OPD	*o*-phenylenediamine
PAPS	peroxidase–antibody–peroxidase smear
PAS	periodic acid–Schiff
PPCA	protective protein cathepsin A
Pse	pseudaminic acid (for abbreviations of derivatives of Pse, see Table 4)
QM	quantum mechanical
RDE	receptor-destroying enzyme (sialidase)
ROS	reactive oxygen species
SDS-PAGE	sodium dodecyl sulfate–polyacrylamide gel electrophoresis
Ser	serine
Sia	sialic acid
SNA	*Sambucus nigra* agglutinin
SOAT	sialate *O*-acetyltransferase
Thr	threonine
TLC	thin-layer chromatography
TS	*trans*-sialidase
UDP	uridine diphosphate
UTP	uridine triphosphate
UV	ultraviolet
WGA	wheat germ agglutinin
X-ray	Röntgen radiation

1. INTRODUCTION

Alfred Gottschalk, discussing the state of the glycoprotein research at the "C.N.R.S. Colloque Internationale sur les Glycoconjugués" (later named the 2nd International Symposium on Glycoconjugates), organized by Jean Montreuil in 1973 at the Université des Sciences et Technologies de Lille (Villeneuve d'Ascq, France), stated: "We are not at the end of all progress, but at the beginning; we have but reached the shores of a great unexplored continent."[1] And this certainly held for his favorite topic: sialic acid research. Gottschalk encouraged students interested in research to work in the field of carbohydrate (bio)chemistry/glycobiology, which rapidly expanded in the 1970s. Since then, as this review will show, many questions regarding the chemistry, analysis, biochemistry, and biology of sialic acids could be answered, and the important influences of this family of sugar molecules on and in our lives have been documented in a myriad of publications.

Sialic acids are frequently found in Nature in many molecular forms, and detailed reviews have been available since the 1960s.[2–7] These compounds are regularly present in higher animals like *Echinoderma*, *Hemichorda*, *Cephalocorda*, and *Vertebrata* and could also sporadically be identified in some members of other groups of animals, such as the Platyhelminthes (*Polychoerus carmelensis*), *Cephalopoda*, and *Crustaceae* (*Homarus americanus*). It is, however, not clear whether the sialic acids are produced by the animals themselves or originate from nutrition or microorganisms. Sialic acid concentrations have been measured in subcellular membrane fractions of various cell types, showing highest values in the plasma membrane (65%–70% of the total sialic acid quantity of the cell) and in much lower amounts in the smooth endoplasmic reticulum.[3] Generally, most of these sugars are bound to glycoproteins with the exception of nerve and brain cells, where they prevail on glycolipids. The overall sialic acid level in the various organelles was assumed to be influenced by endogenous sialidase, but no experimental evidence existed. Interestingly, also insects were shown to contain sialic acids and to express corresponding enzymes.[8–11] Sialic acids were found in some viruses, various bacteria, protozoa, and pathogenic fungi. They are absent in plants and in most animals in the evolutionary tree lower than the echinoderms. The various studies show that sialic acids appeared relatively late in evolution, which led to the assumption that acquisition of these monosaccharides accelerated differentiation of the higher animals.[2,4] Deaminated neuraminic acid is often seen in vertebrates and bacteria, but its abundant occurrence in animals is limited to the lower vertebrates.[12] Further details are included in the various sections in this chapter, when applicable.

In naturally occurring sialic acid-containing structures, the sialic acid units play important roles in many physiological and pathological processes via carbohydrate–protein interactions, including cellular recognition and communication, cellular aggregation and development, controlling the lifetimes of glycoconjugates in organisms, mediating bacterial and viral infections, being involved in tumor growth and metastasis, playing a great role in immunology, in the biology of the microbiome, in cell signaling, in reproduction biology, and in neurobiology. Sialic acids, which occur mostly as the terminal part of many glycoconjugates (glycoproteins; glycolipids) and carbohydrate chains (oligosaccharides, capsular and tissue polysialic acids, bacterial lipooligo/polysaccharides) occur in various chemical forms, which can dramatically influence their biology. *O*-Acetyl and *N*-glycolyl groups, in particular, are most effective in determining sialic acid function. Many genes and enzymes control the anabolism and catabolism of sialic acids, which are of importance not only in health but also in disease situations. Increasing information about the regulation of sialic acid expression on the genetic level of various hereditary diseases can be gathered. Sialopharmacy is still in its infancy, but recently more publications show progress in this field. But it was not too long ago that groups of scientists worldwide were trying to elucidate the primary structures and the basic biochemical reactions of these unusual monosaccharides.

Nowadays, the basic structural features of the sialic acid (Sia) family, a subclass of the superfamily of naturally occurring non-2-ulosonic acids (NulOs), are well known (Fig. 1). The mother molecule of the family, neuraminic acid (Neu), which does not occur in free form in nature due to its immediate cyclization to form an internal Schiff base, is a nine-carbon-containing monosaccharide, chemically defined as a combination of a 2-keto-carboxylic acid, a deoxysugar, and an aminosugar. It is systematically named 5-amino-3,5-dideoxy-D-*glycero*-D-*galacto*-non-2-ulosonic acid (Neu; $C_9H_{17}NO_8$) (C7 is the anomeric reference atom). The intramolecular (cyclic) hemiketal form of Neu (C2–O–C6) comprises a pyranose ring in a $^2C_5$ chair conformation with an equatorially oriented glycerol side chain at C6. The amino group at C5 is generally N-acetylated [5-acetamido-3,5-dideoxy-D-*glycero*-D-*galacto*-non-2-ulopyranosonic acid (IUPAC/IUBMB Recommendations 1996/1997[13]); *N*-acetylneuraminic acid, Neu5Ac; $C_{11}H_{19}NO_9$] (for different views, see Fig. 2) or N-glycolylated [5-hydroxyacetamido-3,5-dideoxy-D-*glycero*-D-*galacto*-non-2-ulopyranosonic acid (IUPAC/IUBMB Recommendations 1996/1997[13]); *N*-glycolylneuraminic acid, Neu5Gc; $C_{11}H_{19}NO_{10}$]. Note that the full name for Neu5Ac, as presented in the IUPAC/IUB document 1969/1971,[14] namely, 5-acetamido-3,5-dideoxy-D-*glycero*-D-*galacto*-2-nonulopyranosonic acid, is no longer used. The amino group at C5 can also be replaced by a hydroxyl function, yielding 3-deoxy-D-*glycero*-D-*galacto*-non-2-ulopyranosonic acid (ketodeoxynononic acid, Kdn; $C_9H_{16}O_9$) (IUPAC/IUBMB Recommendations 1996/1997[13]), usually called deaminated neuraminic acid (other names used: 2-keto-3-deoxy-nononic acid, 3-deoxynon-2-ulosonic acid, 2-keto-3-deoxy-D-*glycero*-D-*galacto*-nononic acid). To indicate the absolute configuration, up to the 1980s the trivial names of the mother molecule and the derived members were presented with the configurational prefix D, e.g., D-neuraminic acid, *N*-acetyl-D-neuraminic acid, *N*-glycolyl-D-neuraminic acid. According to the IUPAC/IUBMB Recommendations 1996/1997,[13] this prefix should not be used, as the D-*glycero*-D-*galacto* configuration is implied in the trivial names.

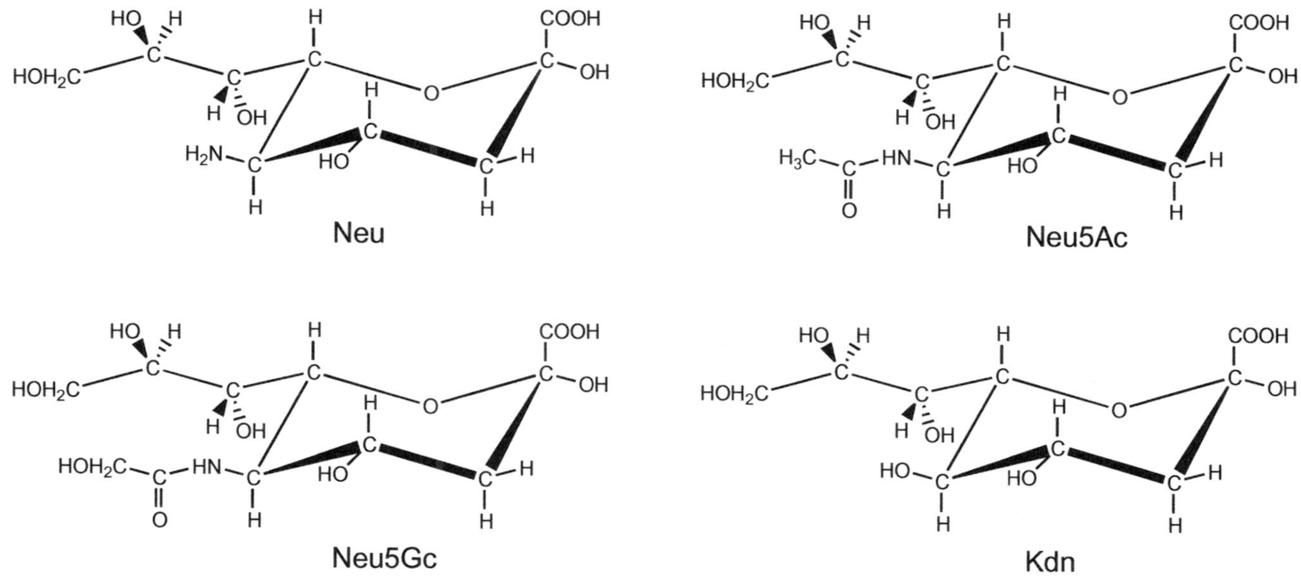

Fig. 1 The “mother” molecule of the family Sialic Acids (Neu), together with the three major “children” (Neu5Ac, Neu5Gc, Kdn), presented in the α-configuration, as occurring in sialic acid-containing carbohydrate chains.

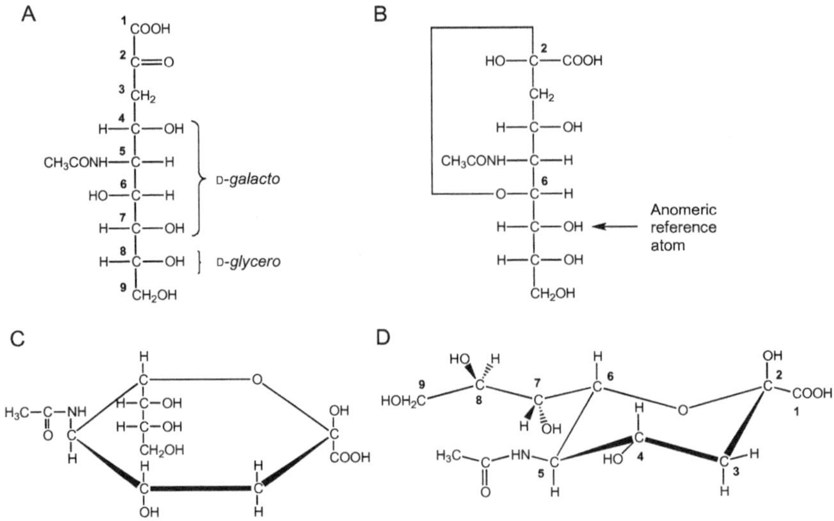

Fig. 2 Chemical structures of *N*-acetylneuraminic acid (5-acetamido-3,5-dideoxy-D-*glycero*-D-*galacto*-non-2-ulosonic acid, Neu5Ac) in different views according to IUPAC/IUBMB recommendations. (A) Neu5Ac in Fischer projection formula, open chain; (B) β-Neu5Ac/*N*-acetyl-β-neuraminic acid/5-acetamido-3,5-dideoxy-D-*glycero*-β-D-*galacto*-non-2-ulopyranosonic acid in Fischer projection formula, cyclic chain, pyranose ring; (C) β-Neu5Ac in Haworth representation, pyranose ring; and (D) β-Neu5Ac in $^2C_5$ chair conformation.

The abbreviations used for sialic acids have a long and controversial history, i.e., for *N*-acetylneuraminic acid: NANA, *N*-AN, NAN, NANS, NeuNAc, AcNeu, and Ac-Neu; for *N*-glycolylneuraminic acid: NGNA, *N*-GN, NGN, NGNS, NeuNGl, NeuNGc, GlNeu, and Gc-Neu; for ketodeoxynononic acid: KDN. At present, the IUPAC/IUBMB recommendations are Neu5Ac (or NeuAc), Neu5Gc (or NeuGc), and Kdn. For neuraminic acid, the abbreviation is Neu, whereas the recommended abbreviation for sialic acid is Sia. A still frequently used old abbreviation for 2-deoxy-2,3-didehydro-*N*-acetylneuraminic acid (Neu2en5Ac) is DANA.

Sialic acids are relatively strong acids, e.g., Neu5Ac has a pK_a value found in the range of 2.0–3.0 in various studies with an average of 2.6. This strong acidity is responsible for processes such as autohydrolysis of sialic acid-containing glycan chains, and for repulsive effects, leading, for example, to the high viscosity of mucins, and for the binding and transport of cationic substances.

In Table 1 a survey of the over 80 known naturally occurring members of the sialic acid family, together with their abbreviations and relevant

Table 1 Survey of Reported Structures of Naturally Occurring Members of the Sialic Acid Family[a]

Name	Abbreviation	References
Neuraminic acid[b]	Neu	15–20
Neuraminic acid 1,5-lactam[c]	Neu1,5lactam	20,21
5-*N*-Acetyl-neuraminic acid	Neu5Ac	6,19,22–33
5-*N*-Acetyl-4-*O*-acetyl-neuraminic acid	Neu4,5Ac_2	6,19,23,24,27,29–34
5-*N*-Acetyl-7-*O*-acetyl-neuraminic acid	Neu5,7Ac_2	6,19,23,24,29–33, 35–38,38a
5-*N*-Acetyl-8-*O*-acetyl-neuraminic acid	Neu5,8Ac_2	6,19,30,32,33,37
5-*N*-Acetyl-9-*O*-acetyl-neuraminic acid	Neu5,9Ac_2	6,19,27,29–33, 36–39
5-*N*-Acetyl-4,9-di-*O*-acetyl-neuraminic acid	Neu4,5,9Ac_3	6,19,27,30,33
5-*N*-Acetyl-7,8-di-*O*-acetyl-neuraminic acid[d]	Neu5,7,8Ac_3	35
5-*N*-Acetyl-7,9-di-*O*-acetyl-neuraminic acid	Neu5,7,9Ac_3	6,19,27,29–33,37, 38,38a,40
5-*N*-Acetyl-8,9-di-*O*-acetyl-neuraminic acid	Neu5,8,9Ac_3	6,19,29–33,37, 38,38a,40
5-*N*-Acetyl-4,7,9-tri-*O*-acetyl-neuraminic acid	Neu4,5,7,9Ac_4	19
5-*N*-Acetyl-7,8,9-tri-*O*-acetyl-neuraminic acid	Neu5,7,8,9Ac_4	6,30–33,37
5-*N*-Acetyl-4,7,8,9-tetra-*O*-acetyl-neuraminic acid	Neu4,5,7,8,9Ac_5	19
5-*N*-Acetyl-4-*O*-glycolyl-neuraminic acid	Neu5Ac4Gc	41
5-*N*-Acetyl-7-*O*-glycolyl-neuraminic acid	Neu5Ac7Gc	41
5-*N*-Acetyl-9-*O*-lactyl-neuraminic acid[e]	Neu5Ac9Lt	6,19,30,32,33,42,43
5-*N*-Acetyl-4-*O*-acetyl-9-*O*-lactyl-neuraminic acid	Neu4,5$Ac_2$9Lt	6,19,30,33,44
5-*N*-Acetyl-7-*O*-acetyl-9-*O*-lactyl-neuraminic acid	Neu5,7$Ac_2$9Lt	19

Continued

Table 1 Survey of Reported Structures of Naturally Occurring Members of the Sialic Acid Family—cont'd

Name	Abbreviation	References
5-*N*-Acetyl-8-*O*-acetyl-9-*O*-lactyl-neuraminic acid	Neu5,8$Ac_2$9Lt	45
5-*N*-Acetyl-8-*O*-methyl-neuraminic acid	Neu5Ac8Me	6,19,25,33,46
5-*N*-Acetyl-4-*O*-acetyl-8-*O*-methyl-neuraminic acid	Neu4,5$Ac_2$8Me	19
5-*N*-Acetyl-9-*O*-acetyl-8-*O*-methyl-neuraminic acid	Neu5,9$Ac_2$8Me	6,19,33,46
5-*N*-Acetyl-9-*O*-methyl-neuraminic acid	Neu5Ac9Me	47
5-*N*-Acetyl-4-*O*-sulfo-neuraminic acid	Neu5Ac4S	48
5-*N*-Acetyl-8-*O*-sulfo-neuraminic acid	Neu5Ac8S	19,32,49,50
5-*N*-Acetyl-4-*O*-acetyl-8-*O*-sulfo-neuraminic acid	Neu4,5$Ac_2$8S	19
5-*N*-Acetyl-9-*O*-phospho-neuraminic acid[f,g]	Neu5Ac9P	51
5-*N*-Acetyl-2-deoxy-2,3-didehydro-neuraminic acid[g,h]	Neu2en5Ac	6,30,33,43,52–54
5-*N*-Acetyl-9-*O*-acetyl-2-deoxy-2,3-didehydro-neuraminic acid[g]	Neu2en5,9Ac_2	55,56
5-*N*-Acetyl-2-deoxy-2,3-didehydro-9-*O*-lactyl-neuraminic acid[g]	Neu2en5Ac9Lt	55,56
5-*N*-Acetyl-2,7-anhydro-neuraminic acid[g,i]	Neu2,7an5Ac	6,33,55,57,58
5-*N*-Acetyl-4,8-anhydro-neuraminic acid[j]	Neu4,8an5Ac	31,32,59,60
5-*N*-Acetyl-neuraminic acid 1,7-lactone	Neu5Ac1,7lactone	19,61,62
5-*N*-Acetyl-9-*O*-acetyl-neuraminic acid 1,7-lactone	Neu5,9$Ac_2$1,7lactone	19

Table 1 Survey of Reported Structures of Naturally Occurring Members of the Sialic Acid Family—cont'd

Name	Abbreviation	References
5-*N*-Acetyl-4,9-di-*O*-acetyl-neuraminic acid 1,7-lactone	Neu4,5,9$Ac_3$1,7lactone	19
1-Tauryl 5-*N*-acetyl-neuraminic amide	Neu5Ac1Tau	63
5-*N*-Glycolyl-neuraminic acid	Neu5Gc	6,19,23,26,27, 29–33,64–66
4-*O*-Acetyl-5-*N*-glycolyl-neuraminic acid	Neu4Ac5Gc	6,27,30–33,41
7-*O*-Acetyl-5-*N*-glycolyl-neuraminic acid	Neu7Ac5Gc	6,30,32,33,37
8-*O*-Acetyl-5-*N*-glycolyl-neuraminic acid	Neu8Ac5Gc	32,55
9-*O*-Acetyl-5-*N*-glycolyl-neuraminic acid	Neu9Ac5Gc	6,19,27,29–33,37
4,7-Di-*O*-acetyl-5-*N*-glycolyl-neuraminic acid	Neu4,7$Ac_2$5Gc	19
4,9-Di-*O*-acetyl-5-*N*-glycolyl-neuraminic acid	Neu4,9$Ac_2$5Gc	19
7,9-Di-*O*-acetyl-5-*N*-glycolyl-neuraminic acid[k]	Neu7,9$Ac_2$5Gc	6,19,30–33,37
8,9-Di-*O*-acetyl-5-*N*-glycolyl-neuraminic acid	Neu8,9$Ac_2$5Gc	6,19,30–33,37
4,7,9-Tri-*O*-acetyl-5-*N*-glycolyl-neuraminic acid	Neu4,7,9$Ac_3$5Gc	19
7,8,9-Tri-*O*-acetyl-5-*N*-glycolyl-neuraminic acid	Neu7,8,9$Ac_3$5Gc	6,30,31,33,37
4,7,8,9-Tetra-*O*-acetyl-5-*N*-glycolyl-neuraminic acid	Neu4,7,8,9$Ac_4$5Gc	19
5-*N*-Glycolyl-9-*O*-lactyl-neuraminic acid	Neu5Gc9Lt	19,32,55
4-*O*-Acetyl-5-*N*-glycolyl-9-*O*-lactyl-neuraminic acid	Neu4Ac5Gc9Lt	19
7-*O*-Acetyl-5-*N*-glycolyl-9-*O*-lactyl-neuraminic acid	Neu7Ac5Gc9Lt	19

Continued

Table 1 Survey of Reported Structures of Naturally Occurring Members of the Sialic Acid Family—cont'd

Name	Abbreviation	References
8-*O*-Acetyl-5-*N*-glycolyl-9-*O*-lactyl-neuraminic acid	Neu8Ac5Gc9Lt	19
4,7-Di-*O*-acetyl-5-*N*-glycolyl-9-*O*-lactyl-neuraminic acid	Neu4,7$Ac_2$5Gc9Lt	19
7,8-Di-*O*-acetyl-5-*N*-glycolyl-9-*O*-lactyl-neuraminic acid	Neu7,8$Ac_2$5Gc9Lt	19
5-*N*-Glycolyl-8-*O*-methyl-neuraminic acid[l]	Neu5Gc8Me	6,31–33,46,67,68
4-*O*-Acetyl-5-*N*-glycolyl-8-*O*-methyl-neuraminic acid	Neu4Ac5Gc8Me	69
7-*O*-Acetyl-5-*N*-glycolyl-8-*O*-methyl-neuraminic acid	Neu7Ac5Gc8Me	32,69
9-*O*-Acetyl-5-*N*-glycolyl-8-*O*-methyl-neuraminic acid	Neu9Ac5Gc8Me	6,31–33,46,55
4,7-Di-*O*-acetyl-5-*N*-glycolyl-8-*O*-methyl-neuraminic acid	Neu4,7$Ac_2$5Gc8Me	69
7,9-Di-*O*-acetyl-5-*N*-glycolyl-8-*O*-methyl-neuraminic acid	Neu7,9$Ac_2$5Gc8Me	31
5-*N*-Glycolyl-9-*O*-methyl-neuraminic acid	Neu5Gc9Me	47,70
5-*N*-Glycolyl-8-*O*-sulfo-neuraminic acid	Neu5Gc8S	19,32,49,71
5-*N*-Glycolyl-9-*O*-sulfo-neuraminic acid	Neu5Gc9S	72
5-*N*-(*O*-Acetyl)glycolyl-neuraminic acid	Neu5GcAc	6,33,73
5-*N*-(*O*-Methyl)glycolyl-neuraminic acid	Neu5GcMe	74
2-Deoxy-2,3-didehydro-5-*N*-glycolyl-neuraminic acid[g]	Neu2en5Gc	6,33,56,75

Table 1 Survey of Reported Structures of Naturally Occurring Members of the Sialic Acid Family—cont'd

Name	Abbreviation	References
9-*O*-Acetyl-2-deoxy-2,3-didehydro-5-*N*-glycolyl-neuraminic acid[g]	Neu2en9Ac5Gc	56
2-Deoxy-2,3-didehydro-5-*N*-glycolyl-9-*O*-lactyl-neuraminic acid[g]	Neu2en5Gc9Lt	55,56
2-Deoxy-2,3-didehydro-5-*N*-glycolyl-8-*O*-methyl-neuraminic acid[g]	Neu2en5Gc8Me	55
2,7-Anhydro-5-*N*-glycolyl-neuraminic acid[g]	Neu2,7an5Gc	33,55,58
2,7-Anhydro-5-*N*-glycolyl-8-*O*-methyl-neuraminic acid[g]	Neu2,7an5Gc8Me	55
4,8-Anhydro-5-*N*-glycolyl-neuraminic acid[j]	Neu4,8an5Gc	32
5-*N*-Glycolyl-neuraminic acid 1,7-lactone	Neu5Gc1,7lactone	19
9-*O*-Acetyl-5-*N*-glycolyl-neuraminic acid 1,7-lactone	Neu9Ac5Gc1,7lactone	19
7-Acetamido-9-*O*-acetyl-7-deoxy-5-*N*-glycolyl-neuraminic acid	Neu9Ac5Gc7NAc	19
7-Acetamido-8,9-di-*O*-acetyl-7-deoxy-5-*N*-glycolyl-neuraminic acid[m]	Neu8,9$Ac_2$5Gc7NAc	19
2-Keto-3-deoxy-nononic acid	Kdn	19,32,76–78
5-*O*-Acetyl-2-keto-3-deoxy-nononic acid	Kdn5Ac	19,32
7-*O*-Acetyl-2-keto-3-deoxy-nononic acid	Kdn7Ac	19,32
8-*O*-Acetyl-2-keto-3-deoxy-nononic acid	Kdn8Ac	79
9-*O*-Acetyl-2-keto-3-deoxy-nononic acid	Kdn9Ac	19,32,80,81

Continued

Table 1 Survey of Reported Structures of Naturally Occurring Members of the Sialic Acid Family—cont'd

Name	Abbreviation	References
4,5-Di-*O*-acetyl-2-keto-3-deoxy-nononic acid	Kdn4,5Ac_2	19
4,7-Di-*O*-acetyl-2-keto-3-deoxy-nononic acid	Kdn4,7Ac_2	19
5,9-Di-*O*-acetyl-2-keto-3-deoxy-nononic acid	Kdn5,9Ac_2	32
7,9-Di-*O*-acetyl-2-keto-3-deoxy-nononic acid	Kdn7,9Ac_2	19,32
8,9-Di-*O*-acetyl-2-keto-3-deoxy-nononic acid	Kdn8,9Ac_2	19
2-Keto-3-deoxy-5-*O*-methyl-nononic acid	Kdn5Me	82
2-Keto-3-deoxy-9-*O*-phospho-nononic acid[g,n]	Kdn9P	83

[a]For a survey of typical biological sources of sialic acids, see refs. 6, 7, 19, and 32.
[b]Present only in bound form and considered to be derived from bound Neu5Ac in an enzymatic N-deacetylation/N-reacetylation cycle.[18] In free open form, Neu is directly converted into its labile internal Schiff base 4-hydroxy-5-[1,2,3,4-tetrahydroxybutyl]-Δ^1-pyrroline-2-carboxylic acid (pH-dependent equilibrium).[15] However, Neu was picked up by GLC–EIMS of its heptafluorobutyrylated methyl ester derivative, and it was suggested that the heptafluorobutyric anhydride acylation is able to disrupt the Schiff base, followed by blocking the free amino group.[19]
[c]Present only in bound form and considered to be derived from bound Neu in an enzymatic dehydration reaction, catalyzed by a so-called sialic acid cyclase. Neu1,5lactam was initially called "cyclic sialic acid."[21]
[d]Detected in bound form in (α2→9)-linked Neu5Ac polysaccharide.
[e]Lactyl = L-Lactyl.
[f]Biosynthetic intermediate to Neu5Ac.
[g]Present only as free form.
[h]Neu2en5Ac, originally known as a synthetic compound with a potent inhibitory effect on various sialidases,[84] may be produced enzymatically from sialoglycoconjugates in tissues.[85] Small amounts are formed by a water elimination side reaction from Neu5Ac during influenza-B-virus-sialidase-catalyzed desialylations of sialoglycoconjugates.[86] Neu2en5Ac can be generated from CMP-Neu5Ac in a non-enzymatic elimination reaction, occurring under physiological and, much faster, under alkaline conditions.[53,56]
[i]Neu2,7an5Ac can be generated from Neu5Ac(α2→3)Gal(β1→-containing sialoglycoconjugates using a sialidase isolated from the leech *Macrobdella decora*.[58,87]
[j]Neu4,8an5Ac does not occur as such in nature. It is assumed to be formed under hydrolytic conditions from bound Neu4,5Ac_2 in a multistep process, before cleavage of the glycosidic linkage.[31,59] Also alkaline treatment of free Neu4,5Ac_2 yielded Neu4,8an5Ac.[31] Alternatively, Neu4,8an5Ac is formed by thermal degradation of sodium *N*-acetylneuraminate.[60] In addition, two other anhydro forms of Neu5Ac were suggested by ref. 32.
[k]Correction in table 3 of ref. 33: Fragments C and D are absent in Neu7,9$Ac_2$5Gc.
[l]Three anhydro derivatives of Neu5Gc8Me were suggested by ref. 32.
[m]Neu4,9$Ac_2$7Am5Gc in the text of ref. 19 is probably a typing error for Neu8,9$Ac_2$7Am5Gc.
[n]Biosynthetic intermediate to Kdn.
In the case of a structure proven by mass spectrometry and/or NMR spectroscopy, reference numbers refer in general to such studies.

references,[15–87] is presented. In the case of Neu, besides *N*-acetyl or *N*-glycolyl groups at C5, the hydroxy groups at C4, C7, C8, and C9 may be free, esterified (acetylated, glycolylated, lactylated, sulfated, phosphorylated) or etherified (methylated). Most free sialic acids occur predominantly in the more thermodynamically stable β-anomeric form (>93%), whereas in nature bound sialic acids, with one exception, possess the α-anomeric configuration. The exception is the nucleotide-bound sialic acid, the activated CMP-donor, e.g., CMP-β-Neu5Ac (Fig. 3C). Crystalline Neu5Ac occurs specifically in the β-anomeric form. The mother molecule Neu and

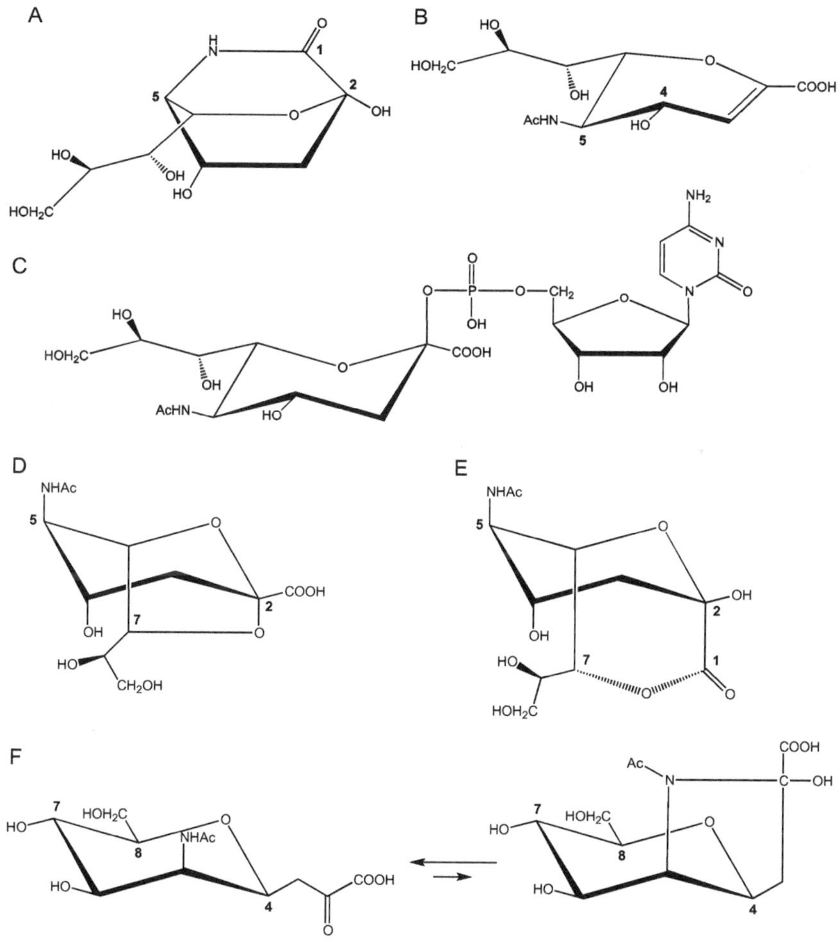

Fig. 3 Chemical structures of (A) Neu1,5lactam ($^{5,2}B$ conformation), (B) Neu2en5Ac ($^{4}H_5$ conformation), (C) CMP-β-Neu5Ac, (D) Neu2,7an5Ac ($^{5}C_2$ conformation), (E) Neu5Ac1,7-lactone ($^{5}C_2$ conformation), and (F) Neu4,8an5Ac ($^{7}C_4$ conformation; two tautomers).

Neu1,5lactam (neuraminic acid 1,5-lactam) (Fig. 3A) only exist in bound form. A number of other members (can) only occur as free form, i.e., Neu5-Ac9P (9-*O*-phospho-*N*-acetylneuraminic acid), (O-substituted) Neu2en5Ac (Fig. 3B), or Neu2en5Gc (2-deoxy-2,3-didehydro-*N*-glycolylneuraminic acid), (O-substituted) Neu2,7an5Ac (2,7-anhydro-*N*-acetylneuraminic acid) (Fig. 3D) or Neu2,7an5Gc (2,7-anhydro-*N*-glycolylneuraminic acid), and Neu4,8an5Ac (4,8-anhydro-*N*-acetylneuramic acid) (Fig. 3F) or Neu4, 8an5Gc (4,8-anhydro-*N*-glycolylneuraminic acid). In the case of Kdn, the hydroxy groups at C4, C5, C7, C8, and C9 may be free, esterified (acetylated, phosphorylated) or etherified (methylated). In Neu5Gc, the hydroxy group at C7 can be replaced by an acetamido group, yielding in fact another variant of Neu, with the full name of 7-acetamido-5-hydroxyacetamido-3,5,7-trideoxy-D-*glycero*-D-*galacto*-non-2-ulosonic acid, so far only found with one or two *O*-acetyl groups.

This chapter will review the fascinating story of the isolation and structural elucidation of the sialic acids and their importance as biological molecules in a historical perspective. It will describe how the founders of the sialic acid field since the mid-1930s, over a period of more than 20 years, were struggling with the chemical structure of these molecules in a period in which mass spectrometry and NMR spectroscopy did not exist. And it will show how sialic acid constituents of glycoproteins, glycolipids, oligosaccharides, and polysaccharides have grown to be considered biologically and medically highly essential factors of the daily life on earth.

Although not discussed in biological and medical detail in this chapter, also the expansion of the more recently well-established "sialic acid-like" subclasses of the superfamily of NulOs, namely, the 5,7-diamino-3,5,7,9-tetradeoxynon-2-ulosonic acids, with the mother molecules pseudaminic acid (Pse), legionaminic acid (Leg), 4-*epi*-legionaminic acid (4eLeg), 8-*epi*-legionaminic acid (8eLeg), acinetaminic acid (Aci), and 8-*epi*-acinetaminic acid (8eAci), so far exclusively found in bacterial polysaccharides and glycoproteins, will be subjects of a limited review.

Over the years several books[88–95] and book chapters (e.g., refs. 6 and 96–114) dealing with the progress in the fields of sialic acid chemistry, biology, medicine, and food science have been published, as well as a number of reviews in scientific journals (e.g., refs. 7, 23, and 115–147) (additional review references are included in the specific sections), but no updated review in a historic perspective about this strongly growing field is available. It is therefore our intention to describe the historical development of the multifaceted sialochemistry/biology field and throw some light also on its present state and future.

2. PAVING THE WAY WITH EHRLICH'S ALDEHYDE AND BIAL'S ORCINOL REAGENTS

It was in 1877, more than 140 years ago, that the German scientist Hoppe-Seyler (1825–1895) described a substance released from epithelial mucins by mild acid treatment at elevated temperature as an acidic compound, being stronger than carbonic acid. When heated under alkaline conditions, the solution turned brown. The compound reduced cupric oxide or bismuth oxide or indigosulfuric acid, contained nitrogen, and was unfermentable by brewer's yeast. Its purification was a continual problem.[148] Similar observations were made by Hoppe-Seyler,[149] Green,[150] Krukenberg,[151] and Zeller,[152] when investigating the mild acid-released material from nests of Asian swiftlets (genus *Collocalia*). For the construction of their nests, these birds use mucin secreted by their salivary glands. The bird's nests are edible and are used in several nutritional and medicinal applications.

Interestingly, Green claimed the formation of crystals of sugar from the released material, a claim that could not be confirmed by Zeller. Citation of Zeller: "Apparently, Green has inorganic material considered as sugar" (translated from the German).

Many years later, collocalia mucin would form an excellent source for the isolation of gram quantities of *N*-acetylneuraminic acid, simply by boiling pulverized edible bird's nest under reflux for 5 h.[153,154]

In 1898, Ehrlich in Germany observed the formation of a purple color after heating mucins with *p*-dimethylaminobenzaldehyde in acid solution.[155,156] In the same year, Müller in Germany also described the formation of a splendid red color when mucus substances, made alkaline and heated slightly, were incubated with acidic *p*-dimethylaminobenzaldehyde at higher temperature.[157] Although initially the mixture of *p*-dimethylaminobenzaldehyde in 50% (v/v) HCl, called Ehrlich's reagent, was frequently applied for the detection of indoles, pyrroles, and related nitrogen-containing compounds, this reaction was going to play an important role in the unraveling of the structure of sialic acids. It turned out that incubation of the newly discovered components, in bound and free form, with Ehrlich's reagent before ("direct" Ehrlich) and after ("indirect" Ehrlich) treatment with alkali yielded a red–violet color (see Section 7.1).

Later, in 1927, Walz in Germany and Levine and Landsteiner in the United States reported that on heating isolated lipid material of bovine spleen,[158] or equine kidney/bovine kidney/bovine brain,[159] respectively, with orcinol in hydrochloric acid containing ferric chloride[158] or copper acetate,[159] a red–violet color was formed. In fact, the orcinol/concd. $HCl/FeCl_3$ mixture, introduced in 1902 by Bial[160] and later called Bial's reagent, was used as a general test for the presence of carbohydrates. A pentose is dehydrated with concd. acid to form furfural, which condenses with orcinol to a product that gives in the presence of ferric ions a bluish-green color. With hexoses, whereby 5-hydroxymethylfurfural is formed, a yellow–brown color is seen. So, the observed red–violet/purple color was quite unusual (see Section 7.1).

2.1 The First Scoutings

The research groups who made the first steps in the discovery and isolation of sialic acids were located in different countries around the world: Gunnar Blix[161] (Fig. 4A) and coworkers at the University of Uppsala (Sweden); Ernst Klenk[162] (Fig. 4B) and coworkers at the University of Tübingen and the University of Cologne (Germany); and Alfred Gottschalk[163,164] (Fig. 4C) and coworkers at the Walter and Eliza Hall Institute in Melbourne (Australia) and the Max-Planck-Institute for Virology in Tübingen (Germany). Interestingly, materials from completely different biological sources played an important role in their journeys of exploration.

In 1936 Blix reported the isolation of a strongly acidic compound from a boiling water solution of bovine submaxillary gland mucin, that he called "**Kohlenhydrat I**."[165] The compound, generated via autohydrolysis and

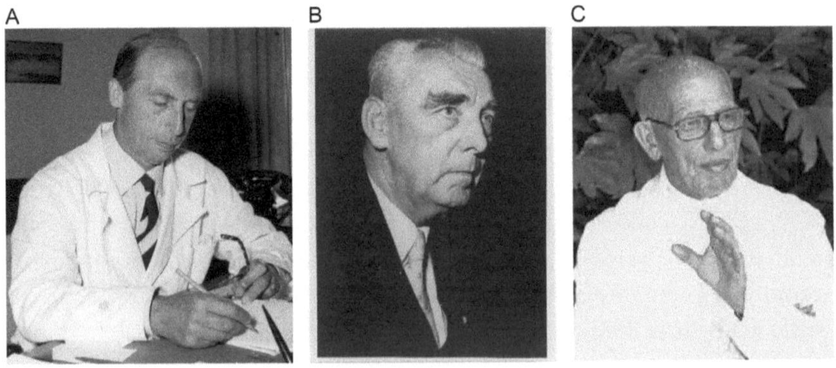

Fig. 4 (A) Gunnar Blix (1894–1981), (B) Ernst Klenk (1896–1971), (C) Alfred Gottschalk (1894–1973). *Panel (A): Reprinted from Lundblad, A. Gunnar Blix and His Discovery of Sialic Acids—Fascinating Molecules in Glycobiology.* Upsala J. Med. Sci. ***2015***, 120, *104–112. With permission of the Board. Panel (B): Photo courtesy of Prof. Dr. Hans-Dieter Klenk. Panel (C): Photo courtesy of one of the authors (R.S.).*

obtained in crystalline form, had reducing power. Elemental analyses yielded $C_{14}H_{24}NO_{11}$ as the preferred chemical formula. Microacetyl determinations suggested the presence of two acetyl groups in one molecule, whereas a negative ninhydrin reaction supposed the presence of one *N*-acetyl group. The product had an mp of 140–150°C (dec.). UV spectroscopy showed the absence of ethylene bonds. The compound turned out to be very sensitive to heating with dilute mineral acid, giving rise to dark humin formation. When heated with Ehrlich's reagent, both with and without alkali pretreatment, an intense red–violet color was formed. Taking into account these and later observations, **a disaccharide structure composed of an N-acetylated hexosamine part and a polyhydroxy acid part with six carbon atoms, being not a hexuronic acid**, was suggested (note that hexosamine was not isolated from the substance).[165–167] An initial suggestion that the hexosamine part should be chondrosamine (galactosamine)[166] was withdrawn when chondrosamine was isolated, separated from "Kohlenhydrat I," and it was stated that both compounds were "probably bound with an easily split linkage."[167] An additional earlier suggestion was that the polyhydroxy acid could be a deoxy-hexuronic acid.[168] Because of the salivary origin of "Kohlenhydrat I" (the Greek name for saliva is "sialon"), Blix called the compound "**sialic acid**."[167]

In further studies, Blix found that the so-called "protagon" lipid fraction of bovine brain tissue gave similar Ehrlich's and Bial's color reactions as found for the mucin-derived compound.[168] A few years earlier, a red color with Bial's reagent had already been mentioned by Klenk for lipid material of the human brain,[169] and the isolation of a nitrogen-containing organic acid in crystalline form, called "**neuraminic acid**" (related to the source: neurons), followed in 1941.[170] Because of the observed lability in acidic medium, leading to dark humin formation, a methanolic-HCl release protocol was chosen. So, the lipid material was treated with 5% methanolic HCl (3 h, 105°C), and the released substance was converted, via its barium salt, into the free acid {elemental analysis: preferred chemical formula $C_{11}H_{21}NO_9$; $[\alpha]^{20}_{D}$ −54.9° (*c* 3.2, H_2O)}. A positive ninhydrin reaction indicated a free amino group, whereas reducing power was not detected. The Bial and Ehrlich colorimetric assays were both strongly positive. Protocols directed to the detection of a hexosamine part in "neuraminic acid" were negative, and a suggestion for **an aliphatic polyoxy-aminodicarboxylic acid** structure was made. When Klenk realized that the isolated substance contained a methoxy group, introduced during the methanolic-HCl treatment (in the 1950s proved with ^{14}C-labeled CH_3OH[171]), he renamed "neuraminic acid" as "**methoxyneuraminic acid**" and kept the name "**neuraminic acid**" for the compound with chemical formula $C_{10}H_{19}NO_9$ ($=C_{11}H_{21}NO_9 - CH_2$).[172] Then, "neuraminic acid" might contain a reactive carbonyl function, which enabled the formation of a glycosidic linkage with other monosaccharides. In the latter study, the term "ganglioside" was introduced for brain glycolipids containing "neuraminic acid."

Besides scientific interest from the mucin and brain glycolipid research areas for these new, but still unknown compounds, an interesting new angle of incidence came from the virology field. At the end of the 1940s, Burnet reported that the ability of various mucoproteins (i.e., serum mucoprotein, ovarian cyst mucoid, chorionic gonadotropin) to act as competitive inhibitors for hemagglutination by heated influenza B virus was lost after pretreatment of the mucoproteins with active virus.[173] Then, Gottschalk and Lind found that incubation of the Melbourne strain of influenza A virus with ovomucin, a protein fraction containing a strong virus hemagglutinin inhibitor, yielded a water-soluble, nitrogen-containing compound with reducing power, that gave, after alkaline treatment, with Ehrlich's reagent, a purple color; furthermore, the Molisch carbohydrate test (concd. H_2SO_4, α-naphthol) was strongly positive.[174] Acid treatment resulted in black insoluble humin formation. Incubation of ovomucin with a *Vibrio cholerae* filtrate yielded a substance with similar properties.[175] In these years, the enzyme-like agent in the filtrates was called "receptor-destroying enzyme" (RDE) (see Section 11.7). Based on the results so far, the enzymatic liberation of a "**carbohydrate–peptide complex**" was suggested, and it was stated that the carbohydrate moiety should be "**an oligosaccharide containing one or more *N*-acyl (acetyl) hexosamine residues linked by an alkali-labile glycosidic linkage to a non-amino-sugar**." Two years later, Gottschalk repeated the incubation experiments with human urinary mucoprotein[176] and influenza B virus and suggested now that the released compound should be an "**N-substituted *iso*glucosamine**" (a fructosamine; an *N*-1-deoxy-1-ketose amino acid or peptide).[177] Interestingly, the presence of a keto function was indicated by a positive Seliwanoff resorcinol/concd. HCl test, an assay whereby ketoses, in contrast to aldoses, rapidly give a red color (ketoses are more rapidly dehydrated than aldoses with concd. acid).

In 1975, Blix described in a letter to Klaus Störiko (Behringwerke AG, Marburg, Germany) about his first contact with Gottschalk: "... I recall how I at the first international biochemist congress in Cambridge 1949 talked with Stacey and Gottschalk, when the latter described the general properties of the carbohydrate component in mucins that inhibited virus haemagglutination and asked for advice from Stacey. It struck me immediately that some of the properties Gottschalk mentioned strongly reminded me about those of sialic acid. When I got home I mailed Gottschalk what we had written about sialic acid and told him that we intended to test the matter closer. Gottschalk obviously did not, at this time, know about sialic acid, which outside Cologne and Uppsala still was of low interest. I had on my part limited knowledge about Burnet's investigations. We agreed later on to study in each of our laboratories the carbohydrate component in the well defined Tamm–Horsfall urinary protein and to publish our results in 1952 simultaneously in Nature *..." (from the archive of Blix's scientific correspondence, available at Uppsala University Library).*[161]

2.2 The Growing Knowledge About the Substance

At the beginning of the 1950s, in their search for more structural details and to elucidate contrasting results, several overlapping studies were published by the Uppsala, Cologne, and Melbourne groups. Step by step the structural relationship among Blix's "sialic acid," Klenk's "neuraminic acid"/"methoxyneuraminic acid," and Gottschalk's "N-substituted *iso*-glucosamine" was growing. Of course, it was also a period of close competition, reflected by several relatively short communications. But, although critical comments on each other's work can be picked up from the literature reports, the groups were also in good contact with each other, exchanging information when necessary.

To start with, Odin (Uppsala group)[178] suggested in a paper published in parallel with Gottschalk's paper about the monosaccharide constituents of human urinary mucoprotein[179] the close chemical relationship between Gottschalk's "N-substituted *iso*glucosamine" and Blix's "sialic acid."

Earlier, the Uppsala group had stated that Klenk's "neuraminic acid" "is probably to be regarded as a degradation product of the native disaccharide compound" (the suggested structure by Blix, composed of an N-acetylated hexosamine part and a polyhydroxy acid part with six carbon atoms).[166] *However, this conclusion was contradicted by Klenk, saying that, based on further research, "there is no reason to doubt that "neuraminic acid" is a native residue of gangliosides" (translated from the German).*[180] *So, it was not a hexosamine-containing cleavage product.*

In further work, the Uppsala group reported the occurrence of "sialic acid" in many biological sources, i.e., egg white, ovomucin, allantoic fluid, ovarian cysts, bovine follicular fluid, cancer fluid, blood serum, plasma glycoproteins, human tears, bovine milk, epithelial mucins, human saliva, meconium, and gangliosides.[178,181–186]

In the same period, the Cologne group prepared "methoxyneuraminic acid" from bovine submaxillary gland mucin and human urinary mucoprotein,[187] following their earlier applied protocol[170] for gangliosides. In subsequent investigations, heating of an aqueous bovine submaxillary gland mucin preparation (2 h, 110°C) yielded, after isolation, a crystalline material with a chemical formula of $C_{12}H_{21}NO_{10}$ and $[\alpha]^{20}{}_{D}$ −31.7° (*c* 2.6, H_2O) that they called "**N-acetylneuraminic acid**."[22] The various chemical reactions, carried out earlier on "sialic acid," gave similar results for "*N*-acetylneuraminic acid" (Ehrlich and Bial reaction: red color; reducing power: positive; ninhydrin reaction: negative), except that "*N*-acetylneuraminic acid" contained only one acetyl group. Treatment of "*N*-acetylneuraminic acid" with 40% NaOH (2 h, 100°C) yielded pyrrole-2-carboxylic acid, a product identified 1 year earlier by Gottschalk[188] in an alkaline hydrolysate of bovine submaxillary gland mucin (see Section 3). In additional studies, crystalline "*N*-acetylneuraminic acid" ($C_{12}H_{21}NO_{10}$) could be obtained from enzymatic incubations of human urinary mucoprotein with influenza virus B-Lee,[189] and of human urinary mucoprotein and bovine submandibular gland mucin with the RDE from *V. cholerae*.[190] It was concluded that both the influenza virus and the RDE are acting as O-glycosidases. Following the methanolyis protocol,[170] crystalline "methoxyneuraminic acid" (neuraminic acid methyl glycoside: $C_{11}H_{21}NO_9$) could be isolated from porcine and equine submaxillary gland mucin and bovine colostrum mucin,[191] from amyloidose liver,[192] from fetuin,[193] and from bovine stroma protein.[194]

The paper of Klenk and Uhlenbruck[191] gives a clear insight into the competition between Klenk ("neuraminic acid") and Blix ("sialic acid") in the slowly developing field. At the end of a long criticism in a footnote (we write 1956), Klenk ends with "As long as the information provided by Blix can not be confirmed by other sources, it must be doubted that a chemically well-defined, uniform compound has been presented here" (translated from the German).

In these years, besides color reactions, elemental analyses, and optical rotations, IR spectroscopy and X-ray analysis were introduced as helpful techniques in comparing different isolates. As an illustration of the application of IR spectroscopy, in Fig. 5 the IR spectra of enzymatically and chemically released "*N*-acetylneuraminic acid" are depicted (indications for carboxyl and amide groups).[190]

Further studies of the Uppsala group focused on bovine, porcine, ovine, and equine submaxillary gland mucins.[23,195] Crystalline "sialic acids" were generated via hydrolytic protocols (1 h, 100°C; pH 3.5–4.0). An overview of some physical parameters (elemental analysis, optical rotation, decomposition range, type of N- and O-substituents, the number of hydroxyl functions) is

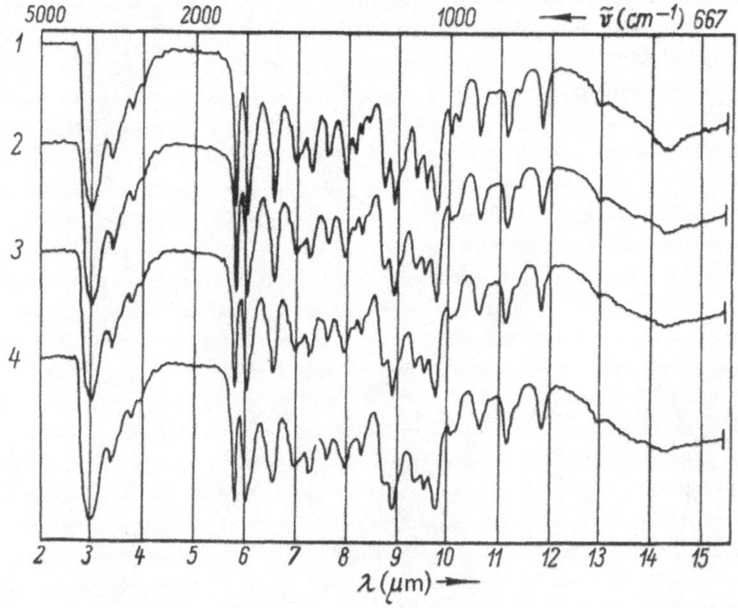

Fig. 5 IR spectra of "*N*-acetylneuraminic acid," enzymatically released (*V. cholerae*) from (1) human urinary mucoprotein, (2) bovine submaxillary gland mucin (crystallization fraction I), and (3) bovine submaxillary gland mucin (crystallization fraction II), and chemically released from (4) bovine submaxillary gland mucin. *Reproduced from Faillard, H. Über die Abspaltung von* N-*Acetyl-neuraminsäure aus Mucinen durch das "Receptor-Destroying-Enzyme" aus* Vibrio cholerae. Hoppe-Seyler's Z. Physiol. Chem. ***1957****,* 307*, 62–86/*308*, 187. Copyright De Gruyter Publishers.*

presented in Table 2. The chemical formulas for "methoxyneuraminic acid," isolated from the mucins and purified "sialic acids" following the methanolic HCl protocol of Klenk, were found to be $C_{10}H_{19}NO_8$ (preferred) or $C_{11}H_{21}NO_9$, being the composition earlier reported by Klenk.[170] Both the Blix and Klenk "methoxyneuraminic acid" samples showed the same X-ray powder patterns. Bial's and Ehrlich's reactions gave red–violet and red–purple colors for all four "sialic acids," respectively. As the "sialic acids" can be reduced with sodium borohydride and subsequent hydrolysis did not produce a reducing substance, whereas the Bial and Ehrlich tests on the reduced compounds were negative, it was stated that, in contrast to earlier conclusions of the Uppsala group,[165] this almost excludes the presence of a glycosidic linkage in "sialic acid." Quantitative determination of hydroxy groups (via an O-acetylation/NaOH titration protocol) showed the presence of five hydroxy groups for the ovine preparation and six hydroxy groups for

Table 2 Some Physical Parameters of Crystalline "Sialic Acids" Isolated From Bovine, Porcine, Ovine, and Equine Submaxillary Gland Mucins[23]

Mucin Source	Elemental Analysis[a]	$[\alpha]^{20}_D$ (5%, H_2O)	Decomposition Range (°C)	*N,O*-Acyl Groups[b]	Hydroxy Groups
Bovine[c]	$C_{13}H_{21}NO_{10}$	$+8° \pm 2°$	134–137	NAc, OAc	n.a.
Ovine	$C_{11}H_{19}NO_9$	$-31° \pm 2°$	185–187	NAc	5
Porcine	$C_{11}H_{19}NO_{10}$	$-32° \pm 2°$	185–187	NGc	6
Equine[c]	$C_{13}H_{21}NO_{10}$	$-59° \pm 2°$	183–187	NAc, OAc	n.a.

[a]Note that in earlier reports it was not realized that during crystallization protocols, wherein methanol was used, partial methyl ester formation took place (all preparations contained one COOH function). The included chemical formulae from this study turned out to be in the end correct.
[b]NAc = *N*-acetyl; OAc = *O*-acetyl; NGc = *N*-glycolyl. Acetyl determination as barium acetate after acid hydrolysis; glycolyl determination as crystalline calcium glycolate after alkaline treatment (see Section 7.1).
[c]The bovine and equine "sialic acids" were suggested to be isomers.
n.a., not available.

the porcine preparation. Amino-group determinations revealed the absence of significant amounts of amino nitrogen, so amino groups are substituted. Periodate studies on bovine "sialic acid" showed the consumption of 1 mol of periodate and no formic acid production (CH_2OH–CHOH–CHOAc– element), whereas the periodate consumptions for the porcine and ovine "sialic acids" were 2 mol of periodate (CH_2OH–CHOH–CHOH– element) and 1 mol of formic acid formation; in all cases 1 mol of formaldehyde was found. As shown in Table 2, the isolated porcine "sialic acid" contained an *N*-glycolyl group, a result confirmed one year later by the Cologne group. Here, "***N*-glycolylneuraminic acid**" was chemically and enzymatically released from porcine submaxillary gland mucin, and a colorimetric method for the quantitative determination of glycolic acid was developed.[64] See also ref. 196. For further discussions on the structure of the presented "sialic acids," see Section 3. To give an impression of the use of X-ray powder diffraction analysis for comparison purposes, Fig. 6 illustrates the diagrams of various "sialic acids."

2.3 The Expanding Network of Researchers

Parallel to the research activities of the Uppsala, Cologne, and Melbourne groups, in the 1950s other reports with information about the existence of such Bial's and Ehrlich's reagent-positive products also appeared. In this context, among others, the groups of Tamio Yamakawa[197] and coworkers at

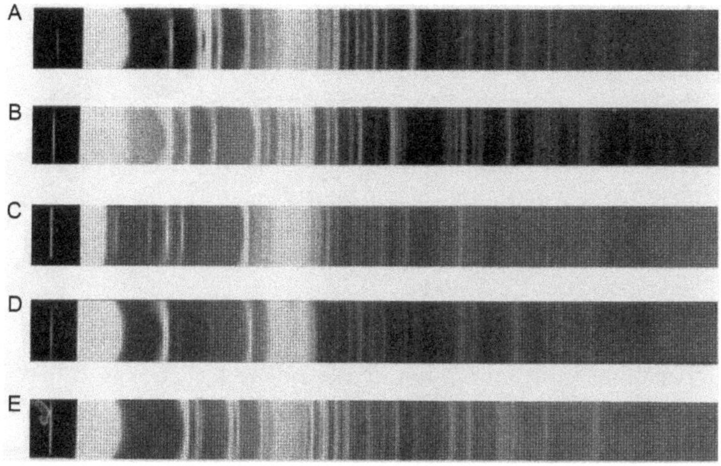

Fig. 6 X-ray powder diagrams of "sialic acid" isolated from submaxillary gland mucins [(A) bovine, (B) ovine, (C) porcine, (D) equine] and "methoxyneuraminic acid" (E). Guinier camera. Nickel-filtered copper K-radiation. *Reproduced from Blix, G.; Lindberg, E.; Odin, L.; Werner, I. Studies on Sialic Acids.* Acta Soc. Med. Upsalien. ***1956***, 61, *1–25. With permission of the Board.*

the University of Tokyo (Japan), Richard Kuhn[198] and coworkers at the Max-Planck-Institute in Heidelberg (Germany), and Paul György[199] and coworkers at the University of Pennsylvania in Philadelphia (USA) should be mentioned.

In 1951, Yamakawa and Suzuki reported the isolation of an unknown crystalline nitrogen-containing polyhydroxy-carboxylic acid (elemental analysis: $C_{10}H_{19}NO_8$) from a methanolysate (8% methanolic HCl, 3 h, 100°C; work-up via a barium salt and treatment with diluted H_2SO_4) of an equine blood stroma glycolipid. The glycolipid was called "hematoside" and the released compound "**hemataminic acid**."[200] Both substances gave a red color with Bial's and Ehrlich's reagents. The ninhydrin test for "hemataminic acid" was positive only in the presence of a trace of alkali. After acid hydrolysis the reducing power changed from negative to positive. As "hemataminic acid" ($[\alpha]^{20}{}_{D}$ −54.2°) was suspected to be a methyl glycoside (and suggested to correspond with Klenk's "methoxyneuraminic acid," $[\alpha]^{20}{}_{D}$ −54.9°,[170] the hypothetical free compound should have the chemical formula $C_9H_{17}NO_8$; it was called "**prehemataminic acid**," described as **a 2-amino-2-deoxy-nonuronic acid** (corresponding with Klenk's "neuraminic acid") (for a proposal of the structure, see Section 3).[201]

*Here, a citation of Yamakawa and Suzuki[201] illustrates the initial controversy with respect to "neuraminic acid" vs "hemataminic acid": "According to a private communication from Dr. Klenk, which we received some days ago, he found chondrosamine in his preparation of ganglioside, and maintained that his 'neuraminic acid' gave, contrary to our '**hemataminic acid**', no reducing hydrolysate after cleavage with hydrochloric acid. For this reason, he was doubtful of the identity of these two substances." However, in the same year, after having analyzed the same material, Klenk withdrew his doubts by saying: "The "prehemataminic acid" occurring therein as a characteristic building block is identical with "neuraminic acid"" (translated from the German).[202]*

From a hydrolysate of equine serum mucoprotein (pH 1–2, 1 h, 80°C), a similar crystalline compound could be isolated that was called "**sero-lactaminic acid**."[203]

In another investigation by Kuhn et al., acid hydrolysis (pH 1, 40 min, 70–80°C) of a mucoprotein fraction of bovine colostrum afforded, after work-up (use of methanol), a low-molecular-mass ninhydrin-negative product, that they called the **methoxy derivative of "lactaminic acid**."[204] The crystalline material had similar Ehrlich and Bial colorimetric properties as reported for "sialic acid" and "methoxyneuraminic acid." However, the paper chromatographic R_f values of the methoxy derivative of "lactaminic acid" in two solvent systems were different from that of "methoxyneuraminic acid." In a subsequent paper, the notation methoxy derivative of "lactaminic acid" was reformulated as "**lactaminic acid methyl ester**" (de-esterification with alkali).[205] Furthermore, "**lactaminic acid**" ($C_{11}H_{19}NO_9$; mp 183–185°C with decomposition; $[\alpha]^{21}{}_D$ −31.0°, *c* 1, H_2O)[206] was established to be identical with Blix's ovine "sialic acid."[23] The same conclusion was drawn by Faillard for "lactaminic acid" in relation to *N*-acetylneuraminic acid.[190] In further screening studies for low-molecular-mass compounds in bovine milk, human milk, and bovine, human, ovine, caprine, and porcine colostrum, thereby demonstrating the presence of Ehrlich- and Bial-positive materials, the Heidelberg group isolated an oligosaccharide from bovine colostrum, which turned out to be *O*-acetyl-lactaminic acid-lactose.[205,207]

György and coworkers reported that nondialyzable preparations of human milk afforded red colors with Bial's and Ehrlich's reagents,[208] and similarities with "sialic acid" and "neuraminic acid" were suggested. The methanolic-HCl protocol was followed to prepare crystalline "methoxy-neuraminic acid" {$[\alpha]^{20}{}_D$ −49.5° (*c* 3.3, H_2O); elemental analysis: $C_{11}H_{21}NO_9$}. In additional reports, a milk oligosaccharide fraction was hydrolyzed under mild conditions (pH 1, 2 h, 80°C), and subsequent work-up, including ion-exchange chromatography, yielded an anhydrous product in crystalline form (no use of methanol), that was called "**gynaminic acid**," a name that referred to human milk.[209,210] The final substance[210] was published with the following analytical parameters: elemental analysis, $C_{11}H_{19}NO_9$; $[\alpha]^{25}{}_D$ −32.1° (*c* 2, H_2O); mp 176–177°C (dec.); positive with Bial's and Ehrlich's reagents; negative with ninhydrin reagent; presence of acetyl group; absence of methoxy group; consumption of 2 mol of periodate; four acetylizable hydroxy groups. In a similar way, "gynaminic acid" was isolated from human meconium. The X-ray powder diagrams of "gynaminic acid" were identical with that of "sialic acid," isolated by the Uppsala group from ovine submaxillary gland mucin.

Other interesting studies from the 1950s by other authors focused on the detection of sialic acids in different biological materials. Those that should be mentioned are those on pooled and fractionated human serum proteins, transudates, urine, cerebrospinal fluid, milk, saliva, and a large variety of animal serum proteins,[211,212] and those on the isolation of neuramin-lactose from rat mammary glands.[213,214]

3. THE FIRST CREATION OF CHEMICAL STRUCTURES

The first creations of a chemical structure were directly influenced by the finding that the Ehrlich reaction should be based on the formation of pyrrole-2-carboxylic acid as reactant for *p*-dimethylaminobenzaldehyde to give the specific red–violet color.

Already in 1949, Hiyama published a first proposal.[215] Bovine sublingual gland mucin was treated with 10% KOH for 10 h at 100°C. After neutralization and work-up, a product was obtained that gave a red color with *p*-dimethylaminobenzaldehyde (absorption max. near 550 nm), and was identified as pyrrole-2-carboxylic acid by comparison with its synthetic analogue. Subjecting Blix's "Kohlenhydrat I"[165] (see Section 2.1), isolated from bovine submaxillary gland mucin, to the same alkaline treatment, also afforded pyrrole-2-carboxylic acid. Based on these results, and taking into account Blix's suggestion that "Kohlenhydrat I" should contain an acetylated hexosamine part (one *N*- and one *O*-acetyl group), a completely substituted pyrrolidine ring with a D-glucosamine part incorporated in its structure ($C_{13}H_{21}NO_{10}$) was tentatively proposed (Scheme 1). The reduction of hot alkaline copper solution by "Kohlenhydrat I" was ascribed to the presence of D-erythrose.

A drawing of a probable structure for the reagent-positive "hemataminic acid" of Ehrlich and Bial was reported in 1952 by Yamakawa and Suzuki (Fig. 7).[201] As mentioned in Section 2.3, "hemataminic acid" was isolated

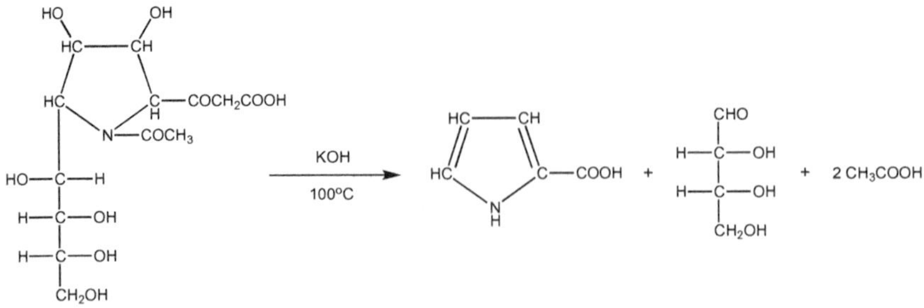

Scheme 1 Proposed structure of "Kohlenhydrat I" with the formation of pyrrole-2-carboxylic acid (and D-erythrose), according to Hiyama, taking into account Blix's findings.[165,215]

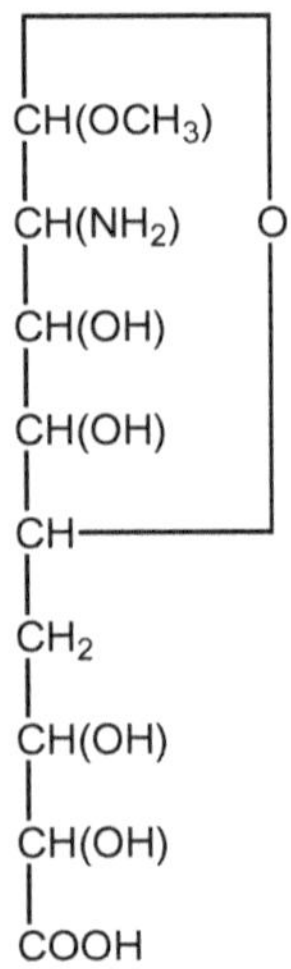

Fig. 7 Structure proposal of Yamakawa and Suzuki for "hemataminic acid" ($C_{10}H_{19}NO_8$), a methoxy derivative of a 2-amino-2-deoxy-nonuronic acid.[201]

from hematoside via the methanolysis protocol and suggested to be identical to Klenk's "methoxyneuraminic acid." The assumed structure was based on a series of observations, e.g.: (i) the gasometric α-amino-carboxyl determination with ninhydrin, used to detect α-amino acids,[216] did not liberate CO_2; (ii) reduction of alkaline ferricyanide or Fehling's reagent after heating with HCl: presence of an aldehyde group; (iii) presence of one methoxy group and a free amino group adjacent to a hydroxy group; (iv) absence of an acetyl group; and (v) consumption of 3 mol of periodate with concomitant formation of 1 mol each of formic acid and glyoxylic acid (CHO–COOH) which was an indication for an α,β-dihydroxy-carboxylic acid.

In 1953, just like Hiyama, Gottschalk also reported that the alkaline hydrolysate of bovine submaxillary gland mucin was shown to contain pyrrole-2-carboxylic acid [comparison with synthetic pyrrole-2-carboxylic acid; formation of DL-proline (pyrrolidine-2-carboxylic acid) on reduction with PtO_2/H_2].[188] Based on the different findings,[174,177,188] it was hypothesized that the substance released with the influenza viral enzyme (Section 2.1) is built up from a pyrrole-2-carboxylic acid, whereby the pyrrole nitrogen is glycosidically linked to an undefined sugar residue (structure **A**) (Fig. 8).[188,217] In bound form, the *N*-glycosyl-pyrrole-2-carboxylic acid **A** was assumed to be connected via an amide linkage with the nitrogen of an adjacent hexosamine (structure **B**), probably D-chondrosamine (D-galactosamine). When incubated with the influenza viral/*V. cholerae* enzymes, and accepting these are amidases, the amide linkage is cleaved, thereby releasing structure **A**. Both nitrogen linkages of structure **B** were expected to be alkali labile, thereby explaining the formation of pyrrole-2-carboxylic acid from the mucoprotein in the Ehrlich's test with *p*-dimethylaminobenzaldehyde.

However, the latter hypothesis was withdrawn in a publication that followed.[218] As UV absorption spectra of the untreated mucoprotein did not give indications for the presence of a pyrrole derivative (absence of a peak at 256 nm), it was suggested that pyrrole-2-carboxylic acid was formed by alkaline degradation (0.05 M Na_2CO_3, 20 min, 100°C) of the enzymatically or chemically released substance from the mucoprotein, in a way as visualized in Scheme 2A. The precursor of pyrrole-2-carboxylic acid was supposed to be pyrroline-4-hydroxy-2-carboxylic acid, stabilized by the engagement of its enolic hydroxy group in an O-glycosidic linkage

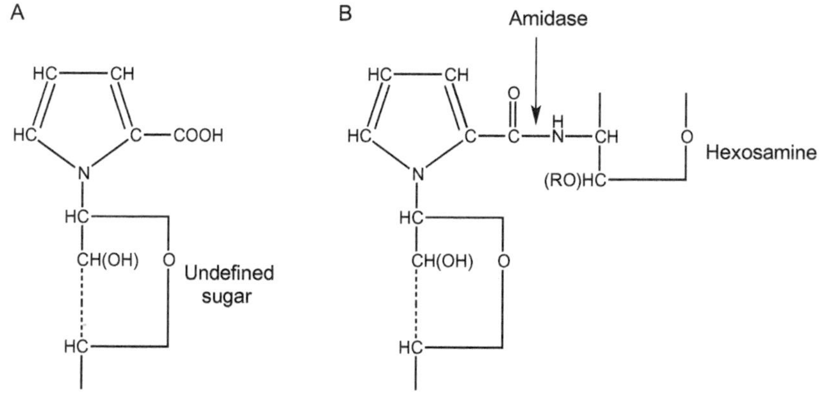

Fig. 8 Gottschalk's *N*-glycosyl-pyrrole-2-carboxylic acid proposal (A, free form; B, bound form) for the structure of the enzymatically released substance from mucoprotein.[188,217]

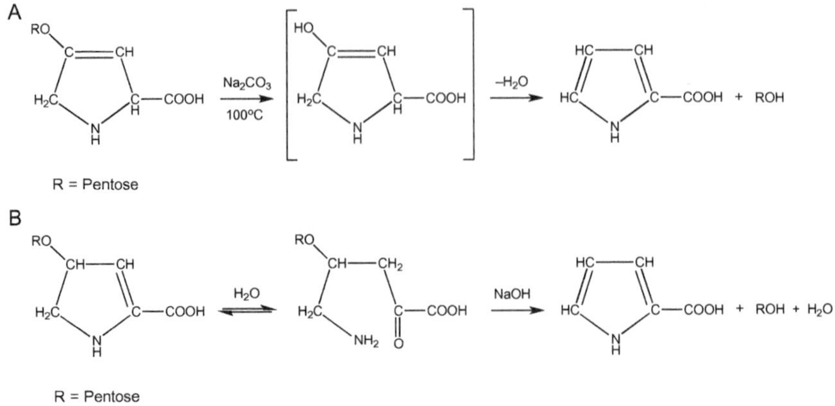

Scheme 2 Gottschalk's *O*-glycosyl-pyrrole-2-carboxylic acid proposals for the structure of the (bio)chemically released substance from mucoprotein with the formation of pyrrole-2-carboxylic acid.[218,219]

apparently with a pentose, being the released substance. The latter compound is very sensitive to alkaline hydrolysis. The choice of position 4 for the hydroxy group was based on the biogenetic relationship to hydroxyproline. In a subsequent paper from 1955, however, an isomeric structure for the *O*-glycosyl-pyrrole-2-carboxylic was assumed (Scheme 2B).[219] In aqueous solution, such a structure will undergo a reversible ring opening, yielding a product that would be unstable to alkali (or acid), and gives, under the expulsion of the sugar, pyrrole-2-carboxylic acid.

> *From a citation of Gottschalk[218] discussing the earlier report of Hiyama[215] about the formation of pyrrole-2-carboxylic acid (see Scheme 1): "It is not known whether pyrrolidines are transformed into pyrroles under the conditions applied, and Hiyama's results cannot therefore be interpreted."*

Then, inspired by the data of the Uppsala and Cologne groups, Gottschalk reinterpreted his earlier structural data[218,219] and formulated a new correlation among his finding of pyrrole-2-carboxylic acid, the bovine "sialic acid" data (Section 2), and the "methoxyneuraminic acid"/"neuraminic acid" data (Section 2). Taking into account the revised chemical formula $C_{13}H_{21}NO_{10}{\cdot}H_2O$ for bovine "sialic acid,"[195] with one *N*- and one *O*-acetyl group, C_9-sugar structures were constructed as depicted in Scheme 3.[220] In this proposal, "neuraminic acid" **c** is presented as an aldol type of condensation product of a 2-amino-2-deoxy-hexose with pyruvic acid. For the cyclization to pyrrole-2-carboxylic acid **d**, the α-keto function, the δ-amino function, and the γ-hydroxyl function of the open

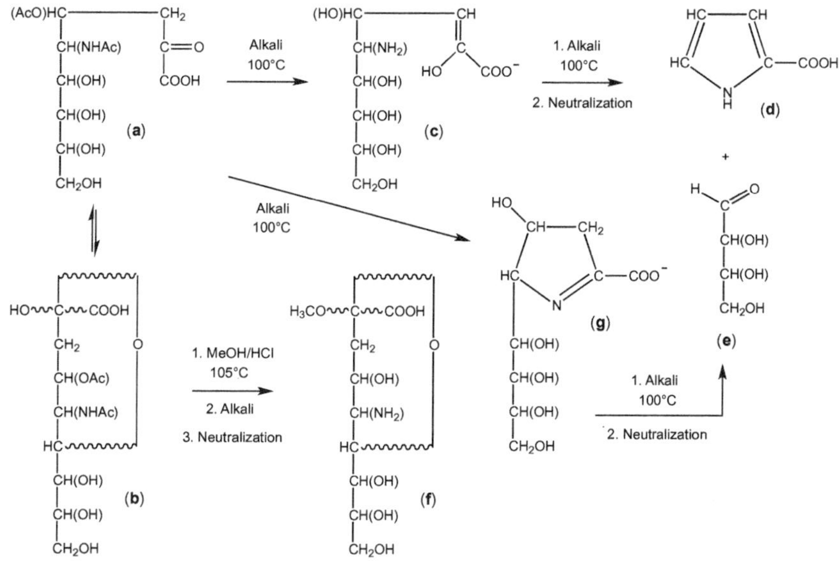

Scheme 3 Reaction scheme to explain Gottschalk's proposal for the structures of "sialic acid" (**a** and **b**) ($C_{13}H_{21}NO_{10}$), "neuraminic acid" (**c**), and "methoxyneuraminic acid" (**f**, $C_{10}H_{19}NO_8$), including the formation of pyrrole-2-carboxylic acid (**d**) and the tetrose (**e**), and a possible intermediate **g**.[220,222]

deacetylated chain (**a** → **c**) are in the right positions relative to the carboxyl group. Blix and coworkers stated that the supposed pyranose ring form is in accordance with their periodate experiments, carried out on bovine, porcine, ovine, and equine "sialic acids."[23] Gottschalk's proposal also included for terminal "*N*-acetylneuraminic acid" an enzymatic release from mucoproteins via a glycosidase instead of an amidase.[97,221] See also ref. 190. A first unraveling of the mechanism behind the enzymatic release had to wait until the 1980s.

The citation of Blix and coworkers (1956) in view of their finding that the O-acetyl group should be on C7 instead of C4 (Scheme 3): "The bovine 'sialic acid' consumes only 1 mole of oxidant per mole, and produces no formic acid. This indicates that the C7 group is acetylated in this substance. The position of the O-acetyl group in Gottschalk's formula is therefore probably wrong."[23] The structure was corrected in 1957.[222]

Experiments, whereby D-glucosamine and pyruvic acid were coupled in alkaline medium (pH 11) for 20 min at 100°C via an aldol condensation, and pyrrole-2-carboxylic acid was formed in trace amounts, formed a further support for this correlation.[220,222] Also a possible other degradation route **a**→**g**→**d** (**g** is an unstable 4-hydroxypyrroline derivative) has been included in Scheme 3.[222] Although the positive aldol condensation with D-glucosamine also fixed in fact the stereochemistry around C5, C6, C7, and C8, these have been left open in Scheme 3 and will be further discussed in Section 4.1. Note that "sialic acid" (**a**), but not "methoxyneuraminic acid" (**f**), can be converted into pyrrole-2-carboxylic acid.

A different proposal for the structure of "methoxyneuraminic acid" appeared in 1956 by Klenk and coworkers.[171] Based on the earlier investigations on "methoxyneuraminic acid," now called "neuraminic acid methyl glycoside," the following observations were taken together[22,170,172,201,208] (see also Section 2): (i) presence of a carboxyl group; (ii) presence of a primary amino group, but not in an α-position to the carboxyl group; (iii) presence of a methoxyl group; (iv) absence of an acetyl group, an acetamido group, or a reducing group; (v) consumption of 3 mol of periodate, and formation of 1 mol of formic acid and 1 mol of formaldehyde in mild periodate oxidation experiments; (vi) IR support for a carboxyl group; (vii) for "*N*-acetylneuraminic acid" it holds that the acetamido group should be situated next to the glycosidic C-atom. Further, using a plethora of chemical reactions, the authors arrived finally at a structure for "methoxyneuraminic acid" ($C_{11}H_{21}NO_9$), as formulated in Fig. 9.

Taking together all structural data so far available, Gottschalk's proposal of a C_9-sugar[220,222] (Fig. 10, Starting Structure) turned out to be the right

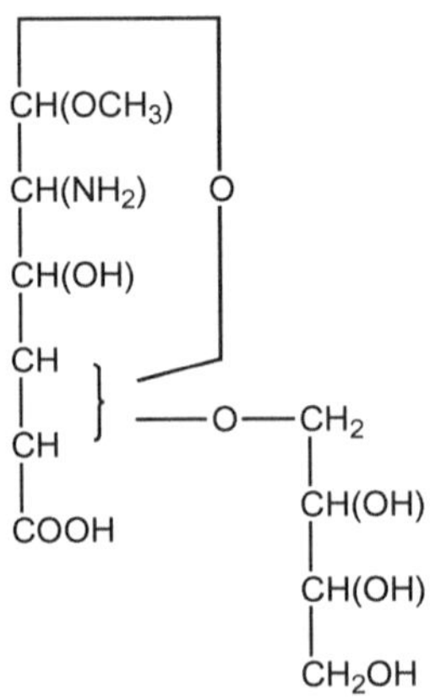

Fig. 9 Structure of "methoxyneuraminic acid" according to Klenk and coworkers; $C_{11}H_{21}NO_9$ (mw 311) was the best fit with the elemental analysis.[171] Note that the ultimately correct chemical formula, $C_{10}H_{19}NO_8$ (mw 281), was pushed aside. Interestingly, molecular weight determination, via cryoscopy in water, yielded 273 and 276 Da.

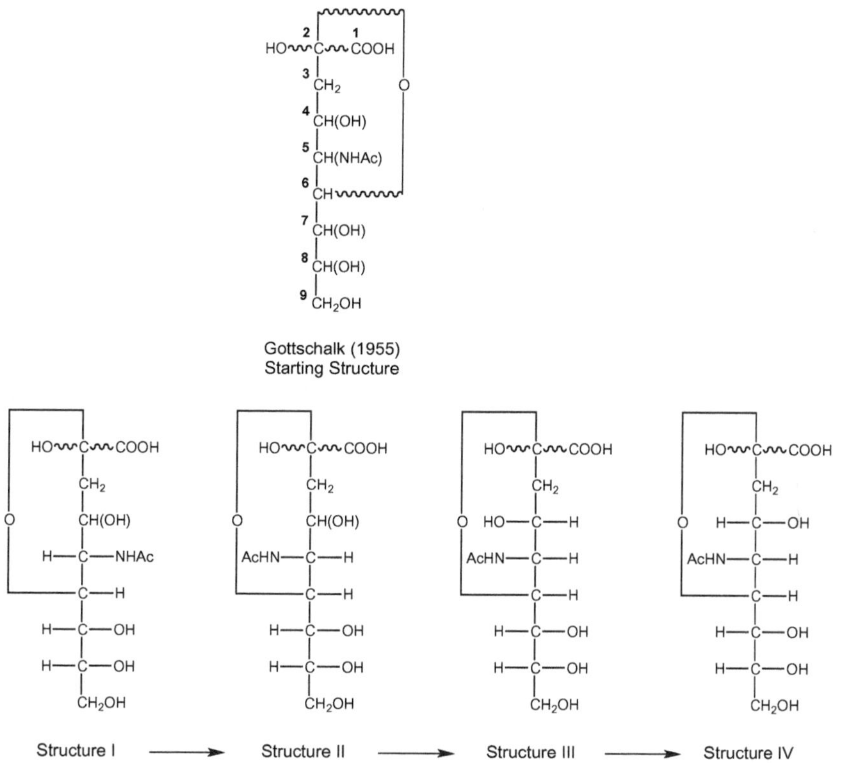

Fig. 10 Survey of the search for the stereochemistry around C4, C5, C6, C7, and C8 of *N*-acetylneuraminic acid in a historical perspective: a path of trial and error (see Section 4).

direction to continue the battle for the chiralities of the carbon atoms in the final structure. The problems with the great variations in elemental analyses (including separate nitrogen determinations) on different preparations of the isolated substances, published over a long period of time, certainly delayed the progress in the field. One of the causes for these variations was due to the frequent use of methanol in work-up procedures, leading to contaminating methyl-esterification of the relatively strong acids. Another complication was mentioned to be the hygroscopicity of the substances. But also the possible presence of closely related family members in specific preparations should not be excluded.

In view of all the foregoing, after more than 20 years of activity in many laboratories, Blix, Gottschalk, and Klenk finally arrived, in 1957, at a joint statement on the chemical relationship of the substances designated over the years by various names, such as "sialic acid," "neuraminic acid," "methoxyneuraminic acid," "lactaminic acid," "sero-lactaminic acid," "gynaminic acid," "hemataminic acid," and "prehemataminic acid." It was formulated as follows: "In order to avoid further confusion, we propose to call the basic, unsubstituted compound **neuraminic acid**. Sialic acid is suggested as group name for the acylated neuraminic acids (for example, *N*-acetylneuraminic acid, *N*-glycolylneuraminic acid, and diacetylneuraminic acids). For the enzyme, which splits the glycosidic linkage joining the terminal sialic acid to the residual oligo- or polysaccharide, the names **neuraminidase** and **sialidase** may be used synonymously. We have agreed to use this nomenclature in future."[223] Looking back, one can say that this joint statement was really a breakthrough after many years of experiments and competition.

Gottschalk wrote in a letter to Klenk (in 1957) about the final phase of the Nature *publication: "... I have just returned the corrected proofs of our joint note to* Nature, *ordered and paid 250 reprints in all and asked the printers to send all reprints to you. I am sorry that I could not consult you with regard to the reprints; however, I had to deal with the proof by return mail and the reprints have to be ordered when returning the proof. I have informed Blix accordingly and asked him to claim reprints (up to 85) from you. I hope you agree with this arrangement ..." (from the archive of Ernst Klenk, held by his son Hans-Dieter Klenk).*

Fifteen years later, Scott, Yamashina, and Jeanloz proposed to change the terminology by replacing the name neuraminic acid by aminosialosonic acid (the reference sugar should be a hypothetical 3-deoxynonulose, called sialose and abbreviated Sia), which means that *N*-acetylneuraminic acid should be named acetylaminosialosonic acid.[224] But the proposal was rejected by the IUPAC-IUB Commission on Biochemical Nomenclature.[225] Only the abbreviation Sia survived, but it is now used as an abbreviation for sialic acids in general.

4. THE BATTLE FOR THE STEREOCHEMISTRY

4.1 The C5, C6, C7, and C8 Chiralities

As mentioned in Section 3, the sialic acid formula (C5–C8 chiralities) of Gottschalk was partly based on an aldol condensation product of D-glucosamine and pyruvic acid, yielding pyrrole-2-carboxylic acid in alkaline medium (Fig. 10, Structure I).[220,222] Support for such a structure was supplied by studies of Kuhn and Brossmer.[206] Heating (90 min, 100°C) of "lactaminic acid" (*N*-acetylneuraminic acid) with $Ni(OAc)_2 \cdot 4H_2O$ in pyridine yielded *N*-acetyl-D-glucosamine (D-GlcNAc), which afforded the chiralities for C5, C6, C7, and C8 (Fig. 10, Structure I), and carbon dioxide [–CH(OH)–CH_2–CO–COOH → –CH(OH)–CH_2–CHO + CO_2]. Because of the anomeric C2 atom, "lactaminic acid" should occur in α- and β-forms; therefore, it was surprising that no mutarotation was observed (see Section 4.3). Also Zilliken and Glick demonstrated that incubation of "gynaminic acid" (*N*-acetylneuraminic acid) with 0.1 M NaOH (5–10 min at 90°C) afforded D-GlcNAc and pyruvic acid.[226] In another study, incubation of bovine submaxillary gland mucin with a *V. cholerae* filtrate, containing the RDE, yielded besides sialic acid, also D-GlcNAc, leading to the suggestion that still another enzyme should be present in the extract.[227] Additional incubation of sialic acid with the same extract gave equimolar amounts of D-GlcNAc and pyruvic acid, and an aldolase mechanism for this splitting was suggested (see Section 12). Finally, *N*-acetylneuraminic acid could be synthesized from D-GlcNAc and oxaloacetic acid at pH 11 in 1%–2% yield (Scheme 4).[228]

In the same period, however, another player in the field, i.e., the group of Saul Roseman[229,230] (Fig. 11) and coworkers at the University of Michigan, Ann Arbor (USA) and later at the Johns Hopkins University, Baltimore (USA), reported enzymatic experiments that were not in agreement with the assigned chirality around C5, as discussed earlier.[231,232] Incubation of *N*-acetyl- and *N*-glycolylneuraminic acid with an enzyme purified from *Clostridium perfringens* (pH 7.2, 3 h, 37°C) yielded in an aldol-type cleavage an *N*-acyl-hexosamine and pyruvic acid. In the case of *N*-acetylneuraminic

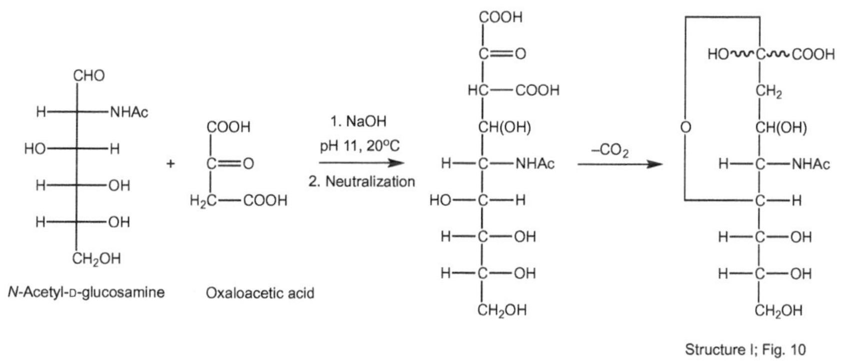

Scheme 4 Synthesis of *N*-acetylneuraminic acid (Structure I; Fig. 10) from *N*-acetyl-D-glucosamine and oxaloacetic acid.[228]

Fig. 11 Saul Roseman (1921–2011). *Photo courtesy of Prof. Dr. Ronald Schnaar.*

acid, the *N*-acyl-hexosamine was identified as *N*-acetyl-D-mannosamine (D-ManNAc), not D-GlcNAc (Fig. 10, Structure II). Note that the D-configuration was established by a chemical degradation reaction of mannosamine, yielding D-arabinose. Note also that D-GlcNAc and D-ManNAc could only be separated by paper chromatography, when using borate-treated paper. The reaction was reversible; thus incubation of D-ManNAc and pyruvate using this enzyme led to the formation of *N*-acetylneuraminic acid (Scheme 5). The enzyme was inactive with D-GlcNAc and *N*-acetyl-D-galactosamine (D-GalNAc). Phosphoenolpyruvate could not replace pyruvate. In view of its activity, the enzyme was called *N*-acetylneuraminic acid aldolase [later called sialate (acylneuraminate)-pyruvate lyase; see Section 12]. Incubation of the enzyme with *N*-glycolyl-D-mannosamine and pyruvic acid afforded *N*-glycolylneuraminic acid.

Pushed by the results of Comb and Roseman, Kuhn and Brossmer repeated the degradation of *N*-acetylneuraminic acid by $Ni(OAc)_2{\cdot}4H_2O$ in hot pyridine and concluded that the primary formation of D-ManNAc was followed by a rapid epimerization to D-GlcNAc.[233] Besides the unexpected epimerization in hot pyridine + $Ni(OAc)_2$ (an epimerization was hardly seen in hot pyridine alone), of which the balance lies strongly on the side of D-GlcNAc, a paper chromatography system used earlier by these authors was not able to separate D-ManNAc and D-GlcNAc. However, using borate-treated paper in a time-course experiment, the initial formation of D-ManNAc, next to D-GlcNAc, was clearly seen. The D-ManNAc ↔ D-GlcNAc reversible epimerization (ratio 1:4) was also observed in aqueous

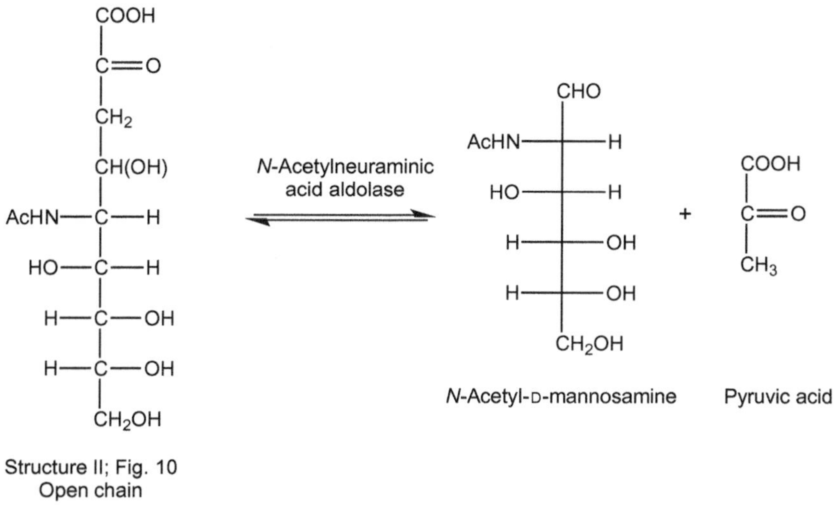

Scheme 5 *N*-Acetylneuraminic acid aldolase-catalyzed cleavage and synthesis of *N*-acetylneuraminic acid (Structure II; Fig. 10) with *N*-acetyl-D-mannosamine and pyruvic acid as counter compounds.[231,232]

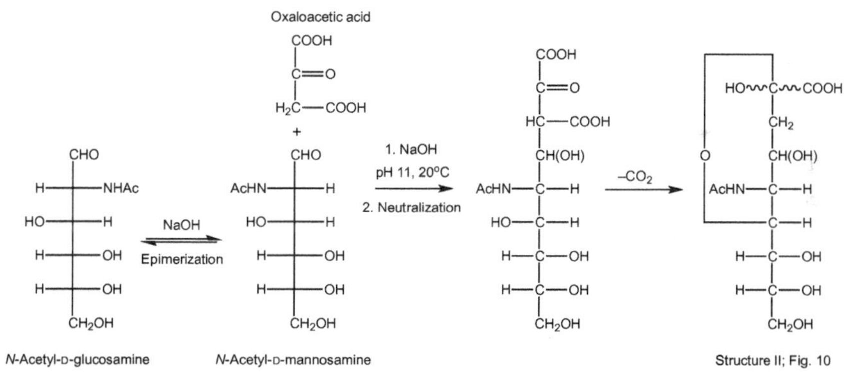

Scheme 6 Synthesis of *N*-acetylneuraminic acid (Structure II; Fig. 10) from D-GlcNAc or D-ManNAc and oxaloacetic acid.[235,236]

alkaline solution (pH 10–11) at room temperature,[233–235] thereby explaining the earlier results of Zilliken and Glick[226] and of Cornforth and coworkers.[228] Repetition of the synthesis of *N*-acetylneuraminic acid from oxaloacetic acid and either D-GlcNAc or D-ManNAc confirmed this view. In both cases *N*-acetylneuraminic acid was formed (Scheme 6; Fig. 10, Structure II).[235,236]

Replacing oxaloacetic acid by its di-*tert*-butyl ester gave higher yields.[237] In a repeated study, incubating *N*-acetylneuraminic acid with a *V. cholerae* filtrate (pH 6.1, 70 h, 37°C), D-ManNAc, but not D-GlcNAc, was found.[238] The earlier observation of splitting into D-GlcNAc and pyruvate[227] was also attributable to a paper chromatography protocol that could not separate D-GlcNAc and D-ManNAc.[238] In another report, the system D-ManNAc/phosphoenolpyruvate/ATP/rat liver extract (not D-GlcNAc or D-GalNAc; not pyruvic acid) was described for the synthesis of *N*-acetylneuraminic acid.[239] Finally, several years later a protocol was described to determine the absolute configuration of *N*-acetylneuraminic acid by converting it into D-arabinose by chemical degradation.[240]

4.2 The C4 Chirality

Also the estimation of the stereochemistry around C4 required a few years, whereby the Heidelberg group of Kuhn and Brossmer played a major role. Reaction of *N*-acetylneuraminic acid with C_2H_5SH/concd. HCl yielded the γ-lactone of *N*-acetylneuraminic acid diethyl dithioacetal. With reference to Hudson's lactone rules,[241] the strong levorotary value of the γ-lactone $\{[\alpha]^{23}{}_{D}\ -83°\ (c\ 1,\ CH_3OH)\}$ was interpreted as belonging to a Fischer projection formula structure with the lactone ring drawn to the left. This meant for *N*-acetylneuraminic acid that the OH group at C4 should project to the left (Fig. 10, Structure III; Scheme 7).[242,243] Using Raney nickel, the diethyl

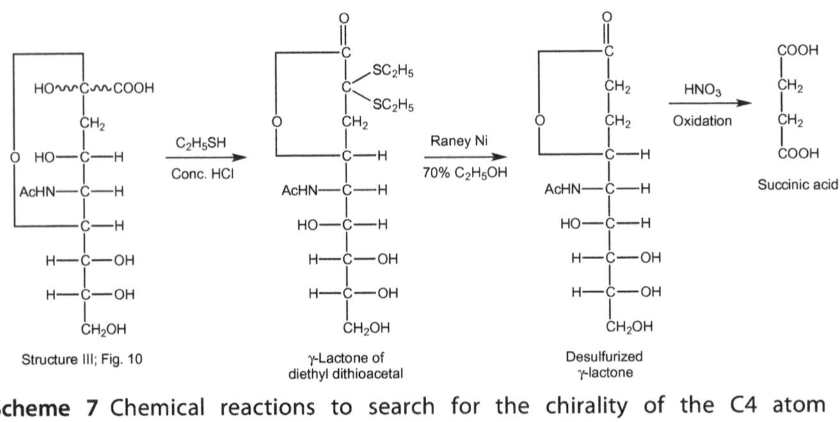

Scheme 7 Chemical reactions to search for the chirality of the C4 atom of *N*-acetylneuraminic acid: The OH group at C4 should project to the *left* in the Fischer projection formula (Structure III; Fig. 10).[243]

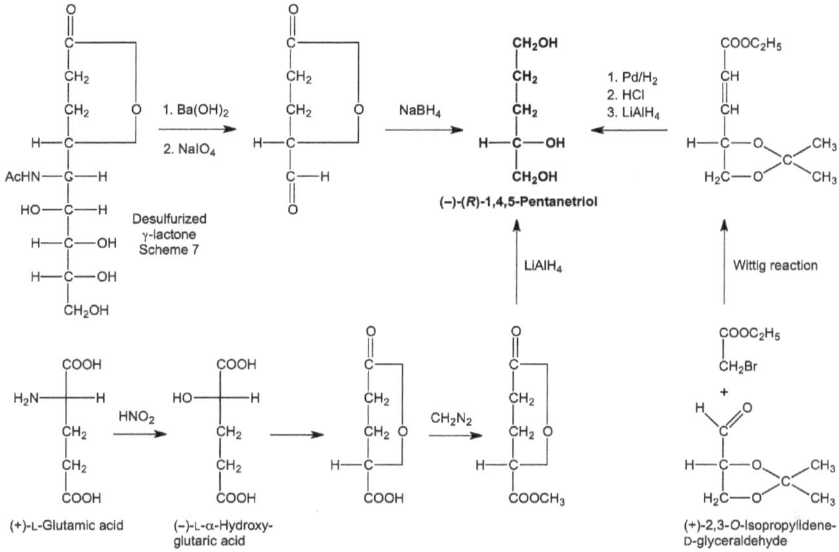

Scheme 8 Three routes to prove the correct chirality around C4 of *N*-acetylneuraminic acid: The OH group at C4 should project to the *right* in the Fischer projection formula (Structure IV; Fig. 10).[244]

dithioacetal could easily be converted into the corresponding levorotating desulfurized γ-lactone. Oxidation of the desulfurized γ-lactone with HNO_3 at 120°C yielded succinic acid.[243]

However, additional experiments along three chemical pathways by Kuhn and Brossmer led to a correction of the stereochemistry around C4 (Scheme 8).[244] First of all, the desulfurized γ-lactone mentioned earlier could be converted into (−)-(*R*)-1,4,5-pentanetriol {$[\alpha]^{20}_{D}$ −18° (pyridine)} (route 1). The same pentanetriol could also be synthesized from (+)-L-glutamic acid via (−)-L-α-hydroxyglutaric acid (route 2), and from the Wittig condensation product of (+)-2,3-*O*-isopropylidene-D-glyceraldehyde and bromoacetic acid ethyl ester (route 3). So, in the Fischer projection formula the OH group at C4 should project to the right instead of to the left (Fig. 10, Structure IV).

A fourth chemical route, described by Kuhn and Baschang[237] also led to an OH group at C4 projected to the right (Scheme 9). Here, ozonolysis of the generated γ-lactone of *N*-acetylneuraminic acid[245] and reductive work-up afforded oxalic acid and 3-acetamido-3-deoxy-D-*glycero*-D-*galacto*-heptose, and not 3-acetamido-3-deoxy-D-*glycero*-D-*talo*-heptose.

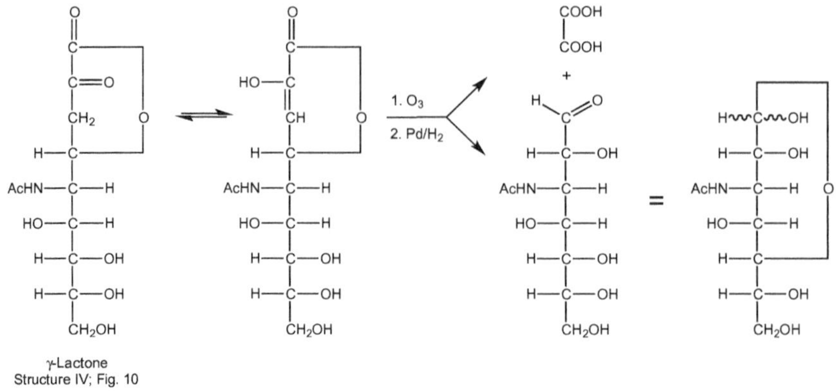

Scheme 9 Synthesis of 3-acetamido-3-deoxy-D-*glycero*-D-*galacto*-heptose from the γ-lactone of *N*-acetylneuraminic acid (Structure IV; Fig. 10).[245]

4.3 The C2 Chirality and the Ring Form/Conformation

From the time that a C_9 structure with a keto group on C2 and an acetamido group at C5 was generally accepted for sialic acid, a pyranose ring form between C2 and C6 was drawn, although still in a Fischer projection formula. A furanose ring form was not possible, and a seven-membered ring was not obvious. As mentioned in Section 3, results from periodate oxidation studies were in agreement with such a pyranose ring form.[23] Further support for a C2–C6 hemiacetal was supplied by detailed inspection of the IR spectra of Blix's bovine, porcine, ovine, and equine sialic acids.[246]

Studies focused on the α,β-anomericity at C2 of sialic acids started with a more detailed characterization oflactaminic acid-lactose (*N*-acetylneuraminyl-lactose) (Section 2.3) by the Heidelberg group. Isolated crystalline *N*-acetylneuraminic acid showed an increasing mutarotation in $(CH_3)_2SO$ as follows: $[\alpha]^{23}_{D}$ −115° (t = 7 min) → −37° (t = 23 min) → −24° (final value) (c 0.2) (extrapolated to t = 0 min, $[\alpha]^{23}_{D}$ should be −150° to −200°). This observation was correlated with the β-form for the crystalline product, which means that the OH group at C2 in the Fischer formula should project to the left.[242] As in H_2O no change in mutarotation was observed, it was concluded that in this solvent the mutarotation was too fast (strong acid) to be followed in the polarimeter. Note that in other studies no mutarotation could be observed in $(CH_3)_2SO$,[247] whereas in H_2O mutarotation was seen.[248] Applying Hudson's isorotation rules,[249] comparison of the optical rotations in $(CH_3)_2SO$ of (*N*-acetylneuraminic acid)-(α,β-lactose) ($[\alpha]^{23}_{D}$ +6°), α,β-lactose ($[\alpha]^{23}_{D}$ +53°), and assumed β-*N*-acetylneuraminic acid

($[\alpha]^{23}{}_{D}$ −150° to −200°; see above), it was hypothesized that the bound *N*-acetylneuraminic acid occurred in the α-form. In case of a β-glycosidic linkage, a negative optical rotation was expected for (*N*-acetylneuraminic acid)-(α,β-lactose).[250,251] But it was stated that for further support pairs of epimeric α- and β-ketosides should be synthesized. Note that the Fischer projection formulae used in these reports (chiralities around C4 and C5) were still not the final ones.

Indeed, in the following years large series of glycosides were synthesized to study the α,β-anomericity at C2 (for a review, see ref. 252), and two examples will be given. Meindl and Tuppy[253] prepared a series of alkyl and aryl glycosides of *N*-acetylneuraminic acid, starting from *N*-acetylneuraminic acid via 2,4,7,8,9-penta-*O*-acetyl-*N*-acetylneuraminic acid and 4,7,8,9-tetra-*O*-acetyl-*N*-acetyl-2-chloro-2-deoxy-neuraminic acid as intermediates (Koenigs–Knorr protocol). It turned out that all glycosides could be cleaved with *V. cholerae* and influenza virus sialidases, and based on the earlier studies of Kuhn and Brossmer, the compounds were assigned as α-ketosides.[253,254] Kuhn and coworkers converted *N*-acetylneuraminic acid methyl ester via 2,4,7,8,9-penta-*O*-acetyl-*N*-acetylneuraminic acid methyl ester and 4,7,8,9-tetra-*O*-acetyl-*N*-acetyl-2-chloro-2-deoxy-neuraminic acid methyl ester as intermediates (Koenigs–Knorr protocol) into *N*-acetylneuraminic acid methyl ester methyl glycoside with $[\alpha]^{20}{}_{D}$ −6° (CH_3OH), and *N*-acetylneuraminic acid, using a CH_3OH/Dowex-50 [H^+] (24 h, boiling under reflux) protocol, into *N*-acetylneuraminic acid methyl ester methyl glycoside with $[\alpha]^{20}{}_{D}$ −46° (CH_3OH).[255] As the first preparation was less levorotatory than the second preparation, the first one was hypothesized to be the α-anomer, and the second one the β-anomer (Hudson's isorotation rules[249]). The first preparation and its free acid form were susceptible to *V. cholerae* sialidase, while the second preparation and its free acid form were resistant.

The application of 100-MHz ^{1}H NMR (nuclear magnetic resonance) spectroscopy (1968) gave conclusive evidence for the occurrence of pyranose ring forms in $^2C_5$ (1*C*) chair conformations.[256,257] Taking the per-O-methylated methyl ester benzyl glycosides α-**3** and β-**3** as examples, ^{1}H NMR analysis in $CDCl_3$ showed axial positions for H3a, H4, H5, and H6 ($J_{3a,4}$, $J_{4,5}$, $J_{5,6}$ >10 Hz) and an equatorial position for H3e ($J_{3e,4}$ ~5 Hz), so equatorial positions for OR^2, NHAc, and the glycerol side chain at C4, C5, and C6, respectively (Fig. 12). Although not discussed,[257] the large *J* coupling constant between H7 and H8 ($J_{7,8}$ ~8 Hz) in the per-O-acetylated methyl ester benzyl glycoside α-**2** is in agreement with

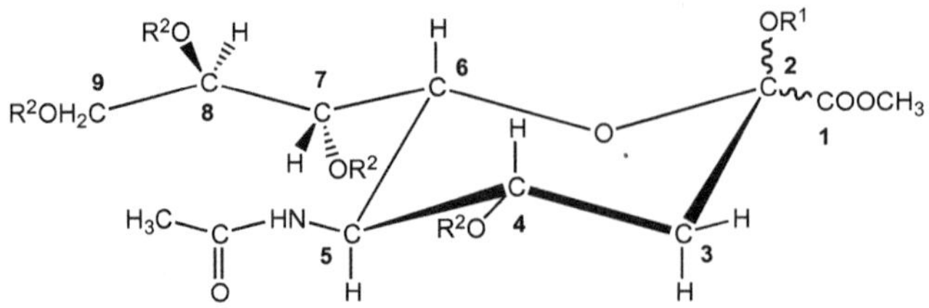

Fig. 12 Conclusive evidence for the $^{2}C_{5}$ chair conformation of the pyranose ring, and the anomeric configuration assignment, of *N*-acetylneuraminic acid methyl ester (derivatives), based on 100-MHz ^{1}H NMR analysis.[257]

a staggered anti orientation of the OR^2 groups at C7 and C8; the *J*-value for the coupling between H6 and H7 is small (~2 Hz). Finally, in the ^{1}H NMR spectrum of methyl ester **1**, recorded in dimethyl sulfoxide-d_6:benzene-d_6 = 8:3, a long-range coupling between H3a and HO2 (*J* 2.1 Hz) indicated that HO2 occurs in the axial orientation, in agreement with a β-configuration (H3a–C–C–HO2 in a W conformation). For a further discussion on the conformation of the glycerol side chain, see Section 6.

With guidance from the observed optical rotation measurements, combined with Hudson's isorotation rules, several studies suggested the α-configuration for bound *N*-acetylneuraminic acid. However, there was a general feeling of caution, given the initially wrong configurational assignment of the hydroxy group at C4 using Hudson's lactone rule.

Conclusive data about the C2 α,β-anomericity were published in 1969 by Yu and Ledeen.[258] For this study, use was made of the earlier synthesized[255] methyl esters of the sialidase-sensitive and sialidase-resistant forms of *N*-acetylneuraminic acid methyl glycoside. Both compounds were subjected to mild periodate oxidation followed by sodium borohydride reduction, and the degradation products that formed were treated with *N,N'*-dicyclohexylcarbodiimide in pyridine (Scheme 10). The C_7 product related to the sialidase-sensitive methyl glycoside did not give rise to lactone formation, whereas the C_7 product related to the sialidase-resistant methyl glycoside gave rise to lactone formation. IR analysis of the isolated lactone fraction [preparative thin-layer chromatography (TLC)] revealed evidence for a δ (major) + γ (minor) lactone mixture. The trimethylsilyl derivative of the major lactone, isolated via gas–liquid chromatography (GLC), was analyzed by electron-impact mass spectrometry (EIMS), and shown to be a δ-lactone. Boat conformations of the lactones were supported by Dreiding modeling. All these findings were in agreement with an axial orientation of the carboxyl function (α-configuration) in the case of the sialidase-susceptible

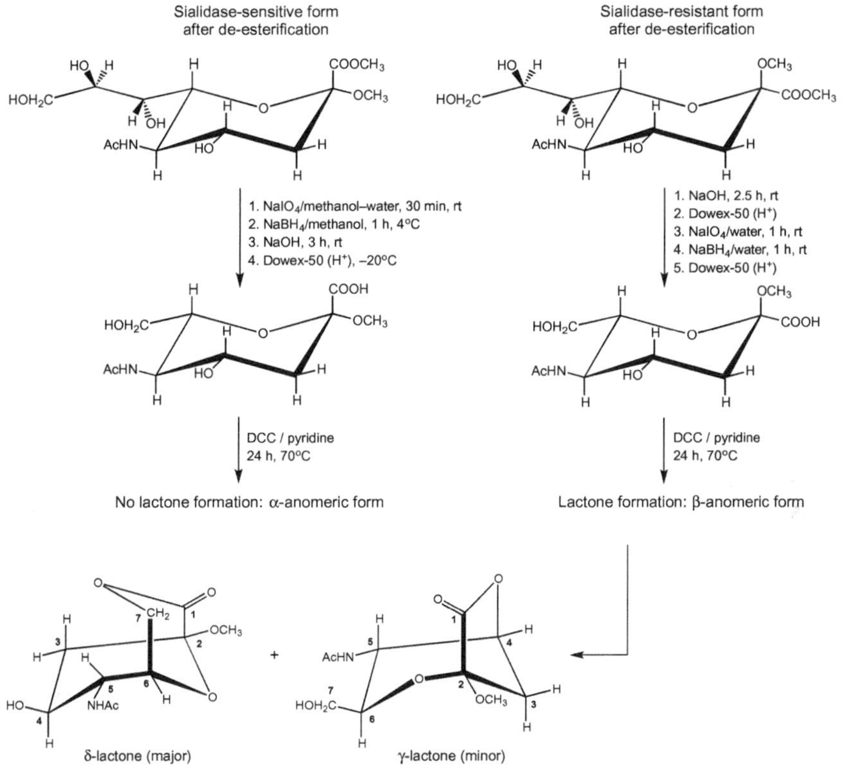

Scheme 10 Elucidation of the stereochemistry around the anomeric carbon atom via a search for lactone formation, by subjecting the two anomeric *N*-acetylneuraminic acid methyl ester methyl glycosides to sequential periodate degradation, borohydride reduction, and treatment with *N,N'*-dicyclohexylcarbodiimide (DCC).[258] The two starting compounds differ, after de-esterification, in their susceptibility to sialidase.

methyl glycoside and with an equatorial orientation of the carboxyl function (β-configuration) in the case of the sialidase-resistant methyl glycoside. So, sialidases (here *V. cholerae* sialidase) were proven to be α-sialidases (see Section 11.7). Note that also the glycerol-side-chain-trimmed α-anomer was susceptible to sialidase, but the β-analogue not susceptible. Note also that *N*-acetyl-α-neuraminic acid methyl ester methyl glycoside was totally inactive with sialidase, which is contrary to results published three years earlier.[255]

4.4 The Replacement of NH_2-C5 by HO-C5

The history with respect to deaminated neuraminic acid (ketodeoxynononic acid, Kdn; Fig. 1) started in 1986.[12] In that year, the groups of Sadako and Yasuo Inoue et al. at the Showa University and the University of Tokyo,

Tokyo (Japan) reported an unknown sialic acid-like component (SiaX) in the monosaccharide GLC analysis of some glycopeptide fractions prepared from *Salmo gairdneri* (rainbow trout) egg polysialoglycoprotein.[259] In subsequent studies,[76] SiaX could be isolated as its methyl ester methyl glycoside. In comparison with permethylated *N*-glycolylneuraminic acid, the EIMS fragmentation pattern of permethylated SiaX supported a structure in which the $N(CH_3)COCH_2OCH_3$ group at C5 was replaced by an OCH_3 group (for EIMS data of sialic acids, see Section 5). Further support for a hydroxy group instead of an *N*-acyl group at C5 was supplied by the ^1H NMR spectrum of the SiaX methyl ester methyl glycoside (for NMR data of sialic acids, see Section 5). The spectrum turned out to be identical with deaminated neuraminic acid methyl ester methyl glycoside, synthesized from *N*-glycolylneuraminic acid.[260] Referring to the earlier work of Roseman et al. (Section 4.1), incubation of free SiaX with *N*-acetylneuraminic acid aldolase yielded D-mannose and pyruvate. Finally, SiaX showed the same absorption spectrum as *N*-acetyl- and *N*-glycolylneuraminic acid in the thiobarbituric acid assay. In conclusion, SiaX could be assigned as 3-deoxy-D-*glycero*-D-*galacto*-non-2-ulosonic acid. For information with respect to the metabolism of Kdn and the enzymatic release of this sialic acid from glycoconjugates, see Sections 11.5.3 and 11.7, respectively.

5. ELUCIDATING THE O-SUBSTITUTION PATTERNS

From the beginning, there were indications for the presence of *O*-acetyl groups in addition to *N*-acetyl or *N*-glycolyl groups in sialic acids. As mentioned in Section 2.1, the first studies, reported in 1936, spoke of the presence of an *O*-acetyl group in the unknown substance, isolated from bovine submandibular gland mucin.[165] It is to be noted that for the release of (O-acylated) sialic acids from sialoglycoconjugates/sialooligosaccharides, the chosen acid hydrolysis conditions must be such that the labile glycosidic linkage between sialic acid and a neighboring monosaccharide is efficiently cleaved, but also that the applied conditions do not cause too much de-esterification. Note that conditions that are too harsh lead to acid- and heat-mediated degradation to chromogens (humin formation). Over the years, the three acid hydrolysis systems most applied are formic acid pH 2.0 (1 h, 70°C) [in later years followed by HCl pH 1.0 (1 h, 80°C)],[24,104] 2 M acetic acid (3 h, 80°C),[261] and 2 M propionic acid (4 h, 80°C)[262]; however, other systems have been advocated. Although these conditions do not lead to significant N-deacylation, O-deacylation has been shown to occur to an extent of approximately 30%–50%. The use of aqueous H_2SO_4 is no longer recommended.[6,33,154] Besides mild acid hydrolysis, in later years sialidases made their appearance.[6,154] As mentioned in Section 2.1, even the acidity of mucin suspensions (pH 3.5–4.0) was low enough to initiate release of sialic acids.[23]

As reported in Section 3, analysis of the sialic acids released from bovine submaxillary gland mucin gave Neu5,7Ac_2, as deduced from periodate oxidation experiments.[23] In additional investigations based on periodate oxidation, besides Neu5,7Ac_2 (consumption of 1 mol of periodate), also Neu5Ac/Neu5Gc (consumption of 2 mol of periodate), and Neu5,7,8/9Ac_3 (no consumption of periodate) were found.[263] In the same report, equine submaxillary gland mucin was shown to contain Neu4,5Ac_2 (consumption of 2 mol of periodate). Also the RDE of *V. cholerae* was used to release *N,O*-acylneuramic acids from bovine submaxillary gland mucin, yielding Neu5Ac, Neu5Gc, and di- and triacetylated Neu5Ac.[264]

For the isolation of sialic acid variants, at the end of the 1960s Faillard and Schauer at the Ruhr-Universität Bochum (Germany) introduced a combination of anion-exchange and cellulose column chromatography.[24,119,154] Subsequently, the positions of the *O*-acetyl groups in the isolated sialic acids were determined by a combined approach as follows:

(i) Polarographic determination of the periodic acid consumption during mild periodate oxidation (pH 3). It was assumed that Neu5Ac, Neu5Gc, and Neu4,5Ac_2 consumed 2 mol of periodate/mol sialic acid, Neu5,7Ac_2 and Neu5,9Ac_2 1 mol of periodate/mol sialic acid, and Neu5,8Ac_2, 0 mol of periodate/mol sialic acid.

(ii) Colorimetric assay using periodic acid/thiobarbituric acid according to Aminoff.[265] The periodate oxidation was carried out at pH 1–2 and 37°C, whereby it was assumed that Neu5Ac, Neu5Gc, Neu4,5Ac_2, and Neu5,9Ac_2, but not Neu5,7Ac_2 and Neu5,8Ac_2, were degraded to β-formylpyruvic acid, the chromogen that forms a red chromophore with thiobarbituric acid (see Section 7.1).[265–269]

(iii) TLC on cellulose and silica-gel layers. Neu5Ac and Neu4,5Ac_2 are discriminated on the basis of different R_f values. Also Neu5,7Ac_2 and Neu5,8Ac_2 have different R_f values (see Section 7.2).[269]

Following this combined protocol, a series of *N,O*-acylneuraminic acids, released from bovine and equine submaxillary gland mucin via mild formic acid hydrolysis (pH 2–2.5, 1 h, 70°C), were characterized (Fig. 13).[24] In case

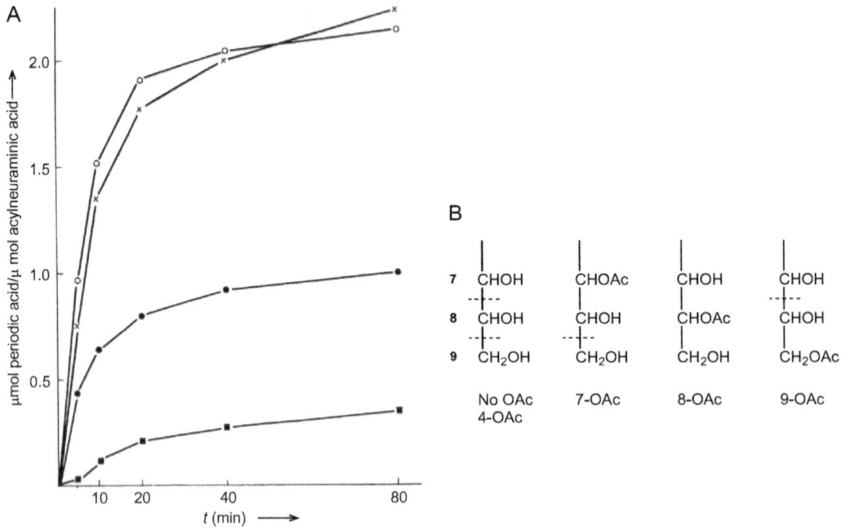

Fig. 13 (A) Polarographic measurement of the periodic acid consumption at 0°C of isomeric *N*-acetyl-mono-*O*-acetyl-neuraminic acids.[24] Neu5Ac, *cross*; Neu4,5Ac$_2$, *open circle*; Neu5,7Ac$_2$, *closed circle*; Neu5,8Ac$_2$, *closed square*. (B) Expected cleavages in the C7–C8–C9 glycerol side chains are indicated with ------. *Panel (A): Reproduced from Schauer, R. Chemistry and Biology of the Acylneuraminic* Acids. Angew. Chem., Int. Ed. Engl. ***1973***, 12, *127–138. Copyright Wiley-VCH Verlag GmbH & Co. KGaA.*

of the bovine source, Neu5Ac, Neu5Gc, Neu5,7Ac$_2$, and Neu5,8Ac$_2$ were identified, and in the case of the equine source, Neu5Ac, Neu5Gc, and Neu4,5Ac$_2$ were identified. Furthermore, it was suggested that the bovine material also contained Neu5,7,8Ac$_3$, and the equine material Neu4,5,?Ac$_3$. The consumption lines of Neu5Ac and Neu4,5Ac$_2$ showed over-consumption after 40 min, due to opening of the pyranose ring. In theory, Neu5,8Ac$_2$ should give no consumption of periodate; the slow consumption of periodate (0.25 mol/mol sialic acid in 40 min) combined with only a partial positive Aminoff test (extinction coefficient 47% of that obtained with free Neu5Ac) was explained as the result of a slow O-deacetylation. In fact, a possible presence of Neu5,9Ac$_2$ was excluded, because the combination of "1 mol periodate consumption and a positive Aminoff test" was not found. The number of *N*-glycolyl groups was estimated using a microanalysis protocol for glycolic acid,[270] and the number of *O*-acetyl groups was determined with the colorimetric Hestrin assay (alkaline hydroxylamine/Fe(ClO$_4$)$_3$; see Section 7.1).[271] Note that all bovine sialic acids could also be released with the sialidases of *V. cholerae* and *C. perfringens*, although at different rates, but equine Neu4,5Ac$_2$ and di-O-acetylated Neu5Ac (presumably one of the O-acetyl groups attached at C4) were resistant to these enzymes (see Section 11.5.2). In the same period, Neu5,8Ac_2, assigned via polarographic periodate oxidation studies, was found to occur in fish brain gangliosides.[272]

A few years later (1975), a novel technology for the analysis of *N*,*O*-acylneuraminic acids, namely, GLC–EIMS, was introduced by Vliegenthart/Kamerling and coworkers at Utrecht University (the Netherlands).[26,27] To this end, the isolated sialic acids were esterified with diazomethane, followed by trimethylsilylation of the free hydroxy groups. In a later protocol, the sialic acids were pertrimethylsilylated.[33] Analysis of the mass spectra suggested strongly that all *N*,*O*-acylneuraminic acids with an earlier identified 8-*O*-acetyl group by periodate oxidation, in fact, had to be assigned with an *O*-acetyl group at C9. This change in assignment was based on mass spectrometric fragmentation patterns that originated from the mass spectral analysis of partially methylated alditol acetates.[273–275] For cleavage of the latter compounds, at the end of the 1960s it had been demonstrated that the charge is preferentially located on an ether oxygen instead of on an ester oxygen atom. Translated to the sialic acid situation, it meant that, in general, cleavage occurs between two trimethylsiloxylated carbon atoms, or between an acetoxylated and a trimethylsiloxylated carbon atom with location of the charge on the ether oxygen, rather than between an acetoxylated and a trimethylsiloxylated carbon atom with location of the charge on the ester oxygen, or between two acetoxylated carbon atoms (Fig. 14A). The mass spectra obtained for the different sialic acid derivatives gave rise to highly specific fragmentation patterns A–H, both for *N*,*O*-acylneuraminic acids and for *N*-acyl-*O*-alkyl-neuraminic acids (Fig. 14B).[6,27,30,33,275] Fragments A and B indicate the molecular mass of the sialic acid derivatives, and thus the number and type of substituents. Fragments C–H contain the information concerning the position of the different substituents. Fragment C only has significant abundance if C7 bears an ether group; when an ester group is present at C7, this fragment is absent or hardly observable. The occurrence of fragment D is dependent on the presence of fragment C. Fragment E is not seen if an *O*-acyl group is attached to C4. Fragment F can be readily formed only if an ether group is attached at C8. As an illustration, the mass spectra of Neu5Ac and Neu5,9Ac_2 are presented in Fig. 15. Additional mass spectra of various members of the sialic acid family can be found in refs. 6, 30, 33, and 104, as well as in the references included in Table 1. For tables of *m*/*z* data of the A–H fragments and GLC retention times of various sialic acid derivatives, see refs. 6, 26, 27, 30, and 33, as well as the references included in Table 1.

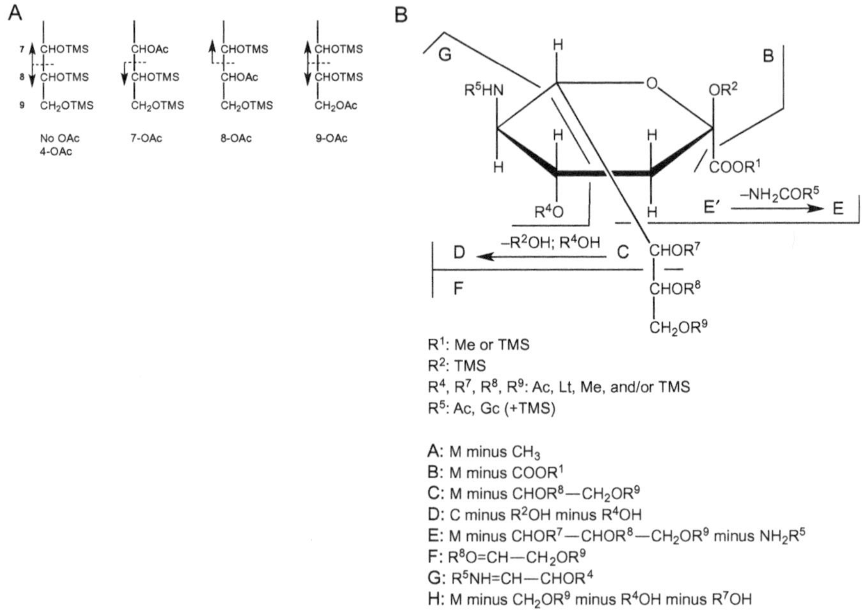

Fig. 14 (A) EI mass spectrometric cleavages in the C7–C8–C9 glycerol side chain of *N,O*-acylneuraminic acids. (B) Fragmentation scheme for the analysis of sialic acid derivatives.[6,27] *Ac*, acetyl; *Lt*, lactyl; *Me*, methyl; *TMS*, trimethylsilyl.

To prove the interpretation of the mass spectra of the 9-O-acetylated sialic acid derivatives, Neu5,9Ac$_2$ and Neu9Ac5Gc were incubated with acylneuraminate pyruvate lyase. The products formed were 6-*O*-acetyl-*N*-acetylmannosamine and 6-*O*-acetyl-*N*-glycolylmannosamine, respectively, as proven by GLC–EIMS.[27] Furthermore, synthetic β-Neu5,9Ac$_2$1,2Me$_2$ and β-Neu4,5,9Ac$_3$1,2Me$_2$, of which the structures were confirmed by ^{1}H and ^{13}C NMR spectroscopy, were subjected to mass spectrometry and periodate oxidation studies. Indeed, both derivatives showed mass spectrometric fragmentation patterns similar to those found for β-Neu5,9Ac$_2$ and β-Neu4,5,9Ac$_3$, and consumed only ~0.15 mol of periodate/mol of sialic acid within 10 min (β-Neu5Ac1,2Me$_2$ consumed 2 mol of periodate/mol sialic acid within 10 min).[276] An explanation for the resistance of 9-O-acetylated sialic acids to periodate oxidation was stated to originate from the staggered anti orientation of the OH groups at C7 and C8, stabilized by some hydrogen-bonding network, that hampers the formation of a diol–periodate complex (see Section 6). It seems to take time to create the complex with both OH groups in a *cis* configuration, as over a period of 48 h a higher periodate consumption (85% of their theoretical value) was

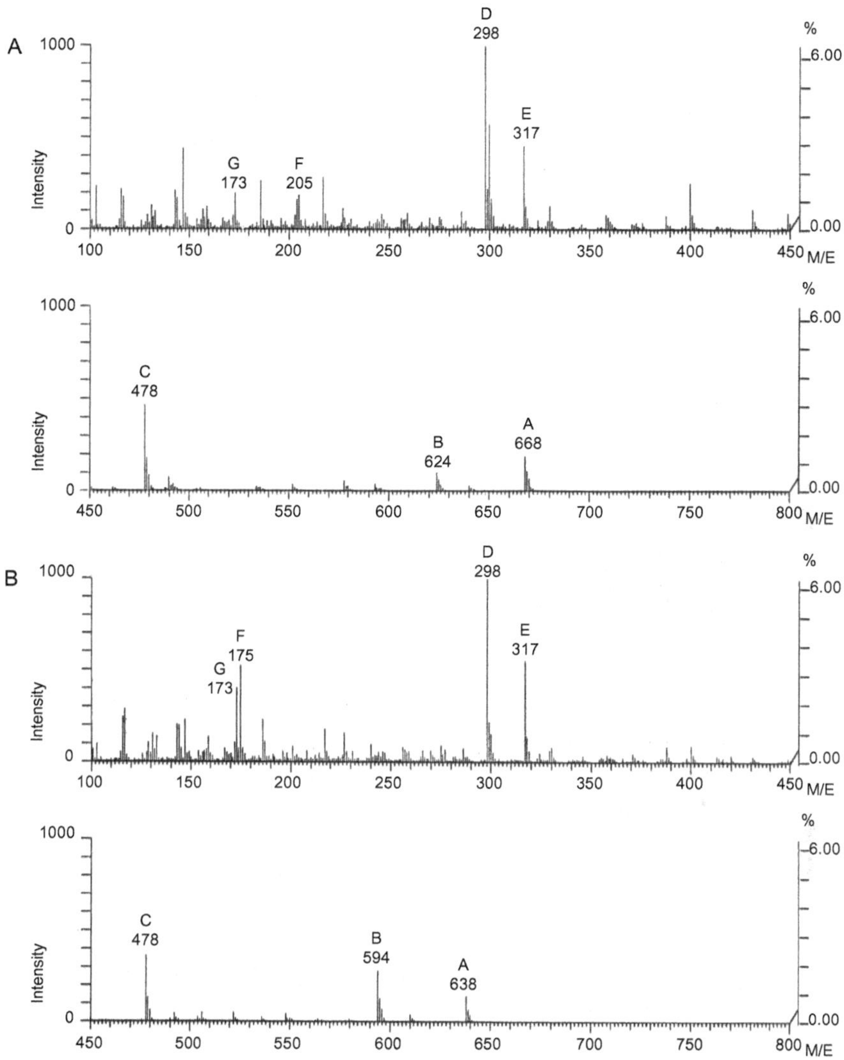

Fig. 15 (A) EI mass spectrum (70 eV) of the per-O-trimethylsilylated methyl ester of β-Neu5Ac. (B) EI mass spectrum (70 eV) of the per-O-trimethylsilylated methyl ester of β-Neu5,9Ac$_2$. *Reproduced from Kamerling, J. P.; Vliegenthart, J. F. G. Gas-Liquid Chromatography and Mass Spectrometry of Sialic Acids. In* Sialic Acids—Chemistry, Metabolism and Function*; Schauer, R., Ed.;* Cell Biology Monographs*, Vol. 10; Springer-Verlag: Vienna, Austria, 1982; pp 95–125. Copyright Springer-Verlag.*

seen for both compounds, only partially due to O-deacetylation. Neu5Ac9Lt behaved in the same way as Neu5,9Ac$_2$ (consumption of only ~0.10 mol of periodate/mol sialic acid within 10 min).[42] Periodate oxidation of Neu5Ac9P yielded glycolaldehyde phosphate, indicating a splitting of the C7–C8 bond, and a positive thiobarbituric acid color reaction (extinction coefficient 78% of that obtained with free Neu5Ac).[51] Also Neu5Gc9S turned out to be easily susceptible to periodate oxidation.[72] In view of the literature protocols followed, it seems that in both cases the oxidation had been carried out for several hours. It should be noted that on treatment of β-Neu5Ac1,2Me_2 with low amounts of periodate, the C8–C9 diol group is preferentially oxidized, as was shown by the high ratio of the aldehyde of the 8-carbon analogue to the aldehyde of the 7-carbon analogue obtained.[277] The aldehyde of the 8-carbon analogue, being a 2-hydroxyaldehyde, is further oxidized if the amount of periodate is sufficient. So, the oxidation of the C7–C8 diol group will be hindered if the formation of the mentioned aldehyde is blocked by the presence of an *O*-acetyl group at C9. See also refs. 6, 30, 277, and 278 for periodate/borohydride degradation studies on the glycoprotein level, generating bound C_8-α-Neu5Ac, C_7-α-Neu5Ac, C_8-α-Neu5Gc, and/or C_7-α-Neu5Gc.

In later years, several other MS protocols, combined with GLC or high-performance liquid chromatography (HPLC), were explored. Underivatized *N,O*-acylneuraminic acids were successfully analyzed by combined HPLC–CIMS (chemical ionization mass spectrometry), although localization of O-substituents based on CIMS data alone was not possible.[279] Although CIMS of pertrimethylsilylated sialic acid derivatives gave less fragmentation than EIMS, GLC–CIMS also furnished suitable information for identification purposes.[104,280] Furthermore, a methodology based on HPLC, coupled with electrospray-ionization mass spectrometry (ESIMS), has been developed.[32] To this end fluorescent sialic acid derivatives were prepared by reaction with 1,2-diamino-4,5-methylenedioxybenzene (DMB) (see Section 7.2). A combination of HPLC retention times and ESI mass spectra allowed the assignment of the type and position of the various substituents.[32,33] A similar approach, but now using matrix-assisted laser-desorption ionization time-of-flight mass spectrometry (MALDI-TOFMS), has also been developed.[281] Analogous to the GLC–EIMS analysis of trimethylsilylated sialic acids, a GLC–EIMS methodology for the analysis of heptafluorobutyrylated (HFB) sialic acid methyl ester derivatives has been worked out.[19] Here, the EI mass spectra are rather complex; a survey of reporter fragment ions and GLC data of specific members of the sialic acid family can be found in refs. 19 and 33. For EI mass spectra of acetylated derivatives, see ref. 25. For fast-atom bombardment mass spectrometry (FABMS) studies of sialic acid derivatives, see ref. 31.

Also ^{1}H NMR spectroscopy has been shown to be an excellent tool in the characterization of *N,O*-acylneuramic acids.[6,29,282] ^{1}H chemical shift values of sialic acids are pH dependent. However, the pH of the solution is also important for the stability of the sialic acids under investigation: autohydrolysis in case of bound sialic acids at low pH; O-deacylation at low and high pH; exchange of H3 in D_2O or tritiated H_2O at high pH[283,284] (see Section 7.3), migration of *O*-acyl groups at neutral pH[38,38a,261] (see Section 11.5.2). With respect to anomeric α:β ratios, interestingly, the ratio of Neu5,7Ac_2 and Neu5,7,9Ac_3 (~23:77) differed strongly from the ratio of the other *N*(*O*)-acylneuraminic acids (~7:93); the 5Ac signal of both Neu5,7Ac_2 anomers resonates at clearly different positions, when compared with the positions of Neu5Ac and Neu5,9Ac_2.[29]

6. DIFFERENT VIEWS ON THE CONFORMATION OF THE GLYCEROL SIDE CHAIN

The studies related to the conformational aspects of sialic acids are mainly attributed to the understanding of the hydrogen-bonding network within α- and β-sialic acids that comes to an expression in the interaction between the "flexible" staggered C7–C8–C9 glycerol side chain and the rigid C2–C3–C4–C5–C6–O6 $^2C_5$ chair conformation.

6.1 X-ray Diffraction Analysis

At the beginning of the 1970s, first crystallographic data appeared on β-methoxyneuraminic acid trihydrate (β-Neu2Me·$3H_2O$),[285] β-*N*-acetylneuraminic acid dihydrate (β-Neu5Ac·$2H_2O$),[286] β-*N*-acetylneuraminic acid methyl ester monohydrate (β-Neu5Ac1Me·H_2O),[287] and at a later stage α-*N*-acetylneuraminic acid methyl ester methyl glycoside (α-Neu5Ac1, 2Me_2).[288] All four substances were found to occur in a $^2C_5$ chair conformation with equatorial groups at C4 (OH), C5 (NH_2 or NHAc), and C6 (glycerol side chain). The first three compounds had an axial OR (R = CH_3 or OH) group at C2 (proof for β-anomer), while the fourth compound had an equatorial OCH_3 group at C2 (proof for α-anomer). The equatorial glycerol side chain of all four compounds was found to occur in an extended, staggered conformation with the OH groups at C7 and C8 in anti orientation. In general, extensive three-dimensional systems of intermolecular hydrogen bonds, also involving H_2O molecules, were seen. In the context of the conformation of the glycerol side chain, intramolecular hydrogen bonds are of interest. For β-Neu5Ac1Me·H_2O, 1 intramolecular hydrogen bond between HO7 and the carbonyl O of $NHCOCH_3$ was detected. In the case of α-Neu5Ac1,2Me_2, intramolecular hydrogen bonds were detected between

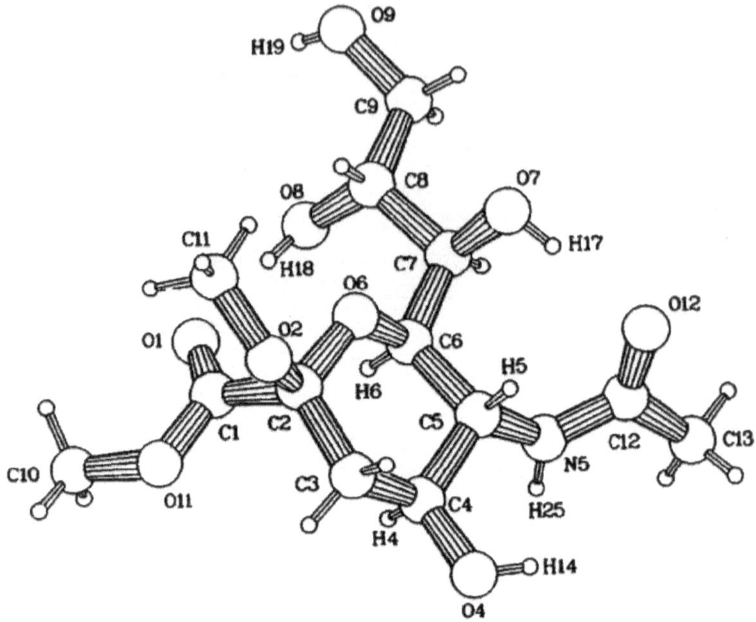

Fig. 16 Molecular conformation of crystalline α-Neu5Ac methyl ester methyl glycoside. *Reproduced from Kooijman, H.; Kroon-Batenburg, L. M. J.; Kroon, J.; Breg, J. N.; de Boer, J. L. Structure of α-D-N-Acetyl-1-O-methylneuraminic Acid Methyl Ester.* Acta Crystallogr. C ***1990***, *46, 407–410. With permission of the International Union of Crystallography.*

HO7 and the carbonyl O of the planar $NHCOCH_3$ group (trans configuration of the peptide bond), and between HO8 and the carbonyl O of the $COOCH_3$ group (Fig. 16). Here, the $^{2}C_5$ chair was shown to be somewhat distorted toward a $^{6}H_5$ half-chair conformation.

6.2 NMR Spectroscopy

The first 60-MHz ^{1}H NMR spectra of Neu5Ac, recorded in D_2O, were published in 1968,[289,290] but these low-resolution spectra did not contain much information (Fig. 17). In fact, the 100-MHz ^{1}H NMR spectrum of β-Neu5Ac1Me, already discussed in Section 4.3, delivered the first useful information with respect to the conformation of sialic acids.[257] In a first high-resolution study, the 270-MHz ^{1}H NMR spectrum of Neu5Ac in D_2O was completely assigned.[28] The wrong assignment of the Neu5Ac sample as α-Neu5Ac was corrected in a later report.[291] In subsequent papers full assignments of various sialic acids were published, as reviewed in refs. 6 and 282. As a typical example, Fig. 18 presents the 500-MHz ^{1}H NMR spectrum of Neu5Ac, dissolved in D_2O, at pD 7, whereas Table 3

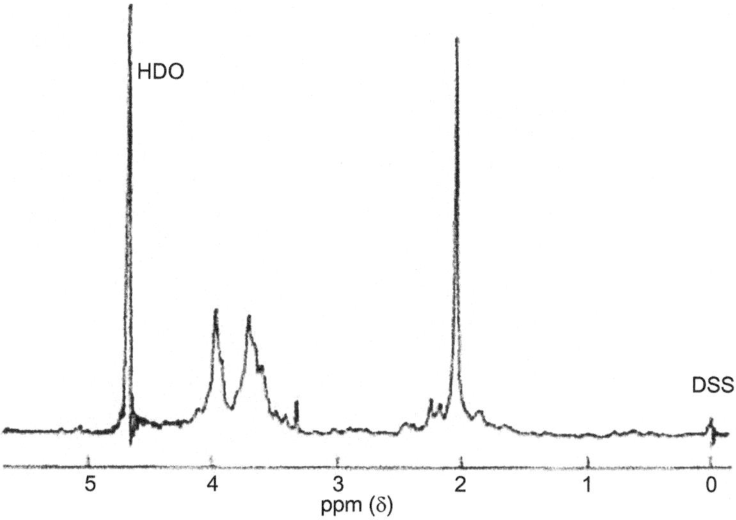

Fig. 17 60-MHz ^{1}H NMR spectrum of Neu5Ac, dissolved in D_2O, recorded at ambient temperature. *Reproduced from Kimura, A.; Tsurumi, K. Studies on Colominic Acid—IV. Conformation and Configuration of the* N*-Acetyl-neuraminic Acid Moiety of Colominic Acid.* Fukushima J. Med. Sci. ***1968****,* 15*, 55–60. With permission of the Board.*

summarizes chemical shifts and coupling constants for α- and β-Neu5Ac (note the $^{3}J_{7,8}$ value of ~9 Hz, reflecting the staggered anti orientation of H7 and H8, and thus of HO7 and HO8 at the glycerol side chain). The spectrum shows a major and a minor set of protons, related to the subspectra of the β- and α-anomer of Neu5Ac, respectively (β:α = 93:7). The positions of the H3e and H3a signals, being highly pH dependent and resonating outside the proton bulk signal, turned out to be of special interest in the structural analysis of sialoglycan chains (see Section 8.6).

The first 22.6- and 25.2-MHz ^{13}C NMR spectra of Neu5Ac (derivatives), recorded in H_2O or D_2O, appeared in the 1970s.[35,291–294] For a review of early ^{13}C NMR studies, including chemical shift data, see ref. 282. As a typical example, in Fig. 19 the 25.2-MHz ^{13}C NMR spectrum of Neu5Ac, dissolved in D_2O, at pD 2, is depicted.

Besides the full interpretation of the ^{1}H and ^{13}C NMR spectra of sialic acids in different solvents, the conformation of the glycerol side chain (average over rotamer states) as part of the whole molecule in D_2O was studied in more detail and compared with the conformation in crystal structures. In a first study,[28] taking into account the values of the vicinal coupling constants $J_{6,7}$, $J_{7,8}$, $J_{8,9}$, and $J_{8,9'}$ (Table 3), the lowest energy conformation of the extended C7–C8–C9 side chain was believed to be as depicted in Fig. 20A. Here, H6 and H7 are placed in a staggered gauche, H7 and H8

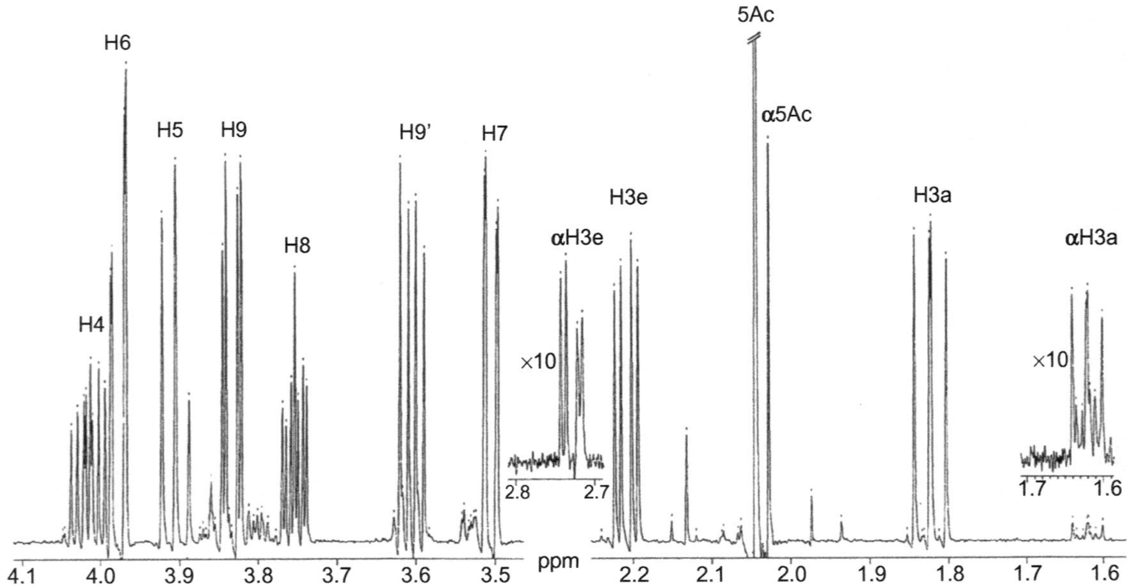

Fig. 18 500-MHz ^{1}H NMR spectrum of Neu5Ac (β:$\alpha = 93$:7), dissolved in D_2O, recorded at pD 7 and 27°C (Table 3), relative to internal acetone in D_2O (δ 2.225 ppm). The H3e and H3a signals of α-Neu5Ac are enlarged 10-fold. *Reproduced from Schauer, R.; Kamerling, J. P. Chemistry, Biochemistry and Biology of Sialic Acids. In* Glycoproteins II; *Montreuil, J.; Vliegenthart, J. F. G.; Schachter, H., Eds.;* New Comprehensive Biochemistry, *Neuberger, A.; van Deenen, L. L. M., Series Eds., Vol. 29b; Elsevier Science B.V.: Amsterdam, The Netherlands, 1997; pp 243–402. Copyright Elsevier Science B.V.*

Table 3 360-MHz ^{1}H Chemical Shifts and Coupling Constants of β- and α-Neu5Ac at Different pH Values, in ppm Relative to Internal Acetone in D_2O (δ 2.225) at 25°C[6,29,282]

Compound	pD	H3a	H3e	H4	H5	H6	H7	H8	H9	H9′	NAc
β-Neu5Ac	1.4	1.880	2.313	4.067	3.931	4.056	3.556	3.750	3.841	3.619	2.053
α-Neu5Ac	1.4	1.705	2.718	n.d.	3.85	3.684	3.53	3.75	3.85	3.62	2.036
β-Neu5Ac	7.0	1.827	2.208	4.024	3.899	3.984	3.514	3.753	3.835	3.608	2.050
α-Neu5Ac	7.0	1.621	2.730	n.d.	n.d.	n.d.	n.d.	n.d.	n.d.	n.d.	2.030
Compound	**pD**	$^{2}J_{3a,3e}$	$^{3}J_{3a,4}$	$^{3}J_{3e,4}$	$^{3}J_{4,5}$	$^{3}J_{5,6}$	$^{3}J_{6,7}$	$^{3}J_{7,8}$	$^{3}J_{8,9}$	$^{3}J_{8,9'}$	$^{2}J_{9,9'}$
β-Neu5Ac	1.4	−13.2	11.8	5.0	10.4	10.7	1.2	9.4	2.8	6.4	−12.4
α-Neu5Ac	1.4	−13.0	11.5	4.5	n.d.	10.5	1.5	9.0	n.d.	6.5	−12.5

n.d., value could not be determined.

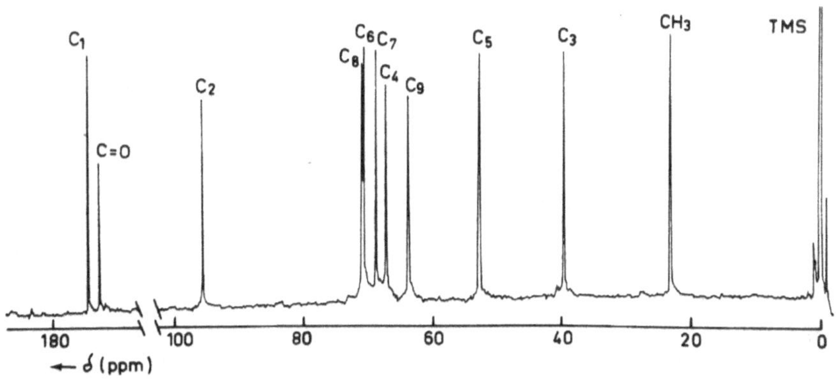

Fig. 19 Proton-noise-decoupled 25.2-MHz ^{13}C NMR spectrum of Neu5Ac, dissolved in D_2O, recorded at pD ~2 and 25°C, relative to external TMS (δ 0 ppm). Only the signals stemming from the β-anomer are clearly observable. For a first report on chemical shifts of both α- and β-Neu5Ac, see ref. 291. *Reproduced from Vliegenthart, J. F. G.; Dorland, L.; van Halbeek, H.; Haverkamp, J. NMR Spectroscopy of Sialic Acids. In* Sialic Acids—Chemistry, Metabolism and Function; *Schauer, R., Ed.;* Cell Biology Monographs, *Vol. 10; Springer-Verlag: Vienna, Austria, 1982; pp 127–172. Copyright Springer-Verlag.*

in a staggered anti, H8 and H9 in a staggered gauche, and H8 and H9′ in a staggered anti orientation. In this spatial model the H of HO7 can form a hydrogen bond with the ring O, and the H of HO8 a hydrogen bond with the O of HO9 (two five-membered ring interactions). In another study, using hard-sphere, exo-anomeric (HSEA) calculations, the latter model (Fig. 20A) got support.[295]

However, the spatial model depicted in Fig. 20A was not supported by results obtained from a ^{13}C spin–lattice relaxation time (T_1) study of the sodium

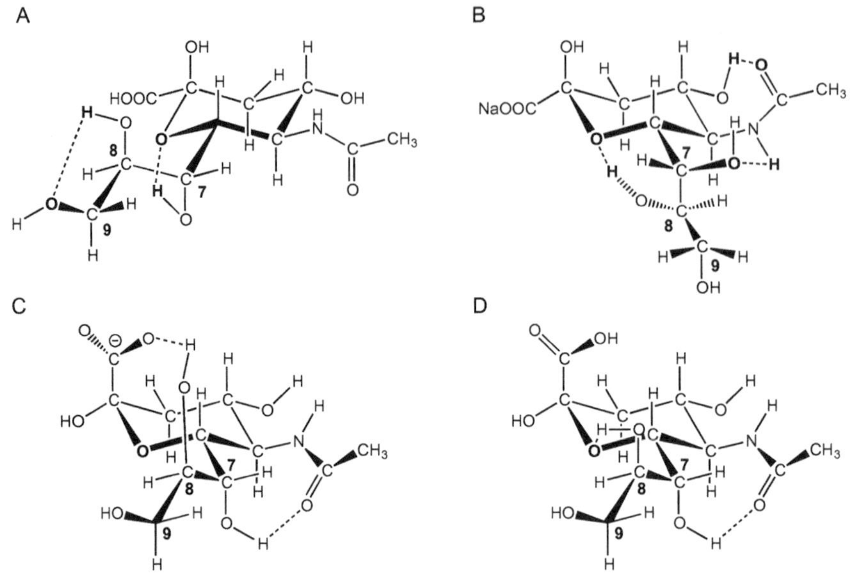

Fig. 20 NMR Molecular conformations in aqueous solution of (A) β-Neu5Ac according to ref. 28, (B) β-Neu5Ac according to refs. 294 and 296, and (C) and (D) α-Neu5Ac according to refs. 297 and 298. For both (A) and (B), as measured in D_2O solution, the H atoms of the OH and NH groups would be replaced by D atoms.

salts of β-Neu5Ac, β-Neu5Ac1Me, β-Neu5Ac2Me, β-Neu5Fo1,2Me$_2$, α-Neu5Ac2Me, β-Neu5Gc, and Neu2en5Ac in D_2O.[294,296] In this case, a conformational model was postulated, consistent with the ^{1}H NMR data and the motional requirements of the T_1 values (Fig. 20B). In this model, intramolecular hydrogen bonds occur between the O of HO7 and the acetamido NH and between the H of HO8 and the ring O, keeping HO7 and HO8 in a fixed staggered anti orientation (two six-membered chair-form ring interactions). Furthermore, the model made a third intramolecular hydrogen bond between the acetamido C=O and the H of HO4 (seven-membered ring interaction) possible. Apparently, the anomeric center is not involved in any hydrogen bond, leading to the same conformation for α- and β-anomers.

Using a new approach to the NMR study of OH and NH protons in aqueous solution (85% H_2O/15% $(CD_3)_2CO$), α-Neu5Ac as part of Neu5Ac-containing trisaccharides was investigated.[297] Careful NOE experiments showed that the strongest NOE was visible between the NH proton and H6. Further NOE contacts were seen between the HO8 proton and H6. Based on a comparison with a study on GM1 ganglioside in $(CD_3)_2SO$,[298] it was concluded that HO8 was involved in a strong intramolecular hydrogen bond with the carboxylic acid function or the ring O, or both, and explained as the main source of rigidity of the glycerol side chain (Fig. 20C and D) (note that in case of β-Neu5Ac only a hydrogen bond between HO8 and the ring O is possible). The investigation on GM1 ganglioside mentioned also a hydrogen bond between the H of HO7 and the acetamido C=O. Both intramolecular hydrogen bonds are also proposed for crystalline α-Neu5Ac1,2Me_2 (Fig. 16).[288]

For NMR complexation studies of Neu5Ac and Neu5Gc (derivatives) with Ca^{2+}, Mn^{2+} or Gd^{3+}, see refs. 66, 291, 292, 299, and 300.

6.3 Molecular Mechanics, Molecular Dynamics, and Ab Initio Calculations

First potential energy calculations of the conformation of β-Neu5Ac suggested that in solution the compound might exist preponderantly in two conformations, differing only in the orientation of the terminal CH_2OH (C9) group of the glycerol side chain.[301] One of these two energy conformations accorded with the crystal structure of β-Neu5Ac·$2H_2O$.[286] Although the two generated conformations were consisted with the earlier reported ^1H and ^{13}C NMR data, there is disagreement with the derived NMR models in Fig. 20A and B. For α-Neu5Ac, two similar minimum energy conformations have been reported.[302] See also ref. 303 for theoretical studies on the conformation of mono- and disialogangliosides.

In a more recent study, density functional theory calculations of the conformation of α-Neu5Ac in the gas phase have been carried out.[304] Out of many possibilities for relatively stable hydrogen bond networks, the most stable structure had an intramolecular hydrogen bond network of O9–H9...O8–H8...O=C1–OH/O7–H7...O=C–NH–C5, which shows similarity with the earlier reported crystal structure of α-Neu5Ac1,2Me_2 (Fig. 16). One of the lower stable structures shows similarity with the model presented in Fig. 20A (H9–O9...H8–O8/O7–H7...O6 (ring)/O4–H4...O=C–NH–C5/O2–H2...O=C1–OH). In addition, α-Neu5Ac–H_2O complexes of the various gas phase conformations were optimized. Here, the following hydrogen bond networks for the optimum complex were seen: O9–H9...OH_2...O8–H8, O8–H8...OH_2(...O=C1–OH)...OH_2(...H2–O2)...O6 (ring), O7–H7...OH_2...O=C–NH–C5, and HO–C1=O...H_2O...HO–C1=O (compare with Fig. 20C).

A first molecular dynamics (MD) simulation of α-Neu5Ac was performed both in vacuum and in water.[305] It was shown that the average conformation of the glycerol side chain was not influenced by the presence of explicit solvent molecules, and the rotation around the C6–C7 and C7–C8 bonds is highly restricted. Hydrogen bond positions were suggested for either

O8–H8...O=C1–OH or O8–H8...O6 (ring), for either O7–H7...O=C–NH–C5 or O7–H7...O6 (ring), and for O4–H4...O=C–NH–C5. The C8–C9 bond was found to be more flexible.

Furthermore, also an MD simulation and quantum mechanical (QM) calculations of α-Neu5Ac directed to the whole molecule have been published.[306] Here, the optimized structure was in good agreement with the crystal structure published for α-Neu5Ac1,2Me_2.[288]

Finally, ab initio calculations (gas phase) were carried out on a series of neuraminic acid derivatives: α-Neu5Ac, α-Neu5Ac1,2Me_2, β-Neu5Ac, β-Neu4,5Ac_2, β-Neu5,9Ac_2, β-Neu5Gc, β-Neu4Ac5Gc, and Neu2en5Ac.[307] The calculations for α-Neu5Ac1,2Me_2 were in close agreement with those of the crystal structure.[288] The calculated structure for α-Neu5Ac is depicted in Fig. 21.

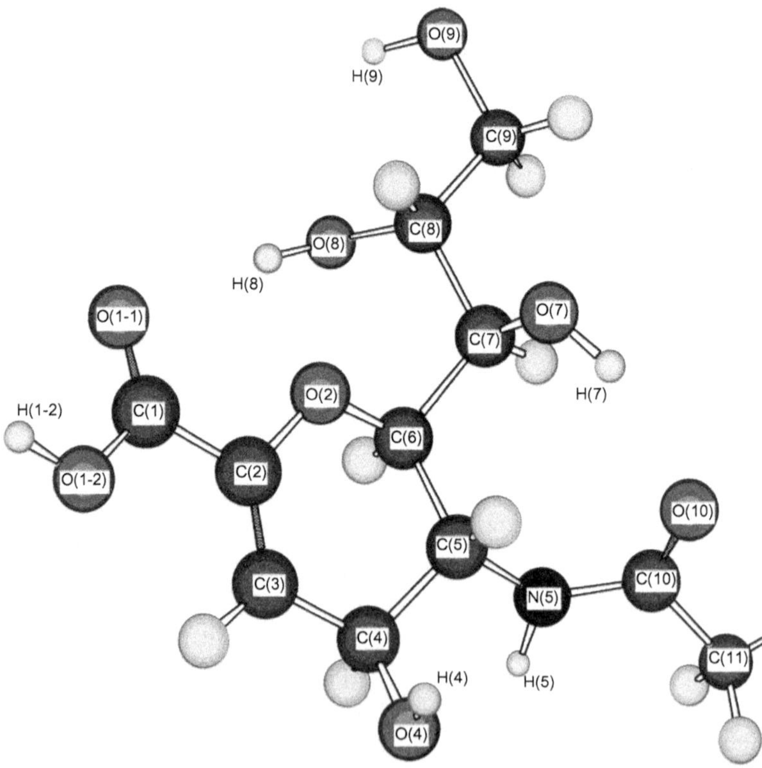

Fig. 21 Ab initio calculated structure of α-Neu5Ac. *Reproduced from van Lenthe, J. H.; den Boer, D. H. W.; Havenith, R. W. A.; Schauer, R.; Siebert, H.-C.* Ab initio *Calculations on Various Sialic Acids Provide Valuable Information About Sialic Acid-Specific Enzymes.* J. Mol. Struct. (Theochem) ***2004****, 677, 29–37. Copyright Elsevier B.V.*

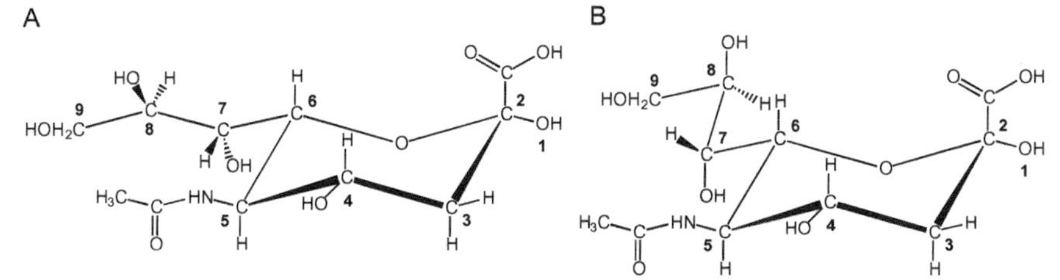

Fig. 22 (A, B) Two different drawings of the structure of α-Neu5Ac.

6.4 Final Remarks

First drawings of the conformation of sialic acids, including the glycerol side chain, stem from the end of the 1960s. These drawings, in later years included in the carbohydrate IUPAC/IUBMB Recommendations 1996/1997,[13] and generally used by many researchers including us, have a glycerol side chain with the H6–C6–C7–C8 part of the chain in an eclipsed conformation. The HO7 and HO8 groups are placed in a staggered anti orientation, whereby the HO7 group points in the direction of the ring O (Fig. 2/Fig. 22A).

Going through the different reports, discussed in Sections 6.1–6.3, it is evident that, although a discussion about the average conformation of the glycerol side chain is highly complex, the drawing as depicted in Fig. 22A does not reflect the most favorable conformation. In contrast, especially models with intramolecular hydrogen bonds between the HO7 and the $NHCOCH_3$ group, and between the HO8 and the ring O or the COOH function (in case of the α-configuration), leading to a strong hydrogen-bonding network between the glycerol side chain and the $^{2}C_{5}$ pyranose ring conformation, pop up. Inspection of the literature shows a great variety in drawing the glycerol side chain. Perhaps, one of the better drawings is shown in Fig. 22B.

7. AN IMPRESSION OF ANALYTICAL TECHNIQUES AS USED IN SIALIC ACID ANALYSIS

7.1 Spectrophotometric Assays

In the early days of the sialic acid history, colorimetry played a major role in the unraveling of sialic acid structures. As demonstrated in Section 2, initially two reagents were applied routinely, namely, Bial's reagent (orcinol/concd. HCl/$FeCl_3$ mixture) and Ehrlich's reagent (*p*-dimethylaminobenzaldehyde/50% HCl). By the 1950s several additional assays were developed; however, the majority of these early methods were found to be unsuitable because of their lack of specificity and/or sensitivity required for sialic acid analysis in biological materials or for enzymatic assays.[6,101,269] In the following, a number of spectrophotometric assays, still in use or used for a long time, will be explained in more detail.

In the Bial reaction,[181,308,309] the sialic acid sample is mixed with the orcinol/~30% HCl/$FeCl_3$ reagent and heated for 15 min at 96°C, whereafter the formed purple to red–violet chromophore is extracted with isoamyl alcohol, and its absorbance measured at 572 nm (for more details, see refs. 101, 104, and 269). Because of the strong acidic conditions used, the method is suitable for the quantification of total amounts of free and glycosidically bound sialic acid. Note that *O*-acyl groups are split off. When applied during the fractionation of biological materials, the Bial assay is therefore typically a reaction to monitor the presence of sialic acids. As other carbohydrates (pentoses, hexoses, uronic acids) can strongly disturb the determination of small amounts of sialic acid, quantification usually requires purified samples. To overcome some of the problems, different variants of the Bial reaction have been explored. With a focus on contaminating 2-ketohexoses, a suitable variant reagent to measure total sialic acid, namely, resorcinol/concd. HCl/$CuSO_4$, was worked out by Svennerholm. It proved to be 30%–50% more sensitive than the Bial assay.[310,311] Here, free or bound sialic acid samples are mixed with the reagent and heated for 15 min at 110°C, and the formed blue–violet chromophore is extracted with amyl alcohol followed by measurement of the absorbance at 580 nm, and also at 450 nm to correct for 2-ketohexoses (for further information, see refs. 269, 312, and 313). Note that Kdn gives no color in the orcinol and resorcinol reactions.[314]

In trying to formulate a reaction mechanism for the Bial and Svennerholm reaction, it was proposed that the strong acidic conditions lead to N-deacetylation, yielding an internal Schiff base formed by the condensation of the free amino group at C5 and the C2 carbonyl group, followed by decarboxylation and dehydration, to yield a pyrrole derivative. The latter chromogen then gives a condensation product with orcinol or resorcinol (Scheme 11).[313,315] When combined with a preceding periodate oxidation step, the sensitivity of the Svennerholm assay can be increased significantly (between three- and sixfold).[312,313]

In the "direct" Ehrlich reaction (without previous treatment with alkali),[165,181] the sialic acid sample is usually mixed with the *p*-dimethylaminobenzaldehyde/~18% HCl reagent and heated for 30 min at 100°C, whereafter the absorbance of the formed red–violet chromophore is measured at 565 nm (for more details, see refs. 88 and 181). In the case of the "indirect" Ehrlich reaction, the test material is first incubated with mild

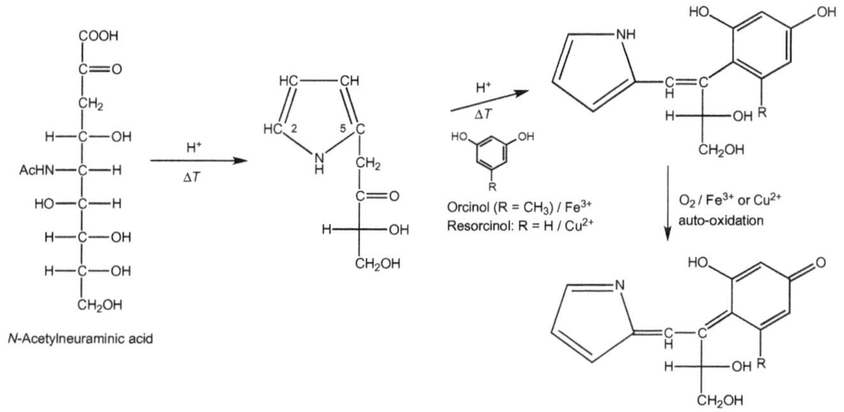

Scheme 11 Possible pyrrole chromogen formation [N-deacetylation, condensation NH_2(C5) and CO(C2), decarboxylation, dehydration] in the Bial (orcinol/concd. HCl/ $FeCl_3$) and Svennerholm (resorcinol/concd. HCl/$CuSO_4$) reaction, taking into account the proposals in refs. 313 and 315.

alkali, e.g., borate buffer at pH 8.5 (45 min, 100°C), followed by the addition of Ehrlich's reagent (20 min, 70°C), whereafter the extinction is determined at 558 nm.[265] To illustrate the chemical background of the color formation in the Ehrlich's test, and taking into account the proposals by different authors, the "direct" and "indirect" Ehrlich pathways for Neu5Ac are presented in Scheme 12. Just like in the Bial/Svennerholm assays, N-deacetylation, internal Schiff base formation, and dehydration play a role. For the formulation of the pathway leading to structure A, the earlier scheme of Tuppy and Gottschalk, combined with information of Treibs and Herrmann, is followed.[100,316] For the formulation of the pathway leading to structure B (including a decarboxylation), the earlier scheme of Wirtz-Peitz, combined with information from Alexander and Butler, is followed.[315,317] Note that the "direct" Ehrlich test does not distinguish free from bound sialic acid. It should be mentioned that the Ehrlich assay as a quantitative method has lost its importance.

The periodic acid/thiobarbituric acid assay, independently developed at about the same time (1959) by Warren and by Aminoff, grew out to the most widely applied quantitative colorimetric method for measuring free sialic acid.[266,267] The methods differ in two ways from each other: (i) the acidity is that of the initial periodate oxidation step (concd. H_3PO_4[266] vs 0.06 M H_2SO_4[265]), and (ii) the use of cyclohexanone[266] or acidic *n*-butanol[265] for the extraction of the chromophore. In the periodate oxidation step (followed by an arsenite step), a chromogen is generated, predicted to be

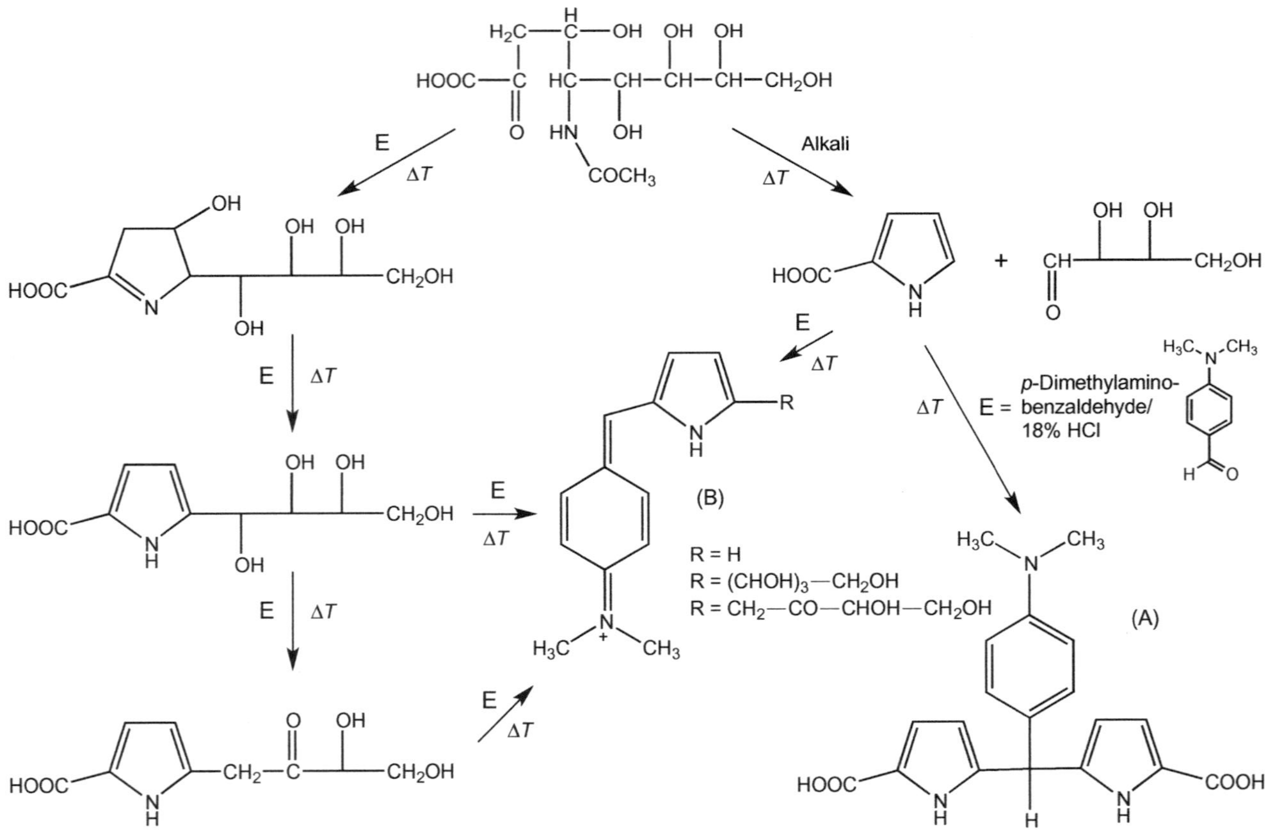

Scheme 12 Possible pyrrole derivatives arising from Neu5Ac, following the "direct" Ehrlich (E = *p*-dimethylaminobenzaldehyde/concd. HCl) and the "indirect" Ehrlich (alkali; E = *p*-dimethylaminobenzaldehyde/concd. HCl) reaction, taking into account the proposals in refs. 100 and 315–317.

β-formylpyruvic acid, which gives a red chromophore with 2-thiobarbituric acid with λ_{max} 549 nm. A proposed reaction mechanism of the assay is presented in Scheme 13.[100,268] In contrast to Warren,[266] it was stated by Paerels and Schut[268] that a conversion of the *N*-acyl group into a free amino group under the acidic periodate oxidation conditions (room temperature; incubation for a relatively short time) is not obvious, because of the high stability of an amide

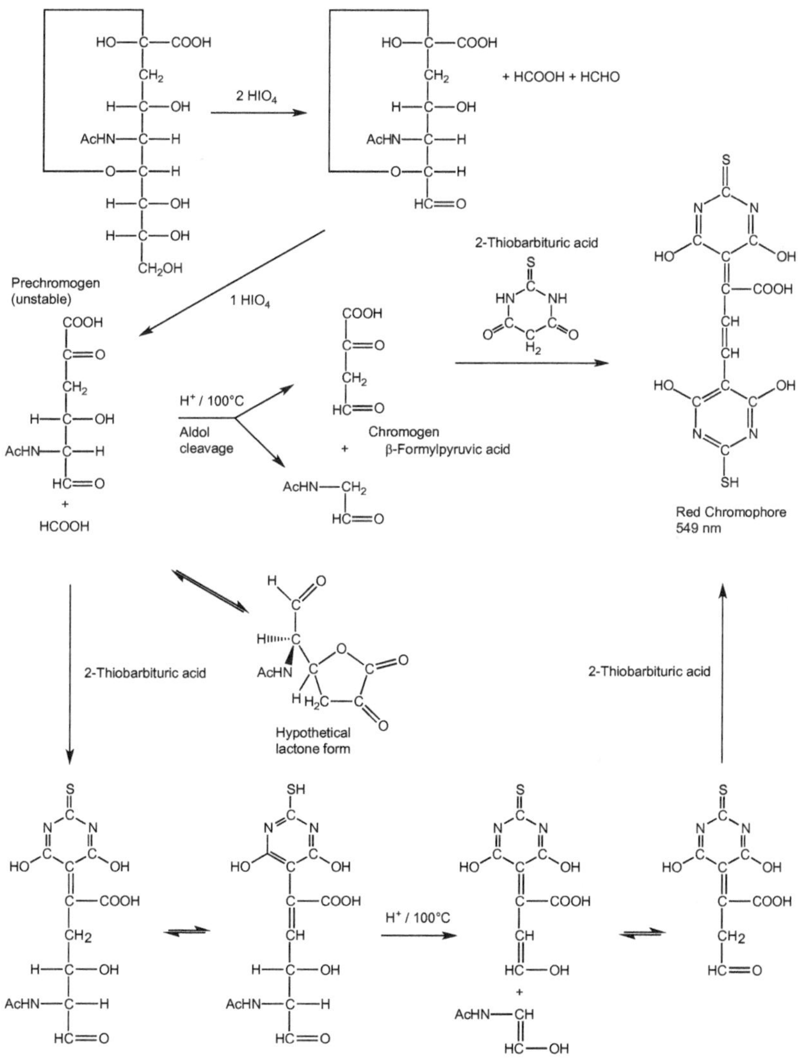

Scheme 13 Proposed reaction mechanisms of the periodic acid/thiobarbituric acid assay.[100,268,319,320]

bond. As sialic acid glycosides did not react in the Warren and Aminoff tests, sialic acid glycosidic linkages are not hydrolyzed under the assay conditions. Furthermore, it was presumed that β-formylpyruvic acid is not a direct product of the periodate oxidation, but the result of an acid-catalyzed C4–C5 aldol splitting of the true oxidation product (for Neu5Ac: formed after consumption of 3 mol of periodate; see the upper part of Scheme 13) during the incubation with the acid solution of 2-thiobarbituric acid at 100°C. The red pigment was earlier prepared in the crystalline form from a derivative of β-formylpyruvic acid, and proven to be a condensation product of 2 mol of 2-thiobarbituric acid and 1 mol of β-formylpyruvic acid.[318] It may be clear that O-substitutions of the glycerol side chain will influence the prechromogen formation (see discussion in Section 5), and a test after O-deacetylation was advised. In case of Neu4,5Ac$_2$, an O-deacetylation at C4 during the reaction with 2-thiobarbituric acid before the aldol splitting was suggested. See the comments on the analysis of Kdn, which directly yields β-formylpyruvic acid.[314]

However, in the 1980s, a positive periodic acid/thiobarbituric acid test with Neu5Ac4Me initiated a new discussion, wherein the postulated prechromogen was suggested to react directly with 1 mol of 2-thiobarbituric acid, followed by an acid-catalyzed reaction (cleavage of the C4–C5 bond) to yield the β-formylpyruvic acid derivative.[319] This derivative can then react with a second mol of 2-thiobarbituric acid to give the final chromophore (lower part Scheme 13). The latter pathway, stating that free β-formylpyruvic acid is not an intermediate, was suggested in the 1970s to explain the positive Warren reaction for 3-deoxy-5-O-methyloctulosonic acid.[320] However, in view of the positive Warren reaction for Neu5Ac4Me, a chromophoric 1:1 condensation product should be expected.

So, applying the periodic acid/thiobarbituric acid methodology on biological samples before and after mild acid hydrolysis, and eventually saponification, the levels of free, bound, and total sialic acid can be determined (for experimental details, see refs. 101, 104, and 269).

In attempts to lower the detection limits of sialic acid, since the 1960s a number of fluorimetric assays have also been developed (for reviews, see refs. 6, 101, 102, and 269). In the following, some typical examples will be presented. Reaction of sialic acid with 3,5-diaminobenzoic acid dihydrochloride in hot dilute HCl yields an intense green fluorescent condensation product (405 nm excitation, 510 nm emission) (free and bound sialic acids react equally).[321] Reaction of sialic acid with pyridoxamine in the presence of zinc acetate and pyridine/methanol for 45 min at 70°C yields a highly fluorescent chelate (395 nm excitation, 470 nm emission; Scheme 14) (free sialic acids).[322,323] The fluorescence of the chromophoric product, obtained from sialic acid in the periodate/2-thiobarbituric acid assay (Scheme 13), is readily detected by excitation at 550 nm and analysis of the emission at 570 nm. Here, special attention has been paid to contamination with 2-deoxyribose.[324] Finally, a different approach has been worked out, whereby free or glycosidically bound sialic acid was subjected to mild periodate oxidation. This method also allowed the discrimination between sialic acids with or without *O*-acyl groups at the glycerol side chain by O-deacetylation of a parallel sample (for a discussion of the mild periodate oxidation of sialic acids, see Section 5). The released formaldehyde, derived from C9 in the case of nonsubstituted HO9 and HO8 functions, is derivatized with acetylacetone in the presence of ammonium acetate, yielding the slightly yellow chromophore 3,5-diacetyl-1,4-dihydro-2,6-dimethylpyridine,[325] of which the fluorescence can be detected by excitation at 410 nm and analysis of the emission at 510 nm (Scheme 15).[326] Using this methodology, it could be demonstrated that rapid O-deacetylation of Neu5,9Ac$_2$ occurred only at pH values below 2 and above 11. When incubated for 10 min at 58°C, no significant loss of *O*-acetyl groups were seen in the range of pH 2–11.[327]

Based on the finding of the acylneuraminate pyruvate lyase (or aldolase)-catalyzed cleavage and synthesis of Neu5Ac with ManNAc and pyruvic acid as counter compounds by Roseman and coworkers (Section 4.1; Scheme 5),[231,232] indirect quantification protocols for sialic acids, via

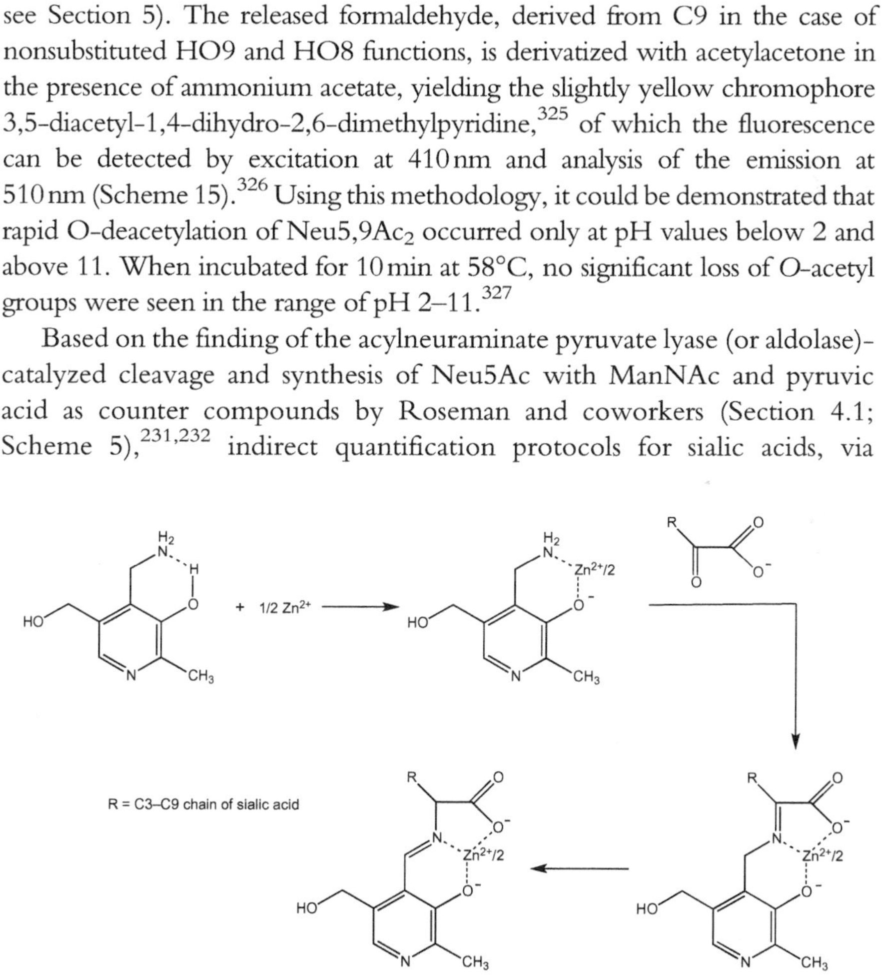

Scheme 14 Proposed reaction mechanism for the transamination of sialic acid with pyridoxamine in the presence of zinc acetate and pyridine/methanol, yielding the fluorescent Zn-chelate.[323]

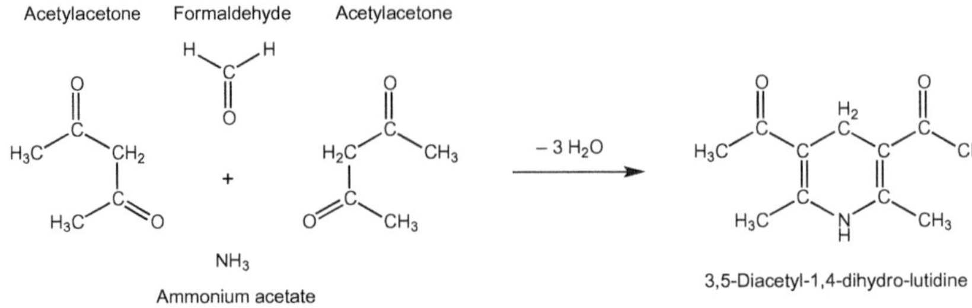

Scheme 15 Reaction of formaldehyde with acetylacetone in the presence of ammonium acetate, yielding the fluorigen 3,5-diacetyl-1,4-dihydro-2,6-dimethylpyridine.[325,326]

estimation of ManNAc or pyruvate, were also developed.[269,328,329] The most sensitive method turned out to be a pyruvate assay, based on the lactic acid dehydrogenase-catalyzed interconversion of pyruvate and lactic acid with the concomitant interconversion of NADH and NAD^+. Coupling of the aldolase-catalyzed cleavage reaction of *N*-acylneuraminic acids to the lactate dehydrogenase/NADH-catalyzed reduction of pyruvate to lactate drives the lyase reaction, otherwise an equilibrium reaction, to completion. Initially, the amount of pyruvate (and thus the amount of sialic acid) was calculated from the amount of oxidized NADH, measured spectrophotometrically by the change in absorbance at 340 nm.[329] In the course of time, various commercial colorimetric and fluorimetric assays have been developed for the estimation of oxidized NADH. Note that the reaction rate of the cleavage by aldolase is influenced by the sialic acid O-substitution pattern, e.g., it has been found that *O*-acyl groups at the sialic acid glycerol side chain lead to a reduction in the reaction rate of 30%–50%, while O-acetylation at C4 gives 90% reduction.[41,330,331]

Originally, acetyl determinations for sialic acids were described using a protocol that included hydrolysis with *p*-toluenesulfonic acid, subsequent distillation to collect acetic acid, conversion into barium acetate, and determination of the Ba^{2+} content.[23,332] In the case of glycolyl groups, sialic acids were treated with 1 M NaOH at 100°C, and the glycolic acid that formed was quantified as crystalline calcium glycolate.[23] At a later stage, among the different colorimetric methods to quantify the *O*-acyl content of sialic acids, the Hestrin assay, as modified by Ludowieg and Dorfman, played an important role.[271,333] Incubation of *N*,*O*-acylneuraminic acids with alkaline hydroxylamine affords acylhydroxamates that form red chromophores with $Fe(ClO_4)_3$, the absorbance of which are measured at 520 nm.[334] *O*-Glycolyl (but also *N*-glycolyl) groups can be quantified using a protocol that includes treatment with ethanolic *p*-toluenesulfonic acid (8 h, 95°C), affording ethyl glycolate, which is collected via distillation in vacuo. After saponification, the amount of sodium glycolate is estimated colorimetrically using the Eegriwe's reagent, 2,7-dihydroxynaphtalene/concd. H_2SO_4.[64,270,334–336] The reddish chromophore that is formed (20 min, 100°C), represents the condensation product of 1 mol of formaldehyde (oxidative decarboxylation of glycolic acid) and 2 mol of 2,7-dihydroxynaphtalene, and yields an absorption at λ_{max} 546 nm. Finally, the configuration and the amount of *O*-lactyl groups in *N,O*-acylneuraminic acids can be measured as lactic acid with L-lactate dehydrogenase and D-lactate dehydrogenase after release with NaOH (1 h, room temperature) and neutralization.[42,43] So far, only *O*-L-lactyl groups have been found.

7.2 Chromatographic Techniques

Just as with structural investigations of many other natural products, in the early days paper chromatography played an important role in the analysis of sialic acids.[88,263,269]

However, at the end of the 1950s, thin-layer chromatography (TLC) on cellulose and silica gel plates slowly surpassed the use of paper. The greater resolution, combined with improved reproducibility, made TLC for many years the technique of choice for screening and (tentative) assignments of sialic acids. Especially, Schauer and coworkers at the Ruhr-Universität Bochum, and later at the Christian-Albrechts-Universität zu Kiel, contributed greatly to the development of TLC technologies. Although several solvents were developed, the most popular flow systems turned out to be propan-1-ol/butan-1-ol/0.1 M HCl (2:1:1, v/v/v) for cellulose sheets (Fig. 23), and propan-1-ol/water (7:3, v/v) for silica gel sheets. For R_f values of different sialic acids, see refs. 6 and 269. For the visualization of the sialic acid members, in most cases, the Bial reagent (orcinol/concd. HCl/$FeCl_3$) was used, yielding typical purple–blue bands. Periodate combined with resorcinol/$CuSO_4$ and periodate combined with arsenite and thiobarbituric acid sprays were also developed. For a further differentiation between *N,O*-acylneuraminic acids with similar R_f values in one direction, a two-dimensional TLC method with intermediate saponification, carried out on cellulose sheets with the solvent system mentioned earlier, was developed. The intermediate exposure to an atmosphere of ammonia vapor over a 5 M NH_4OH solution (2 h, room temperature) generates Neu5Ac and Neu5Gc from their *O*-acyl derivatives.[41,334] A two-dimensional technique

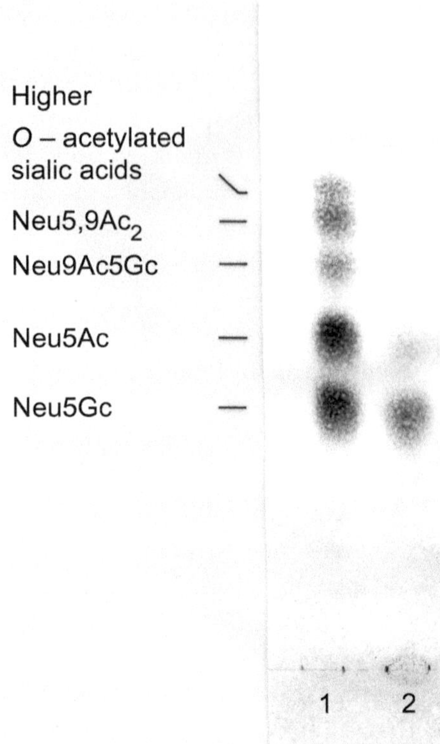

Fig. 23 TLC patterns of sialic acids on a cellulose plate, developed with propan-1-ol/butan-1-ol/0.1 M HCl (2:1:1, v/v/v), and stained with Bial's reagent. Lane 1: sialic acid mixture, released from bovine submandibular gland mucin. Lane 2: standard Neu5Ac and Neu5Gc.

on cellulose sheets was also developed for *O*-acyl group identifications. After a separation in the first dimension using propan-1-ol/butan-1-ol/0.1 M HCl (2:1:1, v/v/v), the lane of migrated sialic acids is sprayed with Hestrin's alkaline hydroxylamine reagent and left for 60 min. Then, the plate is developed in the second dimension using propan-1-ol/10% aq $(NH_4)_2CO_3$/5 M NH_4OH (6:2:1, v/v/v). Visualization of the acylhydroxamates is carried out with a 10% $FeCl_3$ spray.[41,334]

Around the 1970s, GLC (in combination with EIMS) of volatile sialic acid derivatives was introduced.[6,30,33] Major protocols that were explored are based on trimethylsilylation,[260,337–342] trifluoroacetylation,[343] and methylation (in case of partial methylation, followed by trimethylsilylation or acetylation).[344–348] For the application of GLC–EIMS in studying sialic acid O-substitution patterns, see Section 5. The partially methylated derivatives of sialic acids play a role in methylation/linkage analysis studies (GLC–EIMS) of glycan chains with internal sialic acid residues.

As already mentioned in Section 5, the introduction of serial anion-exchange and cellulose column chromatography at the end of the 1960s was a real step forward in the isolation of sialic acid family members on a preparative scale.[24] However, at a later stage the introduction of HPLC using different column materials, elution protocols, and detection techniques replaced this original approach. So, rapid screening of sialic acids in biological materials, affording assignments based on retention times, whereby contaminating substances (e.g., other monosaccharides) did not interfere, could be developed. Also the monitoring of fast transitions between members of the sialic acid family due to migration of substituents, the introduction of substituents, the cleavage of substituents, or other (enzymatic) modifications became possible.

First reports on the application of HPLC for the separation and quantitative analysis of sialic acids stem from the mid-1970s onward. Initially, methods for Neu5Ac were developed using an anion-exchange resin combined with a periodate/arsenite/thiobarbituric acid colorimetric detection assay,[349] a cation-exchange resin combined with UV detection at 206 nm,[350] and an amino column combined with UV detection at 195 nm.[351] Chromophores, obtained by treatment of Neu5Ac and Neu5Gc with periodate/arsenite/thiobarbituric acid, were analyzed on a reversed-phase C_{18} column, monitored at 549 nm.[352] As it was realized that analysis of complex mixtures of sialic acids should require other approaches, the focus was switched to generally applicable methods (for a review, see ref. 6). A first report proposed the use of an anion-exchange resin in borate buffer (pH 8.55), whereby nonderivatized sialic acids were separated as sugar–borate complexes. For the detection, the effluent was mixed with 4,4′-dicarboxy-2,2′-biquinoline disodium salt/$CuSO_4$ (5.4 min, 100°C), and the chromophores (Cu^+/4,4′-dicarboxy-2,2′-biquinoline complexes) that formed were monitored at 570 nm.[353–355] At the same time a different anion-exchange chromatography method was reported, making use of sodium sulfate as mobile phase and UV detection at 195 or 215 nm.[356] Great progress in HPLC approaches was made by the combination of reversed-phase HPLC with fluorometric detection systems and, in a later stage, with additional coupling to MS. To this end fluorescent sialic acid derivatives were prepared by reaction with DMB (373 nm excitation; 448 nm emission) (Scheme 16)[357–359] or *o*-phenylenediamine (OPD) (232 nm excitation; 420 nm emission).[360] In an evaluation of seven different HPLC methods, including high-performacnce anion-exchange

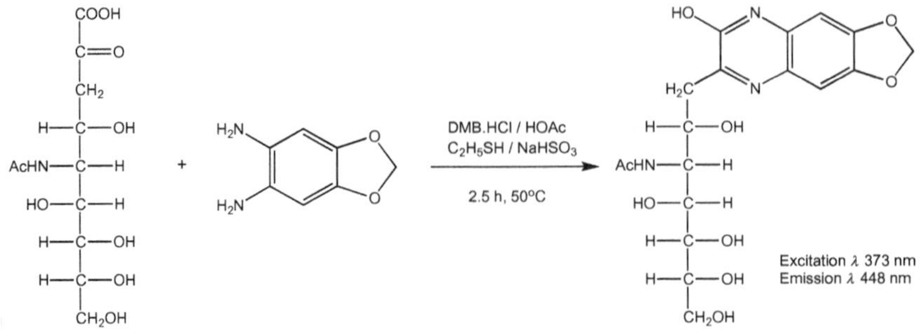

Scheme 16 Reaction of Neu5Ac with 1,2-diamino-4,5-methylenedioxybenzene (DMB) to yield the corresponding fluorescent derivative.[357,358]

chromatography with pulsed amperometric detection (HPAEC–PAD) (see ref. 361), it turned out that no single method is adequate to completely separate and quantitate complex mixtures of sialic acids.[362] For a quantification protocol of Neu5Ac and Neu5Gc by a stable isotope dilution assay using HPLC coupled with ESIMS, see ref. 363. For a more detailed overview of HPLC procedures, see ref. 6.

Around the turn of the century, also capillary electrophoresis (CE) was introduced for the analysis of derivatized and nonderivatized Neu5Ac and Neu5Gc.[364–369]

7.3 Instrumental Techniques

As indicated at several other places in this chapter, since the 1970s, mass spectrometry and NMR spectroscopy have played important roles in the further development of the sialic acid world, and detailed reviews are available.[6,30,282]

In the context of this section, attention will be paid to some NMR studies of sialic acids. As shown in Sections 5 and 11.5.2, the value of ^{1}H NMR spectroscopy for the characterization of *N,O*-acylneuraminic acids and for the migration of *O*-acetyl groups along the glycerol side chain was clearly demonstrated. NMR spectroscopy also proved to be of great value for unraveling of the mechanism of the release of bound sialic acid and for the assignment of the type of glycosidic linkage [e.g., (α2→3), (α2→6), (α2→8); see also Section 8.6]. Following the incubation of Neu5Ac(α2→3 or 6) conjugates with bacterial (*C. perfringens*, *Arthrobacter ureafaciens*, *V. cholerae*, *Bifidobacterium*) and viral (fowl plague virus) sialidases in an NMR tube, NMR spectroscopy showed in a fast process at first instance the formation of free α-Neu5Ac in high amounts, which then mutarotates in a slower process to the equilibrium α-Neu5Ac:β-Neu5-Ac=8:92. The anomerization could be followed easily via the intensities of the H3e, H3a, and *N*-acetyl signals of both anomers (see Section 6.2; Table 3).[370–372] Fortunately, the initial high concentration of α-Neu5Ac made it now possible to follow the mutarotation in more detail than described in Section 4.3. But it should be said that the rapidity of the α-Neu5Ac→β-Neu5Ac mutarotation depends on the pH. At pD 5.4, being the pH for an optimal cleavage of 2-azido-α-Neu5Ac with *C. perfringens* sialidase, the process turned out to be rather slow, but at higher and lower pD values a more rapid establishment of the equilibrium of mutarotation was observed. At pD 1.3 and pD 11.7 the mutarotation was too fast to be measured (no indications for open-chain structures or lactones).[373] For GLC–EIMS studies on the release of Neu5Ac from substrates by sialidases in the presence of $H_2[^{18}O]$, see ref. 374. The fact that the cleavage by these sialidases passed with retention of anomeric configuration received support in a later mechanistic study with the sialidase from influenza virus.[375] These results were in contradiction with an earlier report that Neu5Ac, liberated by the enzymatic cleavage of the α-linked Neu5Ac, leaves the catalytic site of *V. cholerae* sialidase as the β-anomer, which meant an inversion of anomeric configuration in the mechanism.[247] However, such a mechanism was later proposed in case of *Salmonella typhimurium* sialidase.[376,377] With the ability to generate α-Neu5Ac in situ, the aldolase-catalyzed degradation of Neu5Ac to pyruvate and ManNAc (in fact α-ManNAc, followed by a fast mutarotation to α,β-ManNAc) could also be investigated in more detail using ^{1}H NMR spectroscopy.[378,379]

In alkaline D_2O solutions of Neu5Ac at pD 9.0, H3a was completely replaced by D3a, which led to the disappearance of the H3a signal in the ^{1}H NMR spectrum and simpler splitting patterns for the H3e and H4 signals.[284] Further increase of the pD to 12.4 also caused the exchange of H3e for D3e.[283] When using tritiated H_2O, tritium-labeled sialic acids could be prepared. A ^{13}C NMR study of aqueous solutions of Neu5Ac, labeled with ^{13}C at C1, C2, and/or C3 at varying pHs, made the detection and quantification of its acyclic forms possible. Thus, at pH 2 anomerization of β-Neu5Ac leads to α-pyranose:β-pyranose:keto:enol:keto hydrate (gem-diol)=5.8:91.2:0.7:0.5:1.9.[380]

The importance of different MS techniques in the structural analysis of sialic acids and sialoglycans is discussed in Sections 5 and 8.6, respectively.

7.4 Histological Methods

Histochemistry is a most important area of glycobiology research, enabling insight into the chemical architecture of a cell, and especially its surface, the glycocalyx. For this purpose, the cell or tissue remains intact and can be investigated mostly in sliced form by different forms of microscopy. Nowadays, histological experiments are also possible with viable cells. This means that cell biological studies, during which the function of a cell varies, can now be visualized.

Histochemistry has a long history, and the methods applied are manifold. In a short note from 1946 in *Nature*, McManus from the University Museum in Oxford, UK, described the staining of mucins by Schiff's reagent following the action of 0.5% periodic acid.[381] Originally, Schiff's reagent was defined as "a solution of fuchsine decolorized by treatment with sulfur dioxide that gives a useful test for aldehydes because they restore the dye's color." Examples of the staining comprised mucins of goblet cells of the human intestine and bronchus, and also various other tissues from rat and man. A few years later, in 1950 and 1952, Leblond and coworkers used the technology in combination with light microscopy.[382,383] After periodate oxidation, tissues of different animals were stained with fuchsine–sulfurous acid, a technique developed by Hotchkiss.[384] At these times, the nature of the molecules in tissues, oxidized by periodic acid, was not yet known.

Interestingly, in 1957, in a report by Leblond and coworkers[385] describing the staining of glycogen by the periodic acid–Schiff (PAS) technique, sialic acid also came into play as the probable molecule responsible for periodate staining (sialic acid was found in various tissues along with galactose, fucose, and hexosamine). In subsequent years, the PAS technique developed into a powerful technique of sialic acid staining, because of the sensitivity of its glycerol side chain to mild periodate oxidation, thereby generating aldehyde functions. Typical examples are the detection of vicinal hydroxy groups (mostly of sialic acids) in plasma membranes by electron microscopy.[386,387] In an early review,[388] the periodate acid–Schiff technique, together with other methods for the detection of acidic groups/acidic glycoproteins of plasma membranes, e.g., using cationic compounds such as colloidal iron, thorium hydroxide, Alcian blue, and ruthenium red, used earlier for light or electron microscopy, is summarized. The studies confirmed that all cell surfaces seem to be covered by a carbohydrate-rich coat. In a thorough investigation on complex carbohydrates (glycoproteins) in the Golgi apparatus of rat cells, two electron microscopy visualization methods, based on (1) successive treatment of sections with periodic acid, chromic acid, and silver methenamine, and (2) brief treatment of sections with a mixture of chromic acid and phosphotungstic acid, were compared with the PAS technique using light microscopy, and the staining results obtained were shown to be identical/similar.[389]

Soon after the beginning of the use of periodate for histological sialic acid determinations, it was noticed that pretreatment with alkali (0.5% KOH) increased the periodic acid–Schiff (1% PAS) reactivity of, e.g., certain mucins,[390] and the authors confirmed the hypothesis that this KOH–PAS effect was due to the presence of O-acetylated sialic acids.[391,392] It should be noted that ester groups, which prevent aldehyde formation by periodate, were assumed to be located at C8 of the glycerol side chain of sialic acid. Until the mid-1970s, 9-O-acetylated sialic acids were believed to be cleavable between C7 and C8 (see fig. 1 of ref. 391). However, as discussed in Section 5, in the same period (1974), this assumption turned out to be incorrect.[276] So, PAS-resistant sialic acids can be substituted at C9 and not necessarily at C8. $Neu4,5Ac_2$ is fully PAS positive (but sialidase resistant). The specificity of this sialic acid-staining technique can be increased by previous sialidase or acid treatment of tissue slices. The difference of staining intensity with and without alkali treatment enables a rough estimation of the degree of O-acetylation in the glycerol side chain.

In the following years, staining of sialic acid by periodic acid and Schiff or thionine reagents, respectively, was carried out in numerous studies. For an early review by Culling and Reid that describes many details of the histochemical methodologies available for the location and identification of members of the sialic acid family in tissue sections, see ref. 393. It should be mentioned that the Hungarian G. Romhány developed in the 1970s a polarization optical electron microscopic analysis of the spatial arrangement of sialylated cell membrane components by periodate oxidation with the aldehyde bisulfite toluidine blue reaction ("anisotropic ABT reaction"), summarized by Makovitzky and Richter.[394] Also new reagents for staining, making use of periodate oxidation as a first step, were developed, as illustrated by the two following examples. First, a light microscopic histochemical localization of sialic acids (sialoglycoconjugates) in tissue sections of eccrine glands of porcine snout skin could be carried out by employing a method that includes a selective periodate oxidation–phenylhydrazine blockade and a thiocarbohydrazide–silver protein sequence, followed by a physical development procedure.[395,396] Second, for the in situ imaging by confocal laser microscopy and flow cytometric quantification of sialylated glycoproteins on living human gastric normal and cancer cells, the cell lines were treated by mild periodate oxidation (1 mM, 4°C, up to 20 min), and the aldehyde functions generated on the glycoproteins were ligated with the fluorescence tag, fluorescein-5-thiosemicarbazide.[397] Note that the chosen periodate oxidation conditions only introduce aldehyde groups into the sialic acid glycerol side chains of the glycoproteins on the living cell surface.

In the 1990s, staining of sialic acids in tissues or cells also became possible with lectins and antibodies. The specificities of the methods, developed for sialic acid linkage analysis, were higher than those developed using the periodate reaction.[398] Especially, the sialyl(α2→3)-specific *Maackia amurensis* agglutinin (MAA) and the sialyl(α2→6)-specific *Sambucus nigra* agglutinin (SNA) found wide application [and more rarely *Limax flavus* agglutinin (LFA)] in the lectin histochemical localization of glycoconjugates containing Neu5Ac in (α2→3) and (α2→6) linkage, respectively, to Gal and GalNAc. Wheat germ agglutinin (WGA) also interacts with Neu5Ac (not Neu5Gc), but according to a discussion with Jürgen Roth (University of Zurich, Switzerland), the lectin is not suited for sialic acid analysis. For the detection of sialic acid with these lectins, for example, a peroxidase–antibody–peroxidase complex (PAPS), which can be visualized, is in use. In the same period, the labeling of lectins and immunoglobulins with ferritin, and especially colloidal gold particles, used in both light and electron microscopy, got wide attention. Jürgen Roth ("Immunogold Master," "Mr. Compartment"[399]) and collaborators developed, most elegantly and frequently applied, the lectin–gold technique for human and animal tissues, including the Golgi complex.[400–402] Typical examples of staining of normal human colonic mucosa and colon carcinoma with MAA and SNA are shown in Fig. 24. Before these studies, the same research group had developed cationic (poly-L-lysine-coated) colloidal gold particles, which allow the detection of anionic, sialidase-sensitive, sites on erythrocytes.[403]

In more recent years, virolectins have shown to be also powerful reagents for histochemical detection of sialic acids. For example, influenza C viruses bind to 9-O-acetylated sialic acid in tissue slices from rat liver or human colon and are immunologically detected by incubation with rabbit antiserum, followed by treatment with fluorescent anti-rabbit IgG antibody for visualization.[404] Staining is also possible by the esterase activity of the bound viruses. In a similar way, such staining has also been done with a soluble form of influenza C virus hemagglutinin-esterase, wherein the

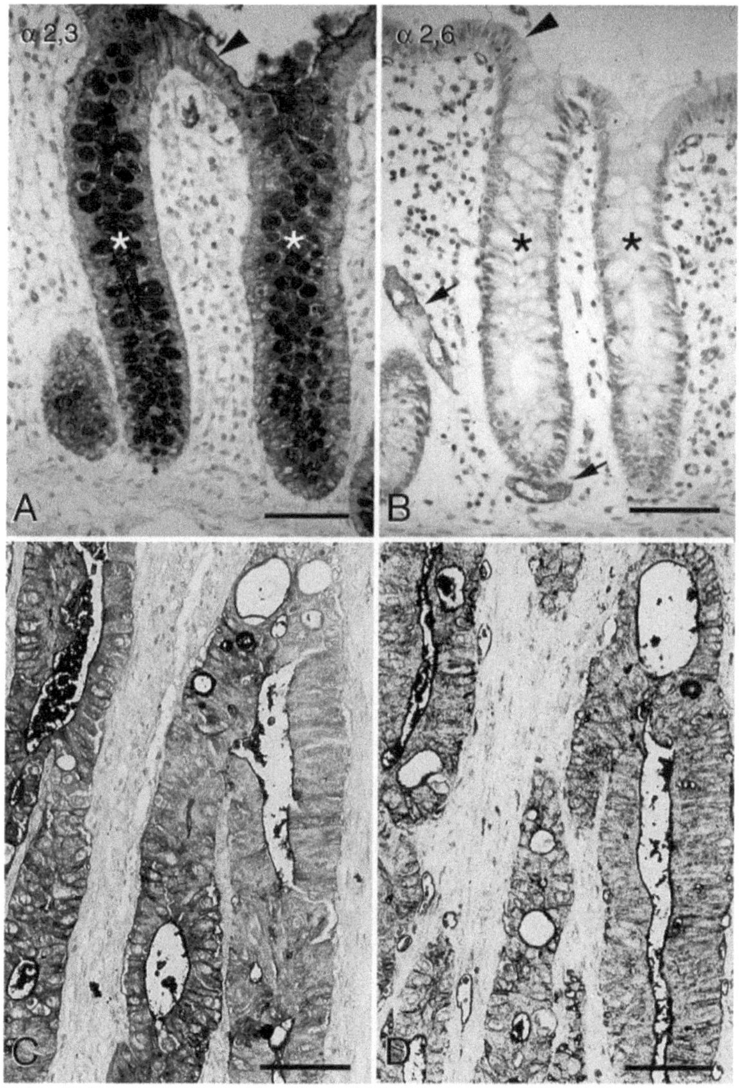

Fig. 24 Detection of sialic acid residues in human colon by silver-intensified lectin–gold labeling of sections from paraffin-embedded tissues. (A) Normal colonic surface (*arrowhead*) and crypt epithelium (*asterisks*) are reactive for (α2→3)-linked sialic acid residues, as detected with the *Maackia amurensis* lectin. (B) In contrast, (α2→6)-linked sialic acid residues are undetectable in the colonic surface (*arrowhead*) and crypt epithelium (*asterisks*) with the *Sambucus nigra* lectin. Note the positive labeling of capillary endothelia (*arrows*). In two adjacent serial sections of a colon carcinoma, both (α2→3)-linked sialic acid residues (C) and (α2→6)-linked sialic acid residues (D) can be detected. Scale bar: 20 μm (A, B), 50 μm (C, D). *Photos courtesy of Prof. Dr. Jürgen Roth.*

C-terminal transmembrane and cytoplasmic domains are replaced by the Fc portion of human IgG. This complex (CHE-Fc) retains both its recognition and enzymatic function. It is detected on the cryosections with alkaline phosphatase-conjugated goat antihuman IgG.[405] Great progress in the identification of 4-mono-, 9-mono, and 7,9-di-O-acetylated sialic acids (Neu5Ac and Neu5Gc) in tissue slices or cells of man and mouse was recently achieved with hemagglutinin-esterase (HE) envelope proteins of nidoviruses (toro- and coronaviruses) and influenza C virus (HEF).[406] For this purpose, HE and HEF ectodomains were fused with the C-terminal domain of human IgG1 or of mouse or bovine IgG2a. The authors showed that O-acetylated sialoglycans occur in a species-, organ-, tissue-, and cell-type-specific way, and that their expression is regulated at the level of the individual cell. They also observed that in several key tissues, especially in the brain, mice and humans share similar 9-*O*-acetyl-sialic acid expression profiles, which suggests an evolutionary and functional conservation. In a similar way, this method with the recombinant soluble virolectins was applied to various tissues of many kinds of animals, including humans, and demonstrated an astonishing variability of the occurrence of the three O-acetylated sialic acids in the tissues studied. While $Neu4,5Ac_2$, for example, was prevalent in horse and guinea pig respiratory tissue, $Neu5,9Ac_2$ and $Neu5,7,9Ac_3$ were found especially in the same tissue of man and mouse.[407] In an early study using a monoclonal antibody bound to gold particles against the *Meningococci* group B capsular polysaccharide, $[\rightarrow 8)Neu5Ac(\alpha 2 \rightarrow]_n$, this polymer was detected in a human Wilms tumor (Fig. 25).[408–410] Treatment with endo-neuraminidase N as control abolished staining. Furthermore, in an immunocytochemical study on ganglioside expression in human breast cancer cells, specific monoclonal antibodies and antimouse IgM or antimouse IgG labeled with Alexa 488 showed the accumulation of gangliosides on the tumor cell surfaces in confocal microscopy.[411]

Finally, at the end of this short overview of the history of histological methods in sialobiology, attention will be paid to some new fluorescence microscopy investigations that allow the manipulation of glycans within their native environment, i.e., to study living cells and the localization, mobility, and lifetime of glycans. Since some sialyltransferases tolerate such large groups as fluorescein at C5 or C9, it was possible to incorporate fluorescent probes via suitably modified CMP-sialic acid donors into cell-surface glycoconjugates[412] (for a general review, see ref. 413) or even enabled the kinetic analyses of the rates of different sialyltransferases within cells.[414] So, it was possible to localize sialyltransferases in the Golgi compartments

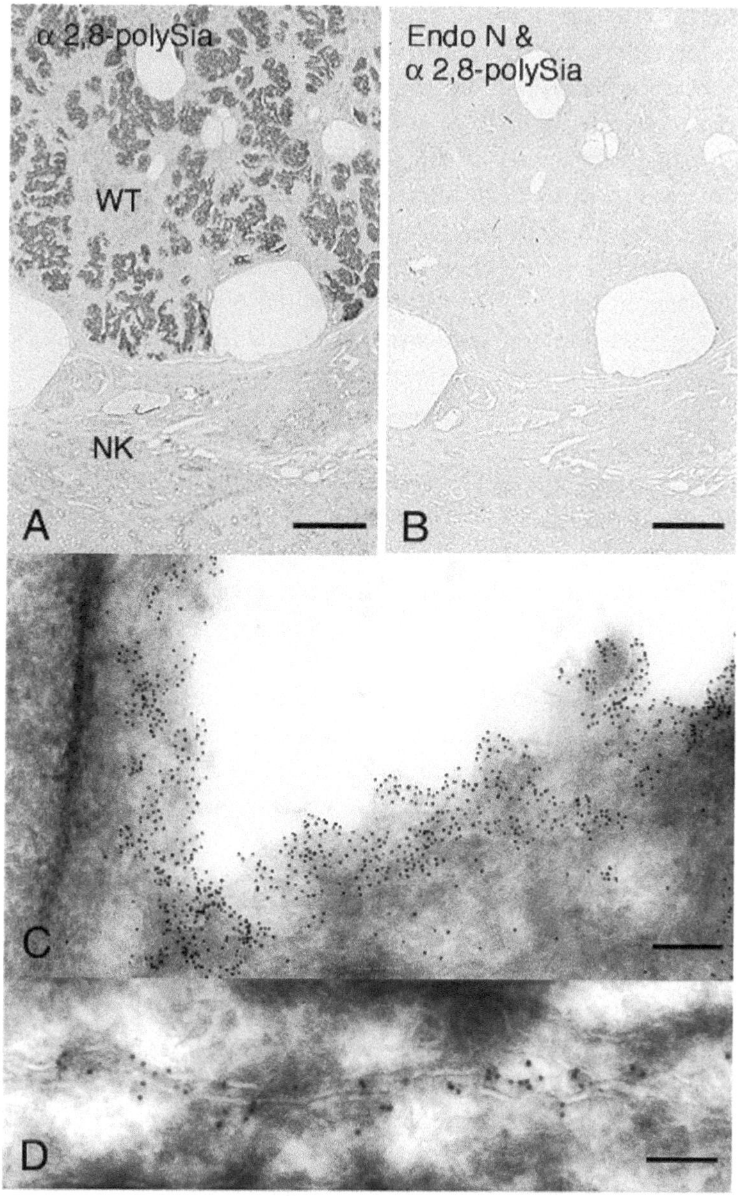

Fig. 25 Reexpression of (α2→8)-linked polysialic acid in human Wilms tumor. (A) Silver-intensified immunogold staining with mAb 735 of a section of paraffin-embedded tissue shows intense labeling in the Wilms tumor (WT). The adjacent normal kidney (NK) is unlabeled. (B) Immunolabeling in a consecutive serial section for (α2→8)-linked polysialic acid is abolished by section pretreatment with endoneuraminidase N (Endo N). (C) Immunogold labeling for (α2→8)-linked polysialic acid of an ultrathin frozen section from a Wilms tumor is very intense at cell-surface regions with a *thick surface coat*. (D) In contrast, immunogold labeling is sparse and patchy in regions of close contact of tumor cells. Scale bar: 300 μm (A, B), 300 nm (C), 150 nm (D). *Photos courtesy of Prof. Dr. Jürgen Roth.*

in rat liver with the CMP-9-fluoresceinyl-Neu5Ac donor substrate.[415] Furthermore, Bertozzi and coworkers introduced the Staudinger ligation for covalent tagging of azido sugars, incorporated biosynthetically into cell-surface glycans. In case of sialic acid, cells are treated with per-O-acetylated *N*-azidoacetyl-D-mannosamine (ManNAz). After entering the cells, the product is O-deacetylated, transformed into the corresponding *N*-azidoacetylneuraminic acid (Neu5Az), subsequently converted into the activated CMP sugar, and finally incorporated into cell-surface glycans on human or other cells.[416] Interestingly, the various biosynthetic enzymes, normally involved in the conversion of ManNAc into bound Neu5Ac (see Section 11.1), tolerate the unnatural substrates. The bound Neu5Az can be visualized using a copper-free click chemistry reaction with, for instance, a difluorinated cyclooctyne (DIFO) reagent, conjugated to an imaging probe, e.g., Alexa Fluor 488. This staining procedure is not toxic and can be carried out in cultured cells and in live mice. Here, a typical example is the study of the sialome during zebrafish development, which showed that the biosynthesis of cell-surface sialosides starts as early as 8.5 h postfertilization.[417] In another example, Neu5Az(α2→3)Lac or Neu5Az(α2→3)Gal was incubated with live trypomastigotes, resulting in Neu5Az(α2→3)Gal-glycoconjugates by means of the parasite *trans*-sialidase (see Section 11.7). For visualization use was made of, for instance, the phosphine–FLAG conjugate, identified by anti-FLAG antibodies. It was learned from this approach that the transfer of sialic acid is mainly to the mucins of the parasites.[418]

8. SIALIC ACIDS AS A PART OF GLYCOPROTEINS, GLYCOLIPIDS, OLIGOSACCHARIDES, AND POLYSACCHARIDES

Although some members of the sialic acid family only occur in free form (see Section 1; Table 1), most members are constituents of glycoproteins, glycolipids, capsular polysaccharides, lipopolysaccharides, and (lipo) oligosaccharides. But also glycosylphosphatidylinositol membrane anchors and keratan sulfates have been shown to contain sialic acid. In the following sections a summary of sialic acid-containing structural elements will be given, as found in first/early reports.

8.1 Glycoprotein N- and O-Glycans

In most eukaryotic glycoproteins the N- and O-glycans are terminated with Neu5Ac or Neu5Gc, or both. First indications stem from the late 1950s.

Over the years, the two sialic acids have been found to be attached to D-galactose (Gal) via (α2→3)[419–423] or (α2→6)[423–425] linkages, to *N*-acetyl-D-galactosamine (GalNAc) via (α2→3)[426–428] or (α2→6)[421,429–432] linkages, and to *N*-acetyl-D-glucosamine (GlcNAc) via an (α2→6) linkage.[433] At the end of the 1970s, mid-1980s, terminal *N*-acylneuraminic acids were also discovered to be (α2→8)-linked[434–436] or (α2→9)-linked[437,438] with internal *N*-acylneuraminic acids, a phenomenon that was already seen in the 1960s for glycolipids (see Section 8.2). Incidentally, terminal Neu5Ac (α2→4)Gal(β1→[439–441] and terminal Neu5Ac(α2→4)GlcNAc(β1→[442] sequences have been published, but discussion about these results does exist. Furthermore, an internal →)Fuc(1→4)Neu5Gc(2→ sequence is known.[443]

Although a long list of naturally occurring *N,O*-acylneuraminic acids is known (Table 1), only for a few glycoproteins, intact glycans with such sialic acids have been published. In short, the terminal sequences comprise Neu4,5-Ac_2(α2→3)Gal(β1→,[444,445] Neu4,5Ac_2(α2→6)Gal(β1→,[445–449] Neu5,9Ac_2 (α2→3)Gal(β1→,[450–454] and Neu5,7,9Ac_3(α2→3)Gal(β1→.[452,453] But also (α2→3) linkages with Gal have been reported for Neu5,7Ac_2, Neu5,8Ac_2, Neu5,7,8Ac_3, and Neu5,8,9Ac_3.[452] Furthermore, [Neu5Ac(α2→8)]$_n$, [Neu5Gc(α2→8)]$_n$, and [Neu5Ac/5Gc(α2→8)]$_n$ sequences with 4Ac, 7Ac, 9Ac, or 9Lt substituents,[455] a Neu5Ac8S(α2→9)[Neu5Ac(α2→9)]$_n$ sequence,[456] and the [Neu5Gc(α2→O5)]$_n$ sequence (linkage between Neu5Gc C2 and the *N*-glycolyl OH function of the next Neu5Gc residue; Fig. 26) should be mentioned.[457] The latter sequence can also be terminated with Neu5Gc9S, i.e., Neu5Gc9S(α2→O5)[Neu5Gc(α2→O5)]$_n$.[72]

The N- and O-glycans can also be terminated by Kdn, and first reports are from the late 1980s and early 1990s. The sialic acid can be (α2→3)-linked with Gal[458,459] and (α2→3)- or (α2→6)-linked with GalNAc.[458,460] Furthermore, Kdn occurs in polysialyl sequences: [Kdn(α2→8)]$_n$ (eventually with 4Ac, 7Ac, or 9Ac substituents),[458,461] and Kdn(α2→8)[Neu5Ac(α2→8)]$_n$, Kdn(α2→8)[Neu5Gc(α2→8)]$_n$, Kdn(α2→8)[Neu5Ac/5Gc(α2→8)]$_n$, and

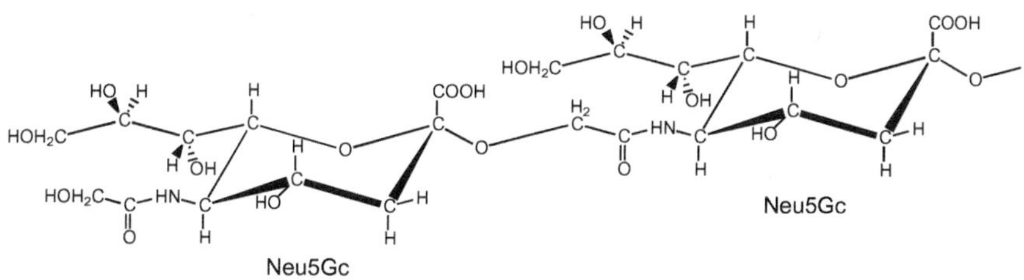

Fig. 26 A terminal Neu5Gc(α2→O5)Neu5Gc sequence.

Kdn9Ac(α2→8)[Neu5Gc(α2→8)]$_n$.[76,80,461] Of interest are the disubstituted internal Kdn units in Fuc(α1→4)[Fuc(α1→5)]Kdn(α2→3)Gal(β1→[462] and in Fuc(α1→4)[Fuc(α1→5)]Kdn(α2→6)GalNAc(α1→,[462,463] and the double Kdn-termination in Kdn(α2→3)[Kdn(α2→6)]GalNAc(α1→,[464] all three occurring in O-glycans.

For extensive reviews of established sialic acid-containing glycoprotein glycan structures, see, e.g., refs. 6 and 465–468. For sialylation of glycosylphosphatidylinositol membrane anchors and keratan sulfates, see ref. 469 and ref. 470, respectively.

8.2 Glycolipids

As in glycoprotein glycans, in glycolipids/gangliosides (first reports from the early 1960s), terminal Neu5Ac and/or Neu5Gc can be coupled with Gal via (α2→3)[471–474] or (α2→6)[475,476] linkages, with GalNAc via (α2→3)[477] or (α2→6)[478] linkages, with GlcNAc via an (α2→6)[479,480] linkage, and with other *N*-acylneuraminic acids via (α2→8)[473,481,482] or (α2→9)[483] linkages. Also Neu5Gc(α2→4)Neu5Gc, Neu5Gc(α2→4)Neu5Ac, Neu5Ac(α2→6)Glc, and Neu5Gc(α2→6)Glc (Glc = D-glucose) elements have been reported.[48,484–486] Furthermore, terminal sequences with Neu itself have been identified: Neu(α2→3)Gal(β1→[16,17,487,488] and Neu(α2→8)Neu5Ac(α2→3)Gal(β1→.[18] In the mid-1960s, it was predicted that gangliosides could occur in the lactonic form (inner esters),[489,490] which a few years later was supported by IR spectroscopy.[491] Besides the inner ester formation between sialic acids (see Section 8.5), such lactone formation was also seen for the Neu5Ac(α2→3)Gal sequence.[492] Here, Neu5Ac(α2→3,1→2)Gal and/or Neu5Ac(α2→3,1→4)Gal elements could be proven by NMR spectroscopy and shown not to be artifacts.[493] More recently, both lactone forms were prepared from sialoglycopeptides (treatment with glacial acidic acid).[494]

In addition to the Neu5Ac/Neu5Gc-glycolipids, from the 1970s onward, intact glycan structures with O-acetylated and O-methylated *N*-acylneuraminic acid residues have been published. Established terminal sequences show

Neu5,9Ac$_2$(α2→3)Gal(β1→,[495]
Neu4Ac5Gc(α2→3)Gal(β1→ (or Neu5GcAc),[496,497]
Neu9Ac5Gc(α2→3)Gal(β1→,[498]
Neu4,9Ac$_2$5Gc(α2→3)Gal(β1→,[499]

Neu5Ac8Me(α2→3)Gal(β1→,[500,501]
Neu5Gc8Me(α2→3)Gal(β1→ (or Neu5GcMe),[74,500]
Neu5Gc8Me(α2→3)GalNAc(β1→,[502]
Neu5Ac9Me(α2→6)Glc(β1→,[503]
Fuc(α1→O5)Neu5Gc(α2→6)Glc(β1→,[504]
Neu4,5Ac$_2$(α2→8)Neu5Ac(α2→3)Gal(β1→,[34]
Neu5,7Ac$_2$(α2→8)Neu5Ac(α2→3)Gal(β1→,[505]
Neu5,9Ac$_2$(α2→8)Neu5Ac(α2→3)Gal(β1→,[505–507]
Neu5,7,9Ac$_3$(α2→8)Neu5Ac(α2→3)Gal(β1→,[505]
Neu5Ac8Me(α2→O5)Neu5Gc(α2→3)Gal(β1→,[508]
[Neu5Gc9Me(α2→O5)]$_n$.[70,509]

Also sequences containing terminal sulfated sialic acid residues have been characterized:

Neu5Ac8S(α2→3)Gal(β1→ and Neu5Gc8S(α2→3)Gal(β1→,[510]
Neu5Ac4S(α2→6)Glc(β1→,[48]
Neu5Ac8S(α2→6)Glc(β1→,[49]
Neu5Gc8S(α2→6)Glc(β1→,[49,71]
Neu5Ac8S/5Gc8S(α2→8)Neu5Ac/5Gc(α2→3)Gal(β1→,[50]
Neu5Ac8S(α2→8)Neu5Ac(α2→6)Glc(β1→.[511]

Furthermore, terminal and internal (branched) elements with sialic acid became known from the late 1970s onward:

Neu5Ac(α2→3)[Neu5Ac(α2→6)]Gal(β1→,[512]
Neu5Ac(α2→3)[Neu5Ac(α2→4)]Gal(β1→8)Neu5Ac(α2→3) GalNAc(β1→,[513]
Neu5Gc8Me(α2→3)[Neu5Gc8Me(α2→6)]GalNAc(β1→,[483]
→6)Gal(β1→4)[Gal(β1→8)]Neu5Gc(α2→3)Gal(β1→,[514]
→6)Gal(β1→4)Neu5Gc(α2→3)Gal(β1→,[515]
→6)Gal(β1→4)Neu5Gc8Me(α2→3)Gal(β1→,[515]
→3/4)Gal(α1→4)Neu5Ac(α2→3)Gal(β1→ [earlier reported with an (α2→4) linkage],[516–518]
→3)Gal(α1→4)Neu5Ac(α2→6)Gal*f*(β1→3)Gal(α1→4) Neu5Ac(α2→3)Gal(β1→,[519]
→3)Gal(α1→4)Neu5Ac8Me(α2→3)Gal(β1→,[501]
→3)Gal(α1→9)Neu5Ac(α2→3)Gal(β1→,[501]
→6)Glc(β1→8)Neu5Gc(α2→6)Glc(β1→,[520]
Fuc(α1→8)Neu5Gc(α2→4)Neu5Ac(α2→,[48]
Fuc(α1→4)Neu5Ac(α2→O5)Neu5Gc(α2→4)Neu5Ac(α2→,[521]
Fuc(α1→O5)Neu5Gc(α2→4)Neu5Ac(α2→ .[522]

Finally, the [Neu5Gc8Me(α2→O5)]$_n$[46,68] and the Neu5Gc8Me(α2→O5)[Neu5Gc(α2→O5)]$_n$[523] sequences, and the Kdn-containing sequences Kdn(α2→3)Gal(β1→,[524] Kdn9Ac(α2→3)Gal(β1→,[525] and Kdn(α2→6)GalNAc(β1→[526] have been established.

In fact, the large variety in sialylated structural elements found so far among glycolipid glycans is certainly influenced by the interest of researchers in the glycolipids of lower animals, such as starfish, sea urchins, sea cucumbers, brittle stars, and feather stars.

For extensive reviews of established sialic acid-containing glycolipid structures, see, e.g., refs. 6 and 527–529.

8.3 Milk Oligosaccharides

Since the late 1950s, neutral and acidic milk oligosaccharides of a large variety of animals have been studied. The ensembles of oligosaccharides differ from species to species, and in the case of human milk over 200 different compounds are expected to be present, of which at the moment 163 have been characterized (up to tridecasaccharides), including 54 sialyloligosaccharides.[530–535]

The established terminal sequences with sialic acid are similar to those found in glycoprotein and glycolipid glycans:

Neu5Ac(α2→3)Gal(β1→,[207,250,251,536]
Neu5Ac(α2→6)Gal(β1→,[536,537]
Neu5Ac(α2→6)GalNAc(β1→,[538]
Neu5Ac(α2→3)GlcNAc(β1→,[539]
Neu5Ac(α2→4)GlcNAc(β1→,[540,541]
Neu5Ac(α2→6)GlcNAc(β1→,[536]
Neu5Ac(α2→6)Glc,[542]
Neu5Gc(α2→3)Gal(β1→,[536]
Neu5Gc(α2→6)Gal(β1→,[543]
Neu5Ac/5Gc(α2→8)Neu5Ac/5Gc(α2→3)Gal(β1→,[536,538,539]
Neu4,5Ac$_2$(α2→3)Gal(β1→,[544,545]
Neu4,5,xAc$_3$(α2→3)Gal(β1→ with x = 7, 8, or 9,[546]
Neu4,5Ac$_2$(α2→6)Gal(β1→,[547] and
Neu5,yAc$_2$(α2→3)Gal(β1→ with y = probably 9.[207,539]

For Neu5,9Ac$_2$ data in human and cow milk carbohydrates, see also ref. 548. Interestingly, besides Neu5Ac and Neu5Gc, porcine milk oligosaccharides (and also glycoproteins and gangliosides) contain very small amounts of Kdn as the building block.[549]

8.4 Urinary Oligosaccharides and Glycopeptides

In normal human urine of different biological origin only small ensembles of sialylated oligosaccharides have been demonstrated to occur, e.g.:

Neu5Ac(α2→3)Gal(β1→4)Glc/GlcNAc/
GlcNAc(α1→P/GalNAc(α1→P,[550–553]
Neu5Ac(α2→6)Gal(β1→4)Glc/GlcNAc/GlcNAc(α1→P,[550,552,553]
Neu5Ac(α2→3)Gal(β1→O)inositols,[553]
Neu5Ac(α2→3)Gal(β1→3)[Neu5Ac(α2→6)]GalNAc,[552] and
Neu5Ac(α2→3)Gal(β1→3)[Neu5Ac(α2→6)]GalNAc(α1→O)Ser.[552]

A disialylated hexasaccharide was found in human pregnancy urine, Neu5Ac (α2→3)Gal(β1→3)[Neu5Ac(α2→6)]GlcNAc(β1→3)Gal(β1→4)Glc.[554]

Studies on rat urine revealed the presence of Neu5Ac/Neu5Gc/ Neu5,9Ac$_2$(α2→3)Gal(β1→4)Glc.[450,551,555]

Several carbohydrate-related inborn errors of metabolism can lead to the accumulation of oligosaccharides in the body (see Sections 13.3 and 13.4). In case of sialidosis/galactosialidosis, (α2→3)- and (α2→6)-sialylated (Neu5Ac) glycoprotein N-glycans (nearly only the expected endo-β-*N*-acetylglucosaminidase-cleaved oligosaccharides) and O-glycans (glycopeptides) are accumulated in the patient's urine, blood serum, amniotic fluid, and ascitic fluid, stored in the placenta, and shown to be present in fibroblasts.[556–569] In a more recent study also relatively small Neu5Ac- and Neu5Ac–Neu5Ac-containing oligosaccharides with a reducing-end hexose were identified. It was suggested that they were derived from glycosphingolipids, whereby the existence of a hitherto unknown endo-glycosylceramidase involved in an alternative glycosphingolipid catabolic pathway was assumed. Unexpectedly, also related sialylated C1-oxidized (aldohexonic acid) versions were detected.[569]

Interestingly, Neu5Ac(α2→6)Man(β1→4)GlcNAc was isolated from the urine of a patient with β-mannosidosis.[570]

Furthermore, in the urine of patients with aspartylglucosaminuria, a series of neutral and acidic glycoasparagines, including Neu5Ac(α2→3/6)Gal(β1→4) GlcNAc(β1→N)Asn, Neu5Ac(α2→3/6)Gal(β1→4)GlcNAc(β1→3)Gal (β1→4)GlcNAc(β1→N)Asn, and Neu5Ac(α2→8)Neu5Ac(α2→3)Gal (β1→4)GlcNAc(β1→N)Asn, are accumulated.[568] Note that the history of elucidating aspartylglucosaminuria carbohydrates showed, besides GlcNAc (β1→N)Asn, initially several proposals for structures, including compounds with terminal Neu5Ac(α2→4)Gal(β1→, Neu5Ac(α2→7)Neu5Ac(α2→3) Gal(β1→, and Neu5Ac(α2→9)Neu5Ac(α2→3)Gal(β1→ elements, that were found to be incorrect in later studies.[571–575]

Patients with Schindler disease type I, type II (Kanzaki), and type III, having an α-*N*-acetylgalactosaminidase deficiency, excrete relatively high concentrations of mono- and disialylated O-glycopeptides and related oligosaccharides in their urine.[568,576–579]

In this context, also pathological abnormalities of (i) lysosomal sialic acid transport (Salla disease) and (ii) a biosynthetic defect in the sialic acid metabolism (CMP-Neu5Ac overproduction; sialuria), described for the first time in 1979 and 1967, respectively, should be mentioned. Depending on the defect, the urinary excretions of Neu5Ac vary from (i) 5- to 10-fold (Salla disease[580,581]) and 200-fold (infantile free sialic acid storage disease[582]) to (ii) 10,000-fold (sialuria; hypersialylation of glycoprotein O-glycans[583–587]) over the normal values, and increased levels in tissues and cells (for further information, see Sections 13.1 and 13.2).

8.5 Capsular Polysaccharides, Lipopolysaccharides, and Lipooligosaccharides

With the first characterized polysialic acid, called colominic acid, in the late 1950s, over the years a continuous flow of structures of sialic acid-containing microbial polysaccharides has appeared. Colominic acid is an (α2→8)-linked homopolymer of Neu5Ac, [→8)Neu5Ac(α2→8)Neu5Ac(α2→]$_n$, isolated from *Escherichia coli* K235.[117,588–590] Such a homopolymer, isolated from *E. coli* K1, can also be partially acetylated at Neu5Ac O7 and/or O9.[36] A polysialic acid from *Neisseria meningitidis* Group C was identified as [→9)-Neu5Ac(α2→9)Neu5Ac(α2→]$_n$ (partly Neu5,7,8Ac$_3$ and Neu5,7Ac$_2$ or Neu5,8Ac$_2$).[35,591] Note that in the late 1950s this homopolymer was suggested to be built up from hexosamine and sialic acid residues.[592] A [→8)Neu5Ac(α2→9)Neu5Ac(α2→]$_n$ polysialic acid was reported to be present in the *E. coli* K92 strain.[593] In several heteropolysaccharides and core regions of lipopolysaccharides (lipooligosaccharides), the (O-acetylated) Neu5Ac(α2→3/6) unit was found in the terminal (side-chain) position.[6,594,595] Interestingly, a microbial decasaccharide revealed the presence of a terminal Neu5,9Ac$_2$(α2→6)Gal(β1→ sequence.[596] Also heteropolysaccharide structures with internal sialic acid, e.g., →4)Neu5Ac(α2→; →4)Neu5,7/9Ac$_2$(α2→; →4)Neu5,9Ac$_2$(α2→; →4)Neu5,7/8,9Ac$_3$(α2→; and →7)Neu5Ac(α2→ units, were established.[6,595,597] Typical examples are [→4)Neu5Ac(α2→6)Gal(α1→]$_n$,[598] [→3)GalNAc(β1→4)Gal(α1→4)Neu5,9Ac$_2$(α2→3)Gal(β1→]$_n$,[39,599] and {→3)GalNAc(β1→7)Neu5Ac(α2→3)[Glc(α1→2)]Glc(β1→}$_n$,[600] and, as core region oligosaccharide of a lipopolysaccharide, Gal(α1→6)Glc(β1→7)Neu5Ac.[601] For Kdn-containing capsular polysaccharides, terminal [Kdn(α2→] and internal

[→9)Kdn5Me(α2→] units have been shown. In case of teichulosonic acids, containing Kdn residues, (β2→4)-linked Kdn backbones, decorated with single monosaccharide side-chain units, have been determined, e.g., {→4)[Glc(β1→8)]Kdn(β2→}$_n$, {→4)[Gal(β1→9)]Kdn(β2→}$_n$, and {→4)[Glc(β1→8)][Glc(β1→9)]Kdn(β2→}$_n$, as well as tetrasaccharides with two Kdn residues, which are (α2→4)-interlinked, e.g., Gal(β1→9)Kdn(α2→4)[Gal(β1→9)]Kdn.[595,602] The Kdn residues in the tetrasaccharide can be acetylated at O7 and O8.[603]

In polysialyl sequences intramolecular lactone formation between sialic acids has been observed.[6] In the case of a Neu5Ac(α2→8)Neu5Ac sequence, lactonization yields a Neu5Ac(α2→8,1→9)Neu5Ac element, whereby the COOH group of one residue reacts with HO9 of an adjacent residue to give a six-membered ring. In a similar way, Neu5Ac(α2→9)Neu5Ac can be converted into Neu5Ac(α2→9,1→8)Neu5Ac.

8.6 General Remarks About the Structural Analysis of Glycans

In the late 1960s and early 1970s, the protocols for the structural analysis of carbohydrates underwent a great change. At first instance, GLC combined with EIMS, developed for the linkage analysis on the monomer level (methylation analysis), caused a revolution in polysaccharide and glycoconjugate characterization.[273,275] And with the introduction of high-resolution ^{1}H and ^{13}C NMR spectroscopy in the mid-1970s, the analysis of glycoprotein glycans, glycolipids, and polysaccharide-derived oligosaccharides came within reach.[275,465,466,468] Concerning sialoglycans, the chemical shifts of H3e and H3a of sialic acids (and other non-2-ulosonic acids) turned out to be highly indicative for the anomeric configuration, the type of linkage with neighboring residues, and the further microenvironment of these acids. Note that the pH of the solution can influence these values (Table 3). For illustration, a few examples at pH 7 will be given. α-Neu5Ac: H3e, δ 2.730/H3a, δ 1.621; β-Neu5Ac: H3e, δ 2.208/H3a, δ 1.827; Neu5Ac(α2→6)Gal(β1→4): H3e, δ 2.67/H3a, δ 1.71–1.72; Neu5Ac(α2→3)Gal(β1→4): H3e, δ 2.75–2.76/H3a, δ 1.80; Neu5Gc(α2→6)Gal(β1→4): H3e, δ 2.69–2.70/H3a, δ 1.73–1.74; Neu5Gc(α2→3)Gal(β1→4): H3e, δ 2.77–2.78/H3a, δ 1.81–1.82; Neu4,5Ac$_2$(α2→6)Gal(β1→4): H3e, δ 2.67–2.68/H3a, δ 1.85; Neu4,5Ac$_2$(α2→3)Gal(β1→4): H3e, δ 2.77/H3a, δ 1.93; Neu5Ac(α2→6)[R→3]GalNAc-ol: H3e, δ 2.72–2.74/H3a, δ 1.69–1.71. For complete tables of H3e/H3a structure–reporter-group NMR data of Neu5Ac/Neu5Gc/Neu4,5Ac$_2$/Neu5,9Ac$_2$-containing structural elements, see refs. 466,468.

Fortunately, the discovery of softer ionization techniques than electron-impact (EIMS) also made mass spectrometry suitable for the analysis of glycoprotein glycans, glycolipids, and (polysaccharide-derived) oligosaccharides. This revolution started with the introduction of fast-atom-bombardment mass spectrometry (FABMS) in the 1980s. However, today, MALDI-TOFMS and ESIMS play the main roles in carbohydrate analysis (for recent reviews, see refs. 275 and 604–610). Even intact glycoproteins can be unraveled now by ESIMS.[611] For a series of papers, specifically focused on mass spectrometry of intact sialoglycans, including information about the stabilization of sialic acid, see refs. 612–618. For the differentiation between (α2→3) and (α2→6) linkages several elegant protocols have been developed. For instance, treatment of Neu5Ac(α2→*x*)Gal(β1→4)-glycans (*x* = 3 or 6) with 4-(4,6-dimethoxy-1,3,5-triazin-2-yl)-4-methyl-morpholinium chloride (carboxylic acid activator) in methanol yielded methyl ester derivatives for (α2→6)-linked Neu5Ac and lactone derivatives for (α2→3)-linked Neu5Ac [Neu5Ac(α2→3, 1→2)Gal(β1→ and/or Neu5Ac(α2→3,1→4)Gal(β1→], providing a difference of 32 Da in mass between both glycosidic linkage types.[617] A very recent example comprises the treatment of Neu5Ac(α2→*x*)Gal(β1→4)-glycans (*x* = 3 or 6) with dimethylamine, 1-ethyl-3-(3-dimethylamino)propyl)carbodiimide (activator)/1-hydroxybenzotriazole (catalyst) in dimethyl sulfoxide, affording dimethylamidate derivatives for (α2→6)-linked Neu5Ac and lactone derivatives for (α2→3)-linked Neu5Ac (difference in mass: 45 Da).[619] Subsequent treatment with ammonium hydroxide had only effect on the lactone ring of the (α2→3)-linked Neu5Ac, resulting in the formation of an amidate derivative (difference in mass: 28 Da).[620]

9. GROWING INTEREST FROM THE ORGANIC AND ENZYMATIC CHEMISTRY

As has been described in Section 4.1, first reports on the organic and enzymatic synthesis of Neu5Ac date from the 1960s, in a period that questions around the different chiralities were still not completely solved. The protocols comprised chemical coupling of oxaloacetic acid (or its di-*tert*-butyl ester) and either D-GlcNAc or D-ManNAc, followed by decarboxylation (Scheme 6),[228,235–237] and enzymatic coupling of pyruvic acid and D-ManNAc in the presence of Neu5Ac aldolase (*N*-acylneuraminate pyruvate lyase) (Scheme 5).[231,232] For the preparation of Neu5Gc, D-ManNAc was replaced by D-ManNGc. Next routes to prepare Neu5Ac enzymatically were the condensation of D-ManNAc with phosphoenolpyruvic acid catalyzed by Neu5Ac synthase (rat liver extract) in the presence of ATP, or by Neu5Ac synthase using nonphosphorylated ManNAc (meningococcal extract).[239,621,622] In the following years, different routes and improvements to obtain higher yields (even at the multikilogram level) were reported.[6,100,623–626] In this context, D-ManNAc was frequently generated from the cheaper D-GlcNAc in an alkaline epimerization process. Note that an excellent biological source for the isolation of large amounts of Neu5Ac was initially the urine of sialuria patients (see Sections 8.4 and 13.1),[584] and later edible bird's nest mucin[153,548] and hen's egg.[627] For the isolation of large amounts of Neu5Gc from holothuroidea gumi, *Cucumaria echinata*, see ref. 628. With the characterization of a long series of *N*,*O*-acylneuraminic acids in the late 1970s and early 1980s (Section 5), synthetic activities in this direction were also started. Via organic synthesis, several free O-acetylated sialic acids became available, e.g., Neu4,5Ac_2, Neu5,9Ac_2, Neu9Ac5Gc, Neu4,5,9Ac_3, and Neu5,7,8,9Ac_4.[629–632] Additionally, a large series of *N*,*O*-acetylneuraminic acid alkyl/aryl ester α-alkyl/α-aryl glycosides were synthesized: Neu4,5Ac_2, Neu5,7Ac_2, Neu5,8Ac_2, Neu5,9Ac_2, Neu5,7,8Ac_3, Neu5,7,9Ac_3, Neu5,8,9Ac_3, Neu4,5,9Ac_3, Neu4,5,8,9Ac_4, and Neu5,7,8,9Ac_4.[633–635] Enzymatic protocols using Neu5Ac aldolase were developed for the synthesis of Neu5,9Ac_2 and Neu5Ac9Lt.[636,637] A number of other naturally occurring sialic acids have also been synthesized along organic chemical or aldolase-catalyzed routes. Examples are as follows: Neu2en5Ac,[638] Neu2en5Gc,[75,638] Neu2en5,9Ac_2,[56] Neu2en5,7,9Ac_3,[56] Neu2en9Ac5Gc,[56] Neu2en5GcAc,[56] Kdn,[639–641] Kdn2en,[642] Neu2,7an5Ac,[643,644] Neu5Ac8Me,[645] Neu5Ac9P,[51] Neu5Ac1,7lactone,[646–648] Neu5Gc1,7lactone,[647] and Kdn1,7lactone.[647] See also refs. 6 and 252. More recently, a highly efficient, one-pot, three-enzyme system with flexible substrate specificity was developed, whereby sialic acid family members are generated as intermediates from corresponding *N*,*O*-acyl-D-mannosamines and *O*-acyl/*O*-alkyl-D-mannoses (catalyst: a sialic acid aldolase) in a protocol that leads via CMP-sialic acids (catalyst: a CMP-sialic acid synthetase) to sialic acid-containing glycosides or sialyl(α2 → 3 or α2 → 6)galacto-glycans (catalyst: a bacterial sialyltransferase).[649–652] Reported examples of intermediates comprise Neu5Ac, Neu5,9Ac_2, Neu5Ac9Lt, Neu5Gc, Neu9Ac5Gc, Neu5GcAc, Neu5GcMe, Kdn, Kdn5Ac, Kdn5Me, Kdn9Ac, and Kdn5,9Ac_2. For the study of biochemical pathways and to understand interactions at the molecular level, isotopically labeled and (stereochemically) modified sialic acids (at C1 and C3 to C9) were also prepared; for reviews, see refs. 6, 106, 108, 252, 626, and 653–660.

The scientific interest in the synthesis of sialo-oligosaccharides dates from the end of the 1960s. The first synthetic compounds comprised two disaccharides with Neu5Ac in a reducing position, i.e., Glc(β1→9)Neu5Ac1Me and Gal(β1→9)Neu5Ac1Me.[661] Furthermore, a series of reducing disaccharides with Neu5Ac in a nonreducing position were synthesized, although in low yields: Neu5Ac(α2→3)Glc, Neu5Ac(α2→3)GlcNAc, Neu5Ac(α2→6)Glc, Neu5Ac(α2→6)Gal, and Neu5Ac(α2→6)GlcNAc.[662] Other interesting disaccharides were nonreducing Neu5Ac(α2↔2α)Neu5Ac and Neu5Ac(α2↔2β)Neu5Ac.[663] Higher yields for Neu5Ac(α2→6)Gal were obtained in the early 1980s,[664] but efforts to synthesize Neu5Ac(α2→6)GlcNAc in larger amounts failed, as mainly Neu5Ac(β2→6)GlcNAc was formed.[665]

It was soon realized that stereoselective chemical sialylation leading to α-glycosidic linkages is one of the most difficult glycosylation reactions in carbohydrate chemistry. First, a neighboring participating group is missing to direct the stereochemical outcome of the coupling (C3 is a deoxy group). Second, the coupling reaction takes place at a sterically hindered, tertiary oxocarbenium-ion intermediate. Third, the deoxy moiety in combination with the electron-withdrawing carboxylic acid group (C1) at the anomeric center (C2) makes sialic acid donors prone to glycal formation. Because it was realized that the availability of labeled sialic acids, sialic acid derivatives, sialic acid O-, N-, C-, and S-glycosides, sialo-analogues, and sialo-oligosaccharides was of the utmost importance for unraveling the secrets of the sialobiology and sialomedicine, the interest in sialochemistry grew enormously. In this context, the creation of a great variety of suitable sialic acid donors and catalysts for organic chemical approaches was developed.[6,107,137,660,666–672] Furthermore, the progress in the biotechnology of carbohydrate enzymes led also to the generation of suitable amounts of stereoselective sialyltransferases/CMP-Neu5Acyl and *trans*-sialidases, which made enzymatic approaches attractive.[107,126,137,139,142,660,669,673–676]

In this way, growing activities focused on a further exploration of the properties of modified sialic acids, the preparation of substrates and inhibitors for sialidases, sialyltransferases, *trans*-sialidases, or sialic acid-converting enzymes, the synthesis of sialo-glycans, and to make compounds accessible for analytical purposes or for studies related to metabolism or biological functions of sialic acids. Over the years, several detailed reviews, describing different aspects of sialic acid organic and enzymatic synthesis, were published (see references in the foregoing paragraphs of this section).

10. STRUCTURAL VARIANTS OF THE SIALIC ACID BACKBONE

A series of naturally occurring nonulosonic acids (NulOs), which do not belong to the classical sialic acid (Sia) family, contain a terminal methyl group as C9 and aminodeoxy functions at C5 and C7, and compared with neuraminic acid (Neu), may exhibit differences in the configuration of the chiral centers at C4, C5, C7, and C8 (Fig. 27). According to IUPAC recommendations, these structural variants of the sialic acid backbone are systematically named 5,7-diamino-3,5,7,9-tetradeoxynon-2-ulosonic acids. In the early days of their discovery, the mid-1980s, they have sometimes been called "bacterial sialic acids." They were found specifically in bacterial lipopolysaccharides and capsular polysaccharides, and since the turn of the century, also in bacterial glycoproteins. Over the years, a number of reviews about these nonulosonic acids have appeared.[114,595,677–681]

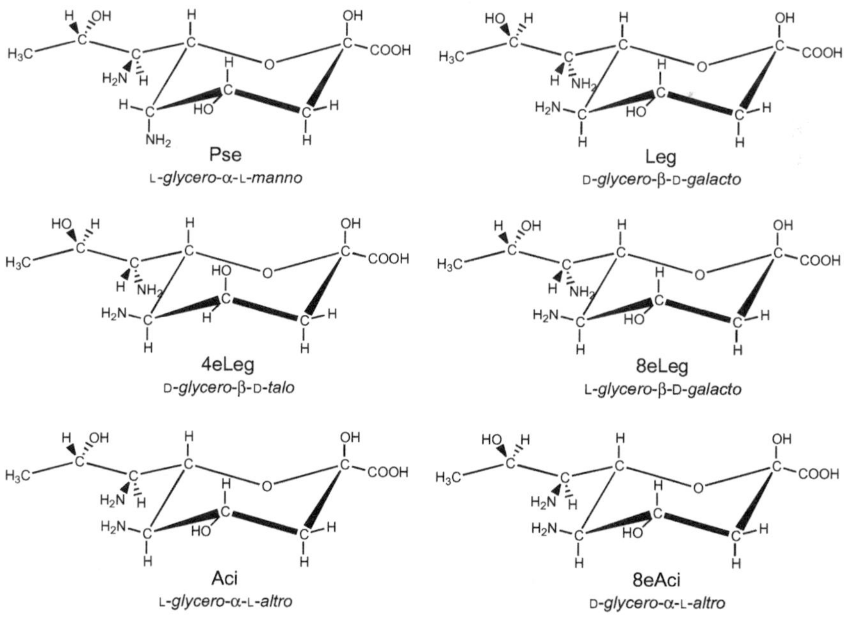

Fig. 27 $^{2}C_{5}$ Chair conformations of α-Pse, β-Leg, β-4eLeg, β-8eLeg, α-Aci, and α-8eAci. Note the difference in orientation of the COOH and OH groups at the anomeric C2 atom.

The first data about the occurrence of these "sialic acid-like" monosaccharides as components of lipopolysaccharides are from 1984 and reported by Knirel and coworkers at the N.D. Zelinsky Institute of Organic Chemistry, Moscow (Russia).[682] In view of the date of their first discovery, it may not be surprising that, in contrast to all the approaches used in the development of the sialic acid field, the analytical techniques had completely changed. Now, from the beginning onward, ^{1}H and ^{13}C NMR spectroscopy and mass spectrometry played major roles in their characterization. Since 2017, six subclasses (six isomers) have been identified, namely,

(i) 5,7-diamino-3,5,7,9-tetradeoxy-L-*glycero*-L-*manno*-non-2-ulosonic acid, called pseudaminic acid (Pse);
(ii) 5,7-diamino-3,5,7,9-tetradeoxy-D-*glycero*-D-*galacto*-non-2-ulosonic acid, called legionaminic acid (Leg);
(iii) 5,7-diamino-3,5,7,9-tetradeoxy-D-*glycero*-D-*talo*-non-2-ulosonic acid, called 4-*epi*-legionaminic acid (4eLeg);
(iv) 5,7-diamino-3,5,7,9-tetradeoxy-L-*glycero*-D-*galacto*-non-2-ulosonic acid, called 8-*epi*-legionaminic acid (8eLeg);
(v) 5,7-diamino-3,5,7,9-tetradeoxy-L-*glycero*-L-*altro*-non-2-ulosonic acid, called acinetaminic acid (Aci); and
(vi) 5,7-diamino-3,5,7,9-tetradeoxy-D-*glycero*-L-*altro*-non-2-ulosonic acid, called 8-*epi*-acinetaminic acid (8eAci).

It should be noted that during the starting period of this research field, some tentative structures were published with misassignment of the chiral configurations. Therefore, to solve these problems, at the end of the 1990s 5,7-di-*N*-acetyl derivatives with the D-*glycero*-D-*galacto*, L-*glycero*-D-*galacto*, D-*glycero*-D-*talo*, L-*glycero*-D-*talo*, D-*glycero*-L-*manno*, L-*glycero*-L-*manno*, D-*glycero*-L-*altro*-, L-*glycero*-L-*altro*, and L-*glycero*-L-*gluco* configuration were synthesized by condensation of suitable 2,4-diacetamido-2,4,6-trideoxyhexoses with oxalacetic acid under basic conditions.[683–685] In contrast to reports from the time of their discovery, presently, the earlier chosen trivial names Pse, Leg, 4eLeg, and 8-Leg have been directly related to the nonsubstituted mother molecules with free amino functions,[595] and the same system was followed for the recently established Aci and 8eAci,[686,687] in a similar way as usual for the Sia family (Neu). As in the case of neuraminic acid, in the full names C7 is the anomeric reference atom. The free sugars exist in the $^{2}C_{5}$ pyranose form as mixtures of α- and β-anomers with a predominance of the thermodynamically more stable anomer having an equatorial carboxyl group. Substitutions of the amino groups at C5 and C7 comprise acetyl (Ac), formyl (Fo), (*R*)-3-hydroxybutyryl (3_{R}Hb), (*S*)-3-hydroxybutyryl (3_{S}Hb), 4-hydroxybutyryl (4Hb), 3,4-dihydroxybutyryl

(3,4Hb), acetimidoyl (Am), *N*-methyl-acetimidoyl (AmMe), *N*,*N*-dimethyl-acetimidoyl ($AmMe_2$), methyl (Me), D-alanyl (Ala), *N*-acetyl-D-alanyl (AlaNAc), *N*-methyl-5-glutamyl (GluNMe), L-glyceryl (Gr), and/or 2,3-di-*O*-methyl-glyceryl (Me_2Gr) groups. Note that only N-acylated derivatives are known; there are no derivatives with free $NH_2$5 or free $NH_2$7 groups. Furthermore, acetyl, *N*-acetyl-glutaminyl (GlnNAc), and glycyl (Gly) groups have been detected at O8, and acetyl groups at O4.

A survey of over 45 known structures[79,602,682,684,686–730] of naturally occurring 5,7-diamino-3,5,7,9-tetradeoxynon-2-ulosonic acid derivatives, together with some related 9-deoxynon-2-ulosonic acids, is presented in Table 4. The symbolic abbreviations for the various derivatives of the 5,7-diamino-3,5,7,9-tetradeoxynon-2-ulosonic acids have been chosen in accordance with the Sia recommendations. The constituents can occur in homo- and heteropolysaccharides, as terminal and internal units, with (α2→ or (β2→ glycosidic linkages. In case of internal residues, the HO4 or HO8 functions can be substituted by another monosaccharide. The same holds for the hydroxy groups of the 3- or 4-hydroxybutyramido substituents at C7. In flagellins, the nonulosonic acids are directly connected to Ser or Thr, e.g., Pse5,7Ac_2(α2→O)Ser.[689] For pilins, an O-linked trisaccharide with Pse7Fo5(3Hb) as the terminal unit has been demonstrated to occur,[703] whereas Leg7Ac5Fo was found to be a terminal unit of an N-linked pentasaccharide of a haloarchaeal virus glycoprotein.[709] For further biochemical information, see Section 11.5.3.

11. UNRAVELING THE METABOLISM OF SIALIC ACIDS

First observations on the metabolism of sialic acids were made at the beginning of the 1940s when sialic acid (at that time an unknown low-molecular-weight substance) was released from cells by a viral or bacterial enzyme, termed the *RDE* (see Section 2.1). This catabolic enzyme was later called neuraminidase or sialidase. Another catabolic enzyme, the acylneuraminate–pyruvate lyase (*N*-acetylneuraminic acid aldolase) from bacteria, was shown to split *N*-acetylneuraminic acid into pyruvate and *N*-acetylmannosamine (see Section 4.1). This serendipitous discovery by the group of Saul Roseman[231,232] at the end of the 1950s turned out to be a milestone in (sialo)glycobiology, leading to Roseman being named "Dr. Saul Serendipity Roseman."[230] The finding of *N*-acetylmannosamine instead of *N*-acetylglucosamine also played a major role in the elucidation of the final structure of *N*-acetylneuraminic acid.

Table 4 Survey of Reported Structures of Naturally Occurring 5,7-Diamino-3,5,7,9-tetradeoxynon-2-ulosonic Acid Derivatives, and Some Related Structures, Which Are Terminal or Internal Units of Bacterial Homo- and Hetero-polysaccharides and Glycoprotein Glycans[595,677,679,681]

Name	Abbreviation	References
**5,7-Diamino-3,5,7,9-tetradeoxy-L-*glycero*-L-*manno*-non-2-ulosonic acid/ pseudaminic acid (Pse)*		
5,7-Di-*N*-acetyl-pseudaminic acid	Pse5,7Ac$_2$	688,689
5,7-Di-*N*-acetyl-8-*O*-acetyl-pseudaminic acid	Pse5,7,8Ac$_3$	689
5,7-Di-*N*-acetyl-8-*O*-glycyl-pseudaminic acid	Pse5,7Ac$_2$8Gly	690
5,7-Di-*N*-glyceryl-pseudaminic acid	Pse5,7Gr$_2$	689
5-*N*-Acetimidoyl-7-*N*-acetyl-pseudaminic acid	Pse7Ac5Am	689,691
5-*N*-Acetimidoyl-7-*N*-acetyl-8-*O*-acetyl-pseudaminic acid	Pse7,8Ac$_2$5Am	692
5-*N*-Acetimidoyl-7-*N*-acetyl-8-*O*-(*N*-acetyl-glutaminyl)-pseudaminic acid	Pse7Ac5Am8GlnNAc	690
5-*N*-Acetyl-7-*N*-formyl-pseudaminic acid	Pse5Ac7Fo	682,693
5-*N*-Acetyl-7-*N*-L-glyceryl-pseudaminic acid	Pse5Ac7Gr	694
5-*N*-Acetyl-7-*N*-[(*R*)-3-hydroxybutyryl]-pseudaminic acid	Pse5Ac7(3$_R$Hb)	682,695,696
5-*N*-Acetyl-7-*N*-[(*R*)-3-hydroxybutyryl]-4-*O*-acetyl-pseudaminic acid	Pse4,5Ac$_2$7(3$_R$Hb)	697
5-*N*-Acetyl-7-*N*-[(*S*)-3-hydroxybutyryl]-pseudaminic acid	Pse5Ac7(3$_S$Hb)	696
5-*N*-Acetyl-7-*N*-(4-hydroxybutyryl)-pseudaminic acid	Pse5Ac7(4Hb)	698
5-*N*-Acetyl-7-*N*-(3,4-dihydroxybutyryl)-pseudaminic acid	Pse5Ac7(3,4Hb)	602
7-*N*-Acetimidoyl-5-*N*-acetyl-pseudaminic acid	Pse5Ac7Am	699
7-*N*-Acetimidoyl-5-*N*-(2,3-di-*O*-methyl-glyceryl)-pseudaminic acid	Pse7Am5Me$_2$Gr	700

Table 4 Survey of Reported Structures of Naturally Occurring 5,7-Diamino-3,5,7,9-tetradeoxynon-2-ulosonic Acid Derivatives, and Some Related Structures, Which Are Terminal or Internal Units of Bacterial Homo- and Hetero-polysaccharides and Glycoprotein Glycans[595,677,679,681]—cont'd

Name	Abbreviation	References
7-*N*-Acetyl-5-*N*-(3-hydroxybutyryl)-pseudaminic acid	Pse7Ac5(3Hb)	701
7-*N*-Acetyl-5-*N*-(2,3-di-*O*-methyl-glyceryl)-pseudaminic acid	Pse7Ac5Me_2Gr	700
7-*N*-Formyl-5-*N*-[(*R*)-3-hydroxybutyryl]-pseudaminic acid	Pse7Fo5(3_RHb)	682,702,703
**5,7-Diamino-3,5,7,9-tetradeoxy-*D*-glycero-*D*-galacto-non-2-ulosonic acid/ legionaminic acid (Leg)*		
5,7-Di-*N*-acetyl-legionaminic acid[a]	Leg5,7Ac_2	684,704,705
5-*N*-Acetimidoyl-7-*N*-acetyl-legionaminic acid[a]	Leg7Ac5Am	684,705
5-*N*-Acetimidoyl-7-*N*-acetyl-8-*O*-acetyl-legionaminic acid[b]	Leg7,8$Ac_2$5Am	684,706
5-*N*-Acetimidoyl-7-*N*-acetyl-5-*N*-methyl-legionaminic acid	Leg7Ac5Am5Me	707
5-*N*-(*N*-Methyl-acetimidoyl)-7-*N*-acetyl-legionaminic acid	Leg7Ac5AmMe	705,707
5-*N*-(*N*,*N*-Dimethyl-acetimidoyl)-7-*N*-acetyl-legionaminic acid	Leg7Ac5AmMe_2	707
5-*N*-Acetimidoyl-7-*N*-acetyl-8-*O*-acetyl-5-*N*-methyl-legionaminic acid	Leg7,8$Ac_2$5Am5Me	706
5-*N*-(*N*,*N*-Dimethyl-acetimidoyl)-7-*N*-acetyl-8-*O*-acetyl-legionaminic acid	Leg7,8$Ac_2$5AmMe_2	706
5-*N*-Acetyl-7-*N*-(*N*-acetyl-D-alanyl)-legionaminic acid	Leg5Ac7AlaNAc	704
5-*N*-Acetyl-7-*N*-(D-alanyl)-legionaminic acid	Leg5Ac7Ala	708
7-*N*-Acetyl-5-*N*-formyl-legionaminic acid	Leg7Ac5Fo	709
7-*N*-Acetyl-5-*N*-[(*S*)-3-hydroxybutyryl]-legionaminic acid[c]	Leg7Ac5(3_SHb)	684

Continued

Table 4 Survey of Reported Structures of Naturally Occurring 5,7-Diamino-3,5,7,9-tetradeoxynon-2-ulosonic Acid Derivatives, and Some Related Structures, Which Are Terminal or Internal Units of Bacterial Homo- and Hetero-polysaccharides and Glycoprotein Glycans[595,677,679,681]—cont'd

Name	Abbreviation	References
7-*N*-Acetyl-5-*N*-(*N*-methyl-5-glutamyl)-legionaminic acid	Leg7Ac5GluNMe	710
**5,7-Diamino-3,5,7,9-tetradeoxy-*D*-glycero-*D*-talo-non-2-ulosonic acid/4-epi-legionaminic acid (4eLeg)*		
5,7-Di-*N*-acetyl-4-*epi*-legionaminic acid	4eLeg5,7Ac_2	711
5,7-Di-*N*-acetyl-8-*O*-acetyl-4-*epi*-legionaminic acid[d]	4eLeg5,7,8Ac_3	684
5-*N*-Acetimidoyl-7-*N*-acetyl-4-*epi*-legionaminic acid	4eLeg7Ac5Am	712
5-*N*-Acetimidoyl-7-*N*-acetyl-8-*O*-acetyl-4-*epi*-legionaminic acid	4eLeg7,8$Ac_2$5Am	712
**5,7-Diamino-3,5,7,9-tetradeoxy-L-glycero-D-galacto-non-2-ulosonic acid/8-epi-legionaminic acid (8eLeg)*		
5,7-Di-*N*-acetyl-8-*epi*-legionaminic acid[e]	8eLeg5,7Ac_2	713–715
5,7-Di-*N*-acetyl-8-*O*-acetyl-8-*epi*-legionaminic acid	8eLeg5,7,8Ac_3	716
5-*N*-Acetimidoyl-7-*N*-acetyl-8-*epi*-legionaminic acid	8eLeg7Ac5Am	717
7-*N*-Acetimidoyl-5-*N*-acetyl-8-*epi*-legionaminic acid	8eLeg5Ac7Am	79
7-*N*-Acetimidoyl-5-*N*-acetyl-8-*O*-acetyl-8-*epi*-legionaminic acid	8eLeg5,8$Ac_2$7Am	79
7-*N*-Acetyl-5-*N*-[(*R*)-3-hydroxybutyryl]-8-*epi*-legionaminic acid[e]	8eLeg7Ac5(3_RHb)	684,713,714
7-*N*-Acetyl-5-*N*-(4-hydroxybutyryl)-8-*epi*-legionaminic acid[e]	8eLeg7Ac5(4Hb)	684
**5,7-Diamino-3,5,7,9-tetradeoxy-L-glycero-L-altro-non-2-ulosonic acid/acinetaminic acid (Aci)*		
5,7-Di-*N*-acetyl-acinetaminic acid	Aci5,7Ac_2	686

Table 4 Survey of Reported Structures of Naturally Occurring 5,7-Diamino-3,5,7,9-tetradeoxynon-2-ulosonic Acid Derivatives, and Some Related Structures, Which Are Terminal or Internal Units of Bacterial Homo- and Hetero-polysaccharides and Glycoprotein Glycans[595,677,679,681]—cont'd

Name	Abbreviation	References
**5,7-Diamino-3,5,7,9-tetradeoxy-D-*glycero*-L-*altro*-non-2-ulosonic acid/8-epi-acinetaminic acid (8eAci)*		
5,7-Di-*N*-acetyl-8-*epi*-acinetaminic acid	8eAci5,7Ac$_2$	687
**Some related 9-deoxynon-2-ulosonic acids*		
5- or 7-Acetamido-,7- or 5-(3-hydroxybutyramido)-5,7,9-trideoxynon-2-ulosonic acid		718
5-Acetamido-7-[(*S*)-3-hydroxybutyramido]-8-amino-3,5,7,8,9-pentadeoxy-L-*glycero*-L-*manno*- or D-*glycero*-L-*manno*-non-2-ulosonic acid		719
5-Acetamidino-3,5,9-trideoxy-L-*glycero*-L-*gluco*-non-2-ulosonic acid (tentatively assigned chirality; trivial name: fusaminic acid)		720
5,7-Diacetamido-8-amino-3,5,7,8,9-pentadeoxy-D-*glycero*-D-*galacto*-non-2-ulosonic acid		721

[a]Initially assigned with the D-*glycero*-L-*galacto*[722] and later with the L-*glycero*-D-*galacto*[713,714,723] configuration.
[b]Initially assigned with the D-*glycero*-L-*galacto*,[724–726] and later with the L-*glycero*-D-*galacto*[713,714] configuration.
[c]Initially assigned with the L-*glycero*-D-*galacto* configuration.[723]
[d]Initially assigned with the L-*glycero*-D-*talo* configuration.[727]
[e]Initially assigned with the D-*glycero*-L-*galacto* configuration.[728–730]
Included literature refers to first reports or revised reports.

In a detailed personal letter handed over to Yuan C. Lee, and recently published by Schnaar and Lee,[731] Roseman describes the full GlcNAc vs ManNAc story, wherein Don Comb played such a great role. With respect to the quantification of N*-acetylhexosamine using the Morgan–Elson color reaction, he memorized that this assay was used for the acylneuraminate–pyruvate lyase degradation of* N*-acetylneuraminic acid. Although the amounts of* N*-acetylneuraminic acid (consumed) and pyruvate (formed) were equal, the amount of formed* N*-acetylhexosamine was only half of the expected amount (reaction time, 3 min). Citing Roseman: "There are two ways to run the Morgan–Elson reaction; one involves heating for 3 min, while the other requires 12 min. There was no way that Don would waste time and heat for 12 min when 3 min was sufficient. But the incredible thing, however, is that both of these sugars [ManNAc and GlcNAc (R.S.)] give the same color when they are heated for 12 min. In other words, the stoichiometry results would have been perfect if he had heated for 12 min and not 3 min. I have often wondered where we would be if Don had heated for 12 min. Would we still think that the Gottschalk structure is correct?" Note that up to this time,* N*-acetylneuraminic acid was thought to be the condensation product of pyruvate and GlcNAc (see Section 4.1).*

For the biosynthesis of sialic acid, it was believed in the beginning that the process is reversibly catalyzed by an enzyme, first observed in extracts of *V. cholerae*, which degrades *N*-acetylneuraminic acid to an *N*-acetylhexosamine and pyruvate.[227] The enzyme also occurred in mammalian tissues and was purified 1700-fold from hog kidney cortex.[328] As said (see Section 4.1), Comb and Roseman found the same enzyme in *C. perfringens* and identified the *N*-acetylhexosamine product as *N*-acetylmannosamine.[231,732] Three arguments, however, were not in favor of this acylneuraminate–pyruvate lyase to be responsible for sialic acid synthesis: (i) in cells, the reaction equilibrium of this enzyme is largely in favor of sialic acid degradation; (ii) the enzyme does not exist in tissues producing large quantities of sialylated mucins[328]; and (iii) the bacteria *V. cholerae* and *C. perfringens* do not contain sialic acid, but express the sialic acid aldolase. Thus, this aldolase is a degradative enzyme, although the reaction is reversible. For an earlier proposed reaction mechanism of the catalysis by this *N*-acetylneuraminic acid lyase, as well as a recent proposal based on crystallography, QM/MM simulation, and mutagenesis, see ref. 6 and ref. 733, respectively.

At this time, at the end of the 1950s, the "golden" and laborious time of the study of the metabolism of sialic acids began.

11.1 The Biosynthesis of *N*-Acetylneuraminic Acid

The biosynthesis of sialic acids, both in animals and in microorganisms, is linked to glycolysis. This central metabolic pathway provides both fructose-6-phosphate and phosphoenol pyruvate as basic substances for the biosynthesis of sialic acids. The chemical steps involved were primarily elucidated in the laboratories of Roseman[734] and Warren.[735] It had been shown in the 1950s by several researchers (see, e.g., refs. 736 and 737) that radioactive D-glucose is converted into D-glucosamine. Moreover, the precursor of D-glucosamine-6-phosphate was investigated in microorganisms[738,739] and in mammals (rat liver).[739,740] Here, the biosynthetic route comprised an amination of D-fructose-6-phosphate (not D-glucose-6-phosphate) with the amido group of L-glutamine catalyzed by L-glutamine-D-fructose-6-phosphate aminotransferase (glucosamine-6-phosphate synthetase), to give D-glucosamine-6-phosphate and L-glutamate.[739] The product D-glucosamine-6-phosphate is then acetylated by an *N*-acetyltransferase, which occurs in many animals and microorganisms, and was first isolated 254-fold from *Neurospora crassa*.[741]

The next step is the transformation of *N*-acetylglucosamine-6-phosphate into *N*-acetylglucosamine-1-phosphate, catalyzed by a phosphoacetylglucosamine mutase, isolated from different sources.[742–744] The sialic acid biosynthesis continues with the formation of UDP-*N*-acetylglucosamine from *N*-acetylglucosamine-1-phosphate and UTP, and the liberation of pyrophosphate by a pyrophosphorylase.[745] UDP-*N*-acetylglucosamine (UDP-GlcNAc) was first detected in rat liver extracts and in yeast.[746] This sugar nucleotide could then, in 1955, be synthesized in extracts of yeast and Rous sarcoma tumors.[747]

The epimerization of *N*-acetylglucosamine (GlcNAc) to *N*-acetylmannosamine (ManNAc) is a key step in sialic acid biosynthesis. In this reaction, the acetylamino group at C2 is epimerized, and the linkage between UDP and the amino sugar is hydrolyzed (UDP-ManNAc could not be detected) (Fig. 28A). It should be noted that this reaction was discovered in rat liver in 1957, but erroneously the reaction product was described as GalNAc.[748] The rectification to ManNAc followed one year later[732]; the responsible enzyme was isolated from rat liver in 1966.[749] The reaction was shown to be irreversible, and the enzyme occurred widely in vertebrate tissues; however, it was not found in bacteria.

For the biosynthesis of sialic acid in vertebrates, ManNAc had to be converted into *N*-acetylmannosamine-6-phosphate (ManNAc6P), and a corresponding kinase was detected in mammalian liver and submandibular glands (see, e.g., ref. 750), as summarized in ref. 751. It can use both ManNAc and *N*-glycolylmannosamine (ManNGc). Many years later (1990s), it was shown by the group of Werner Reutter at the Freie Universität Berlin (Germany) that the conversion of UDP-GlcNAc into ManNAc6P is catalyzed by one bifunctional enzyme, the UDP-*N*-acetylglucosamine 2-epimerase (GNE)/*N*-acetylmannosamine kinase (MNK), a crucial step often discussed between Reutter and one of the authors (R.S.). This enzyme, purified from rat liver, both epimerizes UDP-GlcNAc under release of UDP and phosphorylates the resulting ManNAc by its kinase activity.[752] In the same study, also feedback inhibition of the UDP-*N*-acetylglucosamine 2-epimerase part of the enzyme by CMP-Neu5Ac was demonstrated. In a parallel report, the molecular cloning and functional expression of the GNE/MNK enzyme were published.[753] A selective loss of either the epimerase or the kinase activity was established, via site-directed mutagenesis based on sequence alignments. This technique also enabled the localization of the 2-epimerase activity in the N-terminus and the kinase activity in the C-terminus of the protein.[754] GNE, the root of which is in the Bilateria, was found to have a complex evolutionary origin.[755] It is encoded in the genomes of deuterostomes, acoelomorphs, and *Xenoturbella*, but is absent in protostomes and nonbilaterians.

In order to study the inhibition of the sialic acid biosynthesis, a search for inhibitors of human *N*-acetylmannosamine kinase by high-throughput

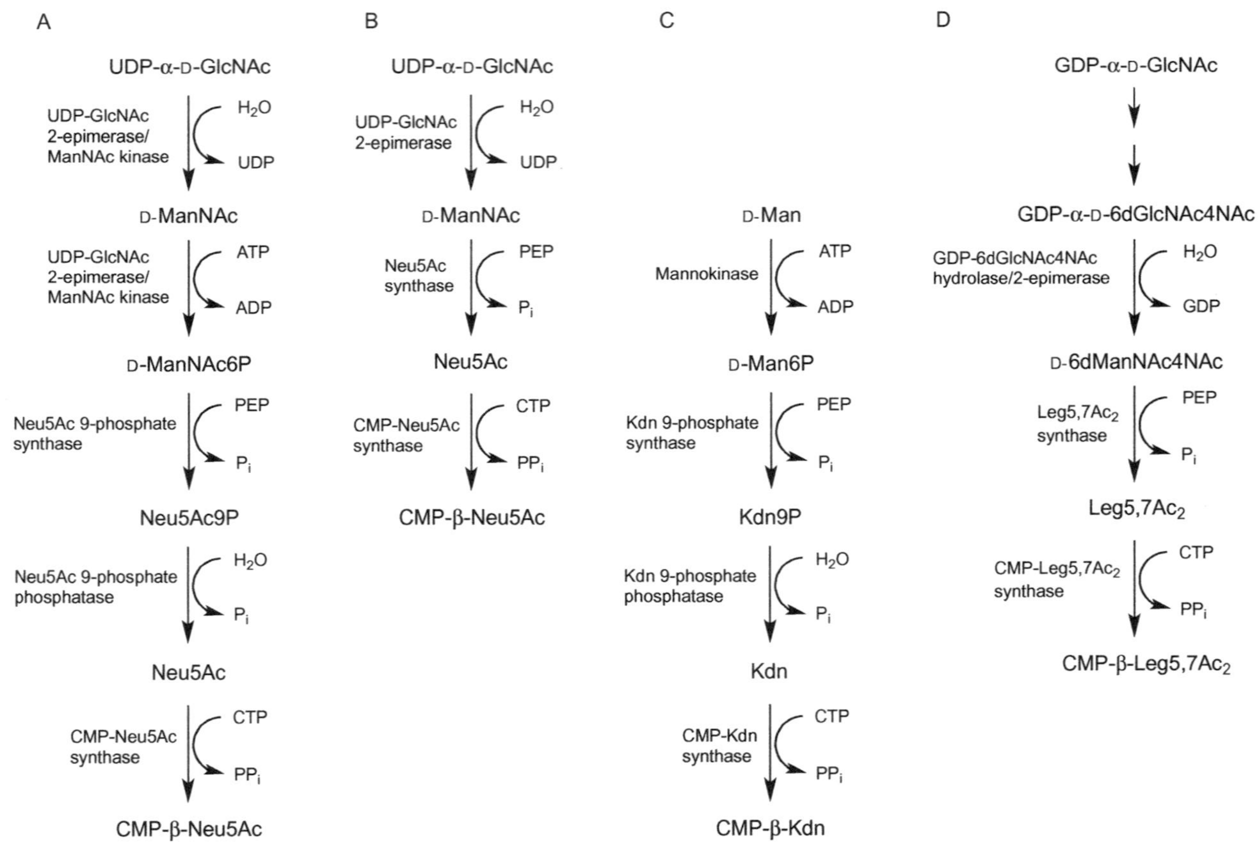

Fig. 28 Metabolism of non-2-ulosonic acids having the D-*glycero*-D-*galacto* configuration. (A) *N*-Acetylneuraminic acid (Neu5Ac) in vertebrates[114]; (B) *N*-acetylneuraminic acid (Neu5Ac) in bacteria[660]; (C) ketodeoxynononic acid (Kdn) in vertebrates[12]; and (D) 5,7-di-*N*-acetyllegionaminic acid ($Leg5,7Ac_2$) in bacteria.[595] *D-6dManNAc4NAc*, 2,4-diacetamido-2,4,6-trideoxy-D-mannose; *GDP-α-D-6dGlcNAc4NAc*, GDP-2,4-diacetamido-2,4,6-trideoxy-α-D-glucose; *Kdn9P*, ketodeoxynononic acid 9-phosphate; *D-ManNAc6P*, *N*-acetyl-D-mannosamine 6-phosphate; *D-Man6P*, D-mannose 6-phosphate; *Neu5Ac9P*, *N*-acetylneuraminic acid 9-phosphate; *PEP*, phosphoenolpyruvate; P_i, phosphate; PP_i, diphosphate. In case of CMP-β-$Leg5,7Ac_2$, also another route, going from UDP-α-D-GlcNAc, via UDP-α-D-6dGlcNAc4NAc to D-6dManNAc4NAc, has been formulated.[595]

screening of drug-like small molecules was carried out,[756] and a few promising compounds binding to the ATP-binding pocket of the enzyme were found. 3-*O*-Methyl-*N*-acetylmannosamine inhibited the same enzyme and correspondingly affected the sialic acid concentration in cells.[757]

In vertebrate tissues, the formation of sialic acid as its 9-phosphate ester (Neu5Ac9P) (Roseman–Warren pathway, Fig. 28A) occurs by condensation of ManNAc6P and phosphoenolpyruvate in an irreversible manner. A synthesizing, anabolic "sialic acid-9-phosphate synthetase" responsible for this reaction was discovered in the early 1960s by the Warren and Roseman groups.[51,621,735] The first sources were rat liver and bovine submandibular gland. The enzyme reaction favors the production of sialic acid.

Before the activation of Neu5Ac by CMP, the 9-phosphate group must be removed. This is achieved by a specific Neu5Ac-9-phosphate phosphatase,[735] first seen in rat liver, and later isolated 800-fold from human erythrocytes (Fig. 28A).[758] This enzyme, completing the Neu5Ac biosynthesis, only occurs in animals. It does not exist in bacteria, where Neu5Ac is formed from nonphosphorylated ManNAc and phosphoenolpyruvate under release of phosphate (Fig. 28B).[759,760]

In summary, at the end of these challenging studies in the 1950s and 1960s, three pathways were described as leading to Neu5Ac:

(i) The Neu5Ac aldolase (lyase) route, which condenses ManNAc and pyruvate and occurs in both animals and bacteria;
(ii) The sialic acid-9-phosphate synthetase route, which condenses ManNAc6P with phosphoenolpyruvate and exists only in animal tissues;
(iii) The bacterial sialic acid synthetase route, condensing ManNAc with phosphoenolpyruvate.

11.2 The Activation of Sialic Acids

The sialylation of glycan chains (see Section 8 for structural elements) requires in general an activated Neu5Ac donor substrate, which is obtained by reaction of Neu5Ac with CTP, resulting in the nucleotide sugar CMP-Neu5Ac and pyrophosphate (Fig. 28A) (for *trans*-sialidases, see Section 11.7). The glycosidic linkage between Neu5Ac and CMP has the β-configuration.[761–763] The responsible enzyme, CMP-sialic acid synthetase, a pyrophosphorylase (EC 2.7.7.43), was first detected in bacteria and purified from *N. meningitidis*.[764] At the same time, it was also found in animal tissues[765,766] and was first purified from pig submandibular glands,[767] as reviewed in ref. 768. The enzyme from different sources was found to react with Neu5Ac, Neu5Gc, and O-acetylated sialic acids, and some of them also with Kdn.[6] Remarkably, in contrast to the other sialic acid-synthesizing enzymes, CMP-Neu5Ac synthetase is located in cell nuclei, first observed by Ed Kean[769] at Case Western Reserve University (Cleveland, USA). From the nucleus, CMP-Neu5Ac is transported into the Golgi, using a carrier-mediated transport system to penetrate liver microsomal vesicles (mice studies).[770] For the cloning and expression of this CMP-sialic acid transporter, see ref. 771. Kean, who devoted most of his scientific life to the investigation of sialic acid activation, discovered also a CMP-Neu5Ac hydrolase in the plasma membranes of rat liver. Its activity was low in other tissues.[772] The role of this hydrolase is still a subject of speculation. Especially in liver, with its most active sialic acid metabolism, the membrane-bound CMP-Neu5Ac hydrolase may regulate the extent of attachment of sialic acid residues to cell membrane glycoconjugates by membrane sialyltransferases.

In the Golgi, CMP-Neu5Ac is used as the substrate for sialyltransferases, which are responsible for sialic acid polymerization (polySia) and sialylation of oligo/polysaccharides, glycoproteins, and glycolipids (see Section 11.3). It additionally serves as a substrate for the *N*-acetyl-hydroxylation reaction leading to Neu5Gc (see Section 11.5.1) and for the O-acetylation of sialic acids (see Section 11.5.2). These modified CMP-sialic acids are also transferred to glycoconjugates in the Golgi. For the synthetic production of CMP-Neu5Ac, see, e.g., refs. 763, 773, and 774.

11.3 Sialic Acid Transfer to Glycans

Studies of the group of sialyltransferases comprise one of the largest and most important fields in sialobiology, and a wealth of papers during the last decades deals with their occurrence, properties, structure, and function. The enzymes are mostly Golgi-membrane-bound. Since several excellent reviews cover this subject, here only some of the milestones will be reported, and the reader is referred to refs. 6, 126, 138, 775, and 776.

In 1963, the first sialyltransferase activities were observed in studies on the biosynthesis of colominic acid[589] and the sialylation of lactose.[777] In the latter case in which details were worked out in Roseman's laboratory, a particulate preparation from the rat mammary gland was incubated with lactose and CMP-[1-^{14}C]Neu5Ac to yield radiolabeled Neu5Ac($\alpha 2 \rightarrow 3$)lactose. In the field of glycolipids, the biosynthesis of GM3 ganglioside was shown from an incubation of lactosyl-ceramide, CMP-[1-^{14}C]Neu5Ac, and a particulate preparation from a 9-day-old embryonic chicken brain.[778]

In Roseman's laboratory the period between 1963 and 1970 was most fruitful, and the detection of four different sialyltransferases for the biosynthesis of glycoproteins and four different sialyltransferases for the biosynthesis of gangliosides was reported by various coworkers (summarized in refs. 230, 779, and 780). These studies suggested that the different sialyltransferases were responsible for the formation of a variety of sialyl glycosidic linkages. The first purification of a sialyltransferase from bovine colostrum by affinity chromatography on CDP-hexanolamine agarose was achieved by the group of Robert Hill at Duke University Medical Center (Durham, NC, USA).[780] The specific activity increased 440,000 times when compared with whole colostrum. The enzyme was pure and consisted of two bands of 43 and 56 kDa, respectively. The enzyme turned out to be a β-D-galactoside α2,6-sialyltransferase.[781] The enzyme was cloned[782] and is cleaved in a whole or post-Golgi compartment.[783] The expression of this ST6Gal-I, regulated by multiple transcriptional promotors, is tissue specific.

Meanwhile, 20 distinct sialyltransferases, which are Golgi type II transmembrane glycosyltransferases, have been identified in murine and human genomes. This superfamily comprises four main sialyltransferase families: (i) an ST3 Gal family with six members (ST3Gal I–VI), whereby sialic acid is ($\alpha 2 \rightarrow 3$)-linked to galactose (Gal); (ii) an ST6 family with two members (ST6Gal I–II), whereby sialic acid is ($\alpha 2 \rightarrow 6$)-linked to Gal; (iii) an ST6GalNAc family with six members (ST6GalNAc I–VI), whereby sialic acid is ($\alpha 2 \rightarrow 6$)-linked to *N*-acetylgalactosamine (GalNAc); and (iv) an ST8Sia family with six members (ST8Sia I–VI), which mediates the transfer of Neu5Ac residues to other Neu5Ac residues of glycoproteins and glycolipids via ($\alpha 2 \rightarrow 8$) linkages.[775] In the latter reference, a phylogenetic approach to animal sialyltransferases and the related genes was undertaken. In this excellent analysis the architecture of these enzymes, which is similar in all vertebrate sialyltransferases, is discussed. Also the precursor genes for animal sialyltransferases are suggested, e.g., for ST6Gal from insects (*Drosophila melanogaster*).[10] There are also over 25 subfamilies in lower vertebrates, orthologs of the 20 mammalian subfamilies, as well as additional subfamilies in fish genomes.[138] For the interesting evolutionary history of the ST8Sia gene family since the early deuterostomes, see ref. 784. Recently, it was shown that the cellular availability of sialic acids regulates the expression of various sialyltransferases at the level of transcription.[785]

Since sialyltransferases are the final enzymes of sialic acid anabolism, they are responsible for the physiologically correct sialylation of glycans in cells, and their activity should be decreased by inhibitors of cellular development or of aberrant oversialylation, such as that which may occur in cancer (see Section 12.3). A natural inhibitor was first purified in 1988 from the 100,000 × g supernatant of a rat-brain homogenate.[786] It is a small, heat-stable protein composed of two components of 14.8 and 22.4 kDa, which inhibits brain α2,6-sialyltransferase noncompetitively. Inhibitory activity was also found in brains from other mammals and in nonbrain tissues. A similar, potent inhibitor was isolated from calf brain, which inhibited (95% by 4 μg of inhibitor) the α2,6-sialyltransferase from ovine submandibular glands.[787] Experiments to chemically modify CMP-Neu5Ac in either its nucleotide or sialic acid moiety in order to gain effective sialyltransferase inhibitors were not successful. Instead, cytidine nucleotides inhibited rat liver α2,6-sialyltransferase (α2,6-ST) and porcine submandibular gland α2,3-sialyltransferase (α2,3-ST) well and at similar rates. The K_i-values of 5′-CMP, 5′-CDP, and 5′-CTP varied between 0.13 and 0.05 mM for α2,6-ST.[788] Probably the strongest sialyltransferase inhibitor presently known is 3-fluoro$_{ax}$-*N*-acetylneuraminic acid (3F$_{ax}$-Neu5Ac), which is supplied to cells or tissues in per-O-acetylated methyl ester form.[789] The hydrophobic molecules penetrate into cells, where they are O-deacetylated. The liberated free 3F$_{ax}$-Neu5Ac more efficiently inhibits α2,3-sialyltransferases and, consequently, hypersialylation of cancer cells. In vivo, delayed growth of inhibitor-treated melanoma cells was observed.

11.4 The Regulation of Sialic Acid Biosynthesis

Such an important biochemical pathway as that of sialic acid synthesis, which results in a great variety of bioactive compounds, must be regulated, since the amount of CMP-sialic acids in a cell may be critical, and influences the level of the sialylation grade in glycoconjugates. The first hint came from studies by Kean.[769] The fact that CMP-sialic acid synthetase is located in cell nuclei of various prominent mammalian tissues suggested that this location is an important control mechanism. Furthermore, two feedback inhibition mechanisms of key enzymes of sialic acid synthesis were detected, i.e., (i) UDP-GlcNAc efficiently inhibits the L-glutamine-D-fructose-6-phosphate aminotransferases, and (ii) CMP-Neu5Ac inhibits UDP-*N*-acetylglucosamine 2-epimerase (GNE).[790] Newer views on the latter enzyme (GNE/MNK) have been discussed in Section 11.1. An example of the lack of the feedback inhibition by CMP-Neu5Ac, required for a physiologically controlled sialic acid metabolism, came from studies of the metabolism of a sialuria patient[583] (see Section 13.1). This young patient excreted gram quantities of free Neu5Ac in urine, as well as Neu2-en5Ac, 2-acetamidoglucal, ManNAc, and GlcNAc.[52,585] 2-Acetamidoglucal represents the putative transition state between UDP-GlcNAc and ManNAc.

11.5 Modified Sialic Acids Different From *N*-Acetylneuraminic Acid

The most frequent natural sialic acid in Nature is Neu5Ac, as is understandable from its biosynthetic pathway. However, from the survey in Table 1 it is also evident that there exist a large number of other sialic acids, variously substituted at the amino and/or hydroxy groups as well as dehydro and anhydro forms. In the following subsections, various derivatives of Neu5Ac will be discussed.

11.5.1 N-Glycolylneuraminic Acid

Leonard Warren, then at the National Institutes of Health, Bethesda, USA, described in 1963 the occurrence of Neu5Gc in different animals.[2] Neu5Gc belongs to the group of the most frequently encountered sialic acids, and it is widely distributed in Nature. Although it has never been found in microorganisms (with the exception of *Trypanosoma cruzi* cultivated in Neu5Gc-containing medium[791]), it is often expressed in species ranging from echinoderms to mammals.[5–7] However, exceptions are known, such as sauropsides (birds and reptiles), and the monotremes, such as platypus and echidna,[792] as well as ferrets, New World monkeys, and man. Very recently, a study appeared about the phylogenetic distribution of the CMP-Neu5Ac hydroxylase (CMAH) gene in 322 deuterostome genomes. It was found that the gene has been lost or inactivated at least 31 times during deuterostome evolution.[793]

Since humans frequently consume Neu5Gc (mainly with red meat), they incorporate a small portion of this sialic acid into their tissues.[794] Neu5Gc is a xeno-antigen and may cause the production of anti-Neu5Gc antibodies. Evidence exists that this can lead to inflammations and tumors, e.g., in the colon.[795] African Great Apes regularly synthesize Neu5Gc in contrast to man.[796] Not much is known about the function of Neu5Gc.[6] Some aberrations in behavior and physiology like retarded wound healing and hearing loss, similar to human pathologies, were observed in Neu5Gc-knockout mice.[797] Neu5Gc may even be toxic. Remarkably, in mammalian brains Neu5Gc exists in only very small amounts. This may be of physiological significance, since experiments leading to a higher expression of Neu5Gc in mouse brain resulted in some neuronal disturbancies.[798] On the other hand, as far as it has been investigated, fetal tissues express more Neu5Gc than adult tissues, as shown in calf and pig.[799,800]

Neu5Gc was first isolated and structurally analyzed by Blix and coworkers,[23] and later it was chemically synthesized by Faillard and Blohm.[801] Elucidation of the biosynthesis of Neu5Gc was a tedious procedure; first attempts go back to 1967.[270,802] Theoretically, the origin of the *N*-glycolyl group is a glycolic acid residue, e.g., from coenzyme A, or it is generated by a direct oxidation of the *N*-acetyl group of the sialic acid precursors GlcNAc and ManNAc, or Neu5Ac itself. Evidence for the formation of glycolyl-CoA from hydroxypyruvate was demonstrated in porcine liver.[803] However, glycolyl-CoA seemed to be a less likely *N*-glycolyl precursor [only traces of O-glycolylated sialic acids have been detected in bovine and equine submandibular glands (Table 1)]. Furthermore, radioactively labeled 3-hydroxypyruvate and serine were not precursors of (radioactive) Neu5Gc.[804] Thus, a transfer of glycolyl groups during Neu5Gc biosynthesis was largely excluded. The following two studies were in agreement with a direct oxidation of the *N*-acetyl group, as had been (theoretically) proposed 10 years earlier[805]: (i) direct hydroxylation of *N*-acetyl anilids to *N*-glycolyl anilids in liver microsomes[806]; (ii) oxidation of UDP-*N*-acetylmuramic acid to UDP-*N*-glycolylmuramic acid before incorporation into the peptide glycan of the bacterial cell wall.[807]

With respect to the suggestion of a direct oxidation of the *N*-acetyl group in the case of Neu5Gc, first experiments with homogenates of porcine submandibular glands did not show the formation of *N*-glycolylglucosamine (GlcNGc) from the sialic acid precursor GlcNAc.[270,802] This pointed to a sialic acid modification at a higher biosynthetic step. Incubation of surviving slices of porcine submandibular glands with *N*-[1-^{14}C]acetyl-glucosamine led to labeling of sialic acids, including Neu5Gc having a radioactive glycolyl group (for analysis of glycolyl groups, see Section 7.1). The direct incorporation of ^{18}O from $^{18}O_2$ gas, resulting in *N*-[^{18}O]glycolyl groups in Neu5Gc, was also shown.[808] Further incubations with [1-^{14}C]acetate, *N*-[1-^{14}C]acetyl-mannosamine, and *N*-[1-^{14}C]acetyl-neuraminic acid led in each case to the identification of *N*-[1-^{14}C]glycolyl-neuraminic acid.[809–811] All these results led to the claim of the existence of an *N*-acetylhydroxylase, involved in Neu5Gc biosynthesis. Most of the newly formed, radioactive Neu5Gc was within the tissue still free and not yet incorporated into glycoconjugates. As shown by the group of one of the authors (R.S.), CMP-Neu5Gc (and CMP-Neu5Ac) is formed in porcine submandibular glands and can act as the donor for the transfer of Neu5Gc to nascent glycoproteins by sialyltransferases.[812] This led to the assumption that Neu5Gc was formed before the transfer to glycoconjugates in the Golgi.

In the same period, using particle-free homogenates from porcine submandibular glands, it was found that the hydroxylation of the *N*-acetyl group of Neu5Ac required, besides of oxygen, the addition of NADPH or ascorbic acid as cofactors.[811] The responsible enzyme was denominated *N*-acetyl-neuraminate:O_2-oxidoreductase, and later Neu5Ac monooxygenase (EC 1.14.18.2). Although activities were found both in the cytosol and in microsomes, they were generally low, whereas glycan-bound Neu5Ac as a substrate was inactive. A later reevaluation of experiments, however, made clear that the positive tissue fractions were contaminated with CMP-Neu5Ac.

Then, in 1988, CMP-Neu5Ac was found to be the real and only substrate for the monooxygenase in porcine submandibular glands, exhibiting highest activity in the soluble protein fraction.[813] The same activity was also detected in the high-speed supernatant fraction of mouse liver.[814] The renamed enzyme, CMP-*N*-acetylneuraminate hydroxylase, was purified from porcine submandibular glands[815] and mouse liver.[816] The molecular weight of the monomeric enzyme from both sources was determined to be 65 kDa. In the same years, the group of Akemi Suzuki at Tokyo Metropolitan Institute of Medical Science (Japan) elucidated the complex mechanism of electron transport required for hydroxylase activity, in which NADH, cytochrome *b*5,[817] and cytochrome *b*5 reductase[818] in mouse liver cytosol were involved. This cytochrome *b*5-dependent electron transport chain is soluble or membrane-bound.[819] The enzyme activity can be stimulated by the addition of iron and is inhibited by iron-binding agents, which suggested the existence of an iron-containing prosthetic group.[820] Electron paramagnetic resonance (EPR) spectroscopy of the primary structure of the hydroxylase revealed that this enzyme is an iron–sulfur protein of the Rieske type (2Fe–2S). Interestingly, it was the first example of such an enzyme found in the cytosol in Eukarya.[821] The Rieske center is located close to the N-terminus of the enzyme protein. For a first molecular cloning of the CMP-Neu5Ac hydroxylase (CMAH) gene of mouse liver, see ref. 822. As the CMAH gene is defective in humans, it became clear why Neu5Gc cannot be synthesized in man.[823] Finally, the group of Ajit Varki at the University of California (San Diego, USA) discovered that the Rieske sulfur–iron center is missing, which is due to an *Alu*-mediated inactivation of the human CMP-Neu5Ac hydroxylase gene.[824,824a]

11.5.2 O-Acetylated Sialic Acids

The existence of O-acetylated sialic acids was assumed very early, when Blix, in 1936, isolated and crystallized sialic acids from bovine submandibular gland mucin under mild acid conditions and found more than one acetyl group per sialic acid molecule ("Kohlenhydrat I")[165] (see Section 2.1). And in a report from 1956, more information about "7-O- and 7,8(9)-O-acetylated *N*-acetylneuraminic acids" after periodate oxidation studies was provided[23,263] (see Section 5). In the same period, O-acetylated sialic acids were also isolated from bovine mucin by incubation with sialidase (*V. cholerae* receptor-destroying enzyme)[264] (see Section 5). Larger quantities of various O-acetylated sialic acids from bovine submandibular gland were obtained by a combination of anion-exchange and cellulose chromatography.[24] The challenges and problems of analyzing O-acetylated sialic acids (Table 1) have already been worked out in Section 5. A turning point in the history of assigning *O*-acyl substitution patterns was a report from the laboratories of both authors (J.P.K. and R.S.) in 1975, describing the determination of "8-*O*-acetyl-*N*-acetylneuraminic acid" as 9-*O*-acetyl-*N*-acetylneuraminic acid (Neu5,9Ac_2). An unexpected misinterpretation of periodate oxidation data[24] was rectified by EIMS[27] (see Section 5). And over the years it became clear that Neu5,9Ac_2 is the most frequent O-acetylated sialic acid occurring in Nature.

The newer analytical techniques made also the study of *O*-acyl migration in sialic acids possible, and Neu4,5Ac_2, Neu5,7Ac_2, Neu5,9Ac_2, Neu5,7,9Ac_3, and Neu5,8,9Ac_3 were subjected to such analyses.[38,38a,101,261] TLC, HPLC, and GLC–EIMS experiments showed that Neu5,7Ac_2, dissolved in pH 7.5 buffer, was almost completely converted into Neu5,9Ac_2 after 30 h at 37°C (only 10% O-deacetylation). At pH 5.0 the substance was relatively stable. This migration phenomenon was studied in more detail by 360-MHz ^{1}H NMR spectroscopy.[38,38a] Using the peak intensities of the *N*-acetyl signals of α,β-Neu5,7Ac_2 and α,β-Neu5,9Ac_2, Fig. 29A illustrates the decrease of Neu5,7Ac_2 and the increase of Neu5,9Ac_2 as a function of the time of incubation in 0.1 M phosphate buffer/D_2O (pD 7.2) at 37°C. In the migration process Neu5,8Ac_2 could not be traced as an intermediate. This could mean that a direct intramolecular transfer Ac7 (secondary ester group) → Ac9 (primary ester group) occurs. But note that so far, an Ac7 → Ac8 → Ac9 mechanism has not been excluded. In Fig. 29B a similar plot is depicted for the conversion of Neu5,7,9Ac_3 to Neu5,8,9Ac_3 (0.1 M phosphate buffer/D_2O, pD 7.5), yielding a molar equilibrium of approximately 1:1. A transient distortion

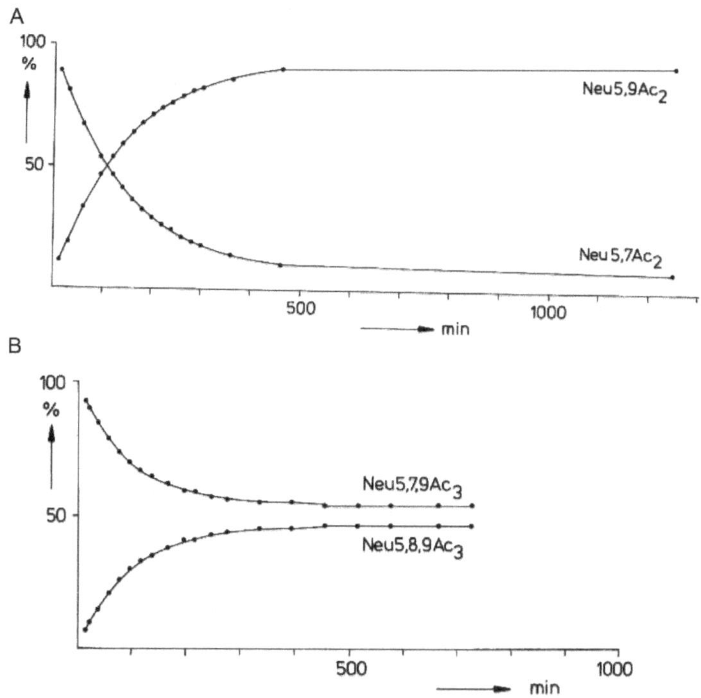

Fig. 29 Time plots for the migration of *O*-acetyl groups in (A) Neu5,7Ac_2 (α:β = 23:77) $\rightarrow$ Neu5,9Ac_2 (α:β = 9:91) and (B) Neu5,7,9Ac_3 (α:β = 22:78) $\rightarrow$ Neu5,8,9Ac_3 (α:β = 3:97) in 0.1 M phosphate, pD 7.2–7.5, at 37°C, as monitored by 360-MHz ^{1}H NMR spectroscopy. As a result of the migration, an anomerization process will also occur. *Reproduced from Kamerling, J. P.; Schauer, R.; Shukla, A. K.; Stoll, S.; van Halbeek, H.; Vliegenthart, J. F. G. Migration of* O*-Acetyl Groups in* N,O*-Acetylneuraminic Acids.* Eur. J. Biochem. ***1987,*** 162, *601–607. Copyright Wiley.*

of the anti orientation of the functional groups at C7 and C8 seems to be necessary to explain the migration of the *O*-acetyl group between the two secondary OH groups (see Section 6). For 4-*O*-acetyl-*N*-acetylneuraminic acid no migration of the *O*-acetyl group was observed.

Application of HPLC and MS techniques revealed that O-acylated sialic acids are most widely distributed, much more than Neu5Gc. They occur in various bacterial species and in vertebrates and invertebrates, from echinoderms to mammals, including those species lacking Neu5Gc, such as monotremes, sauropsides, and humans. For extensive reviews, see refs. 5–7, 792, 825, and 826. In these references, the great variety of biological functions of O-acetylated sialic acids is also described.

The *O*-acetyl groups render the respective sialic acid in most cases completely resistant, as in Neu4,5Ac_2, or partially resistant, as in Neu5,7Ac_2

or $Neu5,9Ac_2$, toward the action of sialidases or sialate lyases. They act as differentiation and tumor antigens, suppress apoptosis, and strongly influence receptor interactions, which also leads to modulation of the immune system. Note that 9-O-acetylation of Neu5Ac and Neu5Gc interferes with the binding of sialoadhesin.[827,827a,828] Increase of virulence by O-acetylation was early observed in *E. coli* K1.[36] Sialic acids, O-acetylated at C4 or at the side chain, are also ligands for some viruses,[826] first studied with influenza C virus. It took some time to demonstrate that the RDE of this virus is a sialate-*O*-acetylesterase, which O-deacetylates $Neu5,9Ac_2$.[829] In contrast, O-acetylation of the sialic acids of mouse erythrocytes inhibits binding of the malaria parasite *Plasmodium falciparum* and infection of the red blood cells.[830]

For a long time, O-acetylated sialic acids appeared as enigmatic molecules in glycobiology, and they were treated as Cinderella molecules, because of their lability, the difficulties in purification, and characterization of the underlying O-acetylating enzyme proteins, as well as the elucidation of their molecular biology. So, it was difficult in the beginning to convince researchers that it may be rewarding to invest time and effort in the study of sialate *O*-acetyltransferases (SOATs), which were rather reluctant to reveal their secrets.

The experiments to gain first insight into the biosynthesis of the O-acetylation of sialic acids were methodologically similar and carried out at about the same time as the study on the Neu5Gc biosynthesis. When surviving slices of submandibular glands from cow and horse were incubated with $[1\text{-}^{14}C]$acetate at various times at 37°C in an atmosphere of oxygen, the radioactivity was almost exclusively incorporated into the *O*-acetyl groups (and *N*-acetyl and *N*-glycolyl groups) of sialic acids within the time of incubation of maximum 4 h.[831] Similar to the Neu5Gc synthesis, the highest specific radioactivity of O-acetylated sialic acids was found in the fraction of free sialic acids, followed by that in the microsomal fractions, and the cytosolic sialic acid of the mature glycoproteins, which were least labeled. These findings were interpreted as O-acetylation also occurs before the sialic acid transfer, either at the hexosamine level (which would not be possible for the horse 4-O-acetylated sialic acid) or at the "free sialic acid" level. In the latter case, free Neu5Ac, Neu5Ac-9-phosphate, or CMP-Neu5Ac were candidates of *O*-acetyltransferase substrates. Indeed, incubation of homogenates of bovine submandibular gland with *N*-$[1\text{-}^{14}C]$acetyl-neuraminic acid and acetyl-CoA led to *N*-$[1\text{-}^{14}C]$acetyl-7- and -8-*O*-acetylneuraminic acid, as shown by radio-TLC analysis and based on our knowledge of sialic acid derivatives at that time (at the end of the 1960s). Hints for the existence of O-acetylated *N*-acetylhexosamines were never obtained, and O-acetylation of mature mucin-bound cytosolic sialic acids did not occur.[832] Therefore, a systematic name for the enzyme was proposed: "acetyl-CoA: *N*-acetylneuraminate-7(8)-*O*-acetyltransferase." As mentioned earlier, the 8-*O*-acetyl position was later corrected to be a 9-*O*-acetyl position, and correspondingly the enzyme was denoted as a 7(9)-*O*-acetyltransferase (Fig. 30). In equine submandibular gland and guinea pig liver, a corresponding acetyl-CoA:sialate-4-*O*-acetyltransferase (EC 2.3.1.44) was detected. Kinetic and other properties of the enzyme were studied in microsomes, prepared from equine submandibular glands, and in an enzyme solubilized from guinea pig liver, respectively.[833,834] In the years that followed, 7 (9)-*O*-acetyltransferase activity was seen in some bacteria and various tissues and cells, ranging from starfish to man.[826]

If sialic acid is modified before the transfer to nascent glycoconjugates, the enzymatic coupling of various sialic acids to CMP should be possible. Indeed, this was shown to be the case for the CMP-sialate synthetase from bovine submandibular glands, which activates Neu5Ac, Neu5Gc, Neu5,7Ac_2,

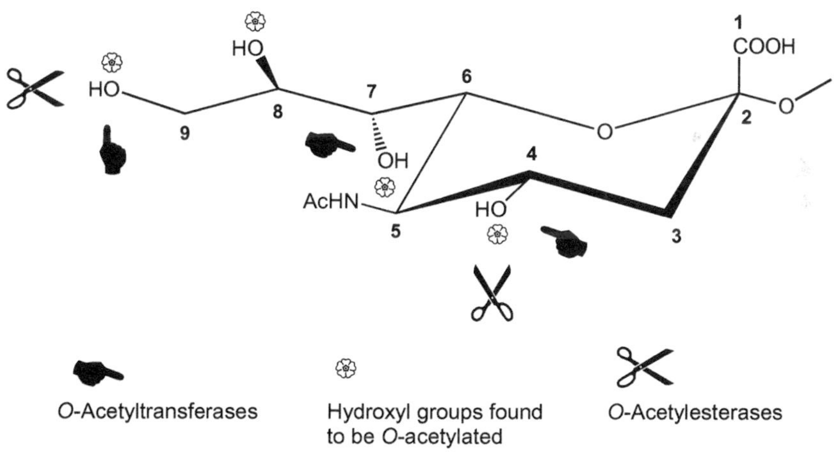

Fig. 30 Metabolism of sialic acid O-acetylation (here Neu5Ac), including a survey of enzymes involved in the transfer and removal of *O*-acetyl groups. The *flowers* indicate the hydroxy groups found to be O-acetylated. The *hands* symbolize the position specificity of the *O*-acetyltransferase activities discovered: sialate-4-*O*-acetyltransferase, sialate-7-*O*-acetyltransferase, and sialate-9-*O*-acetyltransferase. The *scissors* represent the sialate-4-*O*-acetylesterase and sialylate-9-*O*-acetylesterase, involved in the hydrolysis of the ester groups. The *O*-acetyl group at the sialic acid glycerol side chain can migrate between C7 and C9 (via C8?), whereas the 4-*O*-acetyl group seems to be immobile.

and Neu5,8Ac$_2$ (in fact Neu5,9Ac$_2$; see above).[835] Sialyltransferases were also shown to be not absolutely strict toward the kind (structure) of activated sialic acid.[836] Only in 2007, it was established that CMP-Neu5Ac was the correct substrate for *O*-acetyltransferase, and CMP-Neu5,7Ac$_2$ was identified as the primary reaction product. Accordingly, the enzyme's name was then modified into AcCoA-CMP-Neu5Ac 7-*O*-acetyltransferase (EC 2.3.1.45).[837] Interestingly, ab initio calculations gave information concerning possible roles of the different functional groups occurring on sialic acids.[307] Here, the relatively negative charge on the sialic acid side chain O7 supports the hypothesis that in *O*-acetyltransferase reactions this position is the primary insertion site of the acetyl group, and not the less negative oxygen at C9 (Fig. 30). Afterward, the *O*-acetyl group migrates to O9, and the participation of a "migrase" was discussed.[838] Before the 2007 Kiel report,[837] much effort had been undertaken by the group of Ajit Varki to elucidate the mechanism of sialic acid O-acetylation in rat liver.[839,840] A transmembrane transfer of acetyl groups was observed for which histidine and lysine residues were essential. And it was hypothesized that the *O*-acetyltransferase activity might be associated within a membrane-bound complex composed of the acetyl-CoA transporter, the *O*-acetyltransferase, and sialyltransferase activities, and possibly an acetylated intermediate.[841] The same group solubilized acetyl-CoA:polysialic acid *O*-acetyltransferase from K1-positive *E. coli* and studied its basic properties.[842] Unfortunately, it was not possible, in spite of a lot of effort in various laboratories, to isolate and characterize the transferase from this subcellular location, and to get molecular genetic insight on the basis of the amino acid sequence of the enzyme.

However, between 2005 and 2009 several bacterial sialic acid-*O*-acetyltransferase genes from *E. coli* and *Neisseria meningitides* were identified.[843,844] These genes and their encoded proteins do not exhibit significant similarities to animal genes and gene products. *Campylobacter jejuni* also expresses a sialate-*O*-acetyltransferase, which transfers the acetyl groups directly to the O9 position of terminal (α2$\rightarrow$8)-linked sialic acids, as shown by NMR spectroscopy (Fig. 30).[845]

Great progress was achieved by using a rational approach to identify the mammalian sialate-*O*-acetyltransferase (EC 2.3.1.45).[846] To this end, the human genome database was screened for genes that are predicted to possibly encode acetyltransferases and are potentially located in the Golgi membrane. The candidate gene turned out to be CasD1 (capsule structure1 domain containing 1), which is similar to a gene of the fungus *Cryptococcus neoformans* (this yeast contains O-acetylated mannose). Various experiments were described indicating a direct involvement of the human Cas1 including siRNA protein (Cas1p) in the O-acetylation of sialic acids. Expression of this protein and characterization as *O*-acetyltransferase were achieved more recently.[847] CMP-Neu5Ac as substrate of the enzyme was confirmed.

11.5.3 Other Natural Sialic Acids

In addition to the *N*-glycolyl group and *O*-acetyl groups in various positions of the neuraminic acid molecule, other, partially rare substituents were also found, as well as dehydro, anhydro, and lactone derivatives. Since the various functions are often combined, nowadays, over 80 sialic acid derivatives have been identified in Nature (see Table 1; Fig. 31).

A first example is the 9-O-phosphorylated form of Neu5Ac (Neu5Ac9P). This sialic acid only occurs in free form in very low concentrations with high turnover rate as intermediate of the Neu5Ac biosynthesis (see Section 11.1; Fig. 28A). The 9-O-L-lactylation of sialic acids was first seen in 1976, by the identification of Neu5Ac9Lt in bovine submandibular gland mucin[42] and in human serum and saliva.[43] Strikingly, in the latter sources Neu5Ac9Lt was quite often the main *N*,*O*-acylneuraminic acid. 9-O-Lactylation of sialic acid was also found in horse and trout (see Table 1). The function of this sialic acid is not yet known, but it seems to be formed enzymatically, as was studied with a particulate fraction from horse liver.[848]

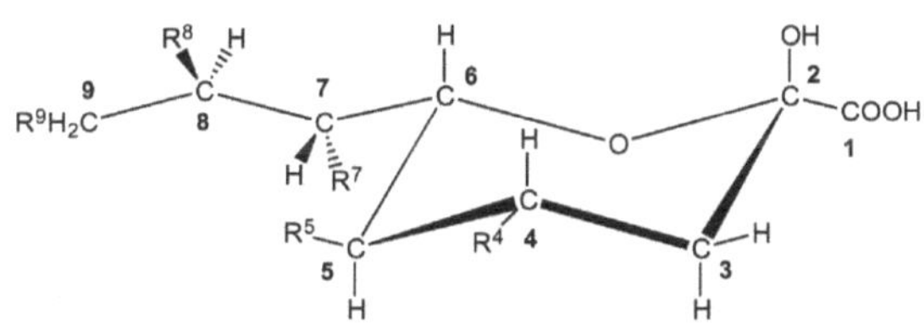

R^5	R^4	R^7	R^8	R^9
NH_2	OH	OH	OH	OH
NH—$COCH_3$	O—$COCH_3$	O—$COCH_3$	O—$COCH_3$	O—$COCH_3$
NH—$COCH_2OH$	O—$COCH_2OH$	O—$COCH_2OH$	O—CH_3	O—$COCH(OH)CH_3$
NH—$COCH_2O$—$COCH_3$	O—SO_3H	NH—$COCH_3$	O—SO_3H	O—CH_3
NH—$COCH_2O$—CH_3				O—SO_3H
OH				O—PO_3H_2
O—$COCH_3$				
O—CH_3				

Tauryl *N*-acetylneuraminic amide is not included in the list above.

Fig. 31 Survey of naturally occurring substituents of the sialic acid backbone. For a full survey of structures, see Table 1.

So far, O-methylated sialic acids in significant quantities have been found only in echinoderms, in various starfish species. The first sialic acid of that type, 8-O-methylated Neu5Gc (Neu5Gc8Me), was seen in the starfish *Asterias forbesi*.[67] The Neu5Ac analogue was later traced.[25] For O-acetylated derivatives of both Neu5Ac8Me and Neu5Gc8Me, see Table 1. *S*-Adenosyl-L-methionine:*N*-acetylneuraminate 8-*O*-methyltransferase (EC 2.1.1.78), involved in the sialic acid methylation, was detected in *Asterias rubens*[46] and purified 22,000-fold from this source.[849] This enzyme methylates both free and bound sialic acids. More recently, a 9-O-methylated variant of Neu5Ac, Neu5Ac9Me, was also found.[47]

In the period between 1976 and 1990 much attention was paid to 8-O-sulfated sialic acids (see Table 1) being present in glycoproteins (see Section 8.1) and glycolipids (see Section 8.2). Here, the group of Kochetkov and coworkers at the Academy of Sciences in Moscow (Russia), working on sialoglycolipids of sea urchin species (Neu5Ac8S, Neu5Gc8S; e.g., ref. 49), was quite active. Neu5Ac8S was also described for a sialoglycosphingolipid from bovine gastric mucosa.[50] With newly developed chemical and immunological techniques, the distribution of Neu5Ac8S and Neu5Gc8S in various cellular sites of sea urchin sperm and eggs was recently shown.[850] Note also the occurrence of Neu5Ac4S in a ganglioside from the sea cucumber *Holothuria pervicax*[48] (see Section 8.2) and Neu5Gc9S in a ganglioside from the sea urchin *Hemicentrotus pulcherrimus*[72] (see Section 8.2). The enzyme responsible for the sulfation of sialic acid is not yet known.

2-Keto-3-deoxynononic acid (Kdn) is a sialic acid in which the amino function at C5 of neuraminic acid (Neu) is substituted by a hydroxy group (Fig. 31). As mentioned in Section 4.4, this substance was discovered in 1986 in cortical alveolar polysialoglycoproteins of rainbow trout eggs,[259] composed mainly of Neu5Gc as sialic acid residue.[76] These sialoglycoproteins were identified 10 years earlier as a novel class of glycoproteins in the unfertilized eggs of rainbow trout.[851] Meanwhile, Kdn is found in all types of glycoconjugates of bacteria and vertebrates, including mammals, in various linkages like Neu5Ac, and in free form in tissues, including cancer tissue. A recent study on the occurrence of free sialic acids in head and neck cancers of the throat revealed relatively high levels of Kdn (2 μg/g tissue) when compared with Neu5Ac and also Neu5Gc. Thus, Kdn was discussed to be of prognostic value, as a biomarker, for detecting some early-stage cancers at biopsy.[852] The occurrence of Kdn in bacteria was first reported in 1989, namely, in the acidic capsular polysaccharide from *Klebsiella ozaenae*, which causes respiratory diseases in man.[77] Kdn was also detected in trout gangliosides[524] and in different organs of the pig, although in smaller amounts than Neu5Ac and Neu5Gc.[800]

For a review from 2006, describing 20 years of history of biological studies on deaminated neuraminic acid, see ref. 12. Focusing here on the metabolism (Fig. 28C), a Kdn-9-phosphate synthase, which condenses mannose-6-phosphate and phosphoenol pyruvate to yield Kdn-9-phosphate, was identified in rainbow trout ovaries and in testes. Thus, the biosynthesis of Kdn is similar to that of Neu5Ac, using Man as the hexose precursor instead of ManNAc. After dephosphorylation, the Kdn that is formed can be linked to CMP (CMP-Kdn synthetase[853]) and transferred to glycans. There exist some enzymes specific for Kdn, including a KDNase (see Section 11.7), but there are also enzymes of the Neu5Ac metabolism that can react with (CMP-)Kdn. For example, a sialyltransferase activity, which catalyzes the transfer of Kdn from CMP-Kdn to the nonreducing termini of the polysialyl chains of polysialoglycoproteins, was found in rainbow trout ovary[854] (see Section 8.1). This capping with Kdn of the polysialyl chains prevents their further elongation.

Information about structural variants of the sialic acid backbone, i.e., the 5,7-diamino-3,5,7,9-tetradeoxynon-2-ulosonic acids pseudaminic acid (Pse), legionaminic acid (Leg), 4-*epi*-legionaminic acid (4eLeg), 8-*epi*-legionaminic acid (8eLeg), acinetaminic acid (Aci), and 8-*epi*-acinetaminic acid (8eAci), has been given in Section 10 and Table 4. For data about homologous biosynthetic pathways for sialic acids and 5,7-diamino-3,5,7,9-tetradeoxynon-2-ulosonic acid derivatives, see refs. 595, 679–681, 686, 699, 705, 708, 716, 855, and 856. As an example, the biosynthetic pathway of 5,7-di-*N*-acetyllegionaminic acid (Leg5,7Ac$_2$),[856] having also the D-*glycero*-D-*galacto* configuration, has been included in Fig. 28D. The biosynthetic pathway of 5,7-di-*N*-acetylpseudaminic acid (L-*glycero*-L-*manno*-configuration) has also been unraveled.[857,858] Here, UDP-α-D-GlcNAc is converted via UDP-β-L-6dAltNAc4NAc and L-6dAltNAc4NAc into Pse5,7Ac$_2$, and then into CMP-α-Pse5,7Ac$_2$ (L-6dAltNAc4NAc = 2,4-diacetamido-2,4,6-trideoxy-L-altrose). For reviews dealing with biological/immunological functions of non-2-ulosonic acids, other than sialic acids, see refs. 595, 679, 681, and 855.

11.5.4 Nonnatural Sialic Acids

The study of the metabolism and biological functions of sialic acids with hexosamine precursors substituted at the amino group with nonnatural functions, inaugurated by Werner Reutter in a visionary way,[859] opened a new area in chemical sialobiology, referred to as metabolic glycoengineering. By the end of the 1960s, Reinhard Brossmer at the Max-Planck-Institut für medizinische Forschung (Heidelberg, Germany) had synthesized, among others, *N*-formyl- and *N*-succinyl-neuraminic acid and had studied the influence of these foreign substituents on sialidase activity[860] (see Section 11.7). At the beginning of the 1980s, the group of Reutter reported that *N*-propionylglucosamine (GlcNProp) and *N*-propionylmannosamine (ManNProp) inhibited the Neu5Ac biosynthesis in a cell-free system of rat liver,[861] and the formation of *N*-propionylneuraminic acid (Neu5Prop) was demonstrated. Following these studies, the uptake of *N*-propionylhexosamines was studied in more detail.[862,863] In rat liver, Neu5Prop biosynthesized from ManNProp was found back in the membrane and serum glycoproteins, as well as in gangliosides. Although some *N*-acetylmannosamine analogues exhibited a degree of toxicity to cells, with this glycoengineering, various modulations of cell functions became possible. For example, human peripheral blood mononuclear cells could be immunomodulated by the replacement of about 10% of their sialic acids with Neu5Prop, after incubation with ManNProp. The cells proliferated faster, produced more IL-2, upregulated their receptor CD25, and expressed the transferrin receptor.[864] Modified sialic acids were found to stimulate the proliferation of astrocytes and microglia cells in the rat central nervous system[865] and modulate virus–receptor interactions.[866] Growth of diploid human fibroblasts with ManNProp, *N*-butyrylmannosamine (ManNBut), and *N*-valerylmannosamine (ManNVar) destroyed their sensitivity to contact inhibition of growth.[867] *N*-Acyl modification of sialic acids seems to be promising in cancer therapy. Treatment of human neuroblastoma cells with ManNProp or ManNVar reduced their sialylation and metastatic potential and increased their sensitivity toward radiation and chemotherapeutics. Furthermore, ManNPent inhibited polysialylation.[868] The formation of cell membrane *N*-azidoacetylneuraminic acid (Neu5Az) after the application of per-*O*-acetylated *N*-azidoacetylmannosamine (ManNAz) also offers many possibilities for cell biological, including histochemical (see Section 7.4), experiments.[413] Recently, this rapidly growing area of the metabolic glycan labeling strategy was excellently and thoroughly reviewed.[869,870] It reports many more applications than mentioned here, and it summarizes that mannosamines with modified aliphatic *N*-acyl functions, like ManNProp, can be used for the investigation of Neu5Ac-dependent biological processes. In contrast, mannosamines with bio-orthogonal modified *N*-acyl groups, like ManNAz, are mainly used to visualize sialylation and sialic acid metabolism in vitro and in vivo.

11.6 The De-esterification of O-Acetylated Sialic Acids

The discovery of more than 20 different naturally occurring O-acetylated sialic acids (see Table 1) implies the necessity of *O*-acetylesterases for the catabolism of these substances, especially, since sialidases and sialate–pyruvate lyases, the next enzymes in the catabolic reaction sequence of bound sialic acids, are strongly influenced in their activity by *O*-acetyl groups. Although 4-O-acetylated sialic acids are resistant toward the action of the latter enzymes [an exception is the fowl plague virus sialidase (see Section 11.7)], no accumulation of sialoglycoconjugates in corresponding tissues, like in the horse, leading to storage diseases, was observed. A first hint for the existence of a "missing link" in the catabolism of O-acetylated sialic acids was obtained in the 1980s when the ganglioside Neu4Ac5Gc-GM3, isolated from horse erythrocytes, was incubated with partially purified equine liver sialidase.[871] Unexpectedly, non-O-acetylated Neu5Gc was slowly released, which was later explained by the possible presence of a 4-*O*-acetylesterase activity in the sialidase preparation.[872] More recently, two esterases could be isolated from the 100,000 × *g* supernatant of horse liver by isoelectric focusing, one O-deacetylating both Neu4,5Ac_2 and Neu5,9Ac_2, and the other only Neu5,9Ac_2.[873] Both enzymes were inactive with 7-O-acetylated sialic acids, as was also observed for other sialate esterases.

Sialate-*O*-acetylesterases (EC 3.1.1.53) were studied in bacteria,[125] in mammalian tissues, for example, rat liver,[841] bovine brain,[874] and horse liver,[873] in corona and toro viruses,[875] and in influenza C viruses. For a summary of the enzyme tests available, see ref. 876. The sialate-*O*-acetylesterase in influenza C virus was one of the first esterases discovered. Although an enzyme destroying the virus receptor had been demonstrated to exist on influenza C viruses in 1950,[877] the specificity of this enzyme and the nature of the cellular receptor for this virus have remained puzzling ever since. Then, in the 1980s, it was found that sialic acid is involved in the receptor function for the virus, because sialidase destroyed a potent hemagglutination inhibitor present in rat serum, i.e., α_1-macroglobulin.[878] The nature of this sialidase-susceptible sialic acid was found to be Neu5,9Ac_2, which lost its

O-acetyl group by incubation of rat α_1-macroglobulin or bovine submandibular gland mucin with influenza C virus.[829] This pointed to the presence of an esterase in the virus, which was thereafter isolated and denoted as sialate 9(4)-*O*-acetylesterase, because it hydrolyzed both Neu4,5Ac_2 and Neu5,9Ac_2.[879] For a further characterization and molecular cloning, see ref. 880. In additional studies, the group of José Cabezas at the Universidad de Salamanca (Spain) found that the influenza C virus esterase had broad substrate specificity, hydrolyzing not only natural substrates containing Neu5,9Ac_2 (e.g., bovine submandibular gland mucin) but also many non-sialic acid *O*-acetyl-containing compounds.[881] This esterase turned out to be part of the hemagglutinin-esterase-fusion (HEF) protein complex of influenza C virus, the crystal structure of which has been elucidated in 1998.[882] In the same period, the HEF protein complex from mouse hepatitis virus strain S was shown to contain a sialate 4-*O*-acetylesterase part.[883] It should be noted that, due to steric hindrance, all these esterases never can act on sialic acid 7-*O*-acetyl groups. Therefore, for O-deacetylation, the 7-*O*-acetyl group must first migrate to position 9 (Fig. 30) in order to get access to the catalytic center of the enzyme.[838] In further investigations, the human sialate-9-*O*-acetylesterase, the gene of which is located on chromosome 11, gained much interest, since it was shown to play a key role in the regulation of B-cell response and the level of O-acetylation in human colon mucosa,[884] and to be involved in immune and autoimmune events.[885–887] In a thorough study of the human sialic acid acetylesterase, including a genomic and phylogenetic analysis, the size of the enzyme and its subcellular localization, the composition of the active center, and other properties have been addressed.[888] One of the most interesting target molecules of this 9-*O*-acetylesterase is the ganglioside 9-O-acetylated GD3, which influences a variety of biological functions, from apoptosis to the growth and metastasis of tumor cells.[826] Its elevated concentration in a variety of tumors was described.

11.7 exo-, endo-, and *trans*-Sialidases

Sialidases are involved in the catabolism of sialic acids. As is clear from the foregoing, sialidases were discovered much earlier than the enzymes involved in the biosynthesis of these sugars (see Section 11.1). In 1942, George Hirst, in the Walter and Eliza Hall Institute of Research in Melbourne (Australia), directed by Frank Macfarlane Burnet (1899–1985), observed a factor eluting influenza viruses from erythrocytes, to which they were bound, resulting in agglutination of the red cells after a period of incubation.[889] Some years later, in the same institute, it was discovered that culture filtrates of various *Clostridia* species made erythrocytes non-agglutinable by influenza virus.[890] They had lost the ligand for virus binding by an enzyme-like activity. In the same year, similar reactions were described with *V. cholerae* filtrates.[175] Additionally, the enzyme-like agent, in the meantime called RDE, was found in various viruses of the mumps influenza group,[891] and later also in pneumococci.[227] As discussed in Section 2.1, the released low-molecular-weight substances turned out to be sialic acids.

The RDE that catalyzes the release of terminal sialic acid residues was initially incorporated in the IUB Nomenclature Recommendations of 1961 with the name of "neuraminidase," as proposed in 1957 by Gottschalk.[892] This name was changed in the IUB Enzyme Nomenclature 1984 edition[893] into "sialidase," but the name "neuraminidase" was kept as an "other name," e.g., viral "sialidases" are quite often called "neuraminidases" because of the use of "N" in their strain designation. In this context, it should be noted that bound neuraminic acid (the mother molecule) is not a substrate for the enzyme. For further enzyme nomenclature discussion, see ref. 894.

After these observations of viral and bacterial sialidases, this enzyme was also discovered in vertebrate tissues, first in both human and bovine plasma,[895] and then in the allantoic membrane of the embryonic chicken,[896] and in several organs of the rat.[897] The following years revealed an explosion of sialidase discoveries in microorganisms and animals. The enzyme turned out to be common in metazoan animals of the deuterostomate lineage from echinoderms to mammals. Various fungi, protozoa, and many pathogenic as well as nonpathogenic bacteria express sialidase. Viral sialidases have been found mainly in orthomyxoviruses, like avian, porcine, equine, and influenza viruses, and in paramyxoviruses to which belong Sendai, Simian, mumps, and Newcastle disease viruses. Presently, influenza virus sialidases, which undergo antigenic drift and are of great pathophysiological significance in influenza, are probably the most studied sialidases. Bacterial sialidases are gaining interest since they are involved in many infectious diseases. For example, *C. perfringens* sialidase, which is involved in the dangerous gas gangrene of muscle tissue, was purified[898] and can be detected by an immunoassay.[899]

The numerous sialidases known are all different regarding their molecular weight, pH dependency (mostly slightly acidic), substrate specificity, sialic acid structure specificity, kinetics, subcellular site, solubility, and susceptibility to inhibitors. They can release α-glycosidically bound sialic acids from N- and O-glycoprotein glycans, gangliosides, oligosaccharides, polysialic acids, and synthetic substrates at rather different rates (Fig. 32A). The isolated bacterial sialidases from *A. ureafaciens*,[900] *C. perfringens*,[898]

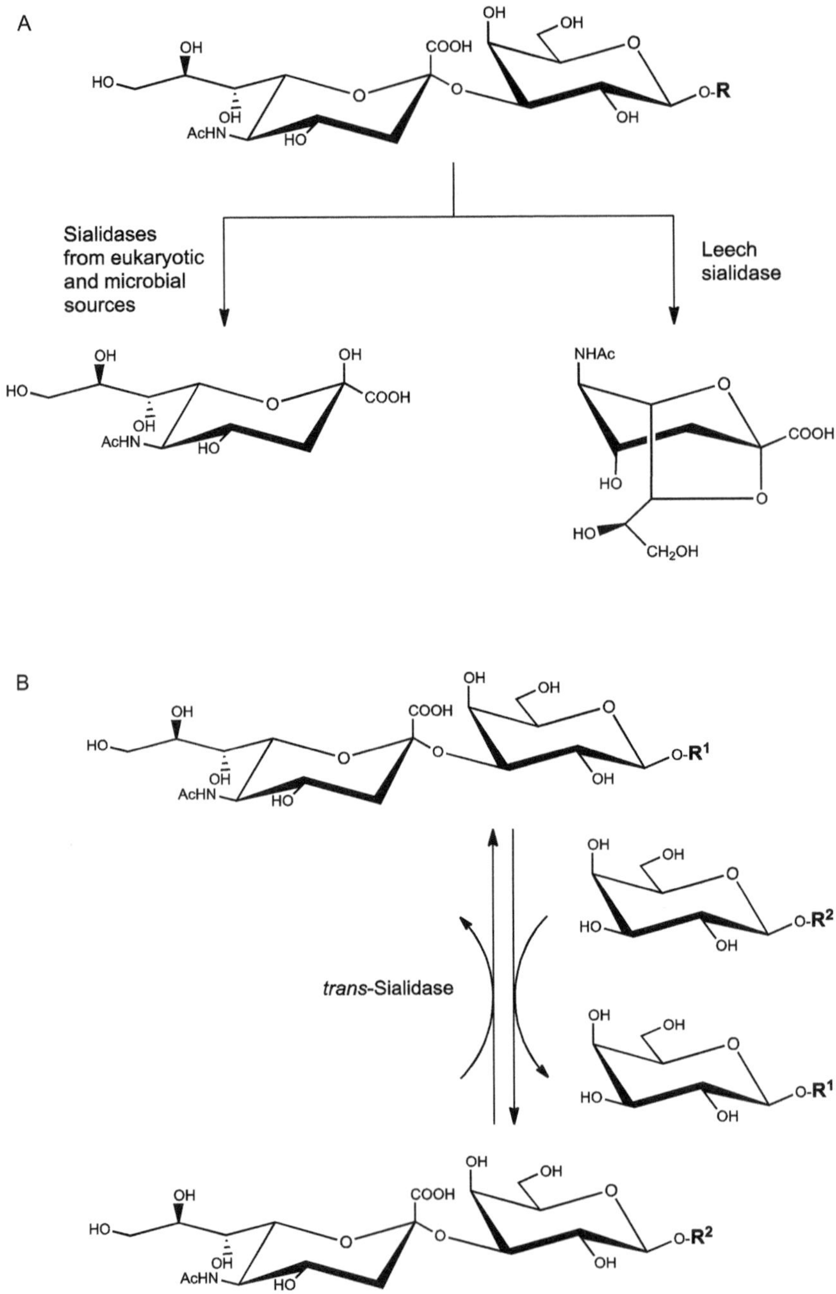

Fig. 32 (A) exo- and endo-Sialidase activities. (B) *trans*-Sialidase activities.

V. cholerae,[901] and *Salmonella typhimurium*,[902] used in structural analysis studies, have similar broad substrate specificities for Neu5Ac. However, a linkage preference [Neu5Ac(α2→6)Gal, Neu5Ac(α2→3)Gal, Neu5Ac(α2→8)Neu5Ac] does exist. The sialidases from Newcastle disease virus[903,904] and *Streptococcus pneumoniae*[905] do not cleave the Neu5Ac(α2→6)Gal linkage. The Newcastle disease virus sialidase is also resistant to the Neu5Ac(α2→6)GalNAc linkage.[904] For more information about the activity of the latter two sialidases on reduced or reductive aminated Neu5Ac(α2→6)GalNAc (open ring form of GalNAc), see refs. 905 and 906. endo-Sialidase, which hydrolyzes (α2→8) linkages in sialic acid polymers, first observed with colominic acid, was detected in the bacteriophage E infecting *E. coli* bacteria.[907] The latter studies comprised rather tedious procedures because the source of the enzyme was feces from elephants kept in a zoo, and the smell during work-up was difficult to be tolerated by the members of the laboratories (Stephan Stirm, personal communication). Nowadays, the gene structures of many isolated sialidases have been analyzed. For reviews over the years, see refs. 6, 89, 113, 676, and 908–915.

Microbial sialidases seem to be inducible by free or bound sialic acids. This was first observed with *V. cholerae*[916,917] and *C. perfringens*[918] sialidases. In the case of the bacterial enzyme, an induction of a permease was observed, enabling the uptake of Neu5Ac into the bacteria. The K_M value of such a (hypothetical) permease reaction was 3.3×10^{-4} M. Furthermore, Neu5Ac lyase activity was also induced in the *Clostridia* by free Neu5Ac. Today, many sialic acid transporters are known,[919] e.g., a first characterized bacterial sialic acid transporter from *E. coli*.[920]

Sialidases are active, although at rather variable rates, with many naturally or synthetically modified α-linked sialic acids.[909,910,921] Two examples from the beginning in the late 1950s are the demonstration that, using *V. cholerae* culture filtrates, O-acetylated sialic acids are released from bovine submandibular mucin at a slower rate than Neu5Ac,[264] and that Neu5Gc is hydrolyzed from the stroma of bovine erythrocytes.[196] All studies revealed that Neu5Ac is released at the highest rate, and Neu5Gc and sialic acids O-acetylated at the glycerol side chain (Neu5,7Ac_2, Neu5,8Ac_2, or Neu5,9Ac_2) are released at slower rates, while Neu4,5Ac_2 shows no activity with most sialidases.[6,24,101] The exception is the sialidase from fowl plague virus, which can slowly hydrolyze the glycosidic bond of Neu4,5Ac_2.[922] It should be noted that the sialic acid lyase behaves toward these naturally substituted sialic acids in a similar way, i.e., it, for example, cannot cleave Neu4,5Ac_2.[330] Using chemically modified sialic acids, it was shown that a smaller function at the sialic acid amino group, like an *N*-formyl group, or a larger one, like an *N*-succinyl substituent, much reduces the *V. cholerae* sialidase activity, when compared with an *N*-acetyl group as in Neu5Ac. While *N*-succinylneuraminic acid is not released at all by the enzyme, the *N*-formyl derivative is slowly hydrolyzed.[860] For more details about this topic, see refs. 6 and 910.

Mammalian sialidases have been investigated very intensively, because they are involved in the regulation of sialic acid degradation, and play a great pathophysiological role in genetic, immunological, and tumor diseases. Among many other researchers, the group of Taeko Miyagi at the Miyagi Cancer Center Research Institute (Miyagi, Japan) has worked very early and intensively on this group of sialidases. Four of them, NEU1–NEU4,[113] are of great importance: (i) lysosomal sialidase NEU1 initiates the degradation of sialoglycoconjugates[923]; (ii) cytosolic sialidase NEU2 exhibits highest activity with gangliosides. It was cloned from rat skeletal muscle[924]; (iii) the plasma membrane-associated sialidase NEU3 is specific for gangliosides. It was cloned from bovine brain[925]; and (iv) sialidase NEU4, bound to the outer mitochondrial membranes via protein–protein interactions, has a wide substrate specificity from glycoproteins to gangliosides and oligosaccharides. It also occurs in the lysosomal lumen. For the cloning of NEU4, see ref. 926.

All these sialidases exert many cell biological functions from cell growth to apoptosis. Being involved in cancer, NEU3 is downregulated in some tumors like colon cancer and acute lymphoblastic leukemia (ALL).[113,927] Importantly, increase of sialidase activity in tumor cells by transfection reduces their aggressiveness. In a recent study with transcripts of the sialidases (NEU1, NEU3, and NEU4) of 11 types of cancer tissues from 170 patients, it was found that NEU1 and NEU3 were upregulated, while NEU4 was downregulated in most cancer types. NEU3 was expressed highest in colorectal and ovarian tumors, and NEU1 in ovarian cancer.[928] The four human sialidase isoenzymes exhibited pronounced differences in their ability to hydrolyze the Neu5,9Ac_2 α-glycosidic linkage. With the exception of NEU4, they hydrolyzed the O-acetylated species slower than Neu5Ac and Neu5Gc.[929]

The mammalian sialidases, including the bacterial and protozoal enzymes, contain a sequence motif of eight amino acids Ser/Thr-X-Asp-X-Gly-X-Thr-Trp/Phe, called the "Asp-box" and a Phe/Tyr-Arg-Ile-Pro motif near the sialidase N-terminus. These motifs, probably involved in the enzymatic cleavage reaction, were observed by comparison of the genes of a variety of mainly Clostridial exo-sialidases and the *T. cruzi trans*-sialidase.[930]

The eukaryotic *trans*-sialidases (TS) are best studied in the pathogenic trypanosomes *T. cruzi*, *Trypanosoma brucei*, and *Trypanosoma congolense*. However, *trans*-sialidase activity was also discovered, for example, in *Corynebacterium diphtheriae*, *Pasteurella multocida*, and *C. jejuni*.[113,139] Trypanosomal *trans*-sialidases are membrane-anchored enzymes that mediate the reversible (α2→3)-sialylation of β-galactose-terminated glycans. In this way, the parasites scavenge sialyl residues from host blood serum components or cells and transfer them to their cell-surface glycans terminated with galactose (Fig. 32B). *trans*-Sialidases can also act as mere sialidases. They are virulence factors in the South American Chagas Disease, the African Sleeping Sickness, and in some African cattle diseases. For a review about their pathomechanism, see ref. 139.

First indications for the existence of such *trans*-sialidases, probably involved in the biology of trypanosomes, stem from 1983, when the group of one of the authors (R.S.) detected Neu5Ac and Neu5Gc in cultivated *T. cruzi*.[791,931] These two sialic acid species probably were not synthesized by the parasites themselves, because their relationship corresponded to that of the culture medium. It was assumed that a sialidase, later localized on the cell membrane of *T. cruzi*,[932] was involved. This assumption was confirmed after the isolation of the first TS from *T. cruzi*,[933,934] followed by TSs from the African species *T. brucei* and *T. congolense*.[935–937] Information about the development of a suitable test system, study of donor and acceptor specificities, reaction mechanism, inhibitors, and biotechnological application of these rare and unique *trans*-sialidases is summarized in refs. 113, 139, and 938. The *T. brucei* group expresses TS during parasite growth in the insect stage, while in *T. cruzi* this enzyme functions mainly in the mammalian host. A wealth of literature, originating especially from South American laboratories, documents the developments during the last 20 years; see ref. 139 for an overview.

As an intramolecular *trans*-sialidase, a sialidase from the leech *Macrobdella decora* can be considered. It catalyzes the formation of 2,7-anhydro-Neu5Ac (Neu2,7an5Ac) from (α2→3)-bound sialic acid of sialyllactose and various glycoconjugates[58,87,939] (Fig. 32A). Whether this sialidase is autochthonous to the leech or of bacterial origin is so far unknown. *S. pneumoniae* sialidase B has also been recognized as an intramolecular *trans*-sialidase.[940] The detection of Neu2,7an5Ac in ear secretion suggests the presence of such an enzyme in human cerumen.[57] Interestingly, gut microbiota are also among the producers of this Neu5Ac derivative.[941]

Although deaminated neuraminic acid (Kdn) is α-glycosidically linked to neighboring residues,[76] the substitution of an *N*-acyl group by a hydroxyl function at C5 blocks nearly completely the action of bacterial exo- or endo-sialidases, known up to the 1990s.[942] A few years later, however, a bacterial sialidase from *Sphingobacterium multivorum* was found to be capable of releasing Kdn (α2→3, α2→6, α2→8), but not *N*-acylneuraminic acids.[943] Therefore, this sialidase was called KDNase. In the same period, a sialidase from the liver of the loach, *Misgurnus fossilis*, released both bound *N*-acetylneuraminic acid and deaminated neuraminic acid residues.[944] The same held for sialidase preparations isolated from the kidney, spleen, and ovary of rainbow trout.[945]

Finally, a unique sialidase was isolated from the hepatopancreas of the oyster species *Crassostrea virginica*.[946] This enzyme was capable of releasing Neu5Gc from the [→O5)Neu5Gc(α2→]$_n$ polysialic acid sequence in a more efficient way than from the [→8)Neu5Ac(α2→]$_n$ or [→8)Neu5Gc (α2→]$_n$ polysialic acid sequences. On the disaccharide level, Neu5Gc (α2→O5)Neu5Gc and also Kdn(α2→8)Kdn were effectively cleaved, whereas Neu5Ac(α2→8)Neu5Ac and Neu5Gc(α2→8)Neu5Gc were poor substrates.

11.7.1 Sialidase Inhibitors

trans-Sialidases and many members of the bacterial and viral sialidases, when involved in diseases, require the availability of inhibitory substances for the study of their mechanism of action and especially for therapeutic use. The best examples are those from influenza viruses, which are absolutely dependent on their sialidase (neuraminidase, symbolized by N) for the infection of host cells. In order to escape the immune defense of the host, they steadily mutate, which, together with the hemagglutinin (H) that also mutates, continuously produce new viruses with variable infection potential. Such mutations are indicated by numbers added to the HN of the viruses, e.g., H1N1.

Early studies on an effective inhibitor of flu were successful in the laboratory of Hans Tuppy and Peter Meindl at the Universität Wien (Vienna, Austria).[84,638] They synthesized various 2-deoxy-2,3-didehydrosialic acids, such as 2-deoxy-2,3-didehydro-*N*-acetylneuraminic acid (Neu2en5Ac) and the corresponding *N*-glycolyl, *N*-propionyl, *N*-butyryl, *N*-benzoyl, and *N*-carbobenzoxy derivatives. Mainly Neu2en5Ac was used for inhibition studies of the propagation of influenza viruses in cell culture. The compound inhibited most viral, bacterial, and mammalian sialidases with an inhibitor constant K_i of about 5 μM, but not trypanosomal *trans*-sialidases. Substitution of the *N*-acetyl group in Neu2en5Ac by an *N*-trifluoroacetyl group led to a twofold better inhibitor.[947] However, these inhibitors were not strong enough to inhibit virus growth in vivo. Neu2en5Ac was also found in human urine and saliva, in the latter source at rather variable quantities up to 2×10^{-5} M, which, in principle, can inhibit influenza virus sialidase and may be of pathophysiological significance.[43]

Zbiral and coworkers at the Universität Wien (Vienna, Austria) synthesized a great variety of Neu2en5Ac variants, among others, 4-amino-Neu2en5Ac,[948] which inhibited viral sialidase about 100-fold stronger than Neu2en5Ac.[6,949] It only weakly inhibited bacterial, e.g., *V. cholerae*, sialidase. A really potent virus sialidase inhibitor was developed when Mark von Itzstein at Monash University, Melbourne (now Griffith University, Brisbane), Australia, substituted this amino group by a guanidino function.[108,949,950] 4-Guanidino-Neu2en5Ac (Fig. 33A) exhibited a K_i-value of 3×10^{-11} M with influenza virus A sialidase. This strong, selective binding to viral sialidases is possible, because a pocket exists in the binding site of viral sialidases near the 4-hydroxy group of Neu5Ac. This pocket also allows slow release of bound Neu4,5Ac$_2$, which is not possible with bacterial and mammalian sialidases.[922] 4-Guanidino-Neu2en5Ac inhibited the growth of a wide range of influenza A and B viruses in a variety of in vitro systems. The resulting drug Relenza® (zanamivir) as a spray is of prophylactic significance in preventing the infection of human bronchial epithelia with influenza viruses and is well suited for medical use.

However, Relenza® cannot enter cells in order to fight against the viruses already present in the tissues. Therefore, a more hydrophobic derivative with similar configuration was synthesized,[951,952] namely, Tamiflu® (oseltamivir, Fig. 33B), which can successfully reduce the symptoms of flu by inhibition of the virus sialidase inside the infected organism.[953,954]

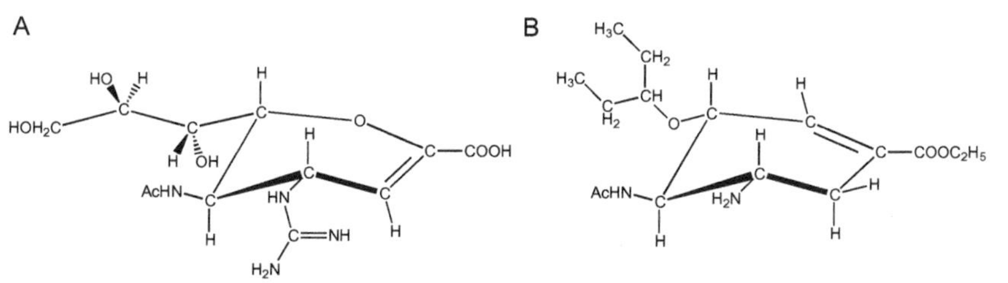

5-Acetamido-2,6-anhydro-4-guanidino-3,4,5-trideoxy-D-*glycero*-D-*galacto*-non-2-enonic acid (4-guanidino-Neu2en5Ac; Zanamivir/Relenza)

Ethyl (*3R,4R,5S*)-4-acetamido-5-amino-3-pentan-3-yloxycyclohexene-1-carboxylate (Oseltamivir/Tamiflu)

Fig. 33 Structures of (A) Relenza®[949] and (B) Tamiflu®.[952]

Because Tamiflu® also has some inhibitory effect on mammalian (human) sialidases, it should not be used for an extended period of time, nor should it be used prophylactically. Since resistance against Tamiflu® has been observed,[955] searches for new and even better flu inhibitors are under way. In a recent publication, such a highly specific inhibitor against influenza viral N1–N9 sialidases, 9-cyclopropyl-carbonylamino-4-guanidino-Neu2en5Ac (cPro-GUN), has been described, which is also highly effective against a Tamiflu®-resistant strain. Furthermore, most importantly, it does not inhibit human sialidases.[956] With such inhibitors, as well as vaccination, it is hoped that such severe and deadly influenza pandemics as had occurred just 100 years ago ("Spanish flu"), claiming many millions of casualties,[957] can be prevented.

12. THE BIOLOGY OF SIALIC ACIDS

In the previous sections, according to the chronological development of the sialic acid field, the history of the sialic acid chemistry and biochemistry was highlighted. Although in recent years the biological roles of sialic acid have been broadly studied in great detail, some important biological discoveries originate from the beginning of the 1940s. As worked out in Sections 2.1 and 11.7, for the progress in sialic acid chemistry, the discoveries in the Walter and Eliza Hall Institute of Research in Melbourne, i.e., the enzymatic release (using virus and microbial filtrates) of a low-molecular-weight substance (later identified as sialic acid) by an RDE (later identified as sialidase), were of great importance. Here, the biological site of this finding will be further cleared up. As mentioned earlier, George Hirst observed that at 4°C influenza viruses bound to chicken erythrocytes, leading to agglutination of the red cells.[889] However, the viruses eluted from the erythrocytes after increasing the temperature to 37°C. Remarkably, the erythrocytes became inagglutinable, but the viruses remained functional. It was concluded that the viruses attach to a receptor substance at the red cell surface, which is inactivated by a viral enzyme. Then, Frank Burnet showed that such an enzyme also existed in the culture filtrate of *V. cholerae.*[175] A mucoprotein of hen egg white (ovomucin) inhibited the reversible virus–erythrocyte binding, a property that was destroyed when the ovomucin was first incubated with influenza virus.[97,174] Thus, the main property of sialic acid was identified in these early days, namely, to act in biological recognition as a receptor, but in due respect to its structure, a ligand. Virologists even today consider sialic acids as receptors that accept viruses. However, sialic acid as a sugar should be called a ligand as it interacts per definition (see below) with a proteinaceous receptor. The first "receptor" function of sialic acids just described represents the binding to influenza virus hemagglutinin, the history and properties of which are described in detail in refs. 958 and 959.

The cell biology, pathophysiology, and pharmacology of sialic acids are a huge and fast developing area, of which only a few highlights can be discussed here. An early review describing the role of glycans and sialic acids was published in 1972.[99] When studying the biological events in which sialic acids are involved, a dual role of these sugars can be recognized, i.e., they can mask recognition sites or they are directly recognized by a lectin-like receptor, for which they represent a ligand. This all mainly happens on the surface of the cell, the so-called glycocalyx.

12.1 The "Sweet Coat of the Cell"

In the 1960s, a study on the migration of cells on electrophoresis led to the discovery of the electronegative charge of cell membranes.[960,961] In fact, this finding contributed to the discovery of the glycocalyx. The term "glycocalyx" was coined by Bennett in 1963 (as "sweet husk") and describes the glycoproteins and glycolipids linked to the external lamina of cells.[962] The carbohydrate composition of the glycocalyx is rather variable and depends on the cell type and function. It represents a sensory system that interacts with the environment. Sialic acids are predominant components at this outer surface, and mainly here they act as biological masks or receptors.

12.2 The Masking Role of Sialic Acids in Serum Glycoproteins and Blood Cells

Sialic acids keep serum glycoproteins and blood cells fit, which is one of the most dominant roles of these monosaccharides. In fact, they regulate the lifetimes of soluble serum glycoproteins and cells, have great significance in all aspects of cell biology, and are most prominent in immunology and in disease states such as cancer (for reviews, see refs. 6 and 963). The sialobiology area was greatly stimulated when Ashwell and Morell in 1974 described in a review that desialylated glycoproteins of rabbit blood serum were rapidly cleared from the circulation and taken up by hepatocytes.[964] This process was mediated by a galactose-specific receptor on the hepatocytes, which recognized galactose residues of the glycoprotein glycans, demasked by the loss of terminal sialic acids (cleavage of Neu5Ac(α2$\rightarrow$3/6)Gal linkage). The first report, describing the fate of desialylated serum glycoproteins, is from 1971.[965] The properties, molecular biology, and biosynthesis of this Ca^{2+}-dependent (galactose) receptor have been intensively investigated, e.g., see ref. 966.

In other studies, it was observed that sialidase treatment of erythrocytes drastically decreased their lifetime.[967] In man it is reduced from about 120 days to a few hours.[968,969] A similar phenomenon was also observed for rabbit erythrocytes.[970] The clearance of asialo-erythrocytes is similar to the sequestration of asialo-glycoproteins, although the erythrocytes are degraded in Kupffer cells. The adhesion of asialo-erythrocytes is mediated by lectin-like receptors on Kupffer cells, which recognize the unmasked galactosyl residues.[969,971,972] Adhesion can be inhibited by glycans that possess galactose in a terminal position. This binding and phagocytosis of sialidase-treated (rat) erythrocytes occurs by a mechanism independent of opsonins.[973] Using rat peritoneal macrophages and desialylated rat erythrocytes, the attachment of the erythrocytes to the macrophages and partial phagocytosis could be clearly demonstrated (Fig. 34A).[974,975] The galactose receptor of rat peritoneal macrophages was identified by photoaffinity labeling as a protein of 42 kDa.[976] It adds to the rapidly increasing family of lectins recognizing galactose and other sugars, as reviewed in ref. 977.

Lymphocytes also alter dramatically their "homing" behavior after sialidase treatment. It was found that enzyme-treated rat thoracic duct lymphocytes accumulated in liver instead of the spleen and lymph nodes.[978,979] However, this loss of orientation was reversible after about 1 day, because the cells recovered, probably by resynthesis of their sialic acid coat. Correspondingly, binding studies with sialidase-treated rat lymphocytes and homologous peritoneal macrophages also showed reversible binding to these nucleated cells (in contrast to the erythrocytes), but no phagocytosis (Fig. 34B and C).[980,981] The interaction was inhibited by galactosides, showing the involvement of the galactose receptor. Also in this in vitro experiment the lymphocytes continuously were released from the macrophages within 1 day. This reversible binding may be due to rapid resynthesis of membrane sialic acids, as observed with human lymphocytes[982] and with rat liver plasma membranes.[983] The turnover rate of sialic acid in a rat liver cell membrane glycoprotein was 33 h, whereas that of the corresponding protein part was 70–78 h.[983]

Sialidase-treated thrombocytes from rats were also rapidly cleared from the circulation.[984] The interaction of desialylated rat thrombocytes with homologous peritoneal macrophages also occurs in a galactose-dependent manner (Fig. 34D).[73,975]

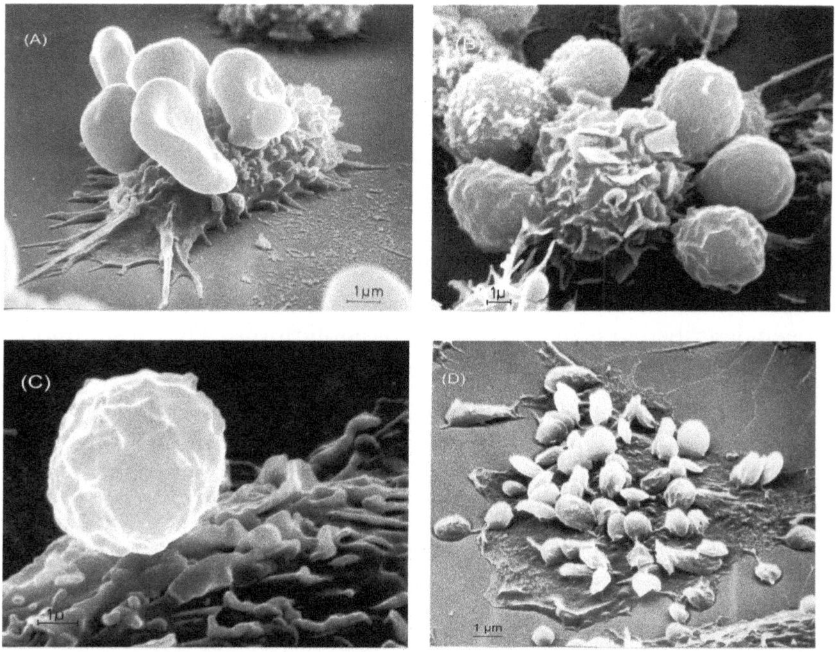

Fig. 34 Binding of sialidase-treated rat erythrocytes, lymphocytes, and thrombocytes to homologues rat peritoneal macrophages by the galactose-specific acceptor. (A) Scanning electron microscopy of erythrocyte–macrophage interaction. After prolonged incubation, the erythrocytes were ingested; (B) scanning electron microscopy of cultured, sialidase-treated lymphocytes bound as a rosette by a peritoneal macrophage; (C) scanning electron microscopy of the characteristic binding of a single sialidase-treated lymphocyte by a rat peritoneal macrophage. In contrast to sialidase-treated, bound erythrocytes, no deformation of phagocytosis is observed. (D) Scanning electron microscopy of sialidase-treated thrombocytes to a macrophage adherent to a Petri dish. *Panels (A) and (D): Reproduced from Schauer, R. Sialic Acids Regulate Cellular and Molecular Recognition. In* Carbohydrates—Synthetic Methods and Applications in Medicinal Chemistry; *Ogura, H.; Hasegawa, A.; Suami, T., Eds.; Kodansha Ltd.: Tokyo, Japan, 1992; pp 340–354. Copyright Wiley-VCH Verlag GmbH & Co. KGaA.*

The various experiments described in this section have shown that sialic acids are of vital importance in the regulation of the lifetimes of these blood components and to maintain homeostasis. The behavior of the fully sialylated glycoproteins and cells is called "self" when they are intact and tolerated by the organism. As soon as sialic acid is lost, the molecules and cells become "nonself" and no longer fit to the chemical architecture required for normal "behavior"; hence, they are recognized by the defense system of the organism and are eventually destroyed. This defense system is part of the immune system; therefore, the dual system "self/nonself" plays a vital role in immunology (see Section 12.4). It functions as an inhibitory "self-signal" to the vertebrate immune system.[985] Also in cancer biology, the masking role of sialic acids is of great significance, as will be discussed in Section 12.3.

12.3 Sialic Acids in Tumors

This is one of the most important areas of sialoglycobiology, and the great variety in the phenomena of tumor growth in which sialic acids are involved has been described in several thousands of publications. The most important feature is the altered, mostly elevated, sialylation of the surface of tumor cells. This is caused by α2,3-, α2,6-, and α2,8-sialyltransferases, which are often more expressed in malignant cells compared to the corresponding normal cells. This high sialylation often leads to accelerated cancer progression including cell–cell repulsion, altered binding of cancer cells to the extracellular matrix, enhanced migration, docking to the vascular endothelium, and invasion that facilitates the formation of metastases and leads to poor prognosis of the disease. An elevated expression of the selectin ligands SLe^{a} and SLe^{x} in a variety of tumor types has been observed. Higher sialylation increases the resistance of tumor cells to apoptosis. Some tumors also exhibit more Neu5Gc and changes in sialic acid O-acetylation or Kdn concentration in comparison to the normal tissues.

Many reviews have appeared before, of which only a few will be listed here. For older reviews, see refs. 986–996. From more recent overviews, refs. 997–999 are herein selected. Here, only the very beginning—according to the aim of this chapter—of tumor sialoglycobiology will be detailed on the basis of publications available. These show the long and tedious way researchers had to go to reach our present profound knowledge on cancer biology. However, many questions still have to be answered, especially in the field of molecular biology and immunology, and, of course, for therapy.

The year 1956 may be considered as the beginning of tumor sialoglycobiology, when Ambrose and coworkers discovered that tumor cells from hamster kidneys have a much greater electrophoretic mobility than normal homologous kidney cells.[1000] In addition, mouse sarcoma cells migrated faster on electrophoresis when they acquired more malignant properties.[1001] At that time, the nature of the molecules responsible for the electrophoretic mobility was not yet known. However, treatment of erythrocytes, as a model, with trypsin released a "sialomucopeptide" which contained terminally linked sialic acids. This was the first demonstration of a sialoglycopeptide at the periphery of a cell.[960] The existence of sialylated carbohydrates on the surface of normal and malignant mammalian cells was nicely demonstrated by using a Hale staining technique and comparison with sialidase treatment.[1002] Most interesting experiments were carried out by the groups of Leonard Warren and Mary C. Glick in the beginning of the 1970s, when they demonstrated a marked difference in cell-surface glycosylation/sialylation of normal and virus-transformed baby hamster kidney (BHK) cells by a double-label technique with [^{14}C]- or [^{3}H]-L-fucose. The radioactive pronase-released glycoproteins from the two cell types were mixed and fractionated on Sephadex G-50. The eluting radioactive glycopeptides of both biological sources showed different profiles.[1003,1004] In a publicaton[1004] by Warren and coworkers, an elevated (up to 11 times) sialyltransferase activity in the transformed BHK cells, when compared to the normal cells, was reported; the sialic acid content was much higher in the transformed cells. Also invasive tumors of human breast and colon expressed elevated levels of glycoprotein sialyltransferases.[1005] Since then, by using various techniques, increased sialylation and altered glycosylation have been observed in many tumors, mostly derived from ectodermal tissues, for example, in the most prevalent breast and colon cancers. A wealth of information about tumor biology was obtained in the early days of this research from the study of blood serum from tumor patients; an increase of sialic acids during tumor growth in humans and mice was identified.[1006–1008] The increase of serum sialic acid was later recognized to be due to high levels of sialylation of both glycoproteins and glycolipids.[1009] In the case of β-myeloma proteins, three to six times more sialic acid was found in blood serum.[1010]

Another milestone in our understanding of tumor biology was the discovery that the four types of known human sialidases behave in different manners during carcinogenesis. Probably the first observation of higher lysosomal sialidase and lower cytosolic sialidase levels was made in rat hepatomas, when compared to that of normal liver.[1011] More recently, human sialidase was established to be a cancer marker, which is downregulated in lysosomes, promoting the growth and metastatic ability of tumor cells. The enzyme is upregulated in the plasma membrane, which suppresses apoptosis of cancer cells.[1012] As described in Section 11.7, the human sialidases NEU1–NEU4 are expressed differently in various tumor tissues.[928] In general, transformed cell lines express higher activities of membrane-bound sialidase that preferably desialylates gangliosides.[1013] Furthermore, the SLe^x

epitope, which is often more expressed on tumor cells and interacts with selectins, facilitates metastasis (see below). Its biosynthesis has been studied in human lung carcinoma cells.[1014]

Tumor cells often express modified Neu5Ac, such as Neu5Gc, O-acetylated sialic acid, or Kdn, at concentrations different from the corresponding normal cells. A well-studied example are human colon cells, the mucin of which is highly O-acetylated in the healthy state. On tumor development, in the adenoma–carcinoma sequence, mucin O-acetylation gradually decreased with the lowest value in adenocarcinoma. This leads to a higher concentration of SLe^x in mucins, which had been masked before by O-acetylation in the metastatic cells, thus facilitating metastasis.[1015] The *O*-acetyltransferase activity declined, also.[1016] Another example are melanoma and basalioma tissues in which gangliosides with O-acetylated sialic acids, mostly 9-*O*-Ac-GD3, are much more expressed than in normal skin.[1017] For example, in gangliosides from basalioma, up to 56 times higher concentration of $Neu5,9Ac_2$ was determined. O-Acetylated gangliosides, which also occur in other tumors of neuroectodermal origin, represent markers for neoplastic development. Interestingly, they are also universal markers for rapidly growing normal cells and tissues. In spinalioma (squamous cell carcinoma) and in skin nevi, sialic acid O-acetylation was not increased (J. Gierthmühlen and R. Schauer, unpublished results). A ganglioside containing sialylated lacto-*N*-fucopentaose II turned out to be an antigen associated with human gastrointestinal cancer cells.[1018] During the last 30 years, more knowledge about gangliosides has been gathered, especially about the signaling mechanisms, particularly in cancer cells, regulated by these sialoglycolipids. Antitumor therapy addressing these molecules has been gradually put into practice.[1019]

N-Glycolylneuraminic acid is also a type of sialic acid, which deserves attention in tumor biology.[794] While Neu5Gc is almost completely absent from normal human tissues (only about 0.01% Neu5Gc of total sialic acid) (see Section 11.5.1), this value is often higher in tumor tissues, especially in their gangliosides (for example, 0.5% of total lipid-bound sialic acid). Elevated levels of Neu5Gc were described as chemical or tumor-associated markers in glycolipids of human breast carcinoma and other tumors and in avian lymphoma cell lines.[1020–1022] They are called Hanganutziu–Deicher (HD) antigens (see Section 12.4). HD antigens occur in blood serum in rather variable amounts from normal individuals to cancer patients, where the values are often increased. With regard to the origin of Neu5Gc, a dietary source is assumed, as was first studied with rats.[1023] Tumor cells may be able to take up more exogenous Neu5Gc than normal cells. There is also some evidence, which, however, needs more solidification, that small amounts of Neu5Gc can also be produced in an endogeneous metabolic pathway (H.-G. Hanisch, personal communication). Hydroxylase activity was, however, never observed. Finally, an important aspect of the xenoantigen Neu5Gc is its immunological and inflammatory role. There is evidence that these processes may lead to malignancies.[1024]

12.4 The Role of Sialic Acids in Immunology

In this section some historical aspects of sialic acids playing a role in immunology will be summarized. The beginning of the classical immunology goes back to 1900, when Landsteiner discovered the ABO blood group system.[1025] For a review, see ref. 1026. The system was originally defined in terms of the antigenic substances occurring at the surface of red blood cells. At that time the carbohydrate nature of the antigens was not yet known. In the following years, more than 100 red cell antigens were analyzed including the Lewis system and the M and N antigens.[1026–1028] The largest group of these glycans are human erythrocyte autoantigens that react with the corresponding antibodies. They can also occur on cells of human tissues and in animals. The antibodies are monoclonal in patients with chronic lymphoproliferation. The cold agglutinins mainly react with erythrocytes, best at 0°C. This can lead via complement activation to intravasal erythrocyte degradation, which leads to anemia. The antigens corresponding with the monoclonal anti-Gd, -Fl, -Li, -Lud, -Pr, -Sa, -Ss, and -Vo are sialylated and can be inactivated by sialidase. As an example, the abbreviation Gd means "ganglioside-dependent." Gangliosides with Neu5Gc are Gd-inactive. Sialic acids, linked ($\alpha2\rightarrow3$) or ($\alpha2\rightarrow6$) at the glycophorin glycans, can also influence the Pr antigen specificity.[1029] For comprehensive reviews about this complex topic, see refs. 1030–1034.

The analysis of the sugar components of the human blood group substances was started in the late 1940s.[1035,1036] Highly purified A, B, H, and Le^a samples from human ovarian cyst fluids were shown to contain L-fucose, D-galactose, *N*-acetyl-D-glucosamine, *N*-acetyl-D-galactosamine, and *N*-acetylneuraminic acid.[1026] Neu5Ac was isolated from Le^a blood group substance and crystallized.[1037] The structural analysis of these macromolecular substances, which occur in blood, but also other human fluids, took many years. This and the early studies of the biosynthesis of blood group substances are thoroughly described in ref. 1026. Regarding the glycosylation of immunoglobulins, as early as 1934 covalently linked carbohydrates were reported to occur on serum proteins with high antitoxin titer after immunizing horses against diphtheria toxin.[1038] For a review, see ref. 1039. But it was until 1967 that sialic acid was described to be present in human IgG.[1040] The variable sialic acid content altered the electrophoretic mobility of this immunoglobulin. In one of the first attempts to elucidate the three-dimensional structure of human γG1, four possible ways of joining the Fab region to a nearby Fc region were shown.[1041]

A later milestone in the glycobiology of immunoglobulins was the discovery of changing sialylation and its influence on function. About 30 different glycan structures, with and without sialic acids, linked to the Fc region of the immunoglobulin are known,[1042] and voluminous reviews on the impact of glycosylation on the biological function and structure of human immunoglobulins have been written.[1043,1044] Immunoglobulin G mediates pro- and antiinflammatory responses through the engagement of its Fc fragment. This difference depends on the sialylation of the Fc core oligosaccharide, and it explains why therapeutic intravenous γ-globulins, if they are sialylated, have antiinflammatory effects in autoimmune diseases.[1045] In a remarkable experiment, it was seen that engineered sialylation of pathogenic antibodies in vivo, at the site of autoimmune inflammation, led to an attenuation of the autoimmune disease.[1046]

The pronounced significance of Neu5Gc, although its nature was not yet known, in immune reactions was observed as early as at the beginning of the last century. M. Hanganutziu in 1924 and H. Deicher in 1926 observed after the therapeutic injection of antisera from horse blood into humans the formation of antibodies, now called "HD antibodies."[1047,1048] About 50 years later, the epitope, recognized by HD antibodies, turned out to be the Neu5Gc residue of gangliosides.[1049,1050] The significance of these antibodies in tumor biology and in nutrition has been discussed in Section 11.5.1. Additionally, monoclonal antibodies have been described that distinguish between Neu5Ac- and Neu5Gc-containing ganglioside GD3.[1051]

Monoclonal antibodies against O-acetylated sialic acids in human colon mucins were obtained by immunization of BALB/c mice with mucosal scrapings from normal human large intestine. Antibody PR.3A5[1052] and antibody 6G4[1053] bound to O-acetylated mucin and were useful in studying the histochemistry of the normal colon, colon cancer, and ulcerative colitis. Immunization of mice with highly purified human colonic mucin afforded the monoclonal antibody MMM-17, which strongly binds to the mucin of goblet cells of the normal colon, but not to those of colon cancer tissue.[1054]

Saponification of the *O*-acetyl groups in the mucin or sialidase treatment of the mucin abolished the binding. O-Acetylated gangliosides are also immunogenic, as was studied with GD3 and GD3 derivatives. Here, monoclonal antibodies were obtained that distinguish between GD3 (CD 60a), 9-*O*-acetyl-GD3 (CD 60b), and 7-*O*-acetyl-GD3 (CD 60c).[1055] It was found in many investigations that the disialoganglioside GD3 is the most frequent carrier of terminal O-acetylated sialic acid. The occurrence and the roles known of O-acetylated GD3 have been reviewed[826] (see also Section 11.5.2). Among the most striking and influential effects of O-acetylation are the inhibition of apoptosis of tumor cells,[1056] the inhibition of the interaction of the siglec sialoadhesin with 9-O-acetylated sialic acid ligands, and the inhibition of the binding of human complement factor H, a negative regulator of the alternative complement pathway, to sialic acids.[1057,1058] Thus, a decrease of O-acetylation may help tumor cells to escape from complement-mediated lysis. A classical example of the protective effect of (unsubstituted) sialic acids from activation of the lytical alternative complement pathway is the channel catfish *Ictalurus punctatus*. Very little bactericidal activity is exerted by the alternative complement pathway against the fish pathogens containing sialic acids, in contrast to a strong response against the nonpathogenic bacteria that lack sialic acid. This phenomenon is sialidase sensitive.[1059]

The assumption of the existence of non-N-acetylated neuraminic acid (Neu) in the gangliosides of some tumors was based on the detection with monoclonal antibodies and pulse-chase experiments.[1060] Note that the first chemical existence for natural Neu was obtained after N-propionylation.[488] N-Deacetylation of capsular polysialic acids provided antibodies protective against *N. meningitidis* groups B and C.[1061] Earlier, antibodies against the virulence factors ($\alpha2\rightarrow8$)- and ($\alpha2\rightarrow9$)-polysialic acids were produced for vaccination against meningococci and *E. coli* K1 after conjugation to tetanus toxoid.[1062] An effective vaccine, applied to small-cell lung cancer patients by the application of N-propionylated group B/($\alpha2\rightarrow8$) polysialic acid, conjugated to keyhole limpet hemocyanin, is known.[1063] Natural polysialic acids are very weak antigens in humans.[1064]

The previous examples have shown that sialic acids are involved in antigenic structures. To these also belong the so-called differentiation and oncodevelopmental antigens of glycoproteins and glycolipids, which are expressed during ontogenesis and also in cancer.[987] However, sialic acids often mask antigenic sites on glycoproteins, glycolipids, and carbohydrate oligo- and polymers and thus represent "antiantigens." Thus, these monosaccharides have a remarkable regulatory role in immunology (see, e.g., refs. 963 and 1065), in a similar way as already discussed for the regulation of the lifetime of blood components and cells, namely, via a dual "self/nonself" system (see Section 12.2). So, the masking effect of sialic acids renders molecules and cells as "self." After the loss of the sialic acid residues, the structures become "nonself" and may be recognized by the immune system, which may lead to autoimmune diseases. A masking role of sialic acids has early been detected for porcine submandibular gland glycoproteins. Sialidase treatment of these mucins revealed H blood group activity, because Neu5Gc had stereospecifically masked the H-determinant Fuc(α1→2).[1066] A classical example is also the masking of the T-antigen Gal(β1→3)GalNAc(α1→O)Ser/Thr and the Tn-antigen GalNAc(α1→O)Ser/Thr of blood cells and some tumors by Neu5Ac in humans.[1032] The Gal(β1→4)[Fuc(α1→3)]GlcNAc antigen was also found to be cryptic in different human tissues.[1067] It is masked to a variable extent by Neu5Ac at the terminal Gal residue of the antigen {Neu5Ac(α2→3)Gal(β1→4)[Fuc(α1→3)]GlcNAc} or at adjacent carbohydrate chains (obscuring the antigen by steric hindrance) and can be made visible by histochemical means by using sialidase. One of the most striking and important examples for the masking of antigenic sites is the immunobarrier between mother and fetus due to the expression of highly sialylated glycoconjugates on syncytiotrophoblast cells.[1068,1069] These molecules prevent the formation of antibodies by the maternal organism against fetal cells. Loss of sialic acids in kidney, e.g., by viral or bacterial infection, may lead to immunological injuries of the glomeruli leading to chronic glomerulonephritis.[1070] Note that tissues rich in sialic acids tend to develop autoimmune diseases.[963] The pronounced masking of tumor antigens by sialic acids has already been discussed in Section 12.3. These observations, together with an old report[1071] about the effect of sialidase to increase the immunogenicity of the Landschütz ascites tumor cells, stimulated attempts to cure cancer by sialidase treatment of tumor cells, reinjection of the cells, and thus vaccination of the patient with these cells.[963] Injection of the cells into various places of the skin ("chessboard vaccination") had been recommended. This method was successful in the treatment of dog mammary tumors,[1072] while in humans the duration of the remission was extended in myelocytic or lymphatic leukemia.[1073] Although some positive results were obtained, this method was not followed further.

Sialic acids on the surface of T cells mediating "delayed-type hypersensitivity" (DTH), which have developed an immune response before, control the entry of the lymphocytes into sites of antigen deposition, where an inflammatory reaction is induced. This could be shown by sialidase treatment of the sensitized T cells that caused a transient reduction of DTH reactions in the periphery. The cells were trapped in the liver, but were released after 24 h, and their biological activity returned.[1074] The involvement of terminal sialic acids on the surface of T cells, in most aspects of T cell fate and function, from cell maturation, differentiation, and migration to cell survival and cell death, was reviewed 27 years later.[1075]

In the context of this section, attention should also be paid to the function of (sialylated) milk oligosaccharides. As mentioned (see Section 8.3 for an updated series of reviews), the extraordinary diversity of these glycans is great, varying from mammal to mammal, and comprising more than 200 different structures in man. They regulate the bacterial colonization of the intestine, prevent the growth of pathogenic bacteria, and feed the microorganisms. This protects especially children from bacterial and viral infections. The variability of the microorganisms, nowadays called "microbiome," also seems to influence the human immune system, although the exact mechanism and the role of sialic acids wait for complete elucidation. The enhancement of the biosynthesis of glycoproteins and glycolipids in the developing brain by milk oligosaccharide feeding and the corresponding stimulation of the intellectual capacities are also under study.[1076,1077]

A striking example of innate, humoral immunity is limulin, a sialic acid-binding lectin from the American horseshoe crab *Limulus polyphemus*. It was isolated and characterized from the hemolymph of this animal, has a molecular weight of 335 kDa, and consists of subunits of each 19 kDa.[1078,1079] Limulin binds to glycosidically linked sialic acids of glycoproteins and gangliosides, but preferably to Neu5Gc-containing gangliosides (i.e., Neu5Gc-GM3), and thus agglutinates erythrocytes and other cells.[1080] A similar lectin was isolated by affinity chromatography from the hemolymph of the Asian horseshoe crab *Tachypleus tridentatus*.[1081] SDS-PAGE showed three protein bands between 27 and 32 kDa. This lectin also agglutinates erythrocytes and preferably binds to sialylated O-glycans of mucins. These lectins of the phylogenetically very ancient horseshoe crabs are considered as potent innate defense systems against parasites with sialic acids or perhaps similar nonulosonic acids.

A most important chapter of modern immunology are the mammalian sialic acid-recognizing lectins, called siglecs, which, in most cases, occur on the surface of immune-competent cells, e.g., B-lymphocytes, and are involved in many immune reactions including autoimmunity.[985,1082,1083] For further details, see Section 12.5.3.

12.5 Proteins Recognizing Sialic Acids

In this section the history of the studies of the interaction of sialic acids as ligands with proteins as receptors (lectins) will be described. Special groups are selectins and siglecs, which will be treated separately. For an overview written by the pioneers of the research on lectins, Nathan Sharon and Halina Lis from The Weizmann Institute of Science (Israel), see ref. 1084.

12.5.1 Plant and Other Lectins, and Virus Hemagglutinins

Lectins were first found almost exclusively in plants.[398,1084–1086] Since they were detected by the agglutination of erythrocytes, they were named "hemagglutinins" or "phytohemagglutinins." As plant lectins are able to distinguish between erythrocytes of different blood types, as early as 1954 Boyd and Shapleigh coined the name "lectin," from the Latin *legere*, "to select."[1087] This definition was later used for all carbohydrate-binding proteins, irrespective of their occurrence.

To go back in history, before 1860, S. Weir Mitchell in Pennsylvania had observed agglutination of pigeon blood by the venom of a rattlesnake.[1088] The first lectin, called "ricin," was discovered in 1888 in the beans of the castor tree (*Ricinus communis*) by Hermann Stillmark at the University of Dorpat (now Tartu) in Estonia.[1089] In fact, it was a mixture of weakly agglutinating protein toxins, which were later found to bind to galactose and *N*-acetylgalactosamine.[1084,1090] Furthermore, Simon Flexner and Hideyo Noguchi,[1091] from the Rockefeller Institute (New York, USA), described in 1902 snake venom agglutinins in some detail in the horseshoe crab (*L. polyphemus*)[1078] and the American lobster *Homarus americanus*. In the first decade of the 20th century also the first bacterial agglutinins were found, which were later identified as lectins. Investigations on blood type-specific lectins stem from the early 1950s, when it was demonstrated for the first time that lectins can bind monosaccharides.[1092]

The already described studies of Hirst in the 1940s on the agglutination of human erythrocytes by influenza virus (see Sections 2.1 and 11.7) can be considered as the first example of the involvement of a lectin/hemagglutinin, a sialic acid-binding agent. Influenza virus hemagglutination was found to be inhibited by a brain lipid fraction, known to contain gangliosides.[1093] In the following decades the structure and pathophysiological significance of the viral hemagglutinin (designated "H" antigen) was most intensively studied, until today, by the group of Hans-Dieter Klenk at Philipps-Universität Marburg (Germany).[958,959,1094] Influenza viruses demonstrate pronounced receptor binding specificity. Human influenza viruses preferentially bind to

Neu5Ac(α2 → 6)Gal sequences, whereas avian influenza viruses prefer Neu5Ac(α2 → 3)Gal as ligand (or receptor in the terminology of virologists).[1095] Competitive inhibitors of the binding of avian or human influenza viruses to sialylated glycans are soluble glycoconjugates carrying (α2 → 3)- or (α2 → 6)-linked sialic acids, respectively. The earlier discussed mucin from the nests of the Chinese swiftlet, genus *Aerodramus collocalia* (collocalia mucoid, edible bird's nest substance),[1096] rich in sialic acids, is a potent inhibitor of the agglutination of myxoviruses.[1097] This may have been the reason that it has been used in Chinese medicine for hundreds of years in order to prevent and cure influenza and its following bronchial diseases. The substance, which is a good substrate for sialidase, has additional health benefits and is therefore highly esteemed in Asia.[1098]

The best known sialic acid-recognizing plant lectins are WGA, accepting Neu5Ac and (β1 → 4)-linked GlcNAc residues, *S. nigra* agglutinin (SNA) from the bark of the elderberry bush, recognizing (α2 → 6)-linked sialic acid to galactose, and *M. amurensis* lectin (MAA), specific for (α2 → 3)-linked sialic acid to galactose.[1084] The fractions of these plant lectins are believed to protect the plants from pathogenic microorganisms and fungi, as well as from herbivorous animals. It was found that the bark lectin SNA provoked toxic effects in animals and created a reaction of avoidance (Fig. 35).[1099] One of the authors (R.S.) observed that during a very cold

Fig. 35 Lectins, for example, the sialic acid-binding *Sambucus nigra* agglutinin (SNA), protect plants against herbivorous and chewing animals. *Reproduced from Peumans, W. J.; Van Damme, E. J. M. Lectins as Plant Defense Proteins.* Plant Physiol. ***1995***, 109, *347–352. Copyright American Society of Plant Biologists.*

winter with much snow, mammals, mainly rabbits, fed on the bark of various bushes, but not on that of the elderberry bush. Both SNA and MAA are used regularly in structural analysis protocols, i.e., lectin affinity chromatography, histochemistry, or lectin microarrays.[275]

Sialic acid-recognizing proteins are manifold in Nature. The first sialic acid-binding protein from vertebrates reported was Complement Factor H, an early response factor of the innate immune system.[1100,1101] In a thorough search about 230 sialic acid-specific lectins in viruses (being the largest group), bacteria, toxins, protozoa, fungi, plants, invertebrates, and vertebrates were described.[1102] In the intervening time, more sialic acid receptors have been detected, especially in the field of virus research, e.g., of the human Noro virus,[1103] the Middle East Respiratory Syndrome Coronavirus (MERS-CoV),[1104] and the Infectious Salmon Anemia Virus (ISAV), the latter binding to Neu4,5Ac_2.[1105] Most of these sialic acid lectins bind to Neu5Ac, but a few only to Neu5Gc or to both sialic acid types. An example of Neu5Gc as virus receptor is the transmissible porcine gastroenteritis coronavirus.[1106] The various lectins may discriminate between (α2→3) and (α2→6) linkages and rarely prefer (α2→8) linkages, like e.g., siglec-7 and -11. Some invertebrate and viral lectins bind to 4- or 9-O-acetylated sialic acids, the best known example being influenza C virus.[826,1102]

Not much is known about how and where influenza C viruses enter the human organism, but the human respiratory tract is the most likely place. In the nasal secreted mucin of only one person studied so far, about 10% of the sialic acids were 9-O-acetylated. In a pool of nasal mucosa probes collected after surgery in 1994, a 9-O-acetylated sialoglycoprotein fraction of 100–130 kDa was detected by Western Blot analysis. This could be the influenza C virus receptor. Since also the ganglioside fraction was found to be O-acetylated, the influenza C virus receptor must be studied further.[1107] Influenza C virus hemagglutinins, specific for (α2→3) and (α2→6) linkages, can differentiate between receptor determinants bearing *N*-acetyl-, *N*-glycolyl-, and O-acetylated sialic acids. This was shown by the agglutination of Neu5Ac, Neu5Gc, or Neu5,9Ac_2 covering erythrocytes.[1108]

Very recently, probes from viral sialic acid-binding proteins were developed to detect 4-*O*-acetyl, 9-*O*-acetyl, and 7,9-di-*O*-acetyl sialic acids on cells and tissues of humans and a great variety of animals.[407] It is astonishing how widely distributed these esterified sialic acids are. The HE proteins were expressed in insect cells and fused to the Fc region of human IgG1 and to a hexahistidine sequence for histochemical or microarray use (see also Section 7.4). Binding of influenza A virus and Siglec-2 (CD22) is inhibited by Neu5Ac O-acetylation.[826]

12.5.2 Selectins

The selectins mediate, as do other cell adhesion molecules, the adhesion of leukocytes from the blood stream to endothelial cells, followed by extravasation into the surrounding tissue. They are thus involved in the initiation of inflammation. Tumor cells can experience the same fate. The basis of this phenomenon is the recognition of terminal sialyl Lewis x {SLe^x; Neu5Ac (α2→3)Gal(β1→4)[Fuc(α1→3)]GlcNAc} and sialyl Lewis a {SLe^a; Neu5Ac (α2→3)Gal(β1→3)[Fuc(α1→4)]GlcNAc} glycan sequences on the mobile cells. The first observations of this complicated process were recorded in the 1880s by Julius Friedrich Cohnheim, a German pathologist, who also worked one year (1868) at Christian-Albrechts-Universität zu Kiel (Germany).[1109,1110] This scientist studied by real-time microscopy the microvessels in the frog and proposed that postcapillary venules at sites of inflammation underwent molecular alterations that mediated the attachment of flowing leukocytes on the vessel internal surface, and subsequent emigration at sites of inflammation. Heinrich (Karl Wilhelm) Schade from the same university also observed about 40 years later the extravasation of white blood cells into inflamed tissues. He assumed that substances penetrating into the blood make the leukocytes "sticky" resulting in adhesion of the cells to the wall of the blood vessels before they extravasate into the tissue guided by chemotaxis.[1111] The molecular basis of these events could not be elucidated in those times. Heinrich Schade is considered the father of molecular pathology, a new area of medical research, in completion with the then existing cellular pathology.

It took a further 30 years before researchers began to study the extensive flux of lymphocytes between blood and lymph ("lymphocyte recirculation").[1110] Here, a "homing molecule" on the surface of lymphocytes was assumed, which directs their traffic to lymph nodes. Electron microscopic studies showed that microvascular structures exist in the lymph nodes, consisting of plump, cuboidal endothelial cells, known as high endothelial venules (HEV),[1112] and lymphocyte adherence to HEV could be demonstrated.[1113] Further studies on the leukocyte extravasation at sites of inflammation showed a dynamic process.[1114] After 2h of injury, postcapillary venules displayed adhesive properties that supported leukocyte binding, followed by endothelial transmigration. At this state of the research, it was not known that carbohydrates were involved in leukocyte extravasation, but soon after these reports the participation of carbohydrates in lymphocyte trafficking was recognized.[978] In many further studies, the HEV receptor of the lymphocytes was shown to be a lectin, whereby sialylated glycans on HEV serve as ligand for this lectin.[1115,1116] Furthermore, a family of sulfated sialofucosylated glycoproteins (sulfated at position 6 of GlcNAc or Gal of SLe^x) was found which serve as L-selectin ligands on HEV, known as "peripheral lymph node addressins."[1117] While L-selectin is expressed on lymphocytes, and functions in the migration of these cells to lymphoid organs, E- and P-selectins are expressed on vascular endothelial cells and mediate the recruitment of neutrophils to sites of inflammation in tissues (Fig. 36). P-Selectin also occurs on lymphocytes. E-Selectin is constitutively expressed on bone marrow microvessels, and human hematopoietic stem cells express high amounts of an E-/L-selectin ligand.[1110] The involvement of selectins in cancer metastasis has drawn many researchers to this field.[1118]

In 1989, the lymph node homing receptor, as well as E- and P-selectins, were cloned.[1119] This protein family was then named "selectins," belonging to the "C-type" lectins. At the same time, studies revealed that, in addition to homing receptors, cell migration is orchestrated by chemoattractants

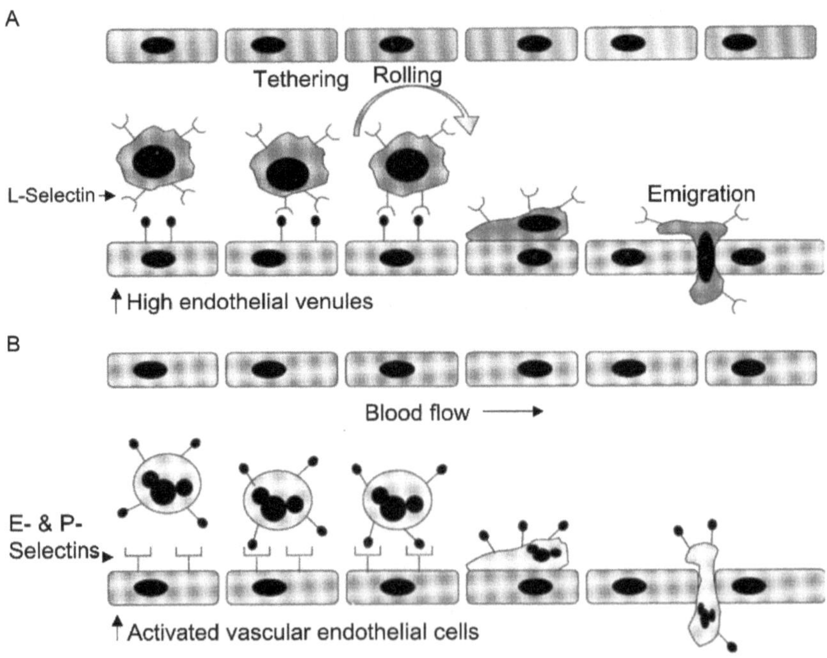

Fig. 36 The selectins play a key role in the control of leukocyte traffic in the body. (A) L-Selectin functions in the migration of lymphocytes to lymphoid organs. (B) E- and P-Selectins mediate the recruitment of neutrophils to sites of inflammation. *Courtesy of Ohad Bairey, and reproduced from Sharon, N.; Lis, H.* Lectins, *2nd ed.; Kluwer Academic Publishers: Dordrecht, The Netherlands, 2003, p 353. Copyright Kluwer Academic Publishers.*

(e.g., chemokines), which may be present within endothelial beds or inducibly expressed at sites of inflammation, and inaugurate intracellular signaling followed by endothelial transmigration.[1110] According to this "multistep paradigm" the function of a homing receptor is not purely to direct cellular trafficking in a tissue-specific manner, but to function as a molecular brake in initiating binding to endothelial cells under hemodynamic shear stress.

The contribution of selectins to the well-being of mammals was shown in genetically engineered mice, lacking selectins, which were prone to severe bacterial infections, and demonstrated the role of these lectins in acute inflammatory responses.[1120] In the rare leukocyte adhesion deficiency type II syndrome (LAD II),[1121] the patients suffer from recurrent bacterial infections as well as from growth and mental retardation. Their neutrophils are deficient in SLe^x and therefore cannot bind E- and P-selectin. The reason of this defect is a decrease of the metabolism of fucose required for the synthesis of Le^x. Correspondingly, supplementation of the patients with fucose efficiently reduced the pathological symptoms.

For further informative reviews, see refs. 129,1122,1123.

12.5.3 Siglecs

Siglecs are a group of the "latest" sialic acid-recognizing lectins discovered. In contrast to the selectins, they are involved in the regulation of inflammation. Due to this pathophysiological function and the manifold members of the siglec family, with a great variety of functions, the number of publications has exploded in the last 20 years. This area is one of the most rapidly evolving in glycobiology research, not only because of the impressive immunological and clinical potential. Therefore, only a short historical introduction to this field is presented, and a number of excellent reviews, which deal with the numerous aspects of siglec occurrence, structure, and biology, will be cited.

The investigations in this area started in 1986, when P. R. Crocker and S. Gordon found a lectin-like hemagglutinin expressed on murine tissue macrophages.[1124] The lectin was called "sialoadhesin (Sn)," a sialic acid-binding receptor that was purified from murine macrophages.[1125] It was shown to be a transmembrane protein with 17 extracellular Ig-like domains, 16 C2-set domains, and 1 unusual V-set domain that contains an intrasheet disulfide bridge.[1126] Sialoadhesin recognizes the sequences Neu5Ac(α2 $\rightarrow$ 3)Gal(β1 $\rightarrow$ 3) GalNAc and Neu5Ac(α2 $\rightarrow$ 3)Gal(β1 $\rightarrow$ 3/4)GlcNAc on glycoproteins and glycolipids.[827,827a,828] The macromolecule is restricted to macrophages and is highest expressed in resident macrophages of hemopoietic and secondary lymphoid tissues. The adhesion mediates interactions with developing myeloid cells in the bone marrow[1125] and lymphocytes in spleen and peripheral lymph nodes.[1127] Recent research has shown that adhesion is also involved both in clearance of sialylated human pathogens and in antigen presentation.[1123]

In the 1990s, a new family of sialic acid-dependent adhesion molecules of the immunoglobulin superfamily, namely, sialoadhesin, CD22, and the myelin-associated glycoprotein (MAG) was defined.[827,827a] The number of extracellular repeats differs much in these three lectins, that of MAG being the shortest (five repeats). After the discovery of Sn, CD22, MAG, and CD33, the group of Ajit Varki proposed the term "siglec," derived from "**si**alic acid," "**i**mmuno**g**lobulin," and "**lec**tin" for each member of the group.[133,133a] Siglecs are now considered a subset of I-type lectins, and they are numbered as follows: Sn (siglec-1), CD22 (siglec-2), CD33 (siglec-3), and MAG (siglec-4). The list of siglecs today[1123] continues up to siglec-16 on macrophages and microglia. Mice express fewer, but partially similar siglecs.

CD22 is a B-cell-specific molecule which binds to the Neu5Ac($\alpha2 \rightarrow 6$)Gal($\beta1 \rightarrow 4$)GlcNAc element on glycoproteins.[1128] It is the best studied member of the siglec family. By cis and trans interactions, CD22 plays a role in B-cell signal transduction,[129] and it regulates immune tolerance.[1123] It carries immune inhibitory domains (ITIM and ITIM-like), as do also the other eight of the human siglecs, numbers 3 and 5–11, which have been shown to suppress immune responses or are assumed to do so.[1129]

The MAG is produced on myelinating oligodendrocytes at Schwann cells.[1130] MAG plays roles in myelination, axonal growth regulation, and signal transduction.[129] It binds to Neu5Ac($\alpha2 \rightarrow 3$)Gal($\beta1 \rightarrow 3$)GalNAc elements. Studies of the binding specificities of MAG and sialoadhesin with synthetic sialylated oligosaccharides demonstrated differences between these two siglecs.[1131] Remarkably, MAG bound Kdn better than Neu5Ac, and sialoadhesin revealed the opposite.

A cDNA encoding CD33, being a differentiation antigen of myeloid progenitor cells, was isolated in 1988.[1132] The antigen belongs to a related subgroup composed of siglec-3 and siglecs-5–13 in primates and similar ones in rodents.[133,133a] This subgroup appears to evolve rapidly. In contrast, siglecs-1, -2, and -4 represent an evolutionarily conserved subgroup.

The roles of siglecs in immunology and human–pathogen interactions are many and varied. Special attention was recently drawn to siglec-8, which seems to function in resolving allergic inflammation, while siglec-9 may dampen neutrophil-mediated inflammation.[1123]

Focusing on the influence of various sialic acids on the binding of siglecs,[133,133a] it was observed that the frequently occurring 9-*O*-acetyl group has a strong negative effect on recognition by human CD22[1133] and mouse sialoadhesin.[828] The masking of the ligand fraction of sialic acid has been observed also with influenza A/B virus hemagglutinin and with selectins. Aberrant O-acetylation of sialic acids, caused by the enzyme pair *O*-acetyltransferase and -esterase, is thought to be involved in autoimmune diseases like asthma.[885] Furthermore, mouse and human Sn prefer Neu5Ac over Neu5Gc. However, murine CD22 strongly prefers Neu5Gc over Neu5Ac, but human CD22, as well as CD33, siglec-5 and siglec-6 accept both types of sialic acids. It can be assumed that these human-specific changes are due to the dramatic alteration in the human "sialome" after the *CMAH* mutation and Neu5Gc loss.[133,133a]

Many of these binding studies with differently linked sialic acids and natural or chemically modified sialic acid derivatives, e.g., Neu5Gc and Neu5,9Ac$_2$, which much helped to understand the biology of siglecs, were, in the beginning of siglec research, based on a close cooperation between Paul Crocker, Jim Paulson, and Sørge Kelm. Paul had the cells available, Jim had purified several sialyltransferases, and Sørge had developed a procedure to transfer various sialic acids to desialylated erythrocytes in order to obtain homogeneous ligands regarding sialic acid nature and linkage.

For a detailed overview of all aspects of "siglectology," see refs. 129, 133, 133a, 134, 776, 1123, and 1134–1139.

13. HEREDITARY SIALIC ACID DISEASES

The last section of this chapter will deal with some diseases with an origin in genetic errors of sialic acid metabolism. Although rare, they clearly show that sialic acids have a decisive impact on health. In the following subsections, the history of the elucidation of sialuria, Salla disease, sialidosis, galactosialidosis, and GNE myopathy, the main diseases known with disturbed sialic acid metabolism, will be described. See also Section 8.4 for more chemical information.

13.1 Sialuria

Sialuria was first reported in 1968, affecting a then 2½-year-old French boy who excreted a 10,000-fold higher amount of sialic acid (Neu5Ac) than normal children.[568,583,584,1140–1142] The Neu5Ac concentration in the urine was so high (excretion of up to 7.2 g Neu5Ac per day) that this monosaccharide could easily be isolated in large quantities from the patient's urine. This helped research on this topic, since Neu5Ac was very expensive at that time, and scientists were interested in a commercial source of the compound. As said in Section 11.4, the underlying metabolic defect in sialuria is the loss of feedback inhibition of the rate-limiting enzyme of sialic acid biosynthesis, i.e., UDP-GlcNAc 2-epimerase (GNE), by CMP-Neu5-Ac. This leads to enhanced enzyme activity and accumulation of Neu5Ac in sialuria fibroblasts.[586,1143] Also the mutations in the epimerase gene have been analyzed.[1142] Sialuria is a very rare disease. The French patient, suffering from sialuria, remained unique for a long time, but other cases, also without lysosomal involvement, but with lower excretion of sialic acids, have been described later.[568] Sialuria patients show mild mental retardation, seizures, and slightly dysmorphic features.

13.2 Salla Disease and Infantile Sialic Acid Storage Disease

Salla disease is an autosomal recessive lysosomal storage disorder, first described in 1978 in four patients[580,1144] presenting severe psychomotor retardation and increased urinary excretion of Neu5Ac, combined with enlarged cellular free Neu5Ac levels. So far, about 150 cases have been reported.[1141,1145] The name "Salla" is derived from a Northern area in Finland from where most of the patients originated. However, two patients with Salla disease were also detected in Turkey.[1146] A similar, but more severe disease than Salla disease, with a much earlier onset, often occurring at birth, was described in 1982 in an infant of German–Swedish origin, and it was called Infantile Sialic Acid Storage Disease (ISSD).[1147] Such patients are, rarely, found in several continents. The pathomechanism of these diseases is an impaired transport of free Neu5Ac across the lysosomal membrane.[1148] The reason for the Neu5Ac accumulation in the lysosomes of various tissues is a defective H^+-driven transport of sialic acid (and D-glucuronic acid) through the lysosomal membrane by a mutation of the integral lysosomal protein Sialin.[1142,1149]

13.3 Sialidosis

Sialidosis belongs to the so-called group of "oligosaccharidoses," which are metabolic disorders characterized by an excessive accumulation of glycoprotein-derived oligosaccharides.[568,1150,1151] Initially, a severe deficiency of neuraminidase in the cultured fibroblasts of a patient, classified as "mucolipidosis I," was seen.[1152,1153] The patient had a neurodegenerative disorder with myoclonus, skeletal changes like in Hurler disease, and cherry-red spots in the macula of the eyes. As reviewed in Section 8.4, excessive quantities of sialyloligosaccharides (sialic acid=Neu5Ac), derived from glycoprotein N- and O-glycans, were found in body fluids and urine. Fibroblasts of the patients exhibit a sialidase defect and accumulate abnormal amounts of sialic acid-containing substances. Besides this group of dysmorphic children with early beginning of the disease, there is a milder type of sialidosis with normal physical appearance of the patients, absent or mild mental retardation, and onset only in adolescence. These patients also have impaired sialidase activity and excrete unusual amounts of sialyloligosaccharides.[1154] These two groups are categorized as type I and type II sialidosis, respectively. In summary, sialidosis is an autosomal recessive lysosomal storage disease, affecting the degradation of glycoproteins. It is caused by a genetic defect of the lysosomal sialidase (Neu 1) gene, first cloned in 1996.[1155] Several mutations are known, which lead to different expressions of the phenotype of this disease.[1156,1157]

13.4 Galactosialidosis

Galactosialidosis is a disease with a combined deficiency of sialidase and β-galactosidase activity, due to a defect of a 32 kDa protein [lysosomal serine carboxypeptidase, protective protein cathepsin A (PPCA)] that shields sialidase and β-galactosidase against lysosomal proteases.[1158] See also refs. 568, 1142, and 1159–1161. Since the result of this disturbed enzyme complex is a reduced sialidase activity, like in sialidosis, there is no significant difference in the pattern of excreted glycoprotein-derived urinary sialooligosaccharides between sialidosis and galactosialidosis.[1162] This suggests that sialidase deficiency, not β-galactosidase deficiency, determines the phenotype of the disease.

13.5 Hereditary Inclusion Body Myopathy (GNE Myopathy)

Hereditary inclusion body myopathy (HIBM) encompasses several syndromes with autosomal recessive or dominant inheritance, and the patients suffer from a progressive course of muscle weakness leading to severe disability. In 2011 about 500 patients with HIBM were known worldwide.[1163] The most common form is the quadriceps muscle-sparing autosomal recessive myopathy,[1164] which was originally described in patients of Persian–Jewish heritage.[1165] The muscle problems usually start in the second to third decade of life with weakness and atrophy of distal lower limb muscles that eventually spreads proximally. The reason for this disease is a reduced activity of the key enzyme UDP-*N*-acetyl-glucosamine-2-epimerase/*N*-acetylmannosamine kinase (GNE/MNK) (see Section 11.1), which leads to impaired Neu5Ac biosynthesis, and to rimmed vacuoles and Congo-Red-positive depositions in fibers in the muscles. So far, 154 mutations have been identified in the gene of GNE that are likely leading to reduced GNE enzyme activities.[1166] It was shown that mutation of methionine 743 to threonine in GNE leads to 30% reduction of enzyme activity and to an aggressive form of GNE myopathy. The introduction of threonine offers an additional phosphorylation/O-GlcNAcylation site. Indeed, the O-GlcNAcylation of the M743T GNE variant was increased. Removal of that residue resulted in almost normal GNE activity. Hereby, the balance of phosphorylation and O-GlcNAcylation is decisively involved in regulation and efficiency of GNE. Evidence is accumulating, especially from experiments with mice, that hyposialylation is underlying the molecular mechanism of this muscle weakness.[1167,1168] Also in humans, a considerable lack of free sialic acids in muscle biopsies of patients, when compared to healthy individuals, has been described.[1169] It was stated that the lack of sialic acids in GNE myopathy contributes to ROS production/oxidative stress followed by muscle weakness and atrophy.[1168] Correspondingly, application of antioxidants had a curative effect.[1168,1170] Furthermore, oral administration to mice of Neu5Ac, sialyllactose, ManNAc, and especially per-O-acetylated ManNAc also decreased the disease symptoms.[1170] In man, intravenous application of high doses of highly sialylated immunoglobulin G (IVIG) also improved the disease symptoms.[1171]

14. CONCLUDING REMARKS

In this chapter, we have presented the historical development of the different areas of sialic acid research. Most impressive was to follow the efforts of those early pioneers to identify the correct structure of the *N*-acylneuraminic acid molecule. To achieve this goal the often cooperative work of dedicated scientists during several decades was necessary, and they had to cope with erroneous conclusions. After the isolation of sialic acids with still unknown structure from bovine submandibular gland mucin, it was by pure chance that a similar substance was discovered by studies with influenza viruses, and that years later also the analysis of the product of a lyase reaction played an important role in solving the stereochemistry of the sialic acid structure. Another difficult and lengthy effort was the development of reliable techniques for the identification of sialic acids and their many derivatives in natural sources, and chemists, synthesizing, for example, sialic acid glycosides, encountered challenges by these compounds that exceeded those experienced from working with "ordinary" sugars.

Sialic acids are among the most striking molecules in higher animals and many microorganisms, and it seems that these monosaccharides play a role in all branches of cell biology. They are involved in probably most diseases, although this is a field where more research should be focused. Sialic acids are prominent molecules on the cell surface, on the glycocalyx, not only in position but also in function, and they can be considered as cytoprotectors. However, in this position, they are often attacked and conquered by pathogens, especially viruses, and they can also be misused by malignant cells. We highlighted the beginnings and discussed the following inquiries into the pathobiochemical functions of sialic acids. The increasing number of publications on this topic shows that in laboratories all over the world efforts are continuing to gather more information and to employ this knowledge, for example, to develop effective therapies for various diseases.

When citing relevant, mostly historical papers in each section, we had to select due to space restrictions. Therefore, many important publications could not be included. And since we could not cover all aspects of sialic acid research in detail, this chapter is accompanied by special reviews on the chemistry, evolution, neurobiology, tumor biology, and virology of these monosaccharides, written by other authors.

With appreciation of the support from numerous colleagues, we are closing this chapter by citing an aphorism from the 1970s by Gerhard Uhlenbruck from the University of Cologne (Germany), who is a devoted glycoimmunologist: "*Neuraminic acid (sialic acid) is a lady whose fascination has not faded away. Again and again stimulating, exciting and good for all kinds of surprises. A lovable molecule, which is enthusiastically courted by many, although the advertising budget never was opulent.*" (Translated from the original German version.)

ACKNOWLEDGMENTS

More than 50 years of sialic acid research brought us in contact with many scientists from all over the world, and—although we could not mention everyone in this review—we acknowledge their contributions and their support. We also thank our numerous associates for their dedication and contributions to the sialoglycoscience field. A special word of thanks goes to Ms Elfriede Schauer, for her continuous assistance in the preparation of this chapter.

REFERENCES

1. Montreuil, J.; Vliegenthart, J. F. G.; Schachter, H. Preface. In: *Glycoproteins II*; Montreuil, J., Vliegenthart, J. F. G., Schachter, H., Eds.; New Comprehensive Biochemistry, Vol. 29b; Elsevier Science B.V: Amsterdam, The Netherlands, 1997; pp v–vi.
2. Warren, L. The Distribution of Sialic Acids in Nature. *Comp. Biochem. Physiol.* **1963**, *10*, 153–171.
3. Warren, L. The Distribution of Sialic Acids Within the Eukaryotic Cell. In: *Biological Roles of Sialic Acid*; Rosenberg, A., Schengrund, C.-L., Eds.; Plenum Press: New York, NY, USA, 1976; pp 103–121.
4. Ng, S.-S.; Dain, J. A. The Natural Occurrence of Sialic Acids. In: *Biological Roles of Sialic Acid*; Rosenberg, A., Schengrund, C.-L., Eds.; Plenum Press: New York, NY, USA, 1976; pp 59–102.
5. Corfield, A. P.; Schauer, R. Occurrence of Sialic Acids. In: *Sialic Acids—Chemistry, Metabolism and Function*; Schauer, R. Ed.; Cell Biology Monographs, Vol. 10; Springer-Verlag: Vienna, Austria, 1982; pp 5–50.
6. Schauer, R.; Kamerling, J. P. Chemistry, Biochemistry and Biology of Sialic Acids. In: *Glycoproteins II*, Montreuil, J., Vliegenthart, J. F. G., Schachter, H., Eds.; New Comprehensive Biochemistry; Neuberger, A., van Deenen, L. L. M., Series Eds.; Vol. 29b; Elsevier Science B.V: Amsterdam, The Netherlands, 1997; pp 243–402.
7. Angata, T.; Varki, A. Chemical Diversity in the Sialic Acids and Related α-Keto Acids: An Evolutionary Perspective. *Chem. Rev.* **2002**, *102*, 439–469.
8. Roth, J.; Kempf, A.; Reuter, G.; Schauer, R.; Gehring, W. J. Occurrence of Sialic Acids in *Drosophila melanogaster. Science* **1992**, *256*, 673–675.
9. Malykh, Y. N.; Krisch, B.; Gerardy-Schahn, R.; Lapina, E. B.; Shaw, L.; Schauer, R. The Presence of *N*-Acetylneuraminic Acid in Malpighian Tubules of Larvae of the Cicada *Philaenus spumarius*. *Glycoconjugate J.* **1999**, *16*, 731–739.
10. Repnikova, E.; Koles, K.; Nakamura, M.; Pitts, J.; Li, H.; Ambavane, A.; Zoran, M. J.; Panin, V. M. Sialyltransferase Regulates Nervous System Function in *Drosophila*. *J. Neurosci.* **2010**, *30*, 6466–6476.
11. Di, W.; Fujita, A.; Hamaguchi, K.; Delannoy, P.; Sato, C.; Kitajima, K. Diverse Subcellular Localizations of the Insect CMP-Sialic Acid Synthetases. *Glycobiology* **2017**, *27*, 329–341.
12. Inoue, S.; Kitajima, K. KDN (Deaminated Neuraminic Acid): Dreamful Past and Exciting Future of the Newest Member of the Sialic Acid Family. *Glycoconjugate J.* **2006**, *23*, 277–290.
13. McNaught, A.D. IUPAC/IUBMB Joint Commission on Biochemical Nomenclature. Nomenclature of Carbohydrates, Recommendations 1996. *Carbohydr. Res.* **1997**, *297*, 1–92.
14. IUPAC Commission on the Nomenclature of Organic Chemistry and IUPAC/IUB Commission on Biochemical Nomenclature. Tentative Rules for Carbohydrate Nomenclature—Part 1, 1969. *Eur. J. Biochem.* **1971**, *21*, 455–477.
15. Gielen, W. Beitrag zur Chemie der Neuraminsäure. *Hoppe Seyler's Z. Physiol. Chem.* **1967**, *348*, 329–333.
16. Hanai, N.; Dohi, T.; Nores, G. A.; Hakomori, S.-I. A Novel Ganglioside, De-*N*-acetyl-GM_3 ($II^3NeuNH_2LacCer$), Acting as a Strong Promoter for Epidermal Growth Factor Receptor Kinase and as a Stimulator for Cell Growth. *J. Biol. Chem.* **1988**, *263*, 6296–6301.
17. Nores, G. A.; Hanai, N.; Levery, S. B.; Eaton, H. L.; Salyan, M. E. K.; Hakomori, S.-I. Synthesis and Characterization of Lyso-GM_3 (II^3Neu5Ac Lactosyl Sphingosine), De-*N*-acetyl-GM_3 (II^3NeuNH_2 Lactosyl Cer), and Related Compounds. *Carbohydr. Res.* **1988**, *179*, 393–410.

18. Sjoberg, E. R.; Chammas, R.; Ozawa, H.; Kawashima, I.; Khoo, K.-H.; Morris, H. R.; Dell, A.; Tai, T.; Varki, A. Expression of De-*N*-acetyl-gangliosides in Human Melanoma Cells Is Induced by Genistein or Nocodazole. *J. Biol. Chem.* **1995**, *270*, 2921–2930.
19. Zanetta, J.-P.; Pons, A.; Iwersen, M.; Mariller, C.; Leroy, Y.; Timmerman, P.; Schauer, R. Diversity of Sialic Acids Revealed Using Gas Chromatography/Mass Spectrometry of Heptafluorobutyrate Derivatives. *Glycobiology* **2001**, *11*, 663–676.
20. Hamada, T.; Hirota, H.; Yokoyama, S.; Otsubo, N.; Ishida, H.; Kiso, M.; Kanamori, A.; Kannagi, R. NMR Analysis of Novel Ganglioside GM4 Analogues Containing De-*N*-acetyl and Lactamized Sialic Acid: Probes for Searching New Ligand Structures for Human L-Selectin. *Magn. Reson. Chem.* **2002**, *40*, 517–523.
21. Mitsuoka, C.; Ohmori, K.; Kimura, N.; Kanamori, A.; Komba, S.; Ishida, H.; Kiso, M.; Kannagi, R. Regulation of Selectin Binding Activity by Cyclization of Sialic Acid Moiety of Carbohydrate Ligands on Human Leukocytes. *Proc. Natl. Acad. Sci. U. S. A.* **1999**, *96*, 1597–1602.
22. Klenk, E.; Faillard, H. Zur Kenntnis der Kohlenhydratgruppen der Mucoproteide. *Hoppe-Seyler's Z. Physiol. Chem.* **1954**, *298*, 230–238.
23. Blix, G.; Lindberg, E.; Odin, L.; Werner, I. Studies on Sialic Acids. *Acta Soc. Med. Upsalien.* **1956**, *61*, 1–25.
24. Schauer, R.; Faillard, H. Zur Wirkungsspezifität der Neuraminidase—Das Verhalten isomerer *N,O*-Diacetyl-neuraminsäureglykoside im Submaxillarismucin von Pferd und Rind bei Einwirkung bakterieller Neuraminidase. *Hoppe-Seyler's Z. Physiol. Chem.* **1968**, *349*, 961–968.
25. Kochetkov, N. K.; Chizhov, O. S.; Kadentsev, V. I.; Smirnova, G. P.; Zhukova, I. G. Mass Spectra of Acetylated Derivatives of Sialic Acids. *Carbohydr. Res.* **1973**, *27*, 5–10.
26. Kamerling, J. P.; Vliegenthart, J. F. G.; Vink, J. Mass Spectrometry of Pertrimethylsilyl Neuraminic Acid Derivatives. *Carbohydr. Res.* **1974**, *33*, 297–306.
27. Kamerling, J. P.; Vliegenthart, J. F. G.; Versluis, C.; Schauer, R. Identification of *O*-Acetylated *N*-Acylneuraminic Acids by Mass Spectrometry. *Carbohydr. Res.* **1975**, *41*, 7–17.
28. Brown, E. B.; Brey, W. S., Jr.; Weltner, W., Jr. Cell-Surface Carbohydrates and Their Interactions—I. NMR of *N*-Acetyl Neuraminic Acid. *Biochim. Biophys. Acta* **1975**, *399*, 124–130.
29. Haverkamp, J.; van Halbeek, H.; Dorland, L.; Vliegenthart, J. F. G.; Pfeil, R.; Schauer, R. High-Resolution ^{1}H-NMR Spectroscopy of Free and Glycosidically Linked *O*-Acetylated Sialic Acids. *Eur. J. Biochem.* **1982**, *122*, 305–311.
30. Kamerling, J. P.; Vliegenthart, J. F. G. Gas-Liquid Chromatography and Mass Spectrometry of Sialic Acids. In: *Sialic Acids—Chemistry, Metabolism and Function*; Schauer, R. Ed.; Cell Biology Monographs, Vol. 10; Springer-Verlag: Vienna, Austria, 1982; pp 95–125.
31. Manzi, A. E.; Dell, A.; Azadi, P.; Varki, A. Studies of Naturally Occurring Modifications of Sialic Acids by Fast-Atom Bombardment-Mass Spectrometry—Analysis of Positional Isomers by Periodate Cleavage. *J. Biol. Chem.* **1990**, *265*, 8094–8107.
32. Klein, A.; Diaz, S.; Ferreira, I.; Lamblin, G.; Roussel, P.; Manzi, A. E. New Sialic Acids From Biological Sources Identified by a Comprehensive and Sensitive Approach: Liquid Chromatography–Electrospray Ionization-Mass Spectrometry (LC-ESI-MS) of SIA Quinoxalinones. *Glycobiology* **1997**, 7, 421–432.
33. Kamerling, J. P.; Gerwig, G. J. Structural Analysis of Naturally Occurring Sialic Acids. In: *Glycobiology Protocols*; Brockhausen, I. Ed.; Methods Mol. Biol., Vol. 347; Humana Press Inc.: Totowa, NJ, USA, 2006; pp 69–91.
34. Niimura, Y.; Ishizuka, I. Unique Disialosyl Gangliosides From Salmon Kidney: Characterization of $V^3\alpha$Fuc,$IV^3\beta$GalNAc,$II^3(\alpha$NeuAc$)_2$-Gg$_4$Cer and Its Analogue With 4-*O*-Acetyl-*N*-acetylneuraminic Acid. *Glycoconjugate J.* **2006**, *23*, 489–499.

35. Bhattacharjee, A. K.; Jennings, H. J.; Kenny, C. P.; Martin, A.; Smith, I. C. P. Structural Determination of the Sialic Acid Polysaccharide Antigens of *Neisseria meningitidis* Serogroups B and C With Carbon 13 Nuclear Magnetic Resonance. *J. Biol. Chem.* **1975**, *250*, 1926–1932.
36. Ørskov, F.; Ørskov, I.; Sutton, A.; Schneerson, R.; Lin, W.; Egan, W.; Hoff, G. E.; Robbins, J. B. Form Variation in *Escherichia coli* K1: Determined by O-Acetylation of the Capsular Polysaccharide. *J. Exp. Med.* **1979**, *149*, 669–685.
37. Reuter, G.; Pfeil, R.; Stoll, S.; Schauer, R.; Kamerling, J. P.; Versluis, C.; Vliegenthart, J. F. G. Identification of New Sialic Acids Derived From Glycoprotein of Bovine Submandibular Gland. *Eur. J. Biochem.* **1983**, *134*, 139–143.
38. Kamerling, J. P.; van Halbeek, H.; Vliegenthart, J. F. G.; Pfeil, R.; Shukla, A. K.; Schauer, R. Migration of *O*-Acetyl Groups in Sialic Acids. *Glycoconjugates*, Proc. 7th Int. Symp., Lund-Ronneby, Sweden, July 17–23, 1983; pp 160–161.
38a. Kamerling, J. P.; Schauer, R.; Shukla, A. K.; Stoll, S.; van Halbeek, H.; Vliegenthart, J. F. G. Migration of *O*-Acetyl Groups in *N*,*O*-Acetylneuraminic Acids. *Eur. J. Biochem.* **1987**, *162*, 601–607.
39. Kogan, G.; Jann, B.; Jann, K. Structure of the *Escherichia coli* O104 Polysaccharide and Its Identity With the Capsular K9 Polysaccharide. *FEMS Microbiol. Lett.* **1992**, *91*, 135–140.
40. Gamian, A.; Romanowska, E.; Ulrich, J.; Defaye, J. The Structure of the Sialic Acid-Containing *Escherichia coli* O104 O-Specific Polysaccharide and Its Linkage to the Core Region in Lipopolysaccharide. *Carbohydr. Res.* **1992**, *236*, 195–208.
41. Buscher, H.-P.; Casals-Stenzel, J.; Schauer, R. New Sialic Acids—Identification of *N*-Glycoloyl-*O*-acetylneuraminic Acids and *N*-Acetyl-*O*-glycoloylneuraminic Acids by Improved Methods for Detection of *N*-Acyl and *O*-Acyl Groups and by Gas–Liquid Chromatography. *Eur. J. Biochem.* **1974**, *50*, 71–82.
42. Schauer, R.; Haverkamp, J.; Wember, M.; Vliegenthart, J. F. G.; Kamerling, J. P. *N*-Acetyl-9-*O*-L-lactylneuraminic Acid, a New Acylneuraminic Acid From Bovine Submandibular Gland. *Eur. J. Biochem.* **1976**, *62*, 237–242.
43. Haverkamp, J.; Schauer, R.; Wember, M.; Farriaux, J.-P.; Kamerling, J. P.; Versluis, C.; Vliegenthart, J. F. G. Neuraminic Acid Derivatives Newly Discovered in Humans: *N*-Acetyl-9-*O*-L-lactoylneuraminic Acid, *N*,9-*O*-Diacetylneuraminic Acid and *N*-Acetyl-2,3-dehydro-2-deoxyneuraminic Acid. *Hoppe-Seyler's Z. Physiol. Chem.* **1976**, *357*, 1699–1705.
44. Reuter, G.; Pfeil, R.; Kamerling, J. P.; Vliegenthart, J. F. G.; Schauer, R. Identification by Gas-Liquid Chromatography-Mass Spectrometry of 4-*O*-Acetyl-9-*O*-lactyl-*N*-acetyl-neuraminic Acid, a New Sialic Acid From Horse Submandibular Gland. *Biochim. Biophys. Acta* **1980**, *630*, 306–310.
45. Bulai, T.; Bratosin, D.; Pons, A.; Montreuil, J.; Zanetta, J.-P. Diversity of the Human Erythrocyte Membrane Sialic Acids in Relation With Blood Groups. *FEBS Lett.* **2003**, *534*, 185–189.
46. Bergwerff, A. A.; Hulleman, S. H. D.; Kamerling, J. P.; Vliegenthart, J. F. G.; Shaw, L.; Reuter, G.; Schauer, R. Nature and Biosynthesis of Sialic Acids in the Starfish *Asterias rubens*—Identification of Sialo-oligomers and Detection of *S*-Adenosyl-L-methionine: *N*-acylneuraminate 8-*O*-Methyltransferase and CPM-*N*-Acetylneuraminate Monooxygenase Activities. *Biochimie* **1992**, *74*, 25–38.
47. Inagaki, M.; Shiizaki, M.; Hiwatashi, T.; Miyamoto, T.; Higuchi, R. Constituents of Crinoidea—5. Isolation and Structure of a New Glycosyl Inositolphosphoceramide-Type Ganglioside From the Feather Star *Comanthina schlegeli*. *Chem. Pharm. Bull.* **2007**, *55*, 1649–1651.
48. Yamada, K.; Harada, Y.; Nagaregawa, Y.; Miyamoto, T.; Isobe, R.; Higuchi, R. Constituents ofHolothuroidea, 7.—Isolation and Structure ofBiologically Active Gangliosides From the Sea Cucumber *Holothuria pervicax*. *Eur. J. Org. Chem.* **1998**, 2519–2525.

49. Kochetkov, N. K.; Smirnova, G. P.; Chekareva, N. V. Isolation and Structural Studies of a Sulfated Sialosphingolipid From the Sea Urchin *Echinocardium cordatum*. *Biochim. Biophys. Acta* **1976**, *424*, 274–283.
50. Slomiany, A.; Kojima, K.; Banas-Gruszka, Z.; Slomiany, B. L. Structure of a Novel Sulfated Sialoglycosphingolipid From Bovine Gastric Mucosa. *Biochem. Biophys. Res. Commun.* **1981**, *100*, 778–784.
51. Roseman, S.; Jourdian, G. W.; Watson, D.; Rood, R. Enzymatic Synthesis of Sialic Acid 9-Phosphates. *Proc. Natl. Acad. Sci. U. S. A.* **1961**, *47*, 958–961.
52. Kamerling, J. P.; Vliegenthart, J. F. G.; Schauer, R.; Strecker, G.; Montreuil, J. Isolation and Identification of 2-Deoxy-2,3-dehydro-*N*-acetylneuraminic Acid From the Urine of a Patient With Sialuria. *Eur. J. Biochem.* **1975**, *56*, 253–258.
53. Beau, J.-M.; Schauer, R.; Haverkamp, J.; Kamerling, J. P.; Dorland, L.; Vliegenthart, J. F. G. Chemical Behaviour of Cytidine 5′-Monophospho-*N*-acetyl-β-D-neuraminic Acid Under Neutral and Alkaline Conditions. *Eur. J. Biochem.* **1984**, *140*, 203–208.
54. Furuhata, K.; Sato, S.; Goto, M.; Takayanagi, H.; Ogura, H. Studies on Sialic Acids—VII. The Crystal and Molecular Structure of *N*-Acetyl-2,3-dehydro-2-deoxyneuraminic Acid. *Chem. Pharm. Bull.* **1988**, *36*, 1872–1876.
55. Schauer, R.; Schröder, C.; Shukla, A. K. New Techniques for the Investigation of Structure and Metabolism of Sialic Acids. *Adv. Exp. Med. Biol.* **1984**, *174*, 75–86.
56. Shukla, A. K.; Schröder, C.; Nöhle, U.; Schauer, R. Natural Occurrence and Preparation of *O*-Acylated 2,3-Unsaturated Sialic Acids. *Carbohydr. Res.* **1987**, *168*, 199–209.
57. Suzuki, M.; Suzuki, A.; Yamakawa, T.; Matsunaga, E. Characterization of 2,7-Anhydro-*N*-acetylneuraminic Acid in Human Wet Cerumen. *J. Biochem.* **1985**, *97*, 509–515.
58. Li, Y.-T.; Nakagawa, H.; Ross, S. A.; Hansson, G. C.; Li, S.-C. A Novel Sialidase Which Releases 2,7-Anhydro-α-*N*-acetylneuraminic Acid From Sialoglycoconjugates. *J. Biol. Chem.* **1990**, *265*, 21629–21633.
59. Pozsgay, V.; Jennings, H.; Kasper, D. L. 4,8-Anhydro-*N*-acetylneuraminic Acid—Isolation From Edible Bird's Nest and Structure Determination. *Eur. J. Biochem.* **1987**, *162*, 445–450.
60. Saito, K.-I.; Sugai, K.; Fujikura, K.; Yamada, N.; Goto, M.; Ban, C.; Hayasaka, E.; Sugiyama, N.; Tomita, K. Thermal Degradation of Sodium *N*-Acetylneuraminate. *Carbohydr. Res.* **1989**, *185*, 307–314.
61. Cebo, C.; Dambrouck, T.; Maes, E.; Laden, C.; Strecker, G.; Michalski, J.-C.; Zanetta, J.-P. Recombinant Human Interleukins IL-1α, IL-1β, IL-4, IL-6, and IL-7 Show Different and Specific Calcium-Independent Carbohydrate-Binding Properties. *J. Biol. Chem.* **2001**, *276*, 5685–5691.
62. Bratosin, D.; Palii, C.; Moicean, A. D.; Zanetta, J.-P.; Montreuil, J. Reduced Diversity of the Human Erythrocyte Membrane Sialic Acids in Polycythemia Vera. Absence of *N*-Glycolylneuraminic Acid and Characterisation of *N*-Acetylneuraminic Acid 1,7-Lactone. *Biochimie* **2007**, *89*, 355–359.
63. Li, Y.-T.; Maskos, K.; Chou, C.-W.; Cole, R. B.; Li, S.-C. Presence of an Unusual GM2 Derivative, Taurine-Conjugated GM2, in Tay–Sachs Brain. *J. Biol. Chem.* **2003**, *278*, 35286–35291.
64. Klenk, E.; Uhlenbruck, G. Über die Abspaltung von *N*-Glykolyl-neuraminsäure (P-Sialinsäure) aus dem Schweine-Submaxillarismucin durch das "Receptor Destroying Enzyme." *Hoppe-Seyler's Z. Physiol. Chem.* **1957**, *307*, 266–271.
65. Handa, S.; Yamakawa, T. The Chemistry of Lipids of Posthemolytic Residue or Stroma of Erythrocytes. XII. Chemical Structure and Chromatographic Behaviour of Hematosides Obtained From Equine and Dog Erythrocytes. *Jpn. J. Exp. Med.* **1964**, *34*, 293–304.

66. Jaques, L. W.; Riesco, B. F.; Weltner, W., Jr. N.M.R. Spectroscopy and Calcium Binding of Sialic Acids: *N*-Glycolylneuraminic Acid and Periodate-Oxidized *N*-Acetylneuraminic Acid. *Carbohydr. Res.* **1980**, *83*, 21–32.
67. Warren, L. *N*-Glycolyl-8-*O*-methylneuraminic Acid—A New Form of Sialic Acid in the Starfish *Asterias forbesi*. *Biochim. Biophys. Acta* **1964**, *83*, 129–132.
68. Smirnova, G. P.; Kochetkov, N. K.; Sadovskaya, V. L. Gangliosides of the Starfish *Aphelasterias japonica*—Evidence for a New Linkage Between Two *N*-Glycolylneuraminic Acid Residues Through the Hydroxy Group of the Glycolic Acid Residue. *Biochim. Biophys. Acta* **1987**, *920*, 47–55.
69. Zanetta, J.-P.; Srinivasan, V.; Schauer, R. Analysis of Monosaccharides, Fatty Constituents and Rare *O*-Acetylated Sialic Acids From Gonads of the Starfish *Asterias rubens*. *Biochimie* **2006**, *88*, 171–178.
70. Arao, K.; Inagaki, M.; Higuchi, R. Constituents of Crinoidea, 2—Isolation and Structure of the Novel Type Gangliosides From the Feather Star *Comanthus japonica*. *Chem. Pharm. Bull.* **2001**, *49*, 695–698.
71. Kubo, H.; Irie, A.; Inagaki, F.; Hoshi, M. Gangliosides From the Eggs of the Sea Urchin, *Anthocidaris crassispina*. *J. Biochem.* **1990**, *108*, 185–192.
72. Kitazume, S.; Kitajima, K.; Inoue, S.; Haslam, S. M.; Morris, H. R.; Dell, A.; Lennarz, W. J.; Inoue, Y. The Occurrence of Novel 9-*O*-Sulfated *N*-Glycolylneuraminic Acid-Capped $\alpha2 \rightarrow 5$-*O*-Glycolyl-Linked Oligo/PolyNeu5Gc Chains in Sea Urchin Egg Cell Surface Glycoprotein—Identification of a New Chain Termination Signal for Polysialyltransferase. *J. Biol. Chem.* **1996**, *271*, 6694–6701.
73. Kluge, A.; Reuter, G.; Lee, H.; Ruch-Heeger, B.; Schauer, R. Interaction of Rat Peritoneal Macrophages With Homologous Sialidase-Treated Thrombocytes In Vitro: Biochemical and Morphological Studies—Detection of *N*-(*O*-Acetyl)glycoloylneuraminic Acid. *Eur. J. Cell Biol.* **1992**, *59*, 12–20.
74. Higuchi, R.; Inukai, K.; Jhou, J. X.; Honda, M.; Komori, T.; Tsuji, S.; Nagai, Y. Biologically Active Glycosides From Asteroidea—XXXI. Glycosphingolipids From the Starfish *Asterias amurensis versicolor* Sladen, 2—Structure and Biological Activity of Ganglioside Molecular Species. *Liebigs Ann. Chem.* **1993**, 359–366.
75. Nöhle, U.; Shukla, A. K.; Schröder, C.; Reuter, G.; Schauer, R.; Kamerling, J. P.; Vliegenthart, J. F. G. Structural Parameters and Natural Occurrence of 2-Deoxy-2,3-didehydro-*N*-glycoloylneuraminic Acid. *Eur. J. Biochem.* **1985**, *152*, 459–463.
76. Nadano, D.; Iwasaki, M.; Endo, S.; Kitajima, K.; Inoue, S.; Inoue, Y. A Naturally Occurring Deaminated Neuraminic Acid, 3-Deoxy-D-*glycero*-D-*galacto*-nonulosonic Acid (KDN)—Its Unique Occurrence at the Nonreducing Ends of Oligosialyl Chains in Polysialoglycoprotein of Rainbow Trout Eggs. *J. Biol. Chem.* **1986**, *261*, 11550–11557.
77. Knirel, Y. A.; Kocharova, N. A.; Shashkov, A. S.; Kochetkov, N. K.; Mamontova, V. A.; Solov'eva, T. F. Structure of the Capsular Polysaccharide of *Klebsiella ozaenae* Serotype K4 Containing 3-Deoxy-D-*glycero*-D-*galacto*-nonulosonic Acid. *Carbohydr. Res.* **1989**, *188*, 145–155.
78. Strecker, G.; Wieruszeski, J.-M.; Michalski, J.-C.; Alonso, C.; Boilly, B.; Montreuil, J. Characterization of Le^x, Le^y and Le^y Antigen Determinants in KDN-Containing *O*-Linked Glycan Chains From *Pleurodeles waltlii* Jelly Coat Eggs. *FEBS Lett.* **1992**, *298*, 39–43.
79. Shashkov, A. S.; Torgov, V. I.; Nazarenko, E. L.; Zubkov, V. A.; Gorshkova, N. M.; Gorshkova, R. P.; Widmalm, G. Structure of the Phenol-Soluble Polysaccharide From *Shewanella putrefaciens* Strain A6. *Carbohydr. Res.* **2002**, *337*, 1119–1127.
80. Iwasaki, M.; Inoue, S.; Troy, F. A. A New Sialic Acid Analogue, 9-*O*-Acetyl-Deaminated Neuraminic Acid, and α-2,8-Linked O-Acetylated Poly(*N*-glycolylneuraminyl) Chains in a Novel Polysialoglycoprotein From Salmon Eggs. *J. Biol. Chem.* **1990**, *265*, 2596–2602.

81. Plancke, Y.; Wieruszeski, J.-M.; Boilly, B.; Strecker, G. Primary Structure of Seven New Acidic Oligosaccharide-Alditols From the Egg Jelly Coats of *Axolotl mexicanum* and *Pleurodeles waltl. Ci. Cult. J. Braz. Assoc. Adv. Sci.* **1994**, *46*, 273–279.
82. Gil-Serrano, A. M.; Rodríguez-Carvajal, M. A.; Tejero-Mateo, P.; Espartero, J. L.; Thomas-Oates, J.; Ruiz-Sainz, J. E.; Buendía-Clavería, A. M. Structural Determination of a 5-*O*-Methyl-Deaminated Neuraminic Acid (Kdn)-Containing Polysaccharide Isolated From *Sinorhizobium fredii. Biochem. J.* **1998**, *334*, 585–594.
83. Angata, T.; Nakata, D.; Matsuda, T.; Kitajima, K.; Troy, F. A., II. Biosynthesis of KDN (2-Keto-3-deoxy-D-*glycero*-D-*galacto*-nononic Acid)—Identification and Characterization of a KDN-9-Phosphate Synthetase Activity From Trout Testis. *J. Biol. Chem.* **1999**, *274*, 22949–22956.
84. Meindl, P.; Tuppy, H. Über 2-Desoxy-2,3-dehydro-sialinsäuren—II. Kompetitive Hemmung der *Vibrio-cholerae*-Neuraminidase durch 2-Desoxy-2,3-dehydro-*N*-acyl-neuraminsäuren. *Hoppe-Seyler's Z. Physiol. Chem.* **1969**, *350*, 1088–1092.
85. Saito, M.; Rosenberg, A. Identification and Characterization of *N*-Acetyl-2,3-didehydro-2-deoxyneuraminic Acid as a Metabolite in Mammalian Brain. *Biochemistry* **1984**, *23*, 3784–3788.
86. Burmeister, W. P.; Henrissat, B.; Bosso, C.; Cusack, S.; Ruigrok, R. W. H. Influenza B Virus Neuraminidase Can Synthesize Its Own Inhibitor. *Structure* **1993**, *1*, 19–26.
87. Chou, M.-Y.; Li, S.-C.; Kiso, M.; Hasegawa, A.; Li, Y.-T. Purification and Characterization of Sialidase L, a NeuAcα2→3Gal-Specific Sialidase. *J. Biol. Chem.* **1994**, *269*, 18821–18826.
88. Gottschalk, A. *The Chemistry and Biology of Sialic Acids and Related Substances*; Cambridge University Press: London, UK, 1960.
89. Rosenberg, A.; Schengrund, C.-L., Eds. *Biological Roles of Sialic Acid*; Plenum Press: New York, NY, USA, 1976.
90. Schauer, R., Ed. *Cell Biology Monographs*, Vol. 10; *Sialic Acids—Chemistry, Metabolism and Function*; Springer-Verlag: Vienna, Austria, 1982.
91. Wiegandt, H., Ed. *New Comprehensive Biochemistry*, Vol. 10; *Glycolipids*; Elsevier Science Publishers: Amsterdam, The Netherlands, 1985.
92. Rosenberg, A., Ed. *Biology of the Sialic Acids*. Plenum Press: New York, NY, USA, 1995.
93. Tiralongo, J.; Martinez-Duncker, I., Eds. *Sialobiology: Structure, Biosynthesis and Function—Sialic Acid Glycoconjugates in Health and Disease*; Bentham Science Publishers: Oak Park, IL, USA, 2013.
94. Gerardy-Schahn, R.; Delannoy, P.; von Itzstein, M., Eds. *SialoGlyco Chemistry and Biology I—Biosynthesis, Structural Diversity and Sialoglycopathologies*; Topics in Current Chemistry, Vol. 366; Springer: Berlin, Germany, 2015.
95. Gerardy-Schahn, R.; Delannoy, P.; von Itzstein, M., Eds. *SialoGlyco Chemistry and Biology II—Tools and Techniques to Identify and Capture Sialoglycans*; Topics in Current Chemistry, Vol. 367; Springer: Berlin, Germany, 2015.
96. Zilliken, F.; Whitehouse, M. W. The Nonulosaminic Acids—Neuraminic Acids and Related Compounds (Sialic Acids). *Adv. Carbohydr. Chem.* **1958**, *13*, 237–263.
97. Gottschalk, A. Neuraminidase: Its Substrate and Mode of Action. *Adv. Enzymol.* **1958**, *20*, 135–146.
98. Blix, G.; Jeanloz, R. W. Sialic Acids and Muramic Acid. In: *The Amino Sugars*; Jeanloz, R. W. Ed.; Vol. IA; Academic Press: New York, USA, 1969; pp 213–265.
99. Faillard, H.; Schauer, R. Glycoproteins as Lubricants, Protective Agents, Carriers, Structural Proteins and as Participants in Other Functions. In: *Glycoproteins—Their Composition, Structure and Function;* Gottschalk, A. Ed.; Vol. 5, Part B, 2nd ed.; B.B.A. Library, Elsevier Publishing Company: Amsterdam, The Netherlands, 1972; pp 1246–1267.

100. Tuppy, H.; Gottschalk, A. The Structure of Sialic Acids and Their Quantitation. In: *Glycoproteins—Their Composition, Structure and Function;* Gottschalk, A. Ed.; Vol. 5, Part A, 2nd ed.; B.B.A. Library, Elsevier Publishing Company: Amsterdam, The Netherlands, 1972; pp 403–449.
101. Schauer, R. Chemistry, Metabolism, and Biological Functions of Sialic Acids. *Adv. Carbohydr. Chem. Biochem.* **1982**, *40*, 131–234.
102. Schauer, R. Analysis of Sialic Acids. *Methods Enzymol.* **1987**, *138*, 132–161.
103. Schauer, R. Sialic Acids: Metabolism of *O*-Acetyl Groups. *Methods Enzymol.* **1987**, *138*, 611–626.
104. Reuter, G.; Schauer, R. Determination of Sialic Acids. *Methods Enzymol.* **1994**, *230*, 168–199.
105. Reuter, G.; Schauer, R. Enzymic Methods of Sialic Acid Analysis. *Methods Carbohydr. Chem.* **1994**, *10*, 29–39.
106. Von Itzstein, M.; Thomson, R. J. The Synthesis of Novel Sialic Acids as Biological Probes. In: *Glycoscience—Synthesis of Oligosaccharides and Glycoconjugates*, Driguez, H.; Thiem, J., Eds.; Topics in Current Chemistry, Springer–Verlag: Berlin, Germany; Vol. 186, 1997; pp 119–170.
107. Boons, G.-J.; Demchenko, A. V. The Chemistry of Sialic Acid. In: *Carbohydrate-Based Drug Discovery*; Wong, C.-H. Ed.; Vol. 1; Wiley–VCH: Weinheim, Germany, 2003; pp 55–102.
108. Thomson, R.; von Itzstein, M. N-Acetylneuraminic Acid Derivatives and Mimetics as Anti-influenza Agents. In: *Carbohydrate-Based Drug Discovery*; Wong, C.-H. Ed.; Vol. 2; Wiley–VCH: Weinheim, Germany, 2003; pp 831–861.
109. Kohla, G.; Schauer, R. Sialic Acids in Gangliosides: Origin and Function. In: *Neuroglycobiology (Molecular and Cellular Neurobiology)*; Fukuda, M., Rutishauser, U., Schnaar, R. L., Eds.; Oxford University Press: Oxford, UK, 2005; pp 133–155.
110. Schauer, R. The Diversity of Sialic Acids and Their Interplay With Lectins. In: *Glycobiology*; Sansom, C., Markman, O., Eds.; Scion Publishing Ltd: Bloxham, UK, 2007; pp 136–149.
111. Miyagi, T.; Yamaguchi, K. Sialic Acids. In: *Comprehensive Glycoscience—From Chemistry to Systems Biology*; Kamerling, J. P., Boons, G.-J., Lee, Y. C., Suzuki, A., Taniguchi, N., Voragen, A. G. J., Eds.; Vol. 3; Elsevier Ltd.: Oxford, UK, 2007; pp 297–323.
112. Varki, A.; Schauer, R. Sialic Acids. In: *Essentials of Glycobiology;* Varki, A., Cummings, R. D., Esko, J. D., Freeze, H. H., Stanley, P., Bertozzi, C. R., Hart, G. W., Etzler, M. E., Eds. 2nd ed.; Cold Spring Harbor Laboratory Press: Cold Spring Harbor, NY, USA, 2009; pp 199–217.
113. Monti, E.; Bonten, E.; D'Azzo, A.; Bresciani, R.; Venerando, B.; Borsani, G.; Schauer, R.; Tettamanti, G. Sialidases in Vertebrates: A Family of Enzymes Tailored for Several Cell Functions. *Adv. Carbohydr. Chem. Biochem.* **2010**, *64*, 403–479.
114. Varki, A.; Schnaar, R. L.; Schauer, R. Sialic Acids and Other Nonulosonic Acids; In: *Essentials of Glycobiology*; Varki, A., Cummings, R. D., Esko, J. D., Stanley, P., Hart, G. W., Aebi, M., Darvill, A., Kinoshita, T., Packer, N. H., Prestegard, J. J., Schnaar, R. L., Seeberger, P. H., Eds, 3rd ed.; Cold Spring Harbor Laboratory Press: Cold Spring Harbor, NY, USA, 2017; pp 179–195.
115. Klenk, E. Chemie und Biochemie der Neuraminsäure. *Angew. Chem.* **1956**, *68*, 349–352.
116. Blix, G. Sialic Acids. In: Proceedings of IVth International Congress of Biochemistry, Vol. 1, Vienna, Austria, 1958; pp 94–106.
117. Barry, G. T. Colominic Acid, a Polymer of *N*-Acetylneuraminic Acid. *J. Exp. Med.* **1958**, *107*, 507–521.
118. Wiegandt, H. Struktur und Funktion der Ganglioside. *Angew. Chem.* **1968**, *80*, 89–98.

119. Schauer, R. Chemistry and Biology of the Acylneuraminic Acids. *Angew. Chem., Int. Ed. Engl.* **1973**, *12*, 127–138.
120. Yamakawa, T. Thus Started Ganglioside Research. *Trends Biochem. Sci.* **1988**, *13*, 452–454.
121. Faillard, H. The Early History of Sialic Acids. *Trends Biochem. Sci.* **1989**, *14*, 237–241.
122. Schauer, R. Biosynthesis and Function of N- and O-Substituted Sialic Acids. *Glycobiology* **1991**, *1*, 449–452.
123. Troy, F. A., II. Polysialylation: From Bacteria to Brains. *Glycobiology* **1992**, *2*, 5–23.
124. Varki, A. Diversity in the Sialic Acids. *Glycobiology* **1992**, *2*, 25–40.
125. Corfield, T. Bacterial Sialidases: Roles in Pathogenicity and Nutrition. *Glycobiology* **1992**, *2*, 509–521.
126. Harduin-Lepers, A.; Recchi, M.-A.; Delannoy, P. 1994, the Year of Sialyltransferases. *Glycobiology* **1995**, *5*, 741–758.
127. Taylor, G. Sialidases: Structures, Biological Significance and Therapeutic Potential. *Curr. Opin. Struct. Biol.* **1996**, *6*, 830–837.
128. Schauer, R. Origin and Biological Role of the Great Chemical Diversity of Natural Sialic Acids. *Trends Glycosci. Glycotechnol.* **1997**, *9*, 315–330.
129. Kelm, S.; Schauer, R. Sialic Acids in Molecular and Cellular Interactions. *Int. Rev. Cytol.* **1997**, *175*, 137–240.
130. Traving, C.; Schauer, R. Structure, Function and Metabolism of Sialic Acids. *Cell. Mol. Life Sci.* **1998**, *54*, 1330–1349.
131. Schauer, R. Achievements and Challenges of Sialic Acid Research. *Glycoconjugate J.* **2000**, *17*, 485–499.
132. Schauer, R. Sialic Acids: Fascinating Sugars in Higher Animals and Man. *Zoology* **2004**, *107*, 49–64.
133. Crocker, P. R.; Clark, E. A.; Filbin, M.; Gordon, S.; Jones, Y.; Kehrl, J. H.; Kelm, S.; Le Douarin, N.; Powell, L.; Roder, J.; Schnaar, R. L.; Sgroi, D. C.; Stamenkovic, I.; Schauer, R.; Schachner, M.; van den Berg, T. K.; van der Merwe, P. A.; Watt, S. M.; Varki, A. Siglecs: A Family of Sialic Acid-Binding Lectins. *Glycobiology* **1998**, *8*, v.
133a. Varki, A.; Angata, T. Siglecs—The Major Subfamily of I-Type Lectins. *Glycobiology* **2006**, *16*, 1R–27R.
134. Crocker, P. R.; Paulson, J. C.; Varki, A. Siglecs and Their Roles in the Immune System. *Nat. Rev. Immunol.* **2007**, 7, 255–266.
135. Schauer, R. Sialic Acids as Regulators of Molecular and Cellular Interactions. *Curr. Opin. Struct. Biol.* **2009**, *19*, 1–8.
136. Wang, B. Sialic Acid Is an Essential Nutrient for Brain Development and Cognition. *Annu. Rev. Nutr.* **2009**, *29*, 177–222.
137. Chen, X.; Varki, A. Advances in the Biology and Chemistry of Sialic Acids. *ACS Chem. Biol.* **2010**, *5*, 163–176.
138. Harduin-Lepers, A. Comprehensive Analysis of Sialyltransferases in Vertebrate Genomes. *Glycobiol. Insights* **2010**, *2*, 29–61.
139. Schauer, R.; Kamerling, J. P. The Chemistry and Biology of Trypanosomal *trans*-Sialidases: Virulence Factors in Chagas Disease and Sleeping Sickness. *ChemBioChem* **2011**, *12*, 2246–2264.
140. Ogura, H. Development of Miracle Medicines From Sialic Acids. *Proc. Jpn. Acad., Ser. B* **2011**, *87*, 328–361.
141. Varki, A.; Gagneux, P. Multifarious Roles of Sialic Acids in Immunity. *Ann. N. Y. Acad. Sci.* **2012**, *1253*, 16–36.
142. Li, Y.; Chen, X. Sialic Acid Metabolism and Sialyltransferases: Natural Functions and Applications. *Appl. Microbiol. Biotechnol.* **2012**, *94*, 887–905.
143. Vimr, E. R. Unified Theory of Bacterial Sialometabolism: How and Why Bacteria Metabolize Host Sialic Acids. *ISRN Microbiol.* **2013**, *2013, 816713*, 26 pages.

144. Schnaar, R. L.; Gerardy-Schahn, R.; Hildebrandt, H. Sialic Acids in the Brain: Gangliosides and Polysialic Acid in Nervous System Development, Stability, Disease, and Regeneration. *Physiol. Rev.* **2014**, *94*, 461–518.
145. Zhang, T.; She, Z.; Huang, Z.; Li, J.; Luo, X.; Deng, Y. Application of Sialic Acid/Polysialic Acid in the Drug Delivery Systems. *Asian J. Pharm. Sci.* **2014**, *9*, 75–81.
146. Stencel-Baerenwald, J. E.; Reiss, K.; Reiter, D. M.; Stehle, T.; Dermody, T. S. The Sweet Spot: Defining Virus-Sialic Acid Interactions. *Nat. Rev. Microbiol.* **2014**, *12*, 739–749.
147. Bauer, J.; Osborn, H. M. I. Sialic Acids in Biological and Therapeutic Processes—Opportunities and Challenges. *Future Med. Chem.* **2015**, 7, 2285–2299.
148. Hoppe-Seyler, F. Allgemeine Biologie—Schleimgewebe. In: *Physiologische Chemie. I. Theil*; Verlag von August Hirschwald: Berlin, Germany, 1877; pp 94–95.
149. Hoppe-Seyler, F. Die Verdauung und Resorption der Nährstoffe—Die Sublingualdrüse und ihr Secret. *Physiologische Chemie. II. Theil.* Verlag von August Hirschwald: Berlin, Germany, 1878; 197–198.
150. Green, J. R. The Edible Bird's-Nest, or Nest of the Java Swift (*Collocalia nidifica*). *J. Physiol.* **1885**, *6*, 40–45.
151. Krukenberg, C. Fr. W. Weitere Mitteilungen über die Hyalogene. *Z. Biol.* **1886**, *4*, 261–271.
152. Zeller, H. Untersuchungen über die essbaren indischen Schwalbennester. *Hoppe-Seyler's Z. Physiol. Chem.* **1913**, *86*, 85–106.
153. Martin, J. E.; Tanenbaum, S. W.; Flashner, M. A Facile Procedure for the Isolation of *N*-Acetylneuraminic Acid From Edible Bird's-Nest. *Carbohydr. Res.* **1977**, *56*, 423–425.
154. Corfield, A. P. Isolation and Purification of Sialic Acids. In: *Sialic Acids—Chemistry, Metabolism and Function*; Schauer, R. Ed.; Cell Biology Monographs, Vol. 10; Springer-Verlag: Vienna, Austria, 1982; pp 51–57.
155. Ehrlich, P. Über die Dimethylamidobenzaldehydreaction. *Med. Woche balneol. Centralz.* **1901**, 151–153.
156. Pröscher, F. Zur Kenntniss der Ehrlich'schen Dimethylamidobenzaldehydreaction. *Hoppe-Seyler's Z. Physiol. Chem.* **1900–1901**, *31*, 520–526.
157. Müller, F. Beiträge zur Kenntnis des Mucins und einiger damit verwandter Eiweissstoffe. *Z. Biol.* **1901**, *42*, 468–564.
158. Walz, E. Über das Vorkommen von Kerasin in der normalen Rindermilz. *Hoppe-Seyler's Z. Physiol. Chem.* **1927**, *166*, 210–222.
159. Levene, P. A.; Landsteiner, K. On Some New Lipoids. *J. Biol. Chem.* **1927**, *75*, 607–612.
160. Bial, M. Die Diagnose der Pentosurie. *Deutsche Med. Wochenschr.* **1902**, *28*, 253–254.
161. Lundblad, A. Gunnar Blix and His Discovery of Sialic Acids—Fascinating Molecules in Glycobiology. *Upsala J. Med. Sci.* **2015**, *120*, 104–112.
162. Faillard, H. Professor Ernst Klenk—Scientific Achievements. *J. Am. Oil Chem. Soc.* **1966**, *43*, 48A/50A/78A.
163. Neuberger, A. Alfred Gottschalk (1894–1973). *Adv. Carbohydr. Chem. Biochem.* **1976**, *33*, 1–10.
164. Carubelli, R. Alfred Gottschalk, His Contributions to the Biochemistry of Sialic Acids and Glycoproteins. *Trends Glycosci. Glycotechnol.* **1992**, *4*, 559–566.
165. Blix, G. Über die Kohlenhydratgruppen des Submaxillarismucins. *Hoppe-Seyler's Z. Physiol. Chem.* **1936**, *240*, 43–54.
166. Blix, G.; Svennerholm, L.; Werner, I. Chondrosamine as a Component of Gangliosides and of Submaxillary Mucin. *Acta Chem. Scand.* **1950**, *4*, 717.
167. Blix, G.; Svennerholm, L.; Werner, I. The Isolation of Chondrosamine From Gangliosides and From Submaxillary Mucin. *Acta Chem. Scand.* **1952**, *6*, 358–362.

168. Blix, G. Einige Beobachtungen über eine hexosaminhaltige Substanz in der Protagonfraktion des Gehirns. *Skand. Arch. Physiol.* **1938**, *80*, 46–51.
169. Klenk, E. Über die Natur der Phosphatide und anderer Lipoide des Gehirns und der Leber bei der Niemann-Pickschen Krankheit. *Hoppe-Seyler's Z. Physiol. Chem.* **1935**, *235*, 24–36.
170. Klenk, E. Neuraminsäure, das Spaltprodukt eines neuen Gehirnlipoids. *Hoppe-Seyler's Z. Physiol. Chem.* **1941**, *268*, 50–58.
171. Klenk, E.; Faillard, H.; Weygand, F.; Schöne, H. H. Untersuchungen über die Konstitution der Neuraminsäure. *Hoppe-Seyler's Z. Physiol. Chem.* **1956**, *304*, 35–52.
172. Klenk, E. Über die Ganglioside, eine neue Gruppe von zuckerhaltigen Gehirnlipoiden. *Hoppe-Seyler's Z. Physiol. Chem.* **1942**, *273*, 76–86.
173. Burnet, F. M. Mucins and Mucoids in Relation to Influenza Virus Action—III. Inhibition of Virus Haemagglutination by Glandular Mucins. *Aust. J. Exp. Biol. Med. Sci.* **1948**, *26*, 371–380.
174. Gottschalk, A.; Lind, P. E. Product of Interaction Between Influenza Virus Enzyme and Ovomucin. *Nature* **1949**, *164*, 232–233.
175. Burnet, F. M.; Stone, J. D. The Receptor-Destroying Enzyme of *Vibrio cholerae*. *Aust. J. Exp. Biol. Med. Sci.* **1947**, *25*, 227–233.
176. Tamm, I.; Horsfall, F. L., Jr. Characterization and Separation of an Inhibitor of Viral Hemagglutination Present in Urine. *Proc. Soc. Exp. Biol. Med.* **1950**, *74*, 108–114.
177. Gottschalk, A. N-Substituted *Iso*glucosamine Released From Mucoproteins by the Influenza Virus Enzyme. *Nature* **1951**, *167*, 845–847.
178. Odin, L. Carbohydrate Residue of a Urine Mucoprotein Inhibiting Influenza Virus Haemagglutination. *Nature* **1952**, *170*, 663–664.
179. Gottschalk, A. Carbohydrate Residue of a Urine Mucoprotein Inhibiting Influenza Virus Haemagglutination. *Nature* **1952**, *170*, 662–663.
180. Klenk, E. Zur Kenntnis der Ganglioside. *Hoppe-Seyler's Z. Physiol. Chem.* **1951**, *288*, 216–220.
181. Werner, I.; Odin, L. On the Presence of Sialic Acid in Certain Glycoproteins and in Gangliosides. *Acta Soc. Med. Upsalien.* **1952**, *57*, 230–241.
182. Odin, L. Sialic Acid in Pseudomyxomatous Gels. *Acta Chem. Scand.* **1955**, *9*, 714–715.
183. Odin, L. Sialic Acid in Human Serum Protein and in Meconium. *Acta Chem. Scand.* **1955**, *9*, 862–864.
184. Odin, L. Sialic Acid in Human Cervical Mucus, in Hog Seminal Gel, and in Ovomucin. *Acta Chem. Scand.* **1955**, *9*, 1235–1237.
185. Blix, G.; Odin, L. Isolation of Sialic Acid From Gangliosides. *Acta Chem. Scand.* **1955**, *9*, 1541–1543.
186. Svennerholm, L. Isolation of Sialic Acid From Brain Gangliosides. *Acta Chem. Scand.* **1955**, *9*, 1033.
187. Klenk, E.; Lauenstein, K. Zur Kenntnis der Kohlenhydratgruppen des Submaxillarismucins und Harnmucoproteids—Die Isolierung von Neuraminsäure als Spaltprodukt. *Hoppe-Seyler's Z. Physiol. Chem.* **1952**, *291*, 147–152.
188. Gottschalk, A. Presence of 2-Carboxy-pyrrole in Mucoproteins and Its Relation to the Viral Enzyme. *Nature* **1953**, *172*, 808.
189. Klenk, E.; Faillard, H.; Lempfrid, H. Über die enzymatische Wirkung von Influenzavirus. *Hoppe-Seyler's Z. Physiol. Chem.* **1955**, *301*, 235–246.
190. Faillard, H. Über die Abspaltung von *N*-Acetyl-neuraminsäure aus Mucinen durch das "Receptor-Destroying-Enzyme" aus *Vibrio cholerae*. *Hoppe-Seyler's Z. Physiol. Chem.* **1957**, *307*. 62–86/*308*, 187.
191. Klenk, E.; Uhlenbruck, G. Über das Vorkommen von Neuraminsäure in den Mucinen von Schweine- und Pferde-Submaxillaris, sowie von Kuh-Colostrum. *Hoppe-Seyler's Z. Physiol. Chem.* **1956**, *305*, 224–231.

192. Klenk, E.; Faillard, H. Über das Vorkommen von Neuraminsäure im Lebereiweiss bei amyloider Degeneration. *Hoppe-Seyler's Z. Physiol. Chem.* **1955**, *299*, 191–192.
193. Klenk, E.; Stoffel, W. Über neuraminsäurehaltige Glykoproteide des Blutserums. *Hoppe-Seyler's Z. Physiol. Chem.* **1955**, *302*, 286–288.
194. Klenk, E.; Stoffel, W. Zur Kenntnis des Zellreceptoren für das Influenzavirus—Über das Vorkommen von Neuraminsäure im Eiweiss des Erythrocytenstromas. *Hoppe-Seyler's Z. Physiol. Chem.* **1956**, *303*, 78–80.
195. Blix, G.; Lindberg, E.; Odin, L.; Werner, I. Sialic Acids. *Nature* **1955**, *175*, 340–341.
196. Klenk, E.; Uhlenbruck, G. Über ein neuraminsäurehaltiges Mucoproteid aus Rindererythrocytenstroma. *Hoppe-Seyler's Z. Physiol. Chem.* **1958**, *311*, 227–233.
197. Suzuki, A. Tamio Yamakawa: Dawn of Glycobiology. *J. Biochem.* **2009**, *146*, 149–156.
198. Baer, H. H. Obituary—Richard Kuhn (1900–1967). *Adv. Carbohydr. Chem. Biochem.* **1969**, *24*, 1–12.
199. Barness, L. A.; Tomarelli, R. M. Paul György (1893–1976): A Bibliographical Sketch. *J. Nutr.* **1979**, *109*, 17–23.
200. Yamakawa, T.; Suzuki, S. The Chemistry of the Lipids of Posthemolytic Residue or Stroma of Erythrocytes—I. Concerning the Ether-Insoluble Lipids of Lyophilized Horse Blood Stroma. *J. Biochem.* **1951**, *38*, 199–212.
201. Yamakawa, T.; Suzuki, S. The Chemistry of the Lipids of Posthemolytic Residue or Stroma of Erythrocytes—II. On the Structure of Hemataminic Acid. *J. Biochem.* **1952**, *39*, 175–183.
202. Klenk, E.; Wolter, H. Über die zuckerhaltigen Lipoide des Erythrocytenstromas vom Pferde. *Hoppe-Seyler's Z. Physiol. Chem.* **1952**, *291*, 259–265.
203. Yamakawa, T.; Suzuki, S. Presence of Sero-Lactaminic Acid and Glucosamine as Constituents of Serum Mucoprotein. *J. Biochem.* **1955**, *42*, 727–736.
204. Kuhn, R.; Brossmer, R.; Schulz, W. Über die prosthetische Gruppe der Mucoproteine des Kuh-Colostrums. *Chem. Ber.* **1954**, *87*, 123–127.
205. Kuhn, R.; Brossmer, R. Isolierung einer nieder-molekularen, durch Influenza-Virus spaltbaren Substanz aus Milch. *Angew. Chem.* **1956**, *68*, 211–212.
206. Kuhn, R.; Brossmer, R. Abbau der Lactaminsäure zu *N*-Acetyl-D-glucosamin. *Chem. Ber.* **1956**, *89*, 2471–2475.
207. Kuhn, R.; Brossmer, R. Über *O*-Acetyl-lactaminsäure-lactose aus Kuh-Colostrum und ihre Spaltbarkeit durch Influenza-Virus. *Chem. Ber.* **1956**, *89*, 2013–2025.
208. Hoover, J. R. E.; Braun, G. A.; György, P. Neuraminic Acid in Mucopolysaccharides of Human Milk. *Arch. Biochem. Biophys.* **1953**, *47*, 216–217.
209. Zilliken, F.; Braun, G. A.; György, P. Gynaminic Acid—A Naturally Occurring Form of Neuraminic Acid in Human Milk. *Arch. Biochem. Biophys.* **1955**, *54*, 564–566.
210. Zilliken, F.; Braun, G. A.; György, P. "Gynaminic Acid" and Other Naturally Occurring Forms of *N*-Acetylneuraminic Acid. *Arch. Biochem. Biophys.* **1956**, *63*, 394–402.
211. Böhm, P.; Baumeister, L. Über die Isolierung der Methoxy-neuraminsäure als Spaltprodukt des Serumeiweisses. *Hoppe-Seyler's Z. Physiol. Chem.* **1955**, *300*, 153–156.
212. Böhm, P.; Baumeister, L. Über das Vorkommen neuraminsäurehaltiger Glykoproteide in Körperflüssigkeiten. *Hoppe-Seyler's Z. Physiol. Chem.* **1956**, *305*, 42–50.
213. Trucco, R. E.; Caputto, R. Neuramin-Lactose, a New Compound Isolated From the Mammary Gland of Rats. *J. Biol. Chem.* **1954**, *206*, 901–909.
214. Heyworth, R.; Bacon, J. S. D. A Nitrogenous Derivative of Lactose From Lactating Rat Mammary Gland. *Biochem. J.* **1957**, *66*, 41–47.
215. Hiyama, N. Biochemical Studies on Carbohydrates. CXXI. On the Direct-Ehrlich Reaction-Giving Structure in Glycidamins. *Tohoku J. Exp. Med.* **1949**, *51*, 319–326.

216. Van Slyke, D. D.; Dillon, R. T.; MacFadyen, D. A.; Hamilton, P. Gasometric Determination of Carboxyl Groups in Free Amino Acids. *J. Biol. Chem.* **1941**, *141*, 627–669.
217. Gottschalk, A. The Influenza Virus Enzyme and Its Mucoprotein Substrate. *Yale J. Biol. Med.* **1954**, *26*, 352–364.
218. Gottschalk, A. The Precursor of 2-Carboxy-pyrrole in Mucoproteins. *Nature* **1954**, *174*, 652–653.
219. Gottschalk, A. 2-Carboxypyrrole: Its Preparation From and Its Precursor in Mucoproteins. *Biochem. J.* **1955**, *61*, 298–307.
220. Gottschalk, A. Structural Relationship Between Sialic Acid, Neuraminic Acid and 2-Carboxy-Pyrrole. *Nature* **1955**, *176*, 881–882.
221. Gottschalk, A. The Linkage of Sialic Acid in Mucoprotein. *Biochim. Biophys. Acta* **1956**, *20*, 560–561.
222. Gottschalk, A. The Synthesis of 2-Carboxypyrrole From D-Glucosamine and Pyruvic Acid: Its Bearing on the Structure of Neuraminic Acid. *Arch. Biochem. Biophys.* **1957**, *59*, 37–44.
223. Blix, F. G.; Gottschalk, A.; Klenk, E. Proposed Nomenclature in the Field of Neuraminic and Sialic Acids. *Nature* **1957**, *179*, 1088.
224. Scott, J. E.; Yamashina, I.; Jeanloz, R. W. A Proposal for a Terminology of 'Sialic Acid' Derivatives. *Biochem. J.* **1982**, *207*, 367–368.
225. Dixon, H. B. F.; Cornish-Bowden, A. Terminology for Sialic Acids. *Biochem. J.* **1983**, *215*, 711.
226. Zilliken, F.; Glick, M. C. Alkalischer Abbau von Gynaminsäure zu Brenztraubensäure und *N*-Acetyl-D-glucosamin. *Naturwissenschaften* **1956**, *43*, 536–537.
227. Heimer, R.; Meyer, K. Studies on Sialic Acid of Submaxillary Mucoid. *Proc. Natl. Acad. Sci. U. S. A.* **1956**, *42*, 728–734.
228. Cornforth, J. W.; Firth, M. E.; Gottschalk, A. The Synthesis of *N*-Acetylneuraminic Acid. *Biochem. J.* **1958**, *68*, 57–61.
229. Schnaar, R. L.; Jourdian, G. W. Obituary Saul Roseman (1921–2011). *Glycobiology* **2011**, *21*, 1393–1394.
230. Basu, S. C. Saul Roseman (1921–2011). *Adv. Carbohydr. Chem. Biochem.* **2013**, *69*, 38–54.
231. Comb, D. G.; Roseman, S. Composition and Enzymatic Synthesis of *N*-Acetylneuraminic Acid (Sialic Acid). *J. Am. Chem. Soc.* **1958**, *80*, 497–499.
232. Comb, D. G.; Roseman, S. The Sialic Acids—I. The Structure and Enzymatic Synthesis of *N*-Acetylneuraminic Acid. *J. Biol. Chem.* **1960**, *235*, 2529–2537.
233. Kuhn, R.; Brossmer, R. Zur Konfiguration der Lactaminsäure—Epimerisierung von *N*-Acetyl-D-glucosamin und *N*-Acetyl-D-mannosamin. *Hoppe-Seyler's Z. Physiol. Chem.* **1958**, *616*, 221–225.
234. Roseman, S.; Comb, D. G. The Hexosamine Moiety of *N*-Acetylneuraminic Acid (Sialic Acid). *J. Am. Chem. Soc.* **1958**, *80*, 3166–3167.
235. Brug, J.; Paerels, G. B. Configuration of *N*-Acetylneuraminic Acid. *Nature* **1958**, *182*, 1159–1160.
236. Carroll, P. M.; Cornforth, J. W. Preparation of *N*-Acetylneuraminic Acid From *N*-Acetyl-D-mannosamine. *Biochim. Biophys. Acta* **1960**, *39*, 161–162.
237. Kuhn, R.; Baschang, G. Aminozucker-Synthesen—XXV. Synthese der Lactaminsäure. *Liebigs Ann. Chem.* **1962**, *659*, 156–163.
238. Brug, J.; Esser, R. J. E.; Paerels, G. B. The Enzymic Cleavage of *N*-Acetylneuraminic Acid. *Biochim. Biophys. Acta* **1959**, *33*, 241–242.
239. Warren, L.; Felsenfeld, H. The Biosynthesis of *N*-Acetylneuraminic Acid. *Biochem. Biophys. Res. Commun.* **1961**, *4*, 232–235.
240. Kisa, F.; Yamada, K.; Miyamoto, T.; Inagaki, M.; Higuchi, R. Determination of the Absolute Configuration of Sialic Acids in Gangliosides From the Sea Cucumber *Cucumaria echinata*. *Chem. Pharm. Bull.* **2007**, *55*, 1051–1052.

241. Hudson, C. S. A Relation Between the Chemical Constitution and the Optical Rotary Power of the Sugar Lactones. *J. Am. Chem. Soc.* **1910**, *32*, 338–346.
242. Kuhn, R.; Brossmer, R. Die Konfiguration der Lactaminsäure. *Angew. Chem.* **1957**, *69*, 534.
243. Kuhn, R.; Brossmer, R. Abbau der Lactaminsäure zu Bernsteinsäure. *Justus Liebigs Ann. Chem.* **1959**, *624*, 137–141.
244. Kuhn, R.; Brossmer, R. Die Konfiguration der Sialinsäuren. *Angew. Chem.* **1962**, *74*, 252–253.
245. Kuhn, R.; Baschang, G. Die Konfiguration der Sialinsäuren am C-atom 4. *Chem. Ber.* **1962**, *95*, 2384–2385.
246. Fischmeister, I. Infrarotspektren von Sialinsäuren und Sialinsäuremethylestern. *Ark. Kemi* **1958**, *13*, 247–258.
247. Holmquist, L.; Östman, B. The Anomeric Configuration of *N*-Acetylneuraminic Acid Released by the Action of *Vibrio cholerae* Neuraminidase. *FEBS Lett.* **1975**, *60*, 327–330.
248. Scheinthal, B. M.; Bettelheim, F. A. Multiple Forms of Sialic Acids. *Carbohydr. Res.* **1968**, *6*, 257–265.
249. Hudson, C. S. The Significance of Certain Numerical Relations in the Sugar Group. *J. Am. Chem. Soc.* **1909**, *31*, 66–86.
250. Kuhn, R.; Brossmer, R. Die Konstitution der Lactaminsäure-lactose–α-Ketosidase-Wirkung von Viren der Influenza-Gruppe. *Angew. Chem.* **1958**, *70*, 25–26.
251. Kuhn, R.; Brossmer, R. Über das durch Viren der Influenza-Gruppe spaltbare Trisaccharid der Milch. *Chem. Ber.* **1959**, *92*, 1667–1671.
252. Vliegenthart, J. F. G.; Kamerling, J. P. Synthesis of Sialic Acids and Sialic Acid Derivatives. In: *Sialic Acids—Chemistry, Metabolism and Function*; Schauer, R. Ed.; Cell Biology Monographs, Vol. 10; Springer-Verlag: Vienna, Austria, 1982; pp 59–76.
253. Meindl, P.; Tuppy, H. Über synthetische Ketoside der N-Acetyl-D-neuraminsäure—1. Mitt.: Darstellung einer Reihe durch Neuraminidase spaltbarer Ketoside. *Monatsh. Chem.* **1965**, *96*, 802–815.
254. Meindl, P.; Tuppy, H. Über synthetische Ketoside der N-Acetyl-D-neuraminsäure—2. Mitt.: Anomere *n*-Amyl- und *n*-Hexyl-ketoside der N-Acetyl-D-neuraminsäure. *Monatsh. Chem.* **1965**, *96*, 816–827.
255. Kuhn, R.; Lutz, P.; MacDonald, D. L. Synthese anomerer Sialinsäure-methylketoside. *Chem. Ber.* **1966**, *99*, 611–617.
256. Jochims, J. C.; Taigel, G.; Seeliger, A.; Lutz, P.; Driesen, H. E. Stereospecific Long-Range Couplings of Hydroxyl Protons of Pyranoses. *Tetrahedron Lett.* **1967**, 4363–4369.
257. Lutz, P.; Lochinger, W.; Taigel, G. Zur Konformation der *N*-Acetyl-neuraminsaüre. *Chem. Ber.* **1968**, *101*, 1089–1094.
258. Yu, R. K.; Ledeen, R. Configuration of the Ketosidic Bond of Sialic Acid. *J. Biol. Chem.* **1969**, *244*, 1306–1313.
259. Kitajima, K.; Inoue, Y.; Inoue, S. Polysialoglycoproteins of *Salmonidae* Fish Eggs—Complete Structure of 200-kDa Polysialoglycoprotein From the Unfertilized Eggs of Rainbow Trout (*Salmo gairdneri*). *J. Biol. Chem.* **1986**, *261*, 5262–5269.
260. Mononen, I. Quantitative Analysis, by Gas–Liquid Chromatography and Mass Fragmentography, of Monosaccharides After Methanolysis and Deamination. *Carbohydr. Res.* **1981**, *88*, 39–50.
261. Varki, A.; Diaz, S. The Release and Purification of Sialic Acids From Glycoconjugates: Methods to Minimize the Loss and Migration of *O*-Acetyl Groups. *Anal. Biochem.* **1984**, *137*, 236–247.
262. Mawhinney, T. P.; Chance, D. L. Hydrolysis of Sialic Acids and *O*-Acetylated Sialic Acids With Propionic Acid. *Anal. Biochem.* **1994**, *223*, 164–167.

263. Blix, G.; Lindberg, E. The Sialic Acids of Bovine and Equine Submaxillary Mucins. *Acta Chem. Scand.* **1960**, *14*, 1809–1814.
264. Faillard, H.; Gebhard, L. Die Wirkung der Neuraminidase auf verschiedene Formen des Rinder-Submaxillarismucins. *Hoppe-Seyler's Z. Physiol. Chem.* **1959**, *317*, 257–268.
265. Aminoff, D. Methods for the Quantitative Estimation of *N*-Acetylneuraminic Acid and Their Application to Hydrolysates of Sialomucoids. *Biochem. J.* **1961**, *81*, 384–392.
266. Warren, L. The Thiobarbituric Acid Assay of Sialic Acids. *J. Biol. Chem.* **1959**, *234*, 1971–1975.
267. Aminoff, D. The Determination of Free Sialic Acid in the Presence of the Bound Compound. *Virology* **1959**, 7, 355–357.
268. Paerels, G. B.; Schut, J. The Mechanism of the Periodate–Thiobarbituric Acid Reaction of Sialic Acids. *Biochem. J.* **1965**, *96*, 787–792.
269. Schauer, R.; Corfield, A. P. Colorimetry and Thin-Layer Chromatography of Sialic Acids. In: *Sialic Acids—Chemistry, Metabolism and Function*; Schauer, R. Ed.; Cell Biology Monographs, Vol. 10; Springer-Verlag: Vienna, Austria, 1982; pp 77–94.
270. Schoop, H. J.; Faillard, H. Ein Verfahren zum spezifischen Nachweis kleinster Glykolylmengen in Glykoproteinen. *Hoppe-Seyler's Z. Physiol. Chem.* **1967**, *348*, 1509–1517.
271. Hestrin, S. The Reaction of Acetylcholine and Other Carboxylic Acid Derivatives With Hydroxylamine, and Its Analytical Application. *J. Biol. Chem.* **1949**, *180*, 249–261.
272. Ishizuka, I.; Kloppenburg, M.; Wiegandt, H. Characterization of Gangliosides From Fish Brain. *Biochim. Biophys. Acta* **1970**, *210*, 299–305.
273. Lönngren, J.; Svensson, S. Mass Spectrometry in Structural Analysis of Natural Carbohydrates. *Adv. Carbohydr. Chem. Biochem.* **1974**, *29*, 41–106.
274. Kamerling, J. P.; Vliegenthart, J. F. G. Carbohydrates. In: *Clinical Biochemistry—Principles, Methods, Applications—Mass Spectrometry*; Lawson, A. M. Ed.; Walter de Gruyter: Berlin, Germany, 1989; pp 175–263.
275. Kamerling, J. P.; Gerwig, G. J. Strategies for the Structural Analysis of Carbohydrates. In: *Comprehensive Glycoscience—From Chemistry to Systems Biology*; Kamerling, J. P., Boons, G.-J., Lee, Y. C., Suzuki, A., Taniguchi, N., Voragen, A. G. J., Eds.; Vol. 2; Elsevier Ltd.: Oxford, UK, 2007; pp 1–68.
276. Haverkamp, J.; Schauer, R.; Wember, M.; Kamerling, J. P.; Vliegenthart, J. F. G. Synthesis of 9-*O*-Acetyl- and 4,9-Di-*O*-acetyl Derivatives of the Methyl Ester of *N*-Acetyl-β-D-neuraminic Acid Methylglycoside—Their Use as Models in Periodate Oxidation Studies. *Hoppe-Seyler's Z. Physiol. Chem.* **1975**, *356*, 1575–1583.
277. McLean, R. L.; Suttajit, M.; Beidler, J.; Winzler, R. J. *N*-Acetylneuraminic Acid Analogues—I. Preparation of the 8-Carbon and 7-Carbon Compounds. *J. Biol. Chem.* **1971**, *246*, 803–809.
278. Reuter, G.; Schauer, R.; Szeiki, C.; Kamerling, J. P.; Vliegenthart, J. F. G. A Detailed Study of the Periodate Oxidation of Sialic Acids in Glycoproteins. *Glycoconjugate J.* **1989**, *6*, 35–44.
279. Shukla, A. K.; Schauer, R.; Schade, U.; Moll, H.; Rietschel, E. T. Structural Analysis of Underivatized Sialic Acids by Combined High-Performance Liquid Chromatography–Mass Spectrometry. *J. Chromatogr.* **1985**, *337*, 231–238.
280. Reuter, G.; Schauer, R. Comparison of Electron and Chemical Ionization Mass Spectrometry of Sialic Acids. *Anal. Biochem.* **1986**, *157*, 39–46.
281. Stehling, P.; Gohlke, M.; Fitzner, R.; Reutter, W. Rapid Analysis of *O*-Acetylated Neuraminic Acids by Matrix Assisted Laser Desorption/Ionization Time-of-Flight Mass Spectrometry. *Glycoconjugate J.* **1998**, *15*, 339–344.

282. Vliegenthart, J. F. G.; Dorland, L.; van Halbeek, H.; Haverkamp, J. NMR Spectroscopy of Sialic Acids. In: *Sialic Acids—Chemistry, Metabolism and Function*; Schauer, R. Ed.; Cell Biology Monographs, Vol. 10; Springer-Verlag: Vienna, Austria, 1982; pp 127–172.
283. Friebolin, H.; Schmidt, H.; Supp, M. ^{1}H-NMR-Untersuchungen an Sialinsäure-Markierung von *N*-Acetyl-D-neuraminsäure mit Deuterium. *Tetrahedron Lett.* **1981**, *22*, 5171–5174.
284. Dorland, L.; Haverkamp, J.; Schauer, R.; Veldink, G. A.; Vliegenthart, J. F. G. The Preparation of Specifically Deuterium- or Tritium-Labeled N-Acetylneuraminic Acid and Cytidine-5′-monophospho-β-N-acetylneuraminic Acid as Precursors for Glycoconjugate Synthesis. *Biochem. Biophys. Res. Commun.* **1982**, *104*, 1114–1119.
285. Biedl, A. Die Kristallstruktur der Methoxy-neuraminsäure. *Naturwissenschaften* **1971**, *58*, 95.
286. Flippen, J. L. The Crystal Structure of β-D-*N*-Acetylneuraminic Acid Dihydrate (Sialic Acid), $C_{11}H_{19}NO_9.2H_2O$. *Acta Crystallogr. B* **1973**, *29*, 1881–1886.
287. O'Connell, A. M. The Crystal Structure of *N*-Acetylneuraminic Acid Methyl Ester Monohydrate. *Acta Crystallogr. B* **1973**, *29*, 2320–2328.
288. Kooijman, H.; Kroon-Batenburg, L. M. J.; Kroon, J.; Breg, J. N.; de Boer, J. L. Structure of α-D-*N*-Acetyl-1-*O*-methylneuraminic Acid Methyl Ester. *Acta Crystallogr. C* **1990**, *46*, 407–410.
289. Kimura, A.; Tsurumi, K. Studies on Colominic Acid—IV. Conformation and Configuration of the N-Acetyl-neuraminic Acid Moiety of Colominic Acid. *Fukushima J. Med. Sci.* **1968**, *15*, 55–60.
290. Chapman, D.; Kamat, V. B.; de Gier, J.; Penkett, S. A. Nuclear Magnetic Resonance Studies of Erythrocyte Membranes. *J. Mol. Biol.* **1968**, *31*, 101–114.
291. Jaques, L. W.; Brown, E. B.; Barrett, J. M.; Brey, W. S., Jr.; Weltner, W., Jr. Sialic Acid—A Calcium-Binding Carbohydrate. *J. Biol. Chem.* **1977**, *252*, 4533–4538.
292. Behr, J.-P.; Lehn, J.-M. The Binding of Divalent Cations by Purified Gangliosides. *FEBS Lett.* **1973**, *31*, 297–300.
293. Eschenfelder, V.; Brossmer, R.; Friebolin, H. ^{13}C-Untersuchungen von anomeren Ketosiden der N-Acetyl-D-neuraminsäure. *Tetrahedron Lett.* **1975**, 3069–3072.
294. Czarniecki, M. F.; Thornton, E. R. Carbon-13 Nuclear Magnetic Resonance Spin-Lattice Relaxation in the *N*-Acylneuraminic Acids—Probes for Internal Dynamics and Conformational Analysis. *J. Am. Chem. Soc.* **1977**, *99*, 8273–8279.
295. Sabesan, S.; Bock, K.; Lemieux, R. U. The Conformational Properties of the Gangliosides G_{M2} and G_{M1} Based on ^{1}H and ^{13}C Nuclear Magnetic Resonance Studies. *Can. J. Chem.* **1984**, *62*, 1034–1045.
296. Czarniecki, M. F.; Thornton, E. R. ^{13}C Spin Lattice Relaxation in the Neuraminic Acids—Evidence for an Unusual Intramolecular Hydrogen-Bonding Network. *J. Am. Chem. Soc.* **1976**, *98*, 1023–1025.
297. Poppe, L.; van Halbeek, H. Nuclear Magnetic Resonance of Hydroxyl and Amido Protons of Oligosaccharides in Aqueous Solution: Evidence for a Strong Intramolecular Hydrogen Bond in Sialic Acid Residues. *J. Am. Chem. Soc.* **1991**, *113*, 363–365.
298. Acquotti, D.; Poppe, L.; Dabrowski, J.; von der Lieth, C.-W.; Sonnino, S.; Tettamanti, G. Three-Dimensional Structure of the Oligosaccharide Chain of GM1 Ganglioside Revealed by a Distance-Mapping Procedure: A Rotating and Laboratory Frame Nuclear Overhauser Enhancement Investigation of Native Glycolipid in Dimethyl Sulfoxide and in Water–Dodecylphosphocholine Solutions. *J. Am. Chem. Soc.* **1990**, *112*, 7772–7778.
299. Czarniecki, M. F.; Thornton, E. R. ^{13}C NMR Chemical Shift Titration of Metal Ion-Carbohydrate Complexes—An Unexpected Dichotomy for Ca^{+2} Binding Between Anomeric Derivatives of *N*-Acetylneuraminic Acid. *Biochem. Biophys. Res. Commun.* **1977**, *74*, 553–558.

300. Daman, M. E.; Dill, K. ^{13}C-N.M.R.-Spectral Study of the Mode of Binding of Mn^{2+} and Gd^{3+} to *N*-Acetyl-α-neuraminic Acid. *Carbohydr. Res.* **1982**, *102*, 47–57.
301. Veluraja, K.; Rao, V. S. R. Theoretical Studies on the Conformation of β-D-*N*-Acetyl Neuraminic Acid (Sialic Acid). *Biochim. Biophys. Acta* **1980**, *630*, 442–446.
302. Rao, V. S. R.; Qasba, P. K.; Balaji, P. V.; Chandrasekaran, R. *Conformation of Carbohydrates.* Harwood Academic Publishers: Amsterdam, The Netherlands, 1998; pp. 70–71.
303. Veluraja, K.; Rao, V. S. R. Theoretical Studies on the Conformation of Monosialogangliosides and Disialogangliosides. *Carbohydr. Polym.* **1983**, *3*, 175–192.
304. Sawada, T.; Hashimoto, T.; Nakano, H.; Shigematsu, M.; Ishida, H.; Kiso, M. Conformational Study of α-*N*-Acetyl-D-neuraminic Acid by Density Functional Theory. *J. Carbohydr. Chem.* **2006**, *25*, 387–405.
305. Mukhopadhyay, C.; Bush, C. A. Molecular Dynamics Simulation of Oligosaccharides Containing *N*-Acetyl Neuraminic Acid. *Biopolymers* **1994**, *34*, 11–20.
306. Priyadarzini, T. R. K.; Subashini, B.; Selvin, J. F. A.; Veluraja, K. Molecular Dynamics Simulation and Quantum Mechanical Calculations on α-D-*N*-Acetylneuraminic Acid. *Carbohydr. Res* **2012**, *351*, 93–97.
307. Van Lenthe, J. H.; den Boer, D. H. W.; Havenith, R. W. A.; Schauer, R.; Siebert, H.-C. Ab Initio Calculations on Various Sialic Acids Provide Valuable Information About Sialic Acid-Specific Enzymes. *J. Mol. Struct. (Theochem)* **2004**, *677*, 29–37.
308. Klenk, E.; Langerbeins, H.; Schumann, E. Über die Verteilung der Neuraminsäure im Gehirn (Mit einer Mikromethode zur quantitativen Bestimmung der Substanz im Nervengewebe). *Hoppe Seyler's Z. Physiol. Chem.* **1941**, *270*, 185–193.
309. Svennerholm, L. Quantitative Estimation of Sialic Acids—I. A Colorimetric Method With Orcinol–Hydrochloric Acid (Bial) Reagent. *Ark. Kemi* **1957**, *10*, 577–596.
310. Svennerholm, L. Quantitative Estimation of Sialic Acids—II. A Colorimetric Resorcinol–Hydrochloric Acid Method. *Biochim. Biophys. Acta* **1957**, *24*, 604–611.
311. Svennerholm, L. Quantitative Estimation of Sialic Acids—III. An Anion-Exchange Resin Method. *Acta Chem. Scand.* **1958**, *12*, 547–554.
312. Jourdian, G. W.; Dean, L.; Roseman, S. The Sialic Acids—XI. A Periodate–Resorcinol Method for the Quantitative Estimation of Free Sialic Acids and Their Glycosides. *J. Biol. Chem* **1971**, *246*, 430–435.
313. Peters, B. P.; Aronson, N. N., Jr. Reactivity of the Sialic Acid Derivative 5-Acetamido-3,5-dideoxy-L-*arabino*-heptulosonic Acid in the Resorcinol and Thiobarbituric Acid Assays. *Carbohydr. Res.* **1976**, *47*, 345–353.
314. Kitajima, K.; Inoue, S.; Kitazume, S.; Inoue, Y. Analytical Methods for Identifying and Quantitating Deamidated Sialic Acid (2-Keto-3-deoxy-D-*glycero*-D-galactonononic Acid) and α2→8-Linked Poly(oligo)nonulosonate Residues in Glycoconjugates. *Anal. Biochem* **1992**, *205*, 244–250.
315. Wirtz-Peitz, F. *Chromogen-Bildung der Neuraminsäure und anderer 2-Keto-3-deoxy-aldonsäuren*; Dissertation, Ruhr-Universität Bochum, 1969.
316. Treibs, A.; Herrmann, E. Über die Ehrlichsche Reaktion der Pyrrole und Indole—IV. Mitteilung über Pyrrolfarbstoffe. *Hoppe-Seyler's Z. Physiol. Chem.* **1955**, *299*, 168–185.
317. Alexander, R. S.; Butler, A. R. Electrophilic Substitution in Pyrroles—Part 1. Reaction With 4-Dimethylaminobenzaldehyde (Ehrlich's Reagent) in Acid Solution. *J. Chem. Soc., Perkin Trans. 2* **1976**, 696–701.
318. Kuhn, R.; Lutz, P. Über Formyl-brenztraubensäure und den Farbstoff der Warren-Reaktion. *Biochem. Z.* **1963**, *338*, 554–560.
319. Beau, J.-M.; Schauer, R. Metabolism of 4-*O*-Methyl-*N*-acetylneuraminic Acid—A Synthetic Sialic Acid. *Eur. J. Biochem* **1980**, *106*, 531–540.
320. Charon, D.; Szabó, L. The Synthesis of 3-Deoxy-5-*O*-methyloctulosonic Acid and Its Behaviour in the Warren Reaction. *Eur. J. Biochem.* **1972**, *29*, 184–187.
321. Hess, H. H.; Rolde, E. Fluorometric Assay of Sialic Acid in Brain Gangliosides. *J. Biol. Chem.* **1964**, *239*, 3215–3220.

322. Murayama, J.-I.; Tomita, M.; Tsuji, A.; Hamada, A. Fluorimetric Assay of Sialic Acids. *Anal. Biochem.* **1976**, *73*, 535–538.
323. Matsushima, Y.; Martell, A. E. Pyridoxal Analogs—X. Zinc(II)-Chelate Catalysis of Transamination in Methanol Solution. *J. Am. Chem. Soc.* **1967**, *89*, 1331–1335.
324. Hammond, K. S.; Papermaster, D. S. Fluorometric Assay of Sialic Acid in the Picomole Range: A Modification of the Thiobarbituric Acid Assay. *Anal. Biochem.* **1976**, *74*, 292–297.
325. Nash, T. The Colorimetric Estimation of Formaldehyde by Means of the Hantzsch Reaction. *Biochem. J.* **1953**, *55*, 416–421.
326. Shukla, A. K.; Schauer, R. Determination of Unsubstituted and 9-*O*-Acetylated Sialic Acids in Erythrocyte Membranes of Different Animals Using a Fluorimetric Method. *Hoppe Seyler's Z. Physiol. Chem.* **1981**, *362*, 236–237.
327. Shukla, A. K.; Schauer, R. Study ofDe-O-acetylation ofN-Acetyl-9-O-acetylneuraminic Acid by a Fluorimetric Method. *Fresenius' Z. Anal. Chem.* **1982**, *311*, 377.
328. Brunetti, P.; Jourdian, G. W.; Roseman, S. The Sialic Acids—III. Distribution and Properties of Animal *N*-Acetylneuraminic Aldolase. *J. Biol. Chem.* **1962**, *237*, 2447–2453.
329. Brunetti, P.; Swanson, A.; Roseman, S. Enzymatic Determination of Sialic Acids: *N*-Acylneuraminic Acid $\rightleftharpoons$ *N*-Acyl-D-mannosamine + Pyruvate. *Methods Enzymol.* **1963**, *6*, 465–473.
330. Schauer, R.; Wember, M.; Wirtz-Peitz, F.; Ferreira do Amaral, C. Studies on the Substrate Specificity of Acylneuraminate Pyruvate-Lyase. *Hoppe-Seyler's Z. Physiol. Chem.* **1971**, *352*, 1073–1080.
331. Corfield, A. P.; Ferreira do Amaral, C.; Wember, M.; Schauer, R. The Metabolism of *O*-Acyl-*N*-acylneuraminic Acids—Biosynthesis of *O*-Acylated Sialic Acids in Bovine and Equine Submandibular Glands. *Eur. J. Biochem.* **1976**, *68*, 597–610.
332. Friedrich, A.; Rapoport, S. Mikromethode für die Acetylbestimmung. *Biochem. Z.* **1932**, *251*, 432–446.
333. Ludowieg, J.; Dorfman, A. A Micromethod for the Colorimetric Determination of *N*-Acetyl Groups in Acid Mucopolysaccharides. *Biochim. Biophys. Acta* **1960**, *38*, 212–218.
334. Schauer, R. Characterization of Sialic Acids. *Methods Enzymol.* **1978**, *50*, 64–89.
335. Eegriwe, E. Reaktionen und Reagenzien zum Nachweis organischer Verbindungen I. *Z. Anal. Chem.* **1932**, *89*, 121–125.
336. Takahashi, K. A Colorimetric Method for Quantitative Determination of Glycolic Acid With 2,7-Dihydroxynaphthalene. *J. Biochem.* **1972**, *71*, 563–565.
337. Sweeley, C. C.; Bentley, R.; Makita, M.; Wells, W. W. Gas–Liquid Chromatography of Trimethylsilyl Derivatives of Sugars and Related Substances. *J. Am. Chem. Soc* **1963**, *85*, 2497–2507.
338. Sweeley, C. C.; Walker, B. Determination of Carbohydrates in Glycolipides and Gangliosides by Gas Chromatography. *Anal. Chem.* **1964**, *36*, 1461–1466.
339. Craven, D. A.; Gehrke, C. W. Quantitative Determination of *N*-Acetylneuraminic Acid by Gas–Liquid Chromatography. *J. Chromatogr.* **1968**, *37*, 414–421.
340. Kamerling, J. P.; Gerwig, G. J.; Vliegenthart, J. F. G.; Clamp, J. R. Characterization by Gas–Liquid Chromatography–Mass Spectrometry and Proton-Magnetic-Resonance Spectroscopy of Pertrimethylsilyl Methyl Glycosides Obtained in the Methanolysis of Glycoproteins and Glycopeptides. *Biochem. J.* **1975**, *151*, 491–495.
341. Casals-Stenzel, J.; Buscher, H.-P.; Schauer, R. Gas–Liquid Chromatography of *N*- and *O*-Acylated Neuraminic Acids. *Anal. Biochem.* **1975**, *65*, 507–524.
342. Sugawara, Y.; Iwamori, M.; Portoukalian, J.; Nagai, Y. Determination of *N*-Acetylneuraminic Acid by Gas Chromatography–Mass Spectrometry With a Stable Isotope as Internal Standard. *Anal. Biochem.* **1983**, *132*, 147–151.

343. Zanetta, J. P.; Breckenridge, W. C.; Vincendon, G. Analysis of Monosaccharides by Gas–Liquid Chromatography of the *O*-Methyl Glycosides as Trifluoroacetate Derivatives—Application to Glycoproteins and Glycolipids. *J. Chromatogr.* **1972**, *69*, 291–304.
344. Bhattacharjee, A. K.; Jennings, H. J. Determination of the Linkages in Some Methylated, Sialic Acid-Containing, Meningococcal Polysaccharides by Mass Spectrometry. *Carbohydr. Res.* **1976**, *51*, 253–261.
345. Rauvala, H.; Kärkkäinen, J. Methylation Analysis of Neuraminic Acids by Gas Chromatography–Mass Spectrometry. *Carbohydr. Res.* **1977**, *56*, 1–9.
346. Van Halbeek, H.; Haverkamp, J.; Kamerling, J. P.; Vliegenthart, J. F. G.; Versluis, C.; Schauer, R. Sialic Acids in Permethylation Analysis: Preparation and Identification of Partially *O*-Methylated Derivatives of Methyl *N*-Acetyl-*N*-methyl-β-D-neuraminate Methyl Glycoside. *Carbohydr. Res.* **1978**, *60*, 51–62.
347. Inoue, S.; Matsumura, G. Identification of the *N*-Glycolylneuraminyl-(2→8)-*N*-glycolylneuraminyl Group in a Trout-Egg Glycoprotein by Methylation Analysis and Gas–Liquid Chromatography–Mass Spectrometry. *Carbohydr. Res.* **1979**, *74*, 361–368.
348. Bruvier, C.; Leroy, Y.; Montreuil, J.; Fournet, B.; Kamerling, J. P. Gas–Liquid Chromatography and Mass Spectrometry of Methyl Ethers of Methyl *N*-Acetyl-*N*-methyl-β-D-neuraminate Methyl Glycoside. *J. Chromatogr.* **1981**, *210*, 487–504.
349. Krantz, M. J.; Lee, Y. C. A Sensitive Autoanalytical Method for Sialic Acids. *Anal. Biochem.* **1975**, *63*, 464–469.
350. Silver, H. K. B.; Karim, K. A.; Gray, M. J.; Salinas, F. A. High-Performance Liquid Chromatography Quantitation of N-Acetylneuraminic Acid in Malignant Melanoma and Breast Carcinoma. *J. Chromatogr. B: Biomed. Sci. Appl.* **1981**, *224*, 381–388.
351. Bergh, M. L. E.; Koppen, P.; van den Eijnden, D. H. High-Pressure Liquid Chromatography of Sialic Acid-Containing Oligosaccharides. *Carbohydr. Res.* **1981**, *94*, 225–229.
352. Powell, L. D.; Hart, G. W. Quantitation of Picomole Levels of *N*-Acetyl- and *N*-Glycolylneuraminic Acids by a HPLC-Adaptation of the Thiobarbituric Acid Assay. *Anal. Biochem.* **1986**, *157*, 179–185.
353. Mopper, K.; Gindler, E. M. A New Noncorrosive Dye Reagent for Automatic Sugar Chromatography. *Anal. Biochem.* **1973**, *56*, 440–442.
354. Sinner, M.; Puls, J. Non-Corrosive Dye Reagent for Detection of Reducing Sugars in Borate Complex Ion-Exchange Chromatography. *J. Chromatogr.* **1978**, *156*, 197–204.
355. Shukla, A. K.; Scholz, N.; Reimerdes, E. H.; Schauer, R. High-Performance Liquid Chromatography of *N,O*-Acylated Sialic Acids. *Anal. Biochem.* **1982**, *123*, 78–82.
356. Shukla, A. K.; Schauer, R. Analysis of N,O-Acylated Neuraminic Acids by High-Performance Liquid Anion-Exchange Chromatography. *J. Chromatogr.* **1982**, *244*, 81–89.
357. Hara, S.; Takemori, Y.; Yamaguchi, M.; Nakamura, M.; Ohkura, Y. Fluorometric High-Performance Liquid Chromatography of *N*-Acetyl- and *N*-Glycolylneuraminic Acids and Its Application to Their Microdetermination in Human and Animal Sera, Glycoproteins, and Glycolipids. *Anal. Biochem.* **1987**, *164*, 138–145.
358. Hara, S.; Yamaguchi, M.; Takemori, Y.; Furuhata, K.; Ogura, H.; Nakamura, M. Determination of Mono-*O*-acetylated *N*-Acetylneuraminic Acids in Human and Rat Sera by Fluorometric High-Performance Liquid Chromatography. *Anal. Biochem.* **1989**, *179*, 162–166.
359. Lin, S.-L.; Inoue, S.; Inoue, Y. Acid-Base Properties of the Reaction Product of Sialic Acid With Fluorogenic Reagent, 1,2-Diamino-4,5-methylenedioxybenzene (DMB). *Carbohydr. Res.* **2000**, *329*, 447–451.
360. Anumula, K. R. Rapid Quantitative Determination of Sialic Acids in Glycoproteins by High-Performance Liquid Chromatography With a Sensitive Fluorescence Detection. *Anal. Biochem.* **1995**, *230*, 24–30.

361. Rohrer, J. S. Analyzing Sialic Acids Using High-Performance Anion-Exchange Chromatography With Pulsed Amperometric Detection. *Anal. Biochem.* **2000**, *283*, 3–9.
362. Manzi, A. E.; Diaz, S.; Varki, A. High-Pressure Liquid Chromatography of Sialic Acids on a Pellicular Resin Anion-Exchange Column With Pulsed Amperometric Detection: A Comparison With Six Other Systems. *Anal. Biochem.* **1990**, *188*, 20–32.
363. Allevi, P.; Femia, E. A.; Costa, M. L.; Cazzola, R.; Anastasia, M. Quantification of *N*-Acetyl- and *N*-Glycolylneuraminic Acids by a Stable Isotope Dilution Assay Using High-Performance Liquid Chromatography–Tandem Mass Spectrometry. *J. Chromatogr. A* **2008**, *1212*, 98–105.
364. Che, F.-Y.; Shao, X.-X.; Wang, K.-Y.; Xia, Q.-C. Characterization of Derivatization of Sialic Acid With 2-Aminoacridone and Determination of Sialic Acid Content in Glycoproteins by Capillary Electrophoresis and High Performance Liquid Chromatography–Ion Trap Mass Spectrometry. *Electrophoresis* **1999**, *20*, 2930–2937.
365. Dong, X.; Xu, X.; Han, F.; Ping, X.; Yuang, X.; Lin, B. Determination of Sialic Acids in the Serum of Cancer Patients by Capillary Electrophoresis. *Electrophoresis* **2001**, *22*, 2231–2235.
366. Taga, A.; Sugimura, M.; Suzuki, S.; Honda, S. Estimation of Sialic Acid in a Sialoglycan and a Sialoglycoprotein by Capillary Electrophoresis With In-Capillary Sialidase Digestion. *J. Chromatogr. A* **2002**, *954*, 259–266.
367. Strousopoulou, K.; Militsopoulou, M.; Stagiannis, K.; Lamari, F. N.; Karamanos, N. K. A Capillary Zone Electrophoresis Method for Determining *N*-Acetylneuraminic Acid in Glycoproteins and Blood Sera. *Biomed. Chromatogr.* **2002**, *16*, 146–150.
368. Lamari, F. N.; Karamanos, N. K. Separation Methods for Sialic Acids and Critical Evaluation of Their Biologic Relevance. *J. Chromatogr. B* **2002**, *781*, 3–19.
369. Ortner, K.; Buchberger, W. Determination of Sialic Acids Released From Glycoproteins Using Capillary Zone Electrophoresis/Electrospray Ionization Mass Spectrometry. *Electrophoresis* **2008**, *29*, 2233–2237.
370. Friebolin, H.; Brossmer, R.; Keilich, G.; Ziegler, D.; Supp, M. ^{1}H-NMR-Spektroskopischer Nachweis der *N*-Acetyl-α-D-neuraminsäure als *primäres* Spaltprodukt der Neuraminidasen. *Hoppe-Seyler's Z. Physiol. Chem.* **1980**, *361*, 697–702.
371. Friebolin, H.; Baumann, W.; Brossmer, R.; Keilich, G.; Supp, M.; Ziegler, D.; von Nicolai, H. ^{1}H-NMR Spectroscopic Evidence for the Release of *N*-Acetyl-α-D-neuraminic Acid as the First Product of Neuraminidase Action. *Biochem. Int.* **1981**, *3*, 321–326.
372. Friebolin, H.; Baumann, W.; Keilich, G.; Ziegler, D.; Brossmer, R.; von Nicolai, H. ^{1}H-NMR-Spektroskopie—Eine aussagekräftige Methode zur Bestimmung der Substratspezifität von Sialidasen. *Hoppe-Seyler's Z. Physiol. Chem.* **1981**, *362*, 1455–1463.
373. Friebolin, H.; Kunzelmann, P.; Supp, M.; Brossmer, R.; Keilich, G.; Ziegler, D. ^{1}H-NMR-Spektroskopische Untersuchungen zur Mutarotation der N-Acetyl-D-neuraminsäure—pH-Abhängigkeit der Mutarotationsgeschwindigkeit. *Tetrahedron Lett.* **1981**, *22*, 1383–1386.
374. Deijl, C. M.; Kamerling, J. P.; Vliegenthart, J. F. G. Release of Sialic Acid From Substrates by Sialidase in the Presence of $H_2[^{18}O]$. *Carbohydr. Res.* **1984**, *126*, 338–342.
375. Chong, A. K. J.; Pegg, M. S.; Taylor, N. R.; von Itzstein, M. Evidence for a Sialosyl Cation Transition-State Complex in the Reaction of Sialidase From Influenza Virus. *Eur. J. Biochem.* **1992**, *207*, 335–343.
376. Guo, X.; Sinnott, M. L. *Salmonella typhimurium* Neuraminidase Acts With Inversion of Configuration. *Biochem. J.* **1993**, *296*, 291–292.
377. Guo, X.; Laver, W. G.; Vimr, E.; Sinnott, M. L. Catalysis by Two Sialidases With the Same Protein Fold but Different Stereochemical Courses: A Mechanistic Comparison of the Enzymes From Influenza A Virus and *Salmonella typhimurium*. *J. Am. Chem. Soc.* **1994**, *116*, 5572–5578.

378. Deijl, C. M.; Vliegenthart, J. F. G. Configuration of Substrate and Products of *N*-Acetylneuraminate Pyruvate-Lyase From *Clostridium perfringens*. *Biochem. Biophys. Res. Commun.* **1983**, *111*, 668–674.
379. Baumann, W.; Freidenreich, J.; Weisshaar, G.; Brossmer, R.; Friebolin, H. Spaltung und Synthese von Sialinsäuren mit Aldolase—^{1}H-NMR-Untersuchungen zur Stereochemie, zur Kinetik und zum Mechanismus. *Biol. Chem. Hoppe-Seyler* **1989**, *370*, 141–149.
380. Klepach, T.; Carmichael, I.; Serianni, A. S. ^{13}C-Labeled *N*-Acetyl-neuraminic Acid in Aqueous Solution: Detection and Quantification of Acyclic Keto, Keto Hydrate, and Enol Forms by ^{13}C NMR Spectroscopy. *J. Am. Chem. Soc.* **2008**, *130*, 11892–11900.
381. McManus, J. F. A. Histological Demonstration of Mucin After Periodic Acid. *Nature* **1946**, *158*, 202.
382. Leblond, C. P. Distribution of Periodic Acid-Reactive Carbohydrates in the Adult Rat. *Am. J. Anat.* **1950**, *86*, 1–49.
383. Leblond, C. P.; Clermont, Y. Spermiogenesis of Rat, Mouse, Hamster and Guinea Pig as Revealed by the "Periodic Acid-Fuchsin Sulfurous Acid" Technique. *Am. J. Anat.* **1952**, *90*, 167–215.
384. Hotchkiss, R. D. A Microchemical Reaction Resulting in the Staining of Polysaccharide Structures in Fixed Tissue Preparations. *Arch. Biochem.* **1948**, *16*, 131–141.
385. Leblond, C. P.; Glegg, R. E.; Eidinger, D. Presence of Carbohydrates With Free 1,2-Glycol Groups in Sites Stained by the Periodic Acid-Schiff Technique. *J. Histochem. Cytochem.* **1957**, *5*, 445–458.
386. Gasic, G.; Gasic, T. Removal of PAS Positive Surface Sugars in Tumor Cells by Glycosidases. *Exp. Biol. Med.* **1963**, *114*, 660–663.
387. Rambourg, A.; Neutra, M.; Leblond, C. P. Presence of a 'Cell Coat' Rich in Carbohydrate at the Surface of Cells in the Rat. *Anat. Rec.* **1966**, *154*, 41–71.
388. Winzler, R. J. Glycoproteins of Plasma Membranes—Chemistry and Function. In: *Glycoproteins—Their Composition, Structure and Function*; Gottschalk, A. Ed.; Vol. 5, Part B, 2nd ed.; B.B.A. Library, Elsevier Publishing Company: Amsterdam, The Netherlands, 1972; pp 1268–1293.
389. Rambourg, A.; Hernandez, W.; Leblond, C. P. Detection of Complex Carbohydrates in the Golgi Apparatus of Rat Cells. *J. Cell Biol.* **1969**, *40*, 395–414.
390. Culling, C. F. A.; Reid, P. E.; Dunn, W. L. The Effect of Saponification Upon Certain Histochemical Reactions of the Epithelial Mucins of the Gastrointestinal Tract. *J. Histochem. Cytochem.* **1971**, *19*, 654–662.
391. Culling, C. F. A.; Reid, P. E.; Clay, M. G.; Dunn, W. L. The Histochemical Demonstration of O-Acylated Sialic Acid in Gastrointestinal Mucins—Their Association With the Potassium Hydroxide–Periodic Acid–Schiff Effect. *J. Histochem. Cytochem.* **1974**, *22*, 826–831.
392. Culling, C. F. A.; Reid, P. E.; Dunn, W. L. A New Histochemical Method for the Identification and Visualization of Both Side Chain Acylated and Nonacylated Sialic Acids. *J. Histochem. Cytochem.* **1976**, *24*, 1225–1230.
393. Culling, C. F. A.; Reid, P. E. Histochemistry of Sialic Acids. In: *Sialic Acids—Chemistry, Metabolism and Function*, Schauer, R., Ed.; Cell Biology Monographs, Vol. 10; Springer-Verlag: Vienna, Austria. 1982; pp 173–193.
394. Makovitzky, J.; Richter, S. The Relevance of the Aldehyde Bisulfite Toluidine Blue Reaction and Its Variants in the Submicroscopic Carbohydrate Research. *Acta Histochem.* **2009**, *111*, 273–291.
395. Ueda, T.; Fujimori, O.; Yamada, K. A New Histochemical Method for Detection of Sialic Acids Using a Physical Development Procedure. *J. Histochem. Cytochem.* **1995**, *43*, 1045–1051.

396. Fukui, K.; Yasui, T.; Gomi, H.; Sugiya, H.; Fujimori, O.; Meyer, W.; Tsukise, A. Histochemical Localization of Sialic Acids and Antimicrobial Substances in Eccrine Glands of Porcine Snout Skin. *Eur. J. Histochem.* **2012**, *56,* e6, (28–34).
397. Zhang, Y.; Yuan, J.; Song, J.; Wang, Z.; Huang, L. An Efficient Method for Selectively Imaging and Quantifying In Situ the Expression of Sialylated Glycoproteins on Living Cells. *Glycobiology* **2013**, *23*, 643–653.
398. Goldstein, I. J.; Winter, H. C.; Poretz, R. D. Plant Lectins: Tools for the Study of Complex Carbohydrates. In: *Glycoproteins II*; Montreuil, J., Vliegenthart, J. F. G., Schachter, H., Eds.; New Comprehensive Biochemistry, Vol. 29b; Elsevier Science B.V: Amsterdam, The Netherlands, 1997; pp 403–474.
399. Sedwick, C. Jürgen Roth: Immunogold Master. *J. Cell Biol.* **2013**, *200*, 686–687.
400. Taatjes, D. J.; Roth, J.; Peumans, W.; Goldstein, I. J. Elderberry Bark Lectin–Gold Techniques for the Detection of Neu5Ac(α2,6)Gal/GalNAc Sequences: Applications and Limitations. *Histochem. J.* **1988**, *20*, 478–490.
401. Sata, T.; Lackie, P. M.; Taatjes, D. J.; Peumans, W.; Roth, J. Detection of the Neu5Ac(α2,3)Gal(β1,4)GlcNAc Sequence With the Leukoagglutinin From *Maackia amurensis*: Light and Electron Microscopic Demonstration of Differential Tissue Expression of Terminal Sialic Acid in α2,3 and α2,6 Linkage. *J. Histochem. Cytochem.* **1989**, *37*, 1577–1588.
402. Sata, T.; Roth, J.; Zuber, C.; Stamm, B.; Heitz, P. U. Expression of α2,6-Linked Sialic Acid Residues in Neoplastic but Not in Normal Human Colonic Mucosa—A Lectin–Gold Cytochemical Study With *Sambucus nigra* and *Maackia amurensis* Lectins. *Am. J. Pathol.* **1991**, *139*, 1435–1448.
403. Skutelsky, E.; Roth, J. Cationic Colloidal Gold—A New Probe for the Detection of Anionic Cell Surface Sites by Electron Microscopy. *J. Histochem. Cytochem.* **1986**, *34*, 693–696.
404. Harms, G.; Reuter, G.; Corfield, A. P.; Schauer, R. Binding Specificity of Influenza C-Virus to Variably *O*-Acetylated Glycoconjugates and Its Use for Histochemical Detection of *N*-Acetyl-9-*O*-acetylneuraminic Acid in Mammalian Tissues. *Glycoconjugate J.* **1996**, *13*, 621–630.
405. Klein, A.; Krishna, M.; Varki, N. M.; Varki, A. 9-*O*-Acetylated Sialic Acids Have Widespread but Selective Expression: Analysis Using a Chimeric Dual-Function Probe Derived From Influenza C Hemagglutinin-Esterase. *Proc. Natl. Acad. Sci. U. S. A.* **1994**, *91*, 7782–7786.
406. Langereis, M. A.; Bakkers, M. J. G.; Deng, L.; Padler-Karavani, V.; Vervoort, S. J.; Hulswit, R. J. G.; van Vliet, A. L. W.; Gerwig, G. J.; de Poot, S. A. H.; Boot, W.; van Ederen, A. M.; Heesters, B. A.; van der Loos, C. M.; van Kuppeveld, F. J. M.; Yu, H.; Huizinga, E. G.; Chen, X.; Varki, A.; Kamerling, J. P.; de Groot, R. J. Complexity and Diversity of the Mammalian Sialome Revealed by Nidovirus Virolectins. *Cell Rep.* **2015**, *11*, 1966–1978.
407. Wasik, B. R.; Barnard, K. N.; Ossiboff, R. J.; Khedri, Z.; Feng, K. H.; Yu, H.; Chen, X.; Perez, D. R.; Varki, A.; Parrish, C. R. Distribution of *O*-Acetylated Sialic Acids Among Target Host Tissues for Influenza Virus. *mSphere* **2017**, *2*, e00379-16 (1–17).
408. Roth, J.; Zuber, C.; Wagner, P.; Taatjes, D. J.; Weisgerber, C.; Heitz, P. U.; Goridis, C.; Bitter-Suermann, D. Reexpression of Poly(sialic Acid) Units of the Neural Cell Adhesion Molecule in Wilms Tumor. *Proc. Natl. Acad. Sci. U. S. A.* **1988**, *85*, 2999–3003.
409. Roth, J.; Zuber, C. Immunoelectron Microscopic Investigation of Surface Coat of Wilms Tumor Cells—Dense Lamina Is Composed of Highly Sialylated Neural Cell Adhesion Molecule. *Lab. Invest.* **1990**, *62*, 55–60.
410. Zuber, C.; Roth, J. The Relationship of Polysialic Acid and the Neural Cell Adhesion Molecule N-CAM in Wilms Tumor and Their Subcellular Distributions. *Eur. J. Cell Biol.* **1990**, *51*, 313–321.

411. Cazet, A.; Groux-Degroote, S.; Teylaert, B.; Kwon, K.-M.; Lehoux, S.; Slomianny, C.; Kim, C.-H.; Le Bourhis, X.; Delannoy, P. G_{D3} Synthase Overexpression Enhances Proliferation and Migration of MDA-MB-231 Breast Cancer Cells. *Biol. Chem.* **2009**, *390*, 601–609.
412. Gross, H. J.; Brossmer, R. Enzymatic Introduction of a Fluorescent Sialic Acid Into Oligosaccharide Chains of Glycoproteins. *Eur. J. Biochem.* **1988**, *177*, 583–589.
413. Goon, S.; Bertozzi, C. R. Metabolic Substrate Engineering as a Tool for Glycobiology. *J. Carbohydr. Chem.* **2002**, *21*, 943–977.
414. Gross, H. J.; Brossmer, R. Characterization of Human Plasma Sialyltransferase Using a Novel Fluorometric Assay. *Clin. Chim. Acta* **1991**, *197*, 237–247.
415. Chammas, R.; McCaffery, J. M.; Klein, A.; Ito, Y.; Saucan, L.; Palade, G.; Farquhar, M. G.; Varki, A. Uptake and Incorporation of an Epitope-Tagged Sialic Acid Donor Into Intact Rat Liver Golgi Compartments. Functional Localization of Sialyltransferase Overlaps With β-Galactosyltransferase but Not With Sialic Acid *O*-Acetyltransferase. *Mol. Biol. Cell* **1996**, 7, 1691–1707.
416. Saxon, E.; Bertozzi, C. R. Cell Surface Engineering by a Modified Staudinger Reaction. *Science* **2000**, *287*, 2007–2010.
417. Dehnert, K. W.; Baskin, J. M.; Laughlin, S. T.; Beahm, B. J.; Naidu, N. N.; Amacher, S. L.; Bertozzi, C. R. Imaging the Sialome During Zebrafish Development With Copper-Free Click Chemistry. *ChemBioChem* **2012**, *13*, 353–357.
418. Lantos, A. B.; Carlevaro, G.; Araoz, B.; Ruiz Diaz, P.; de los Milagros Camara, M.; Buscaglia, C. A.; Bossi, M.; Yu, H.; Chen, X.; Bertozzi, C. R.; Mucci, J.; Campetella, O. Sialic Acid Glycobiology Unveils *Trypanosoma cruzi* Trypomastigote Membrane Physiology. *PLoS Pathog.* **2016**, *12*(1–29), e1005559.
419. Popenoe, E. A. The Linkage of Neuraminic Acid in Orosomucoid. *Biochim. Biophys. Acta* **1959**, *32*, 584–585.
420. Endo, Y.; Yamashita, K.; Tachibana, Y.; Tojo, S.; Kobata, A. Structures of the Asparagine-Linked Sugar Chains of Human Chorionic Gonadotropin. *J. Biochem.* **1979**, *85*, 669–679.
421. Fournet, B.; Fiat, A.-M.; Alais, C.; Jollès, P. Cow κ-Casein: Structure of the Carbohydrate Portion. *Biochim. Biophys. Acta* **1979**, *576*, 339–346.
422. Savage, A. V.; Koppen, P. L.; Schiphorst, W. E. C. M.; Trippelvitz, L. A. W.; van Halbeek, H.; Vliegenthart, J. F. G.; van den Eijnden, D. H. Porcine Submaxillary Mucin Contains α2→3- and α2→6-Linked *N*-Acetyl- and *N*-Glycolyl-neuraminic Acid. *Eur. J. Biochem.* **1986**, *160*, 123–129.
423. Damm, J. B. L.; Bergwerff, A. A.; Hård, K.; Kamerling, J. P.; Vliegenthart, J. F. G. Sialic Acid Patterns in *N*-Linked Carbohydrate Chains—Structural Analysis of the *N*-Acetyl-/*N*-Glycolyl-neuraminic-Acid-Containing *N*-Linked Carbohydrate Chains of Bovine Fibrinogen. *Recl. Trav. Chim. Pays-Bas* **1989**, *108*, 351–359.
424. Jamieson, G. A.; Jett, M.; DeBernardo, S. L. The Carbohydrate Sequence of the Glycopeptide Chains of Human Transferrin. *J. Biol. Chem.* **1971**, *246*, 3686–3693.
425. Breg, J.; van Halbeek, H.; Vliegenthart, J. F. G.; Lamblin, G.; Houvenaghel, M.-C.; Roussel, P. Structure of Sialyl-Oligosaccharides Isolated From Bronchial Mucus Glycoproteins of Patients (Blood Group O) Suffering From Cystic Fibrosis. *Eur. J. Biochem.* **1987**, *168*, 57–68.
426. Inoue, S.; Iwasaki, M.; Matsumura, G. Novel Carbohydrate Structures in Trout Egg Glycoprotein: Occurrence of a Neuraminidase-Resistant *N*-Glycolylneuraminosyl-(2→3)-N-acetylgalactosamine Linkage. *Biochem. Biophys. Res. Commun.* **1981**, *102*, 1295–1301.
427. Iwasaki, M.; Inoue, S. Structures of the Carbohydrate Units of Polysialoglycoproteins Isolated From the Eggs of Four Species of Salmonid Fishes. *Glycoconjugate J.* **1985**, *2*, 209–228.

428. Pfeiffer, G.; Dabrowski, U.; Dabrowski, J.; Stirm, S.; Strube, K.-H.; Geyer, R. Carbohydrate Structure of a Thrombin-Like Serine Protease From *Agkistrodon rhodostoma*—Structure Elucidation of Oligosaccharides by Methylation Analysis, Liquid Secondary-Ion Mass Spectrometry and Proton Magnetic Resonance. *Eur. J. Biochem.* **1992**, *205*, 961–978.
429. Gottschalk, A.; Graham, E. R. B. 6-α-D-Sialyl-*N*-acetylgalactosamine: The Neuraminidase-Susceptible Prosthetic Group of Bovine Salivary Mucoprotein. *Biochim. Biophys. Acta* **1959**, *34*, 380–391.
430. Carlson, D. M.; Blackwell, C. Structures and Immunochemical Properties of Oligosaccharides Isolated From Pig Submaxillary Mucins. *J. Biol. Chem.* **1968**, *243*, 616–626.
431. Van Halbeek, H.; Dorland, L.; Haverkamp, J.; Veldink, G. A.; Vliegenthart, J. F. G.; Fournet, B.; Ricart, G.; Montreuil, J.; Gathmann, W. D.; Aminoff, D. Structure Determination of Oligosaccharides Isolated From A^+, H^+ and A^-H^- Hog-Submaxillary-Gland Mucin Glycoproteins, by 360-MHz ^{1}H-NMR Spectroscopy, Permethylation Analysis and Mass Spectrometry. *Eur. J. Biochem.* **1981**, *118*, 487–495.
432. Spik, G.; Coddeville, B.; Montreuil, J. Comparative Study of the Primary Structures of Sero-, Lacto- and Ovotransferrin Glycans From Different Species. *Biochimie* **1988**, *70*, 1459–1469.
433. Mizuochi, T.; Yamashita, K.; Fujikawa, K.; Kisiel, W.; Kobata, A. The Carbohydrate of Bovine Prothrombin—Occurrence of Galβ1→3GlcNAc Grouping in Asparagine-Linked Sugar Chains. *J. Biol. Chem.* **1979**, *254*, 6419–6425.
434. Finne, J.; Krusius, T.; Rauvala, H. Occurrence of Disialosyl Groups in Glycoproteins. *Biochem. Biophys. Res. Commun.* **1977**, *74*, 405–410.
435. Iwasaki, M.; Inoue, S.; KItajima, K.; Nomoto, H.; Inoue, Y. Novel Oligosaccharide Chains on Polysialoglycoproteins Isolated From Rainbow Trout Eggs—A Unique Carbohydrate Sequence With a Sialidase-Resistant Sialyl Group, DGalNAcβ1→4(NeuGc2→3)DGalNAc. *Biochemistry* **1984**, *23*, 305–310.
436. Finne, J. Polysialic Acid—A Glycoprotein Carbohydrate Involved in Neural Adhesion and Bacterial Meningitis. *Trends Biochem. Sci.* **1985**, *10*, 129–132.
437. Fukuda, M. N.; Dell, A.; Oates, J. E.; Fukuda, M. Embryonal Lactosaminoglycan—The Structure of Branched Lactosaminoglycans With Novel Disialosyl (Sialyl α2→9 Sialyl) Terminals Isolated From PA1 Human Embryonal Carcinoma Cells. *J. Biol. Chem.* **1985**, *260*, 6623–6631.
438. Inoue, S.; Poongodi, G. L.; Suresh, N.; Jennings, H. J.; Inoue, Y. Discovery of an α2,9-PolyNeu5Ac Glycoprotein in C-1300 Murine Neuroblastoma (Clone NB41A3). *J. Biol. Chem.* **2003**, *278*, 8541–8546.
439. Eylar, E. H.; Jeanloz, R. W. Periodate Oxidation of the α_1 Acid Glycoprotein of Human Plasma. *J. Biol. Chem.* **1962**, *237*, 1021–1025.
440. Wrann, M. Methylation Analysis of the Carbohydrate Portion of Fibronectin Isolated From Human Plasma. *Biochem. Biophys. Res. Commun.* **1978**, *84*, 269–274.
441. Takasaki, S.; Yamashita, K.; Suzuki, K.; Iwanaga, S.; Kobata, A. The Sugar Chains of Cold-Insoluble Globulin—A Protein Related to Fibronectin. *J. Biol. Chem.* **1979**, *254*, 8548–8553.
442. Slomiany, A.; Slomiany, B. L. Structures of the Acidic Oligosaccharides Isolated From Rat Sublingual Glycoprotein. *J. Biol. Chem.* **1978**, *253*, 7301–7306.
443. Van der Meer, A.; Kamerling, J. P.; Vliegenthart, J. F. G.; Schmid, K.; Schauer, R. Fucopyranosyl-(1→4)-*N*-glycolylneuraminic Acid, a Constituent of Glycoproteins of the Cuvierian Tubules of the Sea Cucumber *Holothuria forskali* Della Chiaje. *Biochim. Biophys. Acta* **1983**, *757*, 371–376.
444. Inoue, S.; Iwasaki, M.; Ishii, K.; Kitajima, K.; Inoue, Y. Isolation and Structures of Glycoprotein-Derived Free Sialooligosaccharides From the Unfertilized Eggs of

Tribolodon hakonensis, a Dace—Intracellular Accumulation of a Novel Class of Biantennary Disialooligosaccharides. *J. Biol. Chem.* **1989**, *264*, 18520–18526.
445. Van Kuik, J. A.; Hård, K.; Vliegenthart, J. F. G. A ^{1}H NMR Database Computer Program for the Analysis of the Primary Structure of Complex Carbohydrates. *Carbohydr. Res.* **1992**, *235*, 53–68.
446. Damm, J. B. L.; Voshol, H.; Hård, K.; Kamerling, J. P.; Vliegenthart, J. F. G. Analysis of *N*-Acetyl-4-*O*-acetylneuraminic-Acid-Containing N-Linked Carbohydrate Chains Released by Peptide-N^4-(*N*-acetyl-β-glucosaminyl)asparagine Amidase F—Application to the Structure Determination of the Carbohydrate Chains of Equine Fibrinogen. *Eur. J. Biochem.* **1989**, *180*, 101–110.
447. Hanaoka, K.; Pritchett, T. J.; Takasaki, S.; Kochibe, N.; Sabesan, S.; Paulson, J. C.; Kobata, A. 4-*O*-Acetyl-*N*-acetylneuraminic Acid in the *N*-Linked Carbohydrate Structures of Equine and Guinea Pig α_2-Macroglobulins, Potent Inhibitors of Influenza Virus Infection. *J. Biol. Chem.* **1989**, *264*, 9842–9849.
448. Coddeville, B.; Stratil, A.; Wieruszeski, J.-M.; Strecker, G.; Montreuil, J.; Spik, G. Primary Structure of Horse Serotransferrin Glycans—Demonstration That Heterogeneity Is Related to the Number of Glycans and to the Presence of *N*-Acetylneuraminic Acid and *N*-Acetyl-4-*O*-acetylneuraminic Acid. *Eur. J. Biochem.* **1989**, *186*, 583–590.
449. Damm, J. B. L.; Hård, K.; Kamerling, J. P.; van Dedem, G. W. K.; Vliegenthart, J. F. G. Structure Determination of the Major N- and O-Linked Carbohydrate Chains of the β Subunit From Equine Chorionic Gonadotropin. *Eur. J. Biochem.* **1990**, *189*, 175–183.
450. Bergwerff, A. A.; van Oostrum, J.; Asselbergs, F. A. M.; Bürgi, R.; Hokke, C. H.; Kamerling, J. P.; Vliegenthart, J. F. G. Primary Structure of N-Linked Carbohydrate Chains of a Human Chimeric Plasminogen Activator K_2tu-PA Expressed in Chinese Hamster Ovary Cells. *Eur. J. Biochem.* **1993**, *212*, 639–656.
451. Hokke, C. H.; Bergwerff, A. A.; van Dedem, G. W. K.; Kamerling, J. P.; Vliegenthart, J. F. G. Structural Analysis of the Sialylated N- and O-Linked Carbohydrate Chains of Recombinant Human Erythropoietin Expressed in Chinese Hamster Ovary Cells—Sialylation Patterns and Branch Location of Dimeric *N*-Acetyllactosamine Units. *Eur. J. Biochem.* **1995**, *228*, 981–1008.
452. Ylönen, A.; Kalkkinen, N.; Saarinen, J.; Bøgwald, J.; Helin, J. Glycosylation Analysis of Two Cysteine Proteinase Inhibitors From Atlantic Salmon Skin: Di-O-acetylated Sialic Acids Are the Major Sialic Acid Species on N-Glycans. *Glycobiology* **2001**, *11*, 523–531.
453. Llop, E.; Gutiérrez Gallego, R.; Belalcazar, V.; Gerwig, G. J.; Kamerling, J. P.; Segura, J.; Pascual, J. A. Evaluation of Protein N-Glycosylation in 2-DE: Erythropoietin as a Study Case. *Proteomics* **2007**, 7, 4278–4291.
454. Yang, Y.; Liu, F.; Franc, V.; Halim, L. A.; Schellekens, H.; Heck, A. J. R. Hybrid Mass Spectrometry Approaches in Glycoprotein Analysis and Their Usage in Scoring Biosimilarity. *Nat. Commun.* **2016**, *13397*(1 − 10), 7.
455. Sato, C.; Kitajima, K.; Tazawa, I.; Inoue, Y.; Inoue, S.; Troy, F. A., II Structural Diversity in the α2 → 8-Linked Polysialic Acid Chains in Salmonid Fish Egg Glycoproteins—Occurrence of Poly(Neu5Ac), Poly(Neu5Gc), Poly(Neu5Ac, Neu5Gc), Poly(KDN), and Their Partially Acetylated Forms. *J. Biol. Chem.* **1993**, *268*, 23675–23684.
456. Miyata, S.; Sato, C.; Kitamura, S.; Toriyama, M.; Kitajima, K. A Major Flagellum Sialoglycoprotein in Sea Urchin Sperm Contains a Novel Polysialic Acid, an α2,9-Linked Poly-*N*-acetylneuraminic Acid Chain, Capped by an 8-*O*-Sulfated Sialic Acid Residue. *Glycobiology* **2004**, *14*, 827–840.
457. Kitazume, S.; Kitajima, K.; Inoue, S.; Troy, F. A., II.; Cho, J.-W.; Lennarz, W. J.; Inoue, Y. Identification of Polysialic Acid-Containing Glycoprotein in the Jelly Coat of Sea Urchin Eggs—Occurrence of a Novel Type of Polysialic Acid Structure. *J. Biol. Chem.* **1994**, *269*, 22712–22718.

458. Kanamori, A.; Inoue, S.; Iwasaki, M.; Kitajima, K.; Kawai, G.; Yokoyama, S.; Inoue, Y. Deaminated Neuraminic Acid-Rich Glycoprotein of Rainbow Trout Egg Vitelline Envelope—Occurrence of a Novel α-2,8-Linked Oligo(Deaminated Neuraminic Acid) Structure in *O*-Linked Glycan Chains. *J. Biol. Chem.* **1990**, *265*, 21811–21819.
459. Tezuka, T.; Taguchi, T.; Kanamori, A.; Muto, Y.; Kitajima, K.; Inoue, Y.; Inoue, S. Identification and Structural Determination of the KDN-Containing N-Linked Glycan Chains Consisting of Bi- and Triantennary Complex-Type Units of KDN-Glycoprotein Previously Isolated From Rainbow Trout Vitelline Envelopes. *Biochemistry* **1994**, *33*, 6495–6502.
460. Iwasaki, M.; Inoue, S.; Nadano, D.; Inoue, Y. Isolation and Structure of a Novel Deaminated Neuraminic Acid Containing Oligosaccharide Chain Present in Rainbow Trout Egg Polysialoglycoprotein. *Biochemistry* **1987**, *26*, 1452–1457.
461. Terada, T.; Kitazume, S.; Kitajima, K.; Inoue, S.; Ito, F.; Troy, F. A.; Inoue, Y. Synthesis of CMP-Deaminoneuraminic Acid (CMP-KDN) Using the CTP:CMP-3-Deoxynonulosonate Cytidyltransferase From Rainbow Trout Testis—Identification and Characterization of a CMP-KDN Synthetase. *J. Biol. Chem.* **1993**, *268*, 2640–2648.
462. Maes, E.; Wieruszeski, J.-M.; Plancke, Y.; Strecker, G. Structure of Three Kdn-Containing Oligosaccharide-Alditols Released From Oviducal Secretions of *Ambystoma tigrinum*: Characterization of the Carbohydrate Sequence Fuc(α1-5)[Fuc(α1-4)]Kdn(α2-3/6). *FEBS Lett.* **1995**, *358*, 205–210.
463. Strecker, G.; Wieruszeski, J.-M.; Michalski, J.-C.; Alonso, C.; Leroy, Y.; Boilly, B.; Montreuil, J. Primary Structure of Neutral and Acidic Oligosaccharide-Alditols Derived From the Jelly Coat of the Mexican Axolotl—Occurrence of Oligosaccharides With Fucosyl(α1-3)fucosyl(α1-4)-3-deoxy-D-*glycero*-D-*galacto*-nonulosonic Acid and Galactosyl(α1-4)[fucosyl(α1-2)]galactosyl(β1-4)-*N*-acetylglucosamine Sequences. *Eur. J. Biochem.* **1992**, *207*, 995–1002.
464. Kimura, M.; Hama, Y.; Sumi, T.; Asakawa, M.; Rao, B. N. N.; Horne, A. P.; Li, S.-C.; Li, Y.-T.; Nakagawa, H. Characterization of a Deaminated Neuraminic Acid-Containing Glycoprotein From the Skin Mucus of the Loach, *Misgurnus anguillicaudatus*. *J. Biol. Chem.* **1994**, *269*, 32138–32143.
465. Vliegenthart, J. F. G.; Dorland, L.; van Halbeek, H. High-Resolution, ^{1}H-Nuclear Magnetic Resonance Spectroscopy as a Tool in the Structural Analysis of Carbohydrates Related to Glycoproteins. *Adv. Carbohydr. Chem. Biochem.* **1983**, *41*, 209–374.
466. Kamerling, J. P.; Vliegenthart, J. F. G. High-Resolution ^{1}H-Nuclear Magnetic Resonance Spectroscopy of Oligosaccharide-Alditols Released From Mucin-Type *O*-Glycoproteins. *Biol. Magn. Reson.* **1992**, *10*, 1–194.
467. Strecker, G. Amphibian Glycoproteins. In: *Glycoproteins II*; Montreuil, J., Vliegenthart, J. F. G., Schachter, H., Eds.; New Comprehensive Biochemistry, Vol. 29b; Elsevier Science B.V: Amsterdam, The Netherlands, 1997; pp 163–172.
468. Vliegenthart, J. F. G. ^{1}H NMR Structural-Reporter-Group Concepts in Carbohydrate Analysis. In: *Comprehensive Glycoscience—From Chemistry to Systems Biology*; Kamerling, J. P., Kamerling, J. P., Boons, G.-J., Lee, Y. C., Suzuki, A., Taniguchi, N., Voragen, A. G. J., Eds.; Vol. 2; Elsevier Ltd.: Oxford, UK, 2007; pp 133–191.
469. Stahl, N.; Baldwin, M. A.; Hecker, R.; Pan, K.-M.; Burlingame, A. L.; Prusiner, S. B. Glycosylinositol Phospholipid Anchors of the Scrapic and Cellular Prion Proteins Contain Sialic Acid. *Biochemistry* **1992**, *31*, 5043–5053.
470. Huckerby, T. N.; Brown, G. M.; Nieduszynski, I. A. ^{13}C-NMR Spectroscopy of Keratan Sulphates—Assignments for Five Sialylated Pentasaccharides Derived From the Non-reducing Chain Termini of Bovine Articular Cartilage Keratan Sulphate by Keratanase II Digestion. *Eur. J. Biochem.* **1998**, *251*, 991–997.

471. Klenk, E.; Heuer, K. Über die Ganglioside der Hundeerythrozyten. *Dtsch. Z. Verdau. Stoffwechselkr.* **1960**, *20*, 180–183.
472. Kuhn, R.; Wiegandt, H. Die Konstitution der Ganglio-*N*-tetraose und des Gangliosids G_I. *Chem. Ber.* **1963**, *96*, 866–880.
473. Kuhn, R.; Wiegandt, H. Die Konstitution der Ganglioside G_{II}, G_{III} und G_{IV}. *Z. Naturforsch. B* **1963**, *18*, 541–543.
474. Wiegandt, H.; Baschang, G. Die Gewinnung des Zuckeranteiles der Glykosphingolipide durch Ozonolyse und Fragmentierung. *Z. Naturforschg. B* **1965**, *20*, 164–166.
475. Wiegandt, H. Gangliosides of Extraneural Organs. *Hoppe-Seyler's Z. Physiol. Chem.* **1973**, *354*, 1049–1056.
476. Watanabe, K.; Powell, M. E.; Hakomori, S.-I. Isolation and Characterization of Gangliosides With a New Sialosyl Linkage and Core Structures—II. Gangliosides of Human Erythrocyte Membranes. *J. Biol. Chem.* **1979**, *254*, 8223–8229.
477. Watanabe, K.; Hakomori, S.-I. Gangliosides of Human Erythrocytes—A Novel Ganglioside With a Unique *N*-Acetylneuraminosyl-(2→3)-*N*-acetylgalactosamine Structure. *Biochemistry* **1979**, *18*, 5502–5504.
478. Taki, T.; Hirabayashi, Y.; Ishikawa, H.; Ando, S.; Kon, K.; Tanaka, Y.; Matsumoto, M. A Ganglioside of Rat Ascites Hepatoma AH 7974F Cells—Occurrence of a Novel Disialoganglioside (G_{D1a}) With a Unique *N*-Acetylneuraminosyl(α2-6)-*N*-acetylgalactosamine Structure. *J. Biol. Chem.* **1986**, *261*, 3075–3078.
479. Prieto, P. A.; Smith, D. F. A New Ganglioside in Human Meconium Detected by Antiserum Against the Human Milk Sialyloligosaccharide, LS-Tetrasaccharide b. *Arch. Biochem. Biophys.* **1985**, *241*, 281–289.
480. Fukushi, Y.; Nudelman, E.; Levery, S. B.; Higuchi, T.; Hakomori, S.-I. A Novel Disialoganglioside ($IV^3NeuAcIII^6NeuAcLc_4$) of Human Adenocarcinoma and the Monoclonal Antibody (FH9) Defining This Disialosyl Structure. *Biochemistry* **1986**, *25*, 2859–2866.
481. Handa, N.; Handa, S. The Chemistry of Lipids of Posthemolytic Residue or Stroma of Erythrocytes. XIV. Chemical Structure of Glycolipid of Cat Erythrocyte Stroma. *Jpn. J. Exp. Med.* **1965**, *35*, 331–341.
482. Furukawa, K.; Chait, B. T.; Lloyd, K. O. Identification of *N*-Glycolylneuraminic Acid-Containing Gangliosides of Cat and Sheep Erythrocytes—^{252}Cf Fission Fragment Ionization Mass Spectrometry in the Analysis of Glycosphingolipids. *J. Biol. Chem.* **1988**, *263*, 14939–14947.
483. Kochetkov, N. K.; Smirnova, G. P.; Glukhoded, I. S. Gangliosides With Sialic Acid Bound to *N*-Acetylgalactosamine From Hepatopancreas of the Starfish, *Evasterias retifera* and *Asterias amurensis*. *Biochim. Biophys. Acta* **1982**, *712*, 650–658.
484. Kochetkov, N. K.; Zhukova, I. G.; Smirnova, G. P.; Glukhoded, I. S. Isolation and Characterization of a Sialoglycolipid From the Sea Urchin *Strongylocentrotus intermedius*. *Biochim. Biophys. Acta* **1973**, *326*, 74–83.
485. Hoshi, M.; Nagai, Y. Novel Sialosphingolipids From Spermatozoa of the Sea Urchin *Anthocidaris crassispina*. *Biochim. Biophys. Acta* **1975**, *388*, 152–162.
486. Kochetkov, N. K.; Smirnova, G. P.; Glukhoded, I. S. The Structure of Sialoglycolipids From Gonads of the Sea Urchin *Strongylocentrotus nudus*. *Bioorg. Khim.* **1978**, *4*, 1093–1099.
487. Hidari, K. I.-P. J.; Irie, F.; Suzuki, M.; Kon, K.; Ando, S.; Hirabayashi, Y. A Novel Ganglioside With a Free Amino Group in Bovine Brain. *Biochem. J.* **1993**, *295*, 259–263.
488. Sonnenburg, J. L.; van Halbeek, H.; Varki, A. Characterization of the Acid Stability of Glycosidically Linked Neuraminic Acid—Use in Detecting De-*N*-acetyl-gangliosides in Human Melanoma. *J. Biol. Chem.* **2002**, *277*, 17502–17510.

489. Kuhn, R.; Müldner, H. Über Glyko-lipo-sialo-proteide des Gehirns. *Naturwissenschaften* **1964**, *51*, 635–636.
490. Wiegandt, H. Gangliosides. *Ergeb. Physiol. Biol. Chem. Exp. Pharmakol.* **1966**, *57*, 190–222.
491. McCluer, R. H.; Evans, J. E. Ganglioside Inner Esters. *Adv. Exp. Med. Biol.* **1972**, *19*, 95–102.
492. Gross, S. K.; Williams, M. A.; McCluer, R. H. Alkali-Labile, Sodium Borohydride-Reducible Ganglioside Sialic Acid Residues in Brain. *J. Neurochem.* **1980**, *34*, 1351–1361.
493. Terabayashi, T.; Ogawa, T.; Kawanishi, Y. Characterization of Ganglioside G_{M4} Lactones Isolated From the Whale Brain. *J. Biochem.* **1990**, *107*, 868–871.
494. Pudelko, M.; Lindgren, A.; Tengel, T.; Reis, C. A.; Elofsson, M.; Kihlberg, J. Formation of Lactones From Sialylated MUC1 Glycopeptides. *Org. Biomol. Chem.* **2006**, *4*, 713–720.
495. Hirabayashi, Y.; Li, Y.-T.; Li, S.-C. Occurrence of a New Hematoside in the Kidney of Guinea Pig. *FEBS Lett.* **1983**, *161*, 127–130.
496. Hakomori, S.-I.; Saito, T. Isolation and Characterization of a Glycosphingolipid Having a New Sialic Acid. *Biochemistry* **1969**, *8*, 5082–5088.
497. Gasa, S.; Makita, A.; Kinoshita, Y. Further Study of the Chemical Structure of the Equine Erythrocyte Hematoside Containing *O*-Acetyl Ester. *J. Biol. Chem.* **1983**, *258*, 876–881.
498. Yachida, Y.; Tsuchihashi, K.; Gasa, S. Characterization of Novel Mono-*O*-Acetylated GM3s Containing 9-*O*-Acetyl Sialic Acid and 6-*O*-Acetyl Galactose in Equine Erythrocytes. *Glycoconjugate J.* **1996**, *13*, 225–233.
499. Yachida, Y.; Tsuchihashi, K.; Gasa, S. Novel Di-*O*-Acetylated GM3s From Equine Erythrocytes, One Containing 4,9-Di-*O*-acetyl-*N*-glycolylneuraminic Acid and Another Containing 4-*O*-Acetyl-*N*-glycolylneuraminic Acid and 6-*O*-Acetyl-D-galactose. *Carbohydr. Res.* **1997**, *298*, 201–212.
500. Kawatake, S.; Inagaki, M.; Isobe, R.; Miyamoto, T.; Higuchi, R. Isolation and Structure of Monomethylated GM_3-Type Ganglioside Molecular Species From the Starfish *Luidia maculata*. *Chem. Pharm. Bull.* **2002**, *50*, 1386–1389.
501. Pan, K.; Tanaka, C.; Inagaki, M.; Higuchi, R.; Miyamoto, T. Isolation and Structure Elucidation of GM4-Type Gangliosides From the Okinawan Starfish *Protoreaster nodosus*. *Mar. Drugs* **2012**, *10*, 2467–2480.
502. Muralikrishna, G.; Reuter, G.; Peter-Katalinic, J.; Egge, H.; Hanisch, F.-G.; Siebert, H.-C.; Schauer, R. Identification of a New Ganglioside From the Starfish *Asterias rubens*. *Carbohydr. Res.* **1992**, *236*, 321–326.
503. Yamada, K.; Tanabe, K.; Miyamoto, T.; Kusumoto, T.; Inagaki, M.; Higuchi, R. Isolation and Structure of a Monomethylated Ganglioside Possessing Neuritogenic Activity From the Ovary of the Sea Urchin *Diadema setosum*. *Chem. Pharm. Bull.* **2008**, *56*, 734–737.
504. Yamada, K.; Hamada, A.; Kisa, F.; Miyamoto, T.; Higuchi, R. Constituents of Holothuroidea, 13—Structure of Neuritogenic Active Ganglioside Molecular Species From the Sea Cucumber *Stichopus chloronotus*. *Chem. Pharm. Bull.* **2003**, *51*, 46–52.
505. Ren, S.; Scarsdale, J. N.; Ariga, T.; Zhang, Y.; Klein, R. A.; Hartmann, R.; Kushi, Y.; Egge, H.; Yu, R. K. *O*-Acetylated Gangliosides in Bovine Buttermilk—Characterization of 7-*O*-Acetyl, 9-*O*-Acetyl, and 7,9-Di-*O*-acetyl G_{D3}. *J. Biol. Chem.* **1992**, *267*, 12632–12638.
506. Ghidoni, R.; Sonnino, S.; Tettamanti, G.; Baumann, N.; Reuter, G.; Schauer, R. Isolation and Characterization of a Trisialoganglioside From Mouse Brain, Containing 9-*O*-Acetyl-*N*-acetylneuraminic Acid. *J. Biol. Chem.* **1980**, *255*, 6990–6995.

507. Thurin, J.; Herlyn, M.; Hindsgaul, O.; Strömberg, N.; Karlsson, K.-A.; Elder, D.; Steplewski, Z.; Koprowski, H. Proton NMR and Fast-Atom Bombardment Mass Spectrometry Analysis of the Melanoma-Associated Ganglioside 9-*O*-Acetyl-G_{D3}. *J. Biol. Chem.* **1985**, *260*, 14556–14563.
508. Inagaki, M.; Isobe, R.; Higuchi, R. Biologically Active Glycosides From Asteroidea, 39—Glycosphingolipids From the Starfish *Linckia laevigata*, 1—Isolation and Structure of a New Ganglioside Molecular Species. *Eur. J. Org. Chem.* **1999**, 771–774.
509. Arao, K.; Inagaki, M.; Miyamoto, T.; Higuchi, R. Constituents of Crinoidea, 3—Isolation and Structure of a Glycosyl Inositolphosphoceramide-Type Ganglioside With Neuritogenic Activity From the Feather Star *Comanthus japonica*. *Chem. Pharm. Bull.* **2004**, *52*, 1140–1142.
510. Slomiany, B. L.; Kojima, K.; Banas-Gruszka, Z.; Murty, V. L. N.; Galicki, N. I.; Slomiany, A. Characterization of the Sulfated Monosialosyltriglycosylceramid e From Bovine Gastric Mucosa. *Eur. J. Biochem.* **1981**, *119*, 647–650.
511. Ijuin, T.; Kitajima, K.; Song, Y.; Kitazume, S.; Inoue, S.; Haslam, S. M.; Morris, H. R.; Dell, A.; Inoue, Y. Isolation and Identification of Novel Sulfated and Nonsulfated Oligosialyl Glycosphingolipids From Sea Urchin Sperm. *Glycoconjugate J.* **1996**, *13*, 401–413.
512. Kundu, S. K.; Samuelsson, B. E.; Pascher, I.; Marcus, D. M. New Gangliosides From Human Erythrocytes. *J. Biol. Chem.* **1983**, *258*, 13857–13866.
513. Kaneko, M.; Kisa, F.; Yamada, K.; Miyamoto, T.; Higuchi, R. Structure of a New Neuritogenic-Active Ganglioside From the Sea Cucumber *Stichopus japonicus*. *Eur. J. Org. Chem.* **2003**, 1004–1008.
514. Sugita, M. Studies on the Glycosphingolipids of the Starfish, *Asterina pectinifera*—II. Isolation and Characterization of a Novel Ganglioside With an Internal Sialic Acid Residue. *J. Biochem.* **1979**, *86*, 289–300.
515. Sugita, M. Studies on the Glycosphingolipids of the Starfish, *Asterina pectinifera*—III. Isolation and Structural Studies of Two Novel Gangliosides Containing Internal Sialic Acid Residues. *J. Biochem.* **1979**, *86*, 765–772.
516. Kawano, Y.; Higuchi, R.; Komori, T. Biologically Active Glycosides From Asteroidea, XIX. Glycosphingolipids From the Starfish *Acanthaster planci*—4. Isolation and Structure of Five New Gangliosides. *Liebigs Ann. Chem.* **1990**, 43–50.
517. Miyamoto, T.; Inagaki, M.; Isobe, R.; Tanaka, Y.; Higuchi, R.; Iha, M.; Teruya, K. Biologically Active Glycosides From Asteroidea, 36—Re-examination of the Structure of Acanthaganglioside C, and the Identification of Three Minor Acanthagangliosides F, G and H. *Liebigs Ann./Recl.* **1997**, 931–936.
518. Miyamoto, T.; Yamamoto, A.; Wakabayashi, M.; Nagaregawa, Y.; Inagaki, M.; Higuchi, R.; Iha, M.; Teruya, K. Biologically Active Glycosides From Asteroidea, 40—Two New Gangliosides, Acanthagangliosides I and J From the Starfish *Acanthaster planci*. *Eur. J. Org. Chem.* **2000**, 2295–2301.
519. Higuchi, R.; Inoue, S.; Inagaki, K.; Sakai, M.; Miyamoto, T.; Komori, T.; Inagaki, M.; Isobe, R. Biologically Active Glycosides From Asteroidea, 42—Isolation and Structure of a New Biologically Active Ganglioside Molecular Species From the Starfish *Asterina pectinifera*. *Chem. Pharm. Bull.* **2006**, *54*, 287–291.
520. Prokazova, N. V.; Mikhailov, A. T.; Kocharov, S. L.; Malchenko, L. A.; Zvezdina, N. D.; Buznikov, G.; Bergelson, L. D. Unusual Gangliosides of Eggs and Embryos of the Sea Urchin *Strongylocentrotus intermedius*—Structure and Density-Dependence of Surface Localization. *Eur. J. Biochem.* **1981**, *115*, 671–677.
521. Yamada, K.; Harada, Y.; Miyamoto, T.; Isobe, R.; Higuchi, R. Constituents of Holothuroidea, 9—Isolation and Structure of a New Ganglioside Molecular Species From the Sea Cucumber *Holothuria pervicax*. *Chem. Pharm. Bull.* **2000**, *48*, 157–159.

522. Yamada, K.; Matsubara, R.; Kaneko, M.; Miyamoto, T.; Higuchi, R. Constituents of Holothuroidea, 10—Isolation and Structure of a Biologically Active Ganglioside Molecular Species From the Sea Cucumber *Holothuria leucospilota. Chem. Pharm. Bull.* **2001**, *49*, 447–452.
523. Inagaki, M.; Miyamoto, T.; Isobe, R.; Higuchi, R. Biologically Active Glycosides From Asteroidea, 43—Isolation and Structure of a New Neuritogenic-Active Ganglioside Molecular Species From the Starfish *Linckia laevigata. Chem. Pharm. Bull.* **2005**, *53*, 1551–1554.
524. Song, Y.; Kitajima, K.; Inoue, S.; Inoue, Y. Isolation and Structural Elucidation of a Novel Type of Ganglioside, Deaminated Neuraminic Acid (KDN)-Containing Glycosphingolipid, From Rainbow Trout Sperm—The First Example of the Natural Occurrence of KDN-Ganglioside, (KDN)G_{M3}. *J. Biol. Chem.* **1991**, *266*, 21929–21935.
525. Song, Y.; Kitajima, K.; Inoue, S.; Muto, Y.; Kasama, T.; Handa, S.; Inoue, Y. Structure of Novel Gangliosides, Deaminated Neuraminic Acid (KDN)-Containing Glycosphingolipids, Isolated From Rainbow Trout Ovarian Fluid. *Biochemistry* **1993**, *32*, 9221–9229.
526. Song, Y.; Kitajima, K.; Inoue, S.; Khoo, K.-H.; Morris, H. R.; Dell, A.; Inoue, Y. Expression of New KDN-Gangliosides in Rainbow Trout Testis During Spermatogenesis and Their Structural Identification. *Glycobiology* **1995**, *5*, 207–218.
527. Nagai, Y.; Iwamori, M. Cellular Biology of Gangliosides. In: *Biology of the Sialic Acids*; Rosenberg, A. Ed.; Plenum Press: New York, USA, 1995; pp 197–241.
528. Higuchi, R.; Inagaki, M.; Yamada, K.; Miyamoto, T. Biologically Active Gangliosides From Echinoderms. *J. Nat. Med.* **2007**, *61*, 367–370.
529. Yu, R. K.; Yanagisawa, M.; Ariga, T. Glycosphingolipid Structures. In: *Comprehensive Glycoscience—From Chemistry to Systems Biology*; Kamerling, J. P., Boons, G.-J., Lee, Y. C., Suzuki, A., Taniguchi, N., Voragen, A. G. J., Eds.; Vol. 1; Elsevier Ltd.: Oxford, UK, 2007; pp 73–122.
530. Urashima, T.; Asakuma, S.; Messer, M. Milk Oligosaccharides. In: *Comprehensive Glycoscience—From Chemistry to Systems Biology*; Kamerling, J. P., Boons, G.-J., Lee, Y. C., Suzuki, A., Taniguchi, N., Voragen, A. G. J., Eds.; Vol. 4; Elsevier Ltd.: Oxford, UK, 2007; pp 695–724.
531. Urashima, T.; Kitaoka, M.; Asakuma, S.; Messer, M. Milk Oligosaccharides. In: *Advanced Dairy Chemistry—Lactose, Water, Salts and Minor Constituents*; McSweeney, P. L. H., Fox, P. F., Eds.; Vol. 3; Springer Science + Business Media: New York, NY, USA, 2009; pp 295–349.
532. Kobata, A. Structures and Application of Oligosaccharides in Human Milk. *Proc. Jpn. Acad., Ser. B* **2010**, *86*, 731–747.
533. Chen, X. Human Milk Oligosaccharides (HMOS): Structure, Function, and Enzyme-Catalyzed Synthesis. *Adv. Carbohydr. Chem. Biochem.* **2015**, *72*, 113–190.
534. Kobata, A. Structures, Classification, and Biosynthesis of Human Milk Oligosaccharides. In: *Prebiotics and Probiotics in Human Milk—Origins and Functions of Milk-Borne Oligosaccharides and Bacteria*; McGuire, M., McGuire, M. A., Bode, L., Eds.; Elsevier Inc.: Oxford, UK, 2017; pp 17–44.
535. Urashima, T.; Messer, M.; Oftedal, O. T. Oligosaccharides in the Milk of Other Mammals. In: *Prebiotics and Probiotics in Human Milk—Origins and Functions of Milk-Borne Oligosaccharides and Bacteria*; McGuire, M., McGuire, M. A., Bode, L., Eds.; Elsevier Inc.: Oxford, UK, 2017; pp 45–139.
536. Kuhn, R.; Gauhe, A. Bestimmung der Bindungsstelle von Sialinsäureresten in Oligosacchariden mit Hilfe von Perjodat. *Chem. Ber.* **1965**, *98*, 395–413.
537. Kuhn, R. Biochemie der Rezeptoren und Resistenzfaktoren—Von der Widerstandsfähigkeit der Lebewesen gegen Einwirkungen der Umwelt. *Naturwissenschaften* **1959**, *46*, 43–50.

538. Marino, K.; Lane, J. A.; Abrahams, J. L.; Struwe, W. B.; Harvey, D. J.; Marotta, M.; Hickey, R. M.; Rudd, P. M. Method for Milk Oligosaccharide Profiling by 2-Aminobenzamide Labeling and Hydrophilic Interaction Chromatography. *Glycobiology* **2011**, *21*, 1317–1330.
539. Albrecht, S.; Lane, J. A.; Marino, K.; Al Busadah, K. A.; Carrington, S. D.; Hickey, R. M.; Rudd, P. M. A Comparative Study of Free Oligosaccharides in the Milk of Domestic Animals. *Br. J. Nutr.* **2014**, *111*, 1313–1328.
540. Huang, R. T. C. The Structures of Some Neuraminic Acid-Containing Oligosaccharides of Human Milk. *Hoppe-Seyler's Z. Physiol. Chem.* **1971**, *352*, 1645–1652.
541. Wiegandt, H. Bemerkungen zur voranstehenden Arbeit "The Structures of Some Neuraminic Acid-Containing Oligosaccharides of Human Milk"*Hoppe-Seyler's Z. Physiol. Chem.* **1971**, *352*, 1653.
542. Taufik, E.; Fukuda, K.; Senda, A.; Saito, T.; Williams, C.; Tilden, C.; Eisert, R.; Oftedal, O.; Urashima, T. Structural Characterization of Neutral and Acidic Oligosaccharides in the Milks of Strepsirrhine Primates: Greater Galago, Aye-Aye, Coquerel's Sifaka and Mongoose Lemur. *Glycoconjugate J.* **2012**, *29*, 119–134.
543. Veh, R. W.; Michalski, J.-C.; Corfield, A. P.; Sander-Wewer, M.; Gies, D.; Schauer, R. New Chromatographic System for the Rapid Analysis and Preparation of Colostrum Sialyloligosaccharides. *J. Chromatogr.* **1981**, *212*, 313–322.
544. Messer, M. Identification of *N*-Acetyl-4-*O*-acetylneuraminyl-lactose in Echidna Milk. *Biochem. J.* **1974**, *139*, 415–420.
545. Kamerling, J. P.; Dorland, L.; van Halbeek, H.; Vliegenthart, J. F. G.; Messer, M.; Schauer, R. Structural Studies of 4-*O*-Acetyl-α-*N*-acetylneuraminyl-(2→3)-lactose, the Main Oligosaccharide in Echidna Milk. *Carbohydr. Res.* **1982**, *100*, 331–340.
546. Oftedal, O. T.; Nicol, S. C.; Davies, N. W.; Sekii, N.; Taufik, E.; Fukuda, K.; Saito, T.; Urashima, T. Can an Ancestral Condition for Milk Oligosaccharides Be Determined? Evidence From the Tasmanian Echidna (*Tachyglossus aculeatus setosus*). *Glycobiology* **2014**, *24*, 826–839.
547. Urashima, T.; Inamori, H.; Fukuda, K.; Saito, T.; Messer, M.; Oftedal, O. T. 4-*O*-Acetyl-sialic Acid (Neu4,5Ac_2) in Acidic Milk Oligosaccharides of the Platypus (*Ornithorhynchus anatinus*) and Its Evolutionary Significance. *Glycobiology* **2015**, *25*, 683–697.
548. Schauer, R. Sialic Acids as Link to Japanese Scientists. *Proc. Jpn. Acad., Ser. B* **2016**, *92*, 109–120.
549. Jahan, M.; Wynn, P. C.; Wang, B. Molecular Characterization of the Level of Sialic Acids *N*-Acetylneuraminic Acid, *N*-Glycolylneuraminic Acid, and Ketodeoxynonulosonic Acid in Porcine Milk During Lactation. *J. Dairy Sci.* **2016**, *99*, 8431–8442.
550. Huttunen, J. K. Neuraminic Acid-Containing Oligosaccharides of Human Urine: Isolation and Identification of Di-*N*-acetylneuraminyl-3-galactosyl-*N*-acetylgalactosamine, 6-2-*N*-Acetylneuraminyl-lactose, 6-2-*N*-Acetylneuraminyl-*N*-acetyllactosamine and 3-2-*N*-Acetylneuraminyl-lactose. *Ann. Med. Exp. Biol. Fenn.* **1966**, *44*(Suppl. 12), 1–60.
551. Maury, P. Identification of *N*,*O*-Diacetyl-, *N*-Acetyl-, and *N*-Glycolylneuraminyl-(2→3)-lactose in Rat Urine. *Biochim. Biophys. Acta* **1971**, *252*, 472–480.
552. Parkkinen, J.; Finne, J. Isolation and Structural Characterization of Five Major Sialyloligosaccharides and a Sialylglycopeptide From Normal Human Urine. *Eur. J. Biochem.* **1983**, *136*, 355–361.
553. Parkkinen, J.; Finne, J. Isolation of Sialyl Oligosaccharides and Sialyl Oligosaccharide Phosphates From Bovine Colostrum and Human Urine. *Methods Enzymol.* **1987**, *138*, 289–300.
554. Lemonnier, M.; Bourrillon, R. Characterization and Structure of a Sialic Acid-Containing Hexasaccharide Isolated From Human Pregnancy Urine. *Carbohydr. Res.* **1976**, *51*, 99–106.

555. Konradt, T.; Schlichting, A.; May, B.; Veh, R. W. Histochemical Differentiation Between Tissue-Bound 9-*O*-Acetyl-sialic Acid and 7,9-Di-*O*-acetyl-sialic Acid: Periodate Oxidation Studies With Purified 9-*O*-Acetyl-sialyllactose From Rat Urine. *Anat. Anz.* **1987**, *163*, 159–160.
556. Strecker, G.; Hondi-Assah, T.; Fournet, B.; Spik, G.; Montreuil, J.; Maroteaux, P.; Durand, P.; Farriaux, J.-P. Structure of the Three Major Sialyl-Oligosaccharides Excreted in the Urine of Five Patients With Three Distinct Inborn Diseases: "I Cell Disease" and Two New Types of Mucolipidosis. *Biochim. Biophys. Acta* **1976**, *444*, 349–358.
557. Michalski, J. C.; Strecker, G.; Fournet, B.; Cantz, M.; Spranger, J. Structures of Sialyl-Oligosaccharides Excreted in the Urine of a Patient With Mucolipidosis I. *FEBS Lett.* **1977**, *79*, 101–104.
558. Strecker, G.; Peers, M.-C.; Michalski, J.-C.; Hondi-Assah, T.; Fournet, B.; Spik, G.; Montreuil, J.; Farriaux, J.-P.; Maroteaux, P.; Durand, P. Structure of Nine Sialyl-Oligosaccharides Accumulated in Urine of Eleven Patients With Three Different Types of Sialidosis—Mucolipidosis II and Two New Types of Mucolipidosis. *Eur. J. Biochem.* **1977**, *75*, 391–403.
559. Dorland, L.; Haverkamp, J.; Vliegenthart, J. F. G.; Strecker, G.; Michalski, J.-C.; Fournet, B.; Spik, G.; Montreuil, J. 360-MHz ^{1}H Nuclear-Magnetic-Resonance Spectroscopy of Sialyl-Oligosaccharides From Patients With Sialidosis (Mucolipidosis I and II). *Eur. J. Biochem.* **1978**, *87*, 323–329.
560. Kuriyama, M.; Ariga, T.; Ando, S.; Suzuki, M.; Yamada, T.; Miyatake, T. Four Positional Isomers of Sialyloligosaccharides Isolated From the Urine of a Patient With Sialidosis. *J. Biol. Chem.* **1981**, *256*, 12316–12321.
561. Lecat, D.; Lemonnier, M.; Derappe, C.; Lhermitte, M.; van Halbeek, H.; Dorland, L.; Vliegenthart, J. F. G. The Structure of Sialyl-Glycopeptides of the *O*-Glycosidic Type, Isolated From Sialidosis (Mucolipidosis I) Urine. *Eur. J. Biochem.* **1984**, *140*, 415–420.
562. Kuriyama, M.; Ariga, T.; Ando, S.; Suzuki, M.; Yamada, T.; Miyatake, T.; Igata, A. Two Positional Isomers of Sialylheptasaccharides Isolated From the Urine of a Patient With Sialidosis. *J. Biochem.* **1985**, *98*, 1049–1954.
563. Van Pelt, J.; Kamerling, J. P.; Vliegenthart, J. F. G.; Verheijen, F. W.; Galjaard, H. Isolation and Structural Characterization of Sialic Acid-Containing Storage Material From Mucolipidosis I (Sialidosis) Fibroblasts. *Biochim. Biophys. Acta* **1988**, *965*, 36–45.
564. Van Pelt, J.; Kamerling, J. P.; Vliegenthart, J. F. G.; Hoogeveen, A. T.; Galjaard, H. A Comparative Study of the Accumulated Sialic Acid-Containing Oligosaccharides From Cultured Human Galactosialidosis and Sialidosis Fibroblasts. *Clin. Chim. Acta* **1988**, *174*, 325–336.
565. Van Pelt, J.; van Kuik, J. A.; Kamerling, J. P.; Vliegenthart, J. F. G.; van Diggelen, O. P.; Galjaard, H. Storage of Sialic Acid-Containing Carbohydrates in the Placenta of a Human Galactosialidosis Fetus—Isolation and Structural Characterization of 16 Sialyloligosaccharides. *Eur. J. Biochem.* **1988**, *177*, 327–338.
566. Van Pelt, J.; van Bilsen, D. G. J. L.; Kamerling, J. P.; Vliegenthart, J. F. G. Structural Analysis of *O*-Glycosidic Type of Sialyloligosaccharide-Alditols Derived From Urinary Glycopeptides of a Sialidosis Patient. *Eur. J. Biochem.* **1988**, *174*, 183–187.
567. Van Pelt, J.; Hård, K.; Kamerling, J. P.; Vliegenthart, J. F. G.; Reuser, A. J. J.; Galjaard, H. Isolation and Structural Characterization of Twenty-One Sialyloligosaccharides From Galactosialidosis Urine—An Intact *N,N'*-Diacetylchitobiose Unit at the Reducing End of a Diantennary Structure. *Biol. Chem. Hoppe-Seyler* **1989**, *370*, 191–203.
568. Michalski, J.-C. Normal and Pathological Catabolism of Glycoproteins—Catabolic Pathways. In: *Glycoproteins and Disease*; Montreuil, J., Vliegenthart, J. F. G., Schachter, H., Eds.; New Comprehensive Biochemistry, Vol. 30; Elsevier Science B.V: Amsterdam, The Netherlands, 1996; pp 55–97.

569. Bruggink, C.; Poorthuis, B. J. H. M.; Piraud, M.; Froissart, R.; Deelder, A. M.; Wuhrer, M. Glycan Profiling of Urine, Amniotic Fluid and Ascitic Fluid From Galactosialidosis Patients Reveals Novel Oligosaccharides With Reducing End Hexose and Aldohexonic Acid Residues. *FEBS J.* **2010**, *277*, 2970–2986.
570. Van Pelt, J.; Dorland, L.; Duran, M.; Hokke, C. H.; Kamerling, J. P.; Vliegenthart, J. F. G. Sialyl-α2-6-mannosyl-β1-4-*N*-acetylglucosamine, a Novel Compound Occurring in Urine of Patients With β-Mannosidosis. *J. Biol. Chem.* **1990**, *265*, 19685–19689.
571. Pollitt, R. J.; Pretty, K. M. The Glycoasparagines in Urine of a Patient With Aspartylglycosaminuria. *Biochem. J.* **1974**, *141*, 141–146.
572. Sugahara, K.; Funakoshi, S.; Funakoshi, I.; Aula, P.; Yamashina, I. Characterization of One Neutral and Two Acidic Glycoasparagines Isolated From the Urine of Patients With Aspartylglycosylaminuria (AGU). *J. Biochem.* **1976**, *80*, 195–201.
573. Sugahara, K.; Akasaki, M.; Funakoshi, I.; Aula, P.; Yamashina, I. Structural Studies of Glycoasparagines From Urine of a Patient With Aspartylglycosylaminuria (AGU). *FEBS Lett.* **1977**, *78*, 284–286.
574. Sugahara, K.; Akasaki, M.; Funakoshi, I.; Aula, P.; Yamashina, I. Structure of Two Glycoasparagines Isolated From the Urine of Patients With Aspartylglycosylaminuria (AGU). *J. Biochem.* **1977**, *82*, 1499–1501.
575. Irie, F.; Murakoshi, H.; Suzuki, T.; Suzuki, Y.; Kon, K.; Ando, S.; Yoshida, K.; Hirabayashi, Y. Characterization of Four Monosialo and a Novel Disialo Asn N-Glycosides From the Urine of a Patient With Aspartylglycosaminuria. *Glycoconjugate J.* **1995**, *12*, 290–297.
576. Linden, H.-U.; Klein, R. A.; Egge, H.; Peter-Katalinic, J.; Dabrowski, J.; Schindler, D. Isolation and Structural Characterization of Sialic Acid-Containing Glycopeptides of the *O*-Glycosidic Type From the Urine of Two Patients With an Hereditary Deficiency in α-*N*-Acetylgalactosaminidase Activity. *Biol. Chem. Hoppe-Seyler* **1989**, *370*, 661–672.
577. Hirabayashi, Y.; Matsumoto, Y.; Matsumoto, M.; Toida, T.; Iida, N.; Matsubara, T.; Kanzaki, T.; Yokota, M.; Ishizuka, I. Isolation and Characterization of Major Urinary Amino Acid *O*-Glycosides and a Dipeptide *O*-Glycoside From a New Lysosomal Storage Disorder (Kanzaki Disease)—Excessive Excretion of Serine- and Threonine-Linked Glycan in the Patient Urine. *J. Biol. Chem.* **1990**, *265*, 1693–1701.
578. Froesch, M.; Bindila, L.; Zamfir, A.; Peter-Katalinic, J. Sialylation Analysis of *O*-Glycosylated Sialylated Peptides From Urine of Patients Suffering From Schindler's Disease by Fourier Transform Ion Cyclotron Resonance Mass Spectrometry and Sustained Off-Resonance Irradiation Collision-Induced Dissociation. *Rapid Commun. Mass Spectrom.* **2003**, *17*, 2822–2832.
579. Desnick, R. J.; Schindler, D. Schindler Disease: Deficient α-*N*-Acetylgalactosaminidase Activity. In: *The Molecular and Genetic Basis of Neurologic and Psychiatric Disease*, Rosenberg, R. N., DiMauro, S., Paulson, H. L., Ptacek, L., Nestler, E. J., Eds, 4th ed.; Lippincott Williams & Wilkins: Philadelphia, PA, USA, 2008; pp 309–316.
580. Aula, P.; Autio, S.; Raivio, K. O.; Rapola, J.; Thodén, C.-J.; Koskela, S.-L.; Yamashina, I. 'Salla Disease'—A New Lysosomal Storage Disorder. *Arch. Neurol.* **1979**, *36*, 88–94.
581. Renlund, M.; Chester, M. A.; Lundblad, A.; Aula, P.; Raivio, K. O.; Autio, S.; Koskela, S.-L. Increased Urinary Excretion of Free *N*-Acetylneuraminic Acid in Thirteen Patients With Salla Disease. *Eur. J. Biochem.* **1979**, *101*, 245–250.
582. Baumkötter, J.; Cantz, M.; Mendla, K.; Baumann, W.; Friebolin, H.; Gehler, J.; Spranger, J. *N*-Acetylneuraminic Acid Storage Disease. *Hum. Genet.* **1985**, *71*, 155–159.

583. Fontaine, G.; Gaudier, B.; Biserte, G.; Montreuil, J.; Dupont, A.; Farriaux, J.-P.; Strecker, G.; Spik, G.; Puvion, E.; Puvion-Dutilleul, F.; Sezille, G. Elimination Urinaire Permanente d'Acide Sialique Libre chez un Enfant de Trois Ans Atteint de Troubles Cliniques Divers. *Pediatrie* **1967**, *22*, 705–709.
584. Montreuil, J.; Biserte, G.; Strecker, G.; Spik, G.; Fontaine, G.; Farriaux, J.-P. Description d'un Nouveau Type de Méliturie: La Sialurie. *Clin. Chim. Acta* **1968**, *21*, 61–69.
585. Kamerling, J. P.; Strecker, G.; Farriaux, J.-P.; Dorland, L.; Haverkamp, J.; Vliegenthart, J. F. G. 2-Acetamidoglucal, a New Metabolite Isolated From the Urine of a Patient With Sialuria. *Biochim. Biophys. Acta* **1979**, *583*, 403–408.
586. Weiss, P.; Tietze, F.; Gahl, W. A.; Seppala, R.; Ashwell, G. Identification of the Metabolic Defect in Sialuria. *J. Biol. Chem.* **1989**, *264*, 17635–17636.
587. Wopereis, S.; Abd Hamid, U. M.; Critchley, A.; Royle, L.; Dwek, R. A.; Morava, E.; Leroy, J. G.; Wilcken, B.; Lagerwerf, A. J.; Huijben, K. M. L. C.; Lefeber, D. J.; Rudd, P. M.; Wevers, R. A. Abnormal Glycosylation With Hypersialylated O-Glycans in Patients With Sialuria. *Biochim. Biophys. Acta* **2006**, *1762*, 598–607.
588. Barry, G. T.; Goebel, W. F. Colominic Acid, a Substance of Bacterial Origin Related to Sialic Acid. *Nature* **1957**, *179*, 206.
589. Aminoff, D.; Dodyk, F.; Roseman, S. Enzymatic Synthesis of Colominic Acid. *J. Biol. Chem.* **1963**, *238*, PC1177–PC1178.
590. McGuire, E. J.; Binkley, S. B. The Structure and Chemistry of Colominic Acid. *Biochemistry* **1964**, *3*, 247–251.
591. Liu, T.-Y.; Gotschlich, E. C.; Dunne, F. T.; Jonssen, E. K. Studies on the Meningococcal Polysaccharides—II. Composition and Chemical Properties of the Group B and Group C Polysaccharide. *J. Biol. Chem.* **1971**, *246*, 4703–4712.
592. Watson, R. G.; Marinetti, G. V.; Scherp, H. W. The Specific Hapten of Group C (Group II α) Meningococcus—II. Chemical Nature. *J. Immunol.* **1958**, *81*, 337–344.
593. Egan, W.; Liu, T.-Y.; Dorow, D.; Cohen, J. S.; Robbins, J. D.; Gotschlich, E. C.; Robbins, J. B. Structural Studies on the Sialic Acid Polysaccharide Antigen of *Escherichia coli* Strain Bos-12. *Biochemistry* **1977**, *16*, 3687–3692.
594. Dzieciatkowska, M.; Brochu, D.; van Belkum, A.; Heikema, A. P.; Yuki, N.; Houliston, R. S.; Richards, J. C.; Gilbert, M.; Li, J. Mass Spectrometric Analysis of Intact Lipooligosaccharide: Direct Evidence for O-Acetylated Sialic Acids and Discovery of O-Linked Glycine Expressed by *Campylobacter jejuni*. *Biochemistry* **2007**, *46*, 14704–14714.
595. Knirel, Y. A.; Shevelev, S. D.; Perepelov, A. V. Higher Aldulosonic Acids: Components of Bacterial Glycans. *Mendeleev Commun.* **2011**, *21*, 173–182.
596. Deutschmann, R.; Boncheff, A. G.; Daraban, L.; MacInnes, J. I.; Monteiro, M. A. Common Sialylated Glycan in *Actinobacillus suis*. *Glycobiology* **2010**, *20*, 1227–1232.
597. Ali, T.; Weintraub, A.; Widmalm, G. Structural Determination of the O-Antigenic Polysaccharide From the Shiga Toxin-Producing *Escherichia coli* O171. *Carbohydr. Res.* **2006**, *341*, 1878–1883.
598. Bhattacharjee, A. K.; Jennings, H. J.; Kenny, C. P.; Martin, A.; Smith, I. C. P. Structural Determination of the Polysaccharide Antigens of *Neisseria meningitidis* Serogroups Y, W-135, and BO. *Can. J. Biochem.* **1976**, *54*, 1–8.
599. Dutton, G. G. S.; Parolis, H.; Parolis, L. A. S. The Structure of the Neuraminic Acid-Containing Capsular Polysaccharide of *Escherichia coli* Serotype K9. *Carbohydr. Res.* **1987**, *170*, 193–206.
600. Gamian, A.; Kenne, L.; Mieszala, M.; Ulrich, J.; Defaye, J. Structure of the *Escherichia coli* O24 and O56 O-Specific Sialic-Acid-Containing Polysaccharides and Linkage of These Structures to the Core Region in Lipopolysaccharides. *Eur. J. Biochem.* **1994**, *225*, 1211–1220.

601. Krauss, J. H.; Himmelspach, K.; Reuter, G.; Schauer, R.; Mayer, H. Structural Analysis of a Novel Sialic-Acid-Containing Trisaccharide From *Rhodobacter capsulatus* 37b4 Lipopolysaccharide. *Eur. J. Biochem.* **1992**, *204*, 217–223.
602. Tul'skaya, E. M.; Shashkov, A. S.; Streshinskaya, G. M.; Senchenkova, S. N.; Potekhina, N. V.; Kozlova, Y. I.; Evtushenko, L. I. Teichuronic and Teichulosonic Acids of Actinomycetes. *Biochemistry (Moscow)* **2011**, *76*, 736–744.
603. Shashkov, A. S.; Tul'skaya, E. M.; Evtushenko, L. I.; Denisenko, V. A.; Ivanyuk, V. G.; Stomakhin, A. A.; Naumova, I. B.; Stackebrandt, E. Cell Wall Anionic Polymers of *Streptomyces* sp. MB-8, the Causative Agent of Potato Scab. *Carbohydr. Res.* **2002**, *337*, 2255–2261.
604. Dell, A.; Chalabi, S.; Hitchen, P. G.; Jang-Lee, J.; Ledger, V.; North, S. J.; Pang, P.-C.; Parry, S.; Sutton-Smith, M.; Tissot, B.; Morris, H. R.; Panico, M.; Haslam, S. M. Mass Spectrometry of Glycoprotein Glycans: Glycomics and Glycoproteomics. In Comprehensive Glycoscience—From Chemistry to Systems Biology, Kamerling, J. P.; Boons, G.-J.; Lee, Y. C.; Suzuki, A.; Taniguchi, N.; Voragen, A. G. J., Eds.; Elsevier Ltd.: Oxford, UK, 2007; Vol. 2, pp 69–100.
605. Zauner, G.; Deelder, A. M.; Wuhrer, M. Recent Advances in Hydrophilic Interaction Liquid Chromatography (HILIC) for Structural Glycomics. *Electrophoresis* **2011**, *32*, 3456–3466.
606. Leymarie, N.; Zaia, J. Effective Use of Mass Spectrometry for Glycan and Glycopeptide Structural Analysis. *Anal. Chem.* **2012**, *84*, 3040–3048.
607. Wuhrer, M. Glycomics Using Mass Spectrometry. *Glycoconjugate J.* **2013**, *30*, 11–22.
608. Mechref, Y.; Hu, Y.; Desantos-Garcia, J. L.; Hussein, A.; Tang, H. Quantitative Glycomics Strategies. *Mol. Cell. Proteomics* **2013**, *12*, 874–884.
609. Kailemia, M. J.; Ruhaak, L. R.; Lebrilla, C. B.; Amster, I. J. Oligosaccharide Analysis by Mass Spectrometry: A Review of Recent Developments. *Anal. Chem.* **2014**, *86*, 196–212.
610. Banazadeh, A.; Veillon, L.; Wooding, K. M.; Zabet-moghaddam, M.; Mechref, Y. Recent Advances in Mass Spectrometric Analysis of Glycoproteins. *Electrophoresis* **2017**, *38*, 162–189.
611. Yang, Y.; Franc, V.; Heck, A. J. R. Glycoproteomics: A Balance Between High-Throughput and In-Depth Analysis. *Trends Biotechnol.* **2017**, *35*, 598–609.
612. Powell, A. K.; Harvey, D. J. Stabilization of Sialic Acids in *N*-Linked Oligosaccharides and Gangliosides for Analysis by Positive Ion Matrix-Assisted Laser Desorption/Ionization Mass Spectrometry. *Rapid Commun. Mass Spectrom.* **1996**, *10*, 1027–1032.
613. Yamagaki, T.; Nakanishi, H. A New Technique Distinguishing α2-3 Sialyl Linkage From α2-6 Linkage in Sialyllactoses and Sialyl-*N*-acetyllactosamines by Post-Source Decay Fragmentation Method of MALDI-TOF Mass Spectrometry. *Glycoconjugate J.* **1999**, *16*, 385–389.
614. Sekiya, S.; Wada, Y.; Tanaka, K. Derivatization for Stabilizing Sialic Acids in MALDI-MS. *Anal. Chem.* **2005**, *77*, 4962–4968.
615. Galuska, S. P.; Geyer, R.; Mühlenhoff, M.; Geyer, H. Characterization of Oligo- and Polysialic Acids by MALDI-TOF-MS. *Anal. Chem.* **2007**, *79*, 7161–7169.
616. Miura, Y.; Shinohara, Y.; Furukawa, J.-I.; Nagahori, N.; Nishimura, S.-I. Rapid and Simple Solid-Phase Esterification of Sialic Acid Residues for Quantitative Glycomics by Mass Spectrometry. *Chem. Eur. J.* **2007**, *13*, 4797–4804.
617. Wheeler, S. F.; Domann, P.; Harvey, D. J. Derivatization of Sialic Acids for Stabilization in Matrix-Assisted Laser Desorption/Ionization Mass Spectrometry and Concomitant Differentiation of $\alpha(2\rightarrow3)$- and $\alpha(2\rightarrow6)$-Isomers. *Rapid Commun. Mass Spectrom.* **2009**, *23*, 303–312.

618. Galuska, S. P.; Geyer, H.; Bleckmann, C.; Röhrich, R. C.; Maass, K.; Bergfeld, A. K.; Mühlenhoff, M.; Geyer, R. Mass Spectrometric Fragmentation Analysis of Oligosialic and Polysialic Acids. *Anal. Chem.* **2010**, *82*, 2059–2066.
619. De Haan, N.; Reiding, K. R.; Haberger, M.; Reusch, D.; Falck, D.; Wuhrer, M. Linkage-Specific Sialic Acid Derivatization for MALDI-TOF-MS Profiling of IgG Glycopeptides. *Anal. Chem.* **2015**, *87*, 8284–8291.
620. Holst, S.; Heijs, B.; de Haan, N.; van Zeijl, R. J. M.; Briaire-de Bruijn, I. H.; van Pelt, G. W.; Mehta, A. S.; Angel, P. M.; Mesker, W. E.; Tollenaar, R. A.; Drake, R. R.; Bovée, J. V. M. G.; McDonnell, L. A.; Wuhrer, M. Linkage-Specific *In Situ* Sialic Acid Derivatization for N-Glycan Mass Spectrometry Imaging of Formalin-Fixed Paraffin-Embedded Tissues. *Anal. Chem.* **2016**, *88*, 5904–5913.
621. Warren, L.; Felsenfeld, H. *N*-Acetylmannosamine-6-phosphate and *N*-Acetylneuraminic Acid-9-phosphate as Intermediates in Sialic Acid Biosynthesis. *Biochem. Biophys. Res. Commun.* **1961**, *5*, 185–190.
622. Warren, L.; Blacklow, R. S. Biosynthesis of N-Acetylneuraminic Acid and Cytidine-5′-monophospho-N-acetylneuraminic Acid in *Neisseria meningitidis. Biochem. Biophys. Res. Commun.* **1962**, 7, 433–438.
623. Kragl, U.; Kittelmann, M.; Ghisalba, O.; Wandrey, C. *N*-Acetylneuraminic Acid—From a Rare Chemical From Natural Sources to a Multikilogram Enzymatic Synthesis for Industrial Application. *Ann. N. Y. Acad. Sci.* **1995**, *750*, 300–305.
624. Kuboki, A.; Okazaki, H.; Sugai, T.; Ohta, H. An Expeditious Route to *N*-Glycolylneuraminic Acid Based on Enzyme-Catalyzed Reaction. *Tetrahedron* **1997**, *53*, 2387–2400.
625. Maru, I.; Ohnishi, J.; Ohta, Y.; Tsukada, Y. Simple and Large-Scale Production of *N*-Acetylneuraminic Acid From *N*-Acetyl-D-glucosamine and Pyruvate Using *N*-Acyl-D-glucosamine 2-Epimerase and *N*-Acetylneuraminate Lyase. *Carbohydr. Res.* **1998**, *306*, 575–578.
626. Hemeon, I.; Bennet, A. J. Sialic Acid and Structural Analogues: Stereoselective Syntheses. *Synthesis* **2007**, 1899–1926.
627. Juneja, L. R.; Koketsu, M.; Nishimoto, K.; Kim, M.; Yamamoto, T.; Itoh, T. Large-Scale Preparation of Sialic Acid From Chalaza and Egg-Yolk Membrane. *Carbohydr. Res.* **1991**, *214*, 179–186.
628. Sumi, T.; Sallay, I.; Asakawa, M.; Park, S. S.; Miyazaki, M.; Ohba, H. Purification and Identification of N-Glycolylneuraminic Acid (Neu5Gc) From the Holothuroidea Gumi, *Cucumaria echinata. Prep. Biochem. Biotechnol.* **2001**, *31*, 135–146.
629. Ogura, H.; Furuhata, K.; Sato, S.; Anazawa, K.; Itoh, M.; Shitori, Y. Synthesis of 9-*O*-Acyl- and 4-*O*-Acetyl-salic Acids. *Carbohydr. Res.* **1987**, *167*, 77–86.
630. Forstner, M.; Freytag, K.; Paschke, E. A Simple, One-Step Synthesis of *N*-Acetyl-9-*O*-acetylneuraminic Acid by Enzymic Transesterification Mediated by Porcine Pancreas Lipase in Pyridine. *Carbohydr. Res.* **1989**, *193*, 294–295.
631. Hasegawa, A.; Murase, T.; Ogawa, M.; Ishida, H.; Kiso, M. Synthetic Studies on Sialoglycoconjugates—17: Synthesis of 4-*O*-, 9-*O*-, and 4,9-Di-*O*-acetyl-*N*-acetylneuraminic Acids and their Derivatives. *J. Carbohydr. Chem.* **1990**, *9*, 415–428.
632. Hasegawa, A.; Murase, T.; Ogawa, M.; Ishida, H.; Kiso, M. Synthetic Studies on Sialoglycoconjugates—18: Synthesis of 7-*O*-, 7,9-Di-*O*-, and 7,8,9-Tri-*O*-acetyl-*N*-acetyl-neuraminic Acid Derivatives. *J. Carbohydr. Chem.* **1990**, *9*, 429–439.
633. Anazawa, K.; Furuhata, K.; Ogura, H. Synthesis of 7-*O*-Acetyl-*N*-acetylneuraminic Acid Derivative. *Chem. Pharm. Bull.* **1988**, *36*, 4976–4979.
634. Furuhata, K.; Ogura, H. Studies on Sialic Acids—XIX. Syntheses of Partially *O*-Acetylated 4-Methylcoumarin-7-yl 5-Acetamido-3,5-dideoxy-α-D-*glycero*-D-*galacto*-2-nonulopyranosidonic Acids. *Chem. Pharm. Bull.* **1989**, *37*, 2037–2040.

635. Reinhard, B.; Faillard, H. Regioselective Acetylations of Sialic Acid α-Ketosides. *Liebigs Ann. Chem.* **1994**, 193–203.
636. David, S.; Augé, C. Immobilized Enzymes in Preparative Carbohydrate Chemistry. *Pure Appl. Chem.* **1987**, *59*, 1501–1508.
637. Kim, M.-J.; Hennen, W. J.; Sweers, H. M.; Wong, C.-H. Enzymes in Carbohydrate Synthesis: *N*-Acetylneuraminic Acid Aldolase Catalyzed Reactions and Preparation of *N*-Acetyl-2-deoxy-D-neuraminic Acid Derivatives. *J. Am. Chem. Soc.* **1988**, *110*, 6481–6486.
638. Meindl, P.; Tuppy, H. Über 2-Deoxy-2,3-dehydro-sialinsäuren—1. Mitt.: Synthese und Eigenschaften von 2-Deoxy-2,3-dehydro-N-acylneuraminsäuren und deren Methylestern. *Monatsh. Chem.* **1969**, *100*, 1295–1306.
639. Augé, C.; Gautheron, C. The Use of an Immobilized Aldolase in the First Synthesis of a Natural Deaminated Neuraminic Acid. *J. Chem. Soc., Chem. Commun.* **1987**, 859–860.
640. Shirai, R.; Nakamura, M.; Hara, S.; Takayanagi, H.; Ogura, H. Thermal Rearrangement of *N*-Acetyl-*N*-nitrosoneuraminic Acid Derivative: Synthesis of 3-Deoxy-D-nonulosonic Acid (KDN). *Tetrahedron Lett.* **1988**, *29*, 4449–4452.
641. Shirai, R.; Ogura, H. Improved Syntheses of Two 3-Deoxyald-2-ulosonic Acids (KDN, KDO) by Condensation of Oxalacetic Acid With Aldoses Followed by Ni^{2+} Catalyzed Decarboxylation. *Tetrahedron Lett.* **1989**, *30*, 2263–2264.
642. Schreiner, E.; Zbiral, E.; Kleineidam, R. G.; Schauer, R. 2,3-Didehydro-2-deoxysialic Acids Structurally Varied at C-5 and Their Behaviour Towards the Sialidase From *Vibrio cholerae*. *Carbohydr. Res.* **1991**, *216*, 61–66.
643. Ogura, H. Sialic Acid Derivatives as Glycolipids. In: *Carbohydrates—Synthetic Methods and Applications in Medicinal Chemistry*; Ogura, H., Hasegawa, A., Suami, T., Eds.; Kodansha Ltd.: Tokyo, Japan, 1992; pp 282–303.
644. Asressu, K. H.; Wang, C.-C. Concise Synthesis of 2,7-Anhydrosialic Acid Derivatives and Its Application. *Carbohydr. Res.* **2017**, *453–454*, 44–53.
645. Khorlin, A. Y.; Privalova, I. M. Synthesis of *N*-Acetylneuraminic Acid 8-Methyl Ether. *Carbohydr. Res.* **1970**, *13*, 373–377.
646. Colombo, R.; Anastasia, M.; Rota, P.; Allevi, P. The First Synthesis of *N*-Acetylneuraminic Acid 1,7-Lactone. *Chem. Commun. (Cambridge, U. K.)* **2008**, 5517–5519.
647. Allevi, P.; Rota, P.; Scaringi, R.; Colombo, R.; Anastasia, M. Chemoselective Synthesis of Sialic Acid 1,7-Lactones. *J. Org. Chem.* **2010**, *75*, 5542–5548.
648. Rota, P.; Allevi, P.; Anastasia, L. The Sialic Acids Waltz: Novel Stereoselective Isomerization of the 1,7-Lactones of *N*-Acetylneuraminic Acids Into the Corresponding γ-Lactones and Back to the Free Sialic Acids. *Asian J. Org. Chem.* **2015**, *4*, 1315–1321.
649. Yu, H.; Chokhawala, H.; Karpel, R.; Yu, H.; Wu, B.; Zhang, J.; Zhang, Y.; Jia, Q.; Chen, X. A Multifunctional *Pasteurella multocida* Sialyltransferase: A Powerful Tool for the Synthesis of Sialoside Libraries. *J. Am. Chem. Soc.* **2005**, *127*, 17618–17619.
650. Yu, H.; Huang, S.; Chokhawala, H.; Sun, M.; Zheng, H.; Chen, X. Highly Efficient Chemoenzymatic Synthesis of Naturally Occurring and Non-Natural α-2,6-Linked Sialosides: A *P. damsela* α-2,6-Sialyltransferase With Extremely Flexible Donor–Substrate Specificity. *Angew. Chem., Int. Ed.* **2006**, *45*, 3938–3944.
651. Chokhawala, H. A.; Yu, H.; Chen, X. High-Throughput Substrate Specificity Studies of Sialidases by Using Chemoenzymatically Synthesized Sialoside Libraries. *ChemBioChem* **2007**, *8*, 194–201.
652. Chokhawala, H. A.; Huang, S.; Lau, K.; Yu, H.; Cheng, J.; Thon, V.; Hurtado-Ziola, N.; Guerrero, J. A.; Varki, A.; Chen, X. Combinatorial Chemoenzymatic Synthesis and High-Throughput Screening of Sialosides. *ACS Chem. Biol.* **2008**, *3*, 567–576.

653. Zbiral, E. Synthesis of Sialic Acid Analogs and Their Behavior Towards the Enzymes of Sialic Acid Metabolism and Hemagglutinin X-31 of Influenza A-Virus. In: *Carbohydrates—Synthetic Methods and Applications in Medicinal Chemistry*; Ogura, H., Hasegawa, A., Suami, T., Eds.; VCH Verlagsgesellschaft: Weinheim, Germany, 1992; pp 304–339.
654. Brossmer, R.; Gross, H. J. Sialic Acid Analogs and Application for Preparation of Neoglycoconjugates. *Methods Enzymol.* **1994**, *247*, 153–176.
655. Roy, R. Sialoside Mimetics as Anti-inflammatory Agents and Inhibitors of Flu Virus Infections. In: *Carbohydrates in Drug Design*; Witczak, Z. J., Nieforth, K. A., Eds.; Marcel Dekker: New York, NY, USA, 1997; pp 83–135.
656. Von Itzstein, M.; Kiefel, M. J. Sialic Acid Analogues as Potential Antimicrobial Agents. In: *Carbohydrates in Drug Design*; Witczak, Z. J., Nieforth, K. A., Eds.; Marcel Dekker: New York, NY, USA, 1997; pp 39–82.
657. Kiefel, M. J.; von Itzstein, M. Recent Advances in the Synthesis of Sialic Acid Derivatives and Sialylmimetics as Biological Probes. *Chem. Rev.* **2002**, *102*, 471–490.
658. Wilson, J. C.; Kiefel, M. J.; von Itzstein, M. Sialic Acids and Sialylmimetics: Useful Chemical Probes of Sialic Acid-Recognizing Proteins. In: *Handbook of Carbohydrate Engineering*; Yarema, K. J., Ed.; Taylor & Francis CRC Press: Boca Raton, FL, USA, 2005; pp 687–748.
659. De Meo, C.; Priyadarshani, U. C-5 Modifications in *N*-Acetyl-neuraminic Acid: Scope and Limitations. *Carbohydr. Res.* **2008**, *343*, 1540–1552.
660. Cao, H.; Chen, X. General Consideration on Sialic Acid Chemistry. In: *Carbohydrate Microarrays: Methods and Protocols*; Chevolot, Y. Ed.; Methods Mol. Biol, Vol. 808; Springer Science + Business Media: Berlin, Germany, 2012; pp 31–56.
661. Khorlin, A. Y.; Privalova, I. M. Glycosidation of *N*-Acetylneuraminic Acid. *Bull. Acad. Sci. USSR, Div. Chem. Sci.* **1968**, *17*, 217–219.
662. Khorlin, A. Y.; Privalova, I. M.; Bystrova, I. B. The Synthesis of *N*-Acetylneuraminyl-(2→3)- and -(2→6)-hexoses. *Carbohydr. Res.* **1971**, *19*, 272–275.
663. Brossmer, R.; Friebolin, H.; Keilich, G.; Löser, B.; Supp, M. Synthesis of Disaccharides Containing *N*-Acetyl-D-neuraminic Acid. *Hoppe-Seyler's Z. Physiol. Chem.* **1978**, *359*, 1064.
664. Van der Vleugel, D. J. M.; Wassenburg, F. R.; Zwikker, J. W.; Vliegenthart, J. F. G. Synthesis of 6-*O*-(5-Acetamido-3,5-dideoxy-α-D-*glycero*-D-*galacto*-2-nonulopyranosylonic Acid)-D-galactose [6-*O*-(*N*-Acetyl-α-D-neuraminyl)-D-galactose]. *Carbohydr. Res.* **1982**, *104*, 221–233.
665. Van der Vleugel, D. J. M.; Zwikker, J. W.; Vliegenthart, J. F. G.; van Boeckel, S. A. A.; van Boom, J. H. Synthesis of 2-Acetamido-6-*O*-(5-acetamido-3,5-dideoxy-β-D-*glycero*-D-*galacto*-2-nonulopyranosylonic acid)-2-deoxy-D-glucose [2-Acetamido-6-*O*-(*N*-acetyl-β-D-neuraminyl)-2-deoxy-D-glucose]. *Carbohydr. Res.* **1982**, *105*, 19–31.
666. Okamoto, K.; Goto, T. Glycosidation of Sialic Acid. *Tetrahedron* **1990**, *46*, 5835–5857.
667. DeNinno, M. P. The Synthesis and Glycosidation of *N*-Acetylneuraminic Acid. *Synthesis* **1991**, 583–593.
668. Hasegawa, A.; Kiso, M. Systematic Synthesis of Gangliosides Toward the Elucidation and Biomedical Application of Their Biological Functions. In: *Carbohydrates—Synthetic Methods and Applications in Medicinal Chemistry*; Ogura, H., Hasegawa, A., Suami, T., Eds.; Kodansha Ltd.: Tokyo, Japan, 1992 pp 243–266
669. Boons, G.-J.; Demchenko, A. V. Recent Advances in *O*-Sialylation. *Chem. Rev.* **2000**, *100*, 4539–4565.
670. Ress, D. K.; Linhardt, R. J. Sialic Acid Donors: Chemical Synthesis and Glycosylation. *Curr. Org. Synth.* **2004**, *1*, 31–46.

671. De Meo, C.; Boons, G.-J.; Demchenko, A. V. Synthesis of Glycosides of Sialic Acid. In: *Comprehensive Glycoscience—From Chemistry to Systems Biology*; Kamerling, J. P., Boons, G.-J., Lee, Y. C., Suzuki, A., Taniguchi, N., Voragen, A. G. J., Eds.; Vol. 1; Elsevier Ltd.: Oxford, UK, 2007; pp 583–604
672. Adak, A. K.; Yu, C.-C.; Liang, C.-F.; Lin, C.-C. Synthesis of Sialic Acid-Containing Saccharides. *Curr. Opin. Chem. Biol.* **2013**, *17*, 1030–1038.
673. Ichikawa, Y.; Look, G. C.; Wong, C.-H. Enzyme-Catalyzed Oligosaccharide Synthesis. *Anal. Biochem.* **1992**, *202*, 215–238.
674. Taniguchi, N.; Honke, K.; Fukuda, M., Eds.; *Handbook of Glycosyltransferases and Related Genes—Sialyltransferases (Multiauthors)*; Springer-Verlag: Tokyo, Japan, 2002; pp 265–356.
675. Cheng, J.; Huang, S.; Yu, H.; Li, Y.; Lau, K.; Chen, X. Trans-Sialidase Activity of *Photobacterium damsela* α2,6-Sialyltransferase and Its Application in the Synthesis of Sialosides. *Glycobiology* **2010**, *20*, 260–268.
676. Kim, S.; Oh, D.-B.; Kang, H. A.; Kwon, O. Features and Applications of Bacterial Sialidases. *Appl. Microbiol. Biotechnol.* **2011**, *91*, 1–15.
677. Knirel, Y. A.; Kochetkov, N. K. 2,3-Diamino-2,3-dideoxyuronic and 5,7-Diamino-3,5,7,9-tetradeoxynonulosonic Acids: New Components of Bacterial Polysaccharides. *FEMS Microbiol. Rev.* **1987**, *46*, 381–385.
678. Knirel, Y. A.; Kochetkov, N. K. The Structure of the Lipopolysaccharides of Gram-Negative Bacteria—III. The Structure of the O-Antigens. *Biochemistry (Moscow)* **1994**, *59*, 1325–1383.
679. Knirel, Y. A.; Shashkov, A. S.; Tsvetkov, Y. E.; Jansson, P.-E.; Zähringer, U. 5,7-Diamino-3,5,7,9-tetradeoxynon-2-ulosonic Acids in Bacterial Glycopolymers: Chemistry and Biochemistry. *Adv. Carbohydr. Chem. Biochem.* **2003**, *58*, 371–417.
680. Kandiba, L.; Eichler, J. Analysis of Putative Nonulosonic Acid Biosynthesis Pathways in *Archaea* Reveals a Complex Evolutionary History. *FEMS Microbiol. Lett.* **2013**, *345*, 110–120.
681. Zunk, M.; Kiefel, M. J. The Occurrence and Biological Significance of the α-Keto-Sugars Pseudaminic Acid and Legionaminic Acid Within Pathogenic Bacteria. *RSC Adv.* **2014**, *4*, 3413–3421.
682. Knirel, Y. A.; Vinogradov, E. V.; L'vov, V. L.; Kocharova, N. A.; Shashkov, A. S.; Dmitriev, B. A.; Kochetkov, N. K. Sialic Acids of a New Type From the Lipopolysaccharides of *Pseudomonas aeruginosa* and *Shigella boydii*. *Carbohydr. Res.* **1984**, *133*, C5–C8.
683. Tsvetkov, Y. E.; Shashkov, A. S.; Knirel, Y. A.; Backinowsky, L. V.; Zähringer, U. Synthesis of 5,7-Diacetamido-3,5,7,9-tetradeoxy-L-*glycero*-D-*galacto*- and -L-*glycero*-D-*talo*-nonulosonic Acids, Putative Components of Bacterial Lipopolysaccharides. *Mendeleev Commun.* **2000**, *10*, 90–91.
684. Tsvetkov, Y. E.; Shashkov, A. S.; Knirel, Y. A.; Zähringer, U. Synthesis and Identification in Bacterial Lipopolysaccharides of 5,7-Diacetamido-3,5,7,9-tetradeoxy-D-*glycero*-D-*galacto*- and -D-*glycero*-D-*talo*-non-2-ulosonic Acids. *Carbohydr. Res.* **2001**, *331*, 233–237.
685. Tsvetkov, Y. E.; Shashkov, A. S.; Knirel, Y. A.; Zähringer, U. Synthesis and NMR Spectroscopy of Nine Stereoisomeric 5,7-Diacetamido-3,5,7,9-tetradeoxynon-2-ulosonic Acids. *Carbohydr. Res.* **2001**, *335*, 221–243.
686. Kenyon, J. J.; Marzaioli, A. M.; De Castro, C.; Hall, R. M. 5,7-Di-*N*-acetyl-Acinetaminic Acid: A Novel Non-2-ulosonic Acid Found in the Capsule of an *Acinetobacter baumannii* Isolate. *Glycobiology* **2015**, *25*, 644–654.
687. Kenyon, J. J.; Notaro, A.; Hsu, L. Y.; De Castro, C.; Hall, R. M. 5,7-Di-*N*-Acetyl-8-epiacinetaminic Acid: A New Non-2-ulosonic Acid Found in the K73 Capsule Produced by an *Acinetobacter baumannii* Isolate From Singapore. *Sci. Rep.* **2017**, *7*, 11357 (1–6).

688. Staaf, M.; Weintraub, A.; Widmalm, G. Structure Determination of the O-Antigenic Polysaccharide From the Enteroinvasive *Escherichia coli* O136. *Eur. J. Biochem.* **1999**, *263*, 656–661.
689. Thibault, P.; Logan, S. M.; Kelly, J. F.; Brisson, J.-R.; Ewing, C. P.; Trust, T. J.; Guerry, P. Identification of the Carbohydrate Moieties and Glycosylation Motifs in *Campylobacter jejuni* Flagellin. *J. Biol. Chem.* **2001**, *276*, 34862–34870.
690. Schirm, M.; Schoenhofen, I. C.; Logan, S. M.; Waldron, K. C.; Thibault, P. Identification of Unusual Bacterial Glycosylation by Tandem Mass Spectrometry Analyses of Intact Proteins. *Anal Chem.* **2005**, 77, 7774–7782.
691. Kenne, L.; Lindberg, B.; Schweda, E.; Gustafsson, B.; Holme, T. Structural Studies of the O-Antigen From *Vibrio cholerae* O:2. *Carbohydr. Res.* **1988**, *180*, 285–294.
692. Hitchen, P.; Brzostek, J.; Panico, M.; Butler, J. A.; Moris, H. R.; Dell, A.; Linton, D. Modification of the *Campylobacter jejuni* Flagellin Glycan by the Product of the Cj1295 Homopolymeric-Tract-Containing Gene. *Microbiology* **2010**, *156*, 1953–1962.
693. Knirel, Y. A.; Kocharova, N. A.; Shashkov, A. S.; Kochetkov, N. K. The Structure of the *Pseudomonas aeruginosa* Immunotype 6 O-Antigen: Isolation and Identification of 5-Acetamido-3,5,7,9-tetradeoxy-7-formamido-L-*glycero*-L-*manno*-nonulosonic Acid. *Carbohydr. Res.* **1986**, *145*, C1–C4.
694. Vinogradov, E.; Wilde, C.; Anderson, E. M.; Nakhamchik, A.; Lam, J. S.; Rowe-Magnus, D. A. Structure of the Lipopolysaccharide Core of *Vibrio vulnificus* Type Strain 27562. *Carbohydr. Res.* **2009**, *344*, 484–490.
695. Knirel, Y. A.; Vinogradov, E. V.; Shashkov, A. S.; Kochetkov, N. K.; L'vov, V.; Dmitriev, B. A. Identification of 5-Acetamido-3,5,7,9-tetradeoxy-7-[(*R*)-3-hydroxybutyramido]-L-*glycero*-L-*manno*-nonulosonic Acid as a Component of Bacterial Polysaccharides. *Carbohydr. Res.* **1985**, *141*, C1–C3.
696. Gil-Serrano, A. M.; Rodríguez-Carvajal, M. A.; Tejero-Mateo, P.; Espartero, J. L.; Menendez, M.; Corzo, J.; Ruiz-Sainz, J. E.; Buendía-Clavería, A. M. Structural Determination of a 5-Acetamido-3,5,7,9-tetradeoxy-7-(3-hydroxybutyramido)-L-*glycero*-L-*manno*-nonulosonic Acid-Containing Homopolysaccharide Isolated From *Sinorhizobium fredii* HH103. *Biochem. J.* **1999**, *342*, 527–535.
697. Knirel, Y. A.; Vinogradov, E. V.; Shashkov, A. S.; Dmitriev, B. A.; Kochetkov, N. K.; Stanislavsky, E. S.; Mashilova, G. M. Somatic Antigens of *Pseudomonas aeruginosa*—The Structure of O-Specific Polysaccharide Chains of *P. aeruginosa* O10 (Lányi) Lipopolysaccharides. *Eur. J. Biochem.* **1986**, *157*, 129–138.
698. Shashkov, A. S.; Tul'skaya, E. M.; Streshinskaya, G. M.; Senchenkova, S. N.; Avtukh, A. N.; Evtushenko, L. I. New Cell Wall Glycopolymers of the Representatives of the Genus *Kribbella*. *Carbohydr. Res.* **2009**, *344*, 2255–2262.
699. McNally, D. J.; Hui, J. P. M.; Aubry, A. J.; Mui, K. K. K.; Guerry, P.; Brisson, J.-R.; Logan, S. M.; Soo, E. C. Functional Characterization of the Flagellar Glycosylation Locus in *Campylobacter jejuni* 81-176 Using a Focused Metabolomics Approach. *J. Biol. Chem.* **2006**, *281*, 18489–18498.
700. Logan, S. M.; Hui, J. P. M.; Vinogradov, E.; Aubry, A. J.; Melanson, J. E.; Kelly, J. F.; Nothaft, H.; Soo, E. C. Identification of Novel Carbohydrate Modifications on *Campylobacter jejuni* 11168 Flagellin Using Metabolomics-Based Approaches. *FEBS J.* **2009**, *276*, 1014–1023.
701. Reuhs, B. L.; Geller, D. P.; Kim, J. S.; Fox, J. E.; Kolli, V. S. K.; Pueppke, S. G. *Sinorhizobium fredii* and *Sinorhizobium meliloti* Produce Structurally Conserved Lipopolysaccharides and Strain-Specific K Antigens. *Appl. Environ. Microbiol.* **1998**, *64*, 4930–4938.
702. Knirel, Y. A.; Kocharova, N. A.; Shashkov, A. S.; Dmitriev, B. A.; Kochetkov, N. K.; Stanislavsky, E. S.; Mashilova, G. M. Somatic Antigens of *Pseudomonas aeruginosa*—The Structure of O-Specific Polysaccharide Chains of the Lipopolysaccharides From *P. aeruginosa* O5 (Lányi) and Immunotype 6 (Fisher). *Eur. J. Biochem.* **1987**, *163*, 639–652.

703. Castric, P.; Cassels, F. J.; Carlson, R. W. Structural Characterization of the *Pseudomonas aeruginosa* 1244 Pilin Glycan. *J. Biol. Chem.* **2001**, *276*. 26479–26485/36058.
704. Hashii, N.; Isshiki, Y.; Iguchi, T.; Hisatsune, K.; Kondo, S. Structure and Serological Characterization of 5,7-Diamino-3,5,7,9-tetradeoxy-non-2-ulosonic Acid Isolated From Lipopolysaccharides of *Vibrio parahaemolyticus* O2 and O-Untypable Strain KX-V212. *Carbohydr. Res.* **2003**, *338*, 1055–1062.
705. McNally, D. J.; Aubry, A. J.; Hui, J. P. M.; Khieu, N. H.; Whitfield, D.; Ewing, C. P.; Guerry, P.; Brisson, J.-R.; Logan, S. M.; Soo, E. C. Targeted Metabolomics Analysis of *Campylobacter coli* VC167 Reveals Legionaminic Acid Derivatives as Novel Flagellar Glycans. *J. Biol. Chem.* **2007**, *282*, 14463–14475.
706. Kooistra, O.; Lüneberg, E.; Lindner, B.; Knirel, Y. A.; Frosch, M.; Zähringer, U. Complex *O*-Acetylation in *Legionella pneumophila* Serogroup 1 Lipopolysaccharide—Evidence for Two Genes Involved in 8-*O*-Acetylation of Legionaminic Acid. *Biochemistry* **2001**, *40*, 7630–7640.
707. Kooistra, O.; Lüneberg, E.; Knirel, Y. A.; Frosch, M.; Zähringer, U. N-Methylation in Polylegionaminic Acid is Associated With the Phase-Variable Epitope of *Legionella pneumophila* Serogroup 1 Lipopolysaccharide—Identification of 5-(*N*,*N*-Dimethylacetimidoyl)amino- and 5-Acetimidoyl(*N*-methyl)amino-7-acetamido-3,5,7,9-tetradeoxynon-2-ulosonic Acid in the O-Chain Polysaccharide. *Eur. J. Biochem.* **2002**, *269*, 560–572.
708. Li, X.; Perepelov, A. V.; Wang, Q.; Senchenkova, S. N.; Liu, B.; Shevelev, S. D.; Guo, X.; Shashkov, A. S.; Chen, W.; Wang, L.; Knirel, Y. A. Structural and Genetic Characterization of the O-Antigen of *Escherichia coli* O161 Containing a Derivative of a Higher Acidic Diamino Sugar, Legionaminic Acid. *Carbohydr. Res.* **2010**, *345*, 1581–1587.
709. Kandiba, L.; Aitio, O.; Helin, J.; Guan, Z.; Permi, P.; Bamford, D. H.; Eichler, J.; Roine, E. Diversity in Prokaryotic Glycosylation: An Archaeal-Derived N-Linked Glycan Contains Legionaminic Acid. *Mol. Microbiol.* **2012**, *84*, 578–593.
710. Twine, S. M.; Paul, C. J.; Vinogradov, E.; McNally, D. J.; Brisson, J.-R.; Mullen, J. A.; McMullin, D. R.; Jarrell, H. C.; Austin, J. W.; Kelly, J. F.; Logan, S. M. Flagellar Glycosylation in *Clostridium botulinum*. *FEBS J.* **2008**, *275*, 4428–4444.
711. Nazarenko, E. L.; Perepelov, A. V.; Shevchenko, L. S.; Daeva, E. D.; Ivanova, E. P.; Shashkov, A. S.; Widmalm, G. Structure of the O-Specific Polysaccharide From *Shewanella japonica* KMM 3601 Containing 5,7-Diacetamido-3,5,7,9-tetradeoxy-D-*glycero*-D-*talo*-non-2-ulosonic Acid. *Biochemistry (Moscow)* **2011**, *76*, 791–796.
712. Knirel, Y. A.; Senchenkova, S. N.; Kocharova, N. A.; Shashkov, A. S.; Helbig, J. H.; Zähringer, U. Identification of a Homopolymer of 5-Acetamidino-7-acetamido-3,5,7,9-tetradeoxy-D-*glycero*-D-*talo*-nonulosonic Acid in the Lipopolysaccharides of *Legionella pneumophila* Non-1 Serogroups. *Biochemistry (Moscow)* **2001**, *66*, 1035–1041.
713. Knirel, Y. A.; Helbig, J. H.; Zähringer, U. Structure of a Decasaccharide Isolated by Mild Acid Degradation and Dephosphorylation of the Lipopolysaccharide of *Pseudomonas fluorescens* Strain ATCC 49271. *Carbohydr. Res.* **1996**, *283*, 129–139.
714. Edebrink, P.; Jansson, P.-E.; Bøgwald, J.; Hoffman, J. Structural Studies of the *Vibrio salmonicida* Lipopolysaccharide. *Carbohydr. Res.* **1996**, *287*, 225–245.
715. Knirel, Y. A.; Senchenkova, S. N.; Shashkov, A. S.; Shevelev, S. D.; Perepelov, A. V.; Liu, B.; Feng, L.; Wang, L. First Isolation and Identification of a Di-*N*-acyl Derivative of 5,7-Diamino-3,5,7,9-tetradeoxy-L-*glycero*-D-*galacto*-non-2-ulosonic Acid (8-Epilegionaminic Acid). *Adv. Sci. Lett.* **2009**, *2*, 384–387.
716. Perepelov, A. V.; Wang, Q.; Liu, B.; Senchenkova, S. N.; Feng, L.; Shashkov, A. S.; Wang, L.; Knirel, Y. A. Structure of the O-Polysaccharide of *Escherichia coli* O61, Another *E. coli* O-Antigen That Contains 5,7-Diacetamido-3,5,7,9-tetradeoxy-L-*glycero*-D-*galacto*-non-2-ulosonic (Di-*N*-acetyl-8-epilegionaminic) Acid. *J. Carbohydr. Chem.* **2009**, *28*, 463–472.

717. Kilcoyne, M.; Shashkov, A. S.; Senchenkova, S. A.; Knirel, Y. A.; Vinogradov, E. V.; Radziejewska-Lebrecht, J.; Galimska-Stypa, R.; Savage, A. V. Structural Investigation of the O-Specific Polysaccharides of *Morganella morganii* Consisting of Two Higher Sugars. *Carbohydr. Res.* **2002**, *337*, 1697–1702.
718. Corsaro, M. M.; Evidente, A.; Lanzetta, R.; Lavermicocca, P.; Parrilli, M.; Ummarino, S. 5,7-Diamino-5,7,9-trideoxynon-2-ulosonic Acid: A Novel Sugar From a Phytopathogenic *Pseudomonas* Lipopolysaccharide. *Carbohydr. Res.* **2002**, *337*, 955–959.
719. Vinogradov, E.; MacLean, L. L.; Crump, E. M.; Perry, M. B.; Kay, W. W. Structure of the Polysaccharide Chain of the Lipopolysaccharide From *Flexibacter maritimus*. *Eur. J. Biochem.* **2003**, *270*, 1810–1815.
720. Vinogradov, E.; St. Michael, F.; Cox, A. D. The Structure of the LPS O-Chain of *Fusobacterium nucleatum* Strain 25586 Containing Two Novel Monosaccharides, 2-Acetamido-2,6-dideoxy-L-altrose and a 5-Acetimidoylamino-3,5,9-trideoxy-D-*gluco*-non-2-ulosonic Acid. *Carbohydr. Res.* **2017**, *440–441*, 10–15.
721. Mazumder, K.; Choudhury, B. P.; Nair, G. B.; Sen, A. K. Identification of a Novel Sugar 5,7-Diacetamido-8-amino-3,5,7,8,9-pentadeoxy-D-*glycero*-D-*galacto*-non-2-ulosonic Acid Present in the Lipopolysaccharide of *Vibrio parahaemolyticus* O3:K6. *Glycoconjugate J.* **2008**, *25*, 345–354.
722. Nazarenko, E. L.; Shashkov, A. S.; Knirel, Y. A.; Ivanova, E. P.; Ovodov, Y. S. Unusual Acidic Monosaccharides Forming Components of the O-Specific Polysaccharides of Bacteria of the Genus *Vibrio*. *Bioorg. Khim.* **1990**, *16*, 1426–1429.
723. Haseley, S. R.; Wilkinson, S. G. Structural Studies of the Putative O-Specific Polysaccharide of *Acinetobacter baumannii* O24 Containing 5,7-Diamino-3,5,7,9-tetradeoxy-L-*glycero*-D-*galacto*-nonulosonic Acid. *Eur. J. Biochem.* **1997**, *250*, 617–623.
724. Knirel, Y. A.; Rietschel, E. T.; Marre, R.; Zähringer, U. The Structure of the O-Specific Chain of *Legionella pneumophila* Serogroup 1 Lipopolysaccharide. *Eur. J. Biochem.* **1994**, *221*, 239–245.
725. Knirel, Y. A.; Grosskurth, H.; Helbig, J. H.; Zähringer, U. Structures of Decasaccharide and Tridecasaccharide Tetraphosphates Isolated by Strong Alkaline Degradation of *O*-Deacylated Lipopolysaccharide of *Pseudomonas fluorescens* Strain ATCC 49271. *Carbohydr. Res.* **1995**, *279*, 215–226.
726. Helbig, J. H.; Lück, P. C.; Knirel, Y. A.; Witzleb, W.; Zähringer, U. Molecular Characterization of a Virulence-Associated Epitope on the Lipopolysaccharide of *Legionella pneumophila* Serogroup 1. *Epidemiol. Infect.* **1995**, *115*, 71–78.
727. Knirel, Y. A.; Moll, H.; Helbig, J. H.; Zähringer, U. Chemical Characterization of a New 5,7-Diamino-3,5,7,9-tetradeoxynonulosonic Acid Released by Mild Acid Hydrolysis of the *Legionella pneumophila* Serogroup 1 Lipopolysaccharide. *Carbohydr. Res.* **1997**, *304*, 77–79.
728. Knirel, Y. A.; Vinogradov, E. V.; Shashkov, A. S.; Dmitriev, B. A.; Kochetkov, N. K.; Stanislavsky, E. S.; Mashilova, G. M. Somatic Antigens of *Pseudomonas aeruginosa*—The Structure of the O-Specific Polysaccharide Chain of the Lipopolysaccharide From *P. aeruginosa* O13 (Lányi). *Eur. J. Biochem.* **1987**, *163*, 627–637.
729. Vinogradov, E. V.; Shashkov, A. S.; Knirel, Y. A.; Kochetkov, N. K.; Dabrowski, J.; Grosskurth, H.; Stanislavsky, E. S.; Kholodkova, E. V. The Structure of the O-Specific Polysaccharide Chain of the Lipopolysaccharide of *Salmonella arizonae* O61. *Carbohydr. Res.* **1992**, *231*, 1–11.
730. Beynon, L. M.; Richards, J. C.; Perry, M. B. The Structure of the Lipopolysaccharide O Antigen From *Yersinia ruckeri* Serotype 01. *Carbohydr. Res.* **1994**, *256*, 303–317.
731. Schnaar, R. L.; Lee, Y. C. Discoveries of the Structures of Sialic Acid and CMP-Sialic Acid (1957-1960): A Letter From Saul Roseman. *Glycobiology* **2017**, *27*, 513–517.

732. Comb, D. G.; Roseman, S. Enzymic Synthesis of *N*-Acetyl-D-mannosamine. *Biochim. Biophys. Acta* **1958**, *29*, 653–654.
733. Daniels, A. D.; Campeotto, I.; van der Kamp, M. W.; Bolt, A. H.; Trinh, C. H.; Phillips, S. E. V.; Pearson, A. R.; Nelson, A.; Mulholland, A. J.; Berry, A. Reaction Mechanism of *N*-Acetylneuraminic Acid Lyase Revealed by a Combination of Crystallography, QM/MM Simulation, and Mutagenesis. *ACS Chem. Biol.* **2014**, *9*, 1025–1032.
734. Roseman, S. Metabolism of Sialic Acids and D-Mannosamine. *Fed. Proc.* **1962**, *21*, 1075–1083.
735. Warren, L.; Felsenfeld, H. The Biosynthesis of Sialic Acids. *J. Biol. Chem.* **1962**, *237*, 1421–1431.
736. Becker, C. E.; Day, H. G. Utilization of Glucosone and the Synthesis of Glucosamine in the Rat. *J. Biol. Chem.* **1953**, *201*, 795–801.
737. Spiro, R. G. Studies on the Biosynthesis of Glucosamine in the Intact Rat. *J. Biol. Chem.* **1959**, *234*, 742–748.
738. Lowther, D. A.; Rogers, H. J. The Relation of Glutamine to the Synthesis of Hyaluronate or Hyaluronate-Like Substances by Streptococci. *Biochem. J* **1953**, *53*, XXXIX.
739. Ghosh, S.; Blumenthal, H. J.; Davidson, E.; Roseman, S. Glucosamine Metabolism—V. Enzymatic Synthesis of Glucosamine 6-Phosphate. *J. Biol. Chem.* **1960**, *235*, 1265–1273.
740. Pogell, B. M.; Gryder, R. M. Enzymatic Synthesis of Glucosamine 6-Phosphate in Rat Liver. *J. Biol. Chem.* **1957**, *228*, 701–712.
741. Davidson, E. A.; Blumenthal, H. J.; Roseman, S. Glucosamine Metabolism—II. Studies on Glucosamine 6-Phosphate *N*-Acetylase. *J. Biol. Chem.* **1957**, *226*, 125–133.
742. Leloir, L. F.; Cardini, C. E. Enzymes Acting on Glucosamine Phosphates. *Biochim. Biophys. Acta* **1956**, *20*, 33–42.
743. Reissig, J. L. Phosphoacetylglucosamine Mutase of *Neurospora*. *J. Biol. Chem.* **1956**, *219*, 753–767.
744. Carlson, D. M. Phosphoacetylglucosamine Mutase From Pig Submaxillary Gland. *Methods Enzymol.* **1966**, *8*, 179–182.
745. Smith, E. E. B.; Munch-Petersen, A.; Mills, G. T. Uridyl Transferases and the Formation of Uridine Triphosphate: Pyrophosphorolysis of Uridine Diphosphoglucose and 'UDPX' by a Rat Liver Nuclear Fraction. *Nature* **1953**, *172*, 1038–1039.
746. Cabib, E.; Leloir, L. F.; Cardini, C. E. Uridine Diphosphate Acetylglucosamine. *J. Biol. Chem.* **1953**, *203*, 1055–1070.
747. Glaser, L.; Brown, D. H. The Enzymatic Synthesis In Vitro of Hyaluronic Acid Chains. *Proc. Natl. Acad. Sci. U. S. A.* **1955**, *41*, 253–260.
748. Cardini, C. E.; Leloir, L. F. Enzymatic Formation of Acetylgalactosamine. *J. Biol. Chem.* **1957**, *225*, 317–324.
749. Spivak, C. T.; Roseman, S. UDP-*N*-Acetyl-D-glucosamine 2′-Epimerase. *Methods Enzymol.* **1966**, *9*, 612–615.
750. Ghosh, S.; Roseman, S. Enzymatic Phosphorylation of *N*-Acetyl-D-mannosamine. *Proc. Natl. Acad. Sci. U. S. A.* **1961**, *47*, 955–958.
751. McGuire, E. J. Anabolic Reactions Involving Sialic Acids. In: *Biological Roles of Sialic Acid*; Rosenberg, A., Schengrund, C.-L., Eds.; Plenum Press: New York, NY, USA, 1976; pp 123–158.
752. Hinderlich, S.; Stäsche, R.; Zeitler, R.; Reutter, W. A Bifunctional Enzyme Catalyzes the First Two Steps in *N*-Acetylneuraminic Acid Biosynthesis of Rat Liver—Purification and Characterization of UDP-*N*-Acetylglucosamine 2-Epimerase/*N*-Acetylmannosamine Kinase. *J. Biol. Chem.* **1997**, *272*, 24313–24318.

753. Stäsche, R.; Hinderlich, S.; Weise, C.; Effertz, K.; Lucka, L.; Moormann, P.; Reutter, W. A Bifunctional Enzyme Catalyzes the First Two Steps in *N*-Acetylneuraminic Acid Biosynthesis of Rat Liver—Molecular Cloning and Functional Expression of UDP-*N*-acetyl-glucosamine 2-Epimerase/*N*-Acetylmannosamine Kinase. *J. Biol. Chem.* **1997**, *272*, 24319–24324.
754. Effertz, K.; Hinderlich, S.; Reutter, W. Selective Loss of Either the Epimerase or Kinase Activity of UDP-*N*-Acetylglucosamine 2-Epimerase/*N*-Acetylmannosamine Kinase Due to Site-Directed Mutagenesis Based on Sequence Alignments. *J. Biol. Chem.* **1999**, *274*, 28771–28778.
755. De Mendoza, A.; Ruiz-Trillo, I. The Mysterious Evolutionary Origin for the GNE Gene and the Root of Bilateria. *Mol. Biol. Evol.* **2011**, *28*, 2987–2991.
756. Hinderlich, S.; Neuenschwander, M.; Wratil, P. R.; Oder, A.; Lisurek, M.; Nguyen, L. D.; von Kries, J. P.; Hackenberger, C. P. R. Small Molecules Targeting Human *N*-Acetylmannosamine Kinase. *ChemBioChem* **2017**, *18*, 1279–1285.
757. Wratil, P. R.; Rigol, S.; Solecka, B.; Kohla, G.; Kannicht, C.; Reutter, W.; Giannis, A.; Nguyen, L. D. A Novel Approach to Decrease Sialic Acid Expression in Cells by a C-3-Modified *N*-Acetylmannosamine. *J. Biol. Chem.* **2014**, *289*, 32056–32063.
758. Jourdian, G. W.; Swanson, A. L.; Watson, D.; Roseman, S. Isolation of Sialic Acid 9-Phosphatase From Human Erythrocytes. *J. Biol. Chem.* **1964**, *239*, PC2714–PC2716.
759. Blacklow, R. S.; Warren, L. Biosynthesis of Sialic Acids by *Neisseria meningitidis*. *J. Biol. Chem.* **1962**, *237*, 3520–3526.
760. Vimr, E. R.; Kalivoda, K. A.; Deszo, E. L.; Steenbergen, S. M. Diversity of Microbial Sialic Acid Metabolism. *Microbiol. Mol. Biol. Rev.* **2004**, *68*, 132–153.
761. Comb, D. G.; Watson, D. R.; Roseman, S. The Sialic Acids—IX. Isolation of Cytidine 5′-Monophospho-*N*-acetylneuraminic Acid From *Escherichia coli* K-235. *J. Biol. Chem.* **1966**, *241*, 5637–5642.
762. Haverkamp, J.; Spoormaker, T.; Dorland, L.; Vliegenthart, J. F. G.; Schauer, R. Determination of the β-Anomeric Configuration of Cytidine 5′-Monophospho-*N*-acetylneuraminic Acid by ^{13}C NMR Spectroscopy. *J. Am. Chem. Soc.* **1979**, *101*, 4851–4853.
763. Ambrose, M. G.; Freese, S. J.; Reinhold, M. S.; Warner, T. G.; Vann, W. F. ^{13}C NMR Investigation of the Anomeric Specificity of CMP-*N*-Acetylneuraminic Acid Synthetase From *Escherichia coli*. *Biochemistry* **1992**, *31*, 775–780.
764. Warren, L.; Blacklow, R. S. The Biosynthesis of Cytidine 5′-Monophospho-*N*-acetylneuraminic Acid by an Enzyme From *Neisseria meningitidis*. *J. Biol. Chem.* **1962**, *237*, 3527–3534.
765. Roseman, S. Enzymatic Synthesis of Cytidine 5′-Monophospho-sialic Acids. *Proc. Natl. Acad. Sci. U. S. A.* **1962**, *48*, 437–441.
766. Spiro, M. J.; Spiro, R. G. Glycoprotein Biosynthesis: Studies on Thyroglobulin—Thyroid Sialyltransferase. *J. Biol. Chem.* **1968**, *243*, 6520–6528.
767. Kean, E. L.; Roseman, S. The Sialic Acids—X. Purification and Properties of Cytidine 5′-Monophosphosialic Acid Synthetase. *J. Biol. Chem.* **1966**, *241*, 5643–5650.
768. Kean, E. L.; Münster-Kühnel, A. K.; Gerardy-Schahn, R. CMP-Sialic Acid Synthetase of the Nucleus. *Biochim. Biophys. Acta* **2004**, *1673*, 56–65.
769. Kean, E. L. Nuclear Cytidine 5′-Monophosphosialic Acid Synthetase. *J. Biol. Chem.* **1970**, *245*, 2301–2308.
770. Carey, D. J.; Sommers, L. W.; Hirschberg, C. B. CMP-N-Acetylneuraminic Acid: Isolation From and Penetration Into Mouse Liver Microsomes. *Cell* **1980**, *19*, 597–605.
771. Eckhardt, M.; Mühlenhoff, M.; Bethe, A.; Gerardy-Schahn, R. Expression Cloning of the Golgi CMP-Sialic Acid Transporter. *Proc. Natl. Acad. Sci. U. S. A.* **1996**, *93*, 7572–7576.
772. Kean, E. L.; Bighouse, K. J. Cytidine 5′-Monophosphosialic Acid Hydrolase—Subcellular Location and Properties. *J. Biol. Chem.* **1974**, *249*, 7813–7823.

773. Liu, J. L.-C.; Shen, G.-J.; Ichikawa, Y.; Rutan, J. F.; Zapata, G.; Vann, W. F.; Wong, C.-H. Overproduction of CMP-Sialic Acid Synthetase for Organic Synthesis. *J. Am. Chem. Soc.* **1992**, *114*, 3901–3910.
774. Hamamoto, T.; Takeda, S.; Noguchi, T. Enzymatic Synthesis of Cytidine 5′-Monophospho-*N*-acetylneuraminic Acid. *Biosci. Biotechnol. Biochem.* **2005**, *69*, 1944–1950.
775. Harduin-Lepers, A.; Mollicone, R.; Delannoy, P.; Oriol, R. The Animal Sialyltransferases and Sialyltransferase-Related Genes: A Phylogenetic Approach. *Glycobiology* **2005**, *15*, 805–817.
776. Bhide, G. P.; Colley, K. J. Sialylation of N-Glycans: Mechanism, Cellular Compartmentalization and Function. *Histochem. Cell Biol.* **2017**, *147*, 149–174.
777. Jourdian, G. W.; Carlson, D. M.; Roseman, S. The Enzymatic Synthesis of Sialyllactose. *Biochem. Biophys. Res. Commun.* **1963**, *10*, 352–358.
778. Basu, S.; Kaufman, B. Ganglioside Biosynthesis in Embryonic Chicken Brain. *Fed. Proc.* **1965**, *24*, 479.
779. Roseman, S.; Carlson, D. M.; Jourdian, G. W.; McGuire, E. J.; Kaufman, B.; Basu, S.; Bartholomew, B. Animal Sialic Acid Transferases (Sialyl-transferases). *Methods Enzymol.* **1966**, *8*, 354–372.
780. Paulson, J. C.; Beranek, W. E.; Hill, R. L. Purification of a Sialyltransferase From Bovine Colostrum by Affinity Chromatography on CDP-Agarose. *J. Biol. Chem.* **1977**, *252*, 2356–2362.
781. Paulson, J. C.; Rearick, J. I.; Hill, R. L. Enzymatic Properties of β-D-Galactoside α2→6 Sialyltransferase From Bovine Colostrum. *J. Biol. Chem.* **1977**, *252*, 2363–2371.
782. Weinstein, J.; Lee, E. U.; McEntee, K.; Lai, P.-H.; Paulson, J. C. Primary Structure of β-Galactoside α2,6-Sialyltransferase—Conversion of Membrane-Bound Enzyme to Soluble Forms by Cleavage of the NH_2-Terminal Signal Anchor. *J. Biol. Chem.* **1987**, *262*, 17735–17743.
783. Ma, J.; Qian, R.; Rausa, F. M., III; Colley, K. J. Two Naturally Occurring α2,6-Sialyltransferase Forms With a Single Amino Acid Change in the Catalytic Domain Differ in Their Catalytic Activity and Proteolytic Processing. *J. Biol. Chem.* **1997**, *272*, 672–679.
784. Harduin-Lepers, A.; Petit, D.; Mollicone, R.; Delannoy, P.; Petit, J.-M.; Oriol, R. Evolutionary History of the α2,8-Sialyltransferase (ST8Sia) Gene Family: Tandem Duplications in Early Deuterostomes Explain Most of the Diversity Found in the Vertebrate ST8Sia Genes. *BMC Evol. Biol.* **2008**, *8*, 258 (1–18).
785. Bork, K.; Weidemann, W.; Berneck, B.; Kuchta, M.; Bennmann, D.; Thate, A.; Huber, O.; Gnanapragassam, V. S.; Horstkorte, R. The Expression of Sialyltransferases Is Regulated by the Bioavailability and Biosynthesis of Sialic Acids. *Gene Expression Patterns* **2017**, *23–24*, 52–58.
786. Albarracin, I.; Lassaga, F. E.; Caputto, R. Purification and Characterization of an Endogenous Inhibitor of the Sialyltransferase CMP-*N*-Acetylneuraminate: Lactosylceramide α2,6-*N*-Acetylneuraminyltransferase (EC 2.4.99.-). *Biochem. J.* **1988**, *254*, 559–565.
787. Van den Eijnden, D. H.; Schiphorst, W. E. C. M. Specificity and Inhibition of Sialyltransferases. In: *Sialic Acids;* Schauer, R., Yamakawa, T., Eds.; *Proc. Japanese-German Symp. Sialic Acids*; Kieler Verlag Wissenschaft + Bildung: Kiel, Germany, 1988; pp 128–129.
788. Kleineidam, R. G.; Schmelter, T.; Schwarz, R. T.; Schauer, R. Studies on the Inhibition of Sialyl- and Galactosyltransferase. *Glycoconjugate J.* **1997**, *14*, 57–66.
789. Büll, C.; Boltje, T. J.; Wassink, M.; de Graaf, A. M. A.; van Delft, F. L.; den Brok, M. H.; Adema, G. J. Targeting Aberrant Sialylation in Cancer Cells Using a Fluorinated Sialic Acid Analog Impairs Adhesion, Migration, and *In Vivo* Tumor Growth. *Mol. Cancer Ther.* **2013**, *12*, 1935–1946.

790. Kornfeld, S.; Kornfeld, R.; Neufeld, E. F.; O'Brien, P. J. The Feedback Control of Sugar Nucleotide Biosynthesis in Liver. *Proc. Natl. Acad. Sci. U. S. A.* **1964**, *52*, 371–379.
791. Schauer, R.; Reuter, G.; Mühlpfordt, H.; Andrade, A. F. B.; Pereira, M. E. A. The Occurrence of *N*-Acetyl- and *N*-Glycoloylneuraminic Acid in *Trypanosoma cruzi*. *Hoppe-Seyler's Z. Physiol. Chem.* **1983**, *364*, 1053–1057.
792. Schauer, R.; Srinivasan, G. V.; Coddeville, B.; Zanetta, J.-P.; Guérardel, Y. Low Incidence of *N*-Glycolylneuraminic Acid in Birds and Reptiles and Its Absence in the Platypus. *Carbohydr. Res.* **2009**, *344*, 1494–1500.
793. Peri, S.; Kulkarni, A.; Feyertag, F.; Berninsone, P. M.; Alvarez-Ponce, D. Phylogenetic Distribution of CMP-Neu5Ac Hydroxylase (CMAH), the Enzyme Synthetizing the Proinflammatory Human Xenoantigen Neu5Gc. *Genome Biol. Evol.* **2018**, *10*, 207–219.
794. Malykh, Y. N.; Schauer, R.; Shaw, L. *N*-Glycolylneuraminic Acid in Human Tumours. *Biochimie* **2001**, *83*, 623–634.
795. Hedlund, M.; Padler-Karavani, V.; Varki, N. M.; Varki, A. Evidence for a Human-Specific Mechanism for Diet and Antibody-Mediated Inflammation in Carcinoma Progression. *Proc. Natl. Acad. Sci. U. S. A.* **2008**, *105*, 18936–18941.
796. Uhlenbruck, G.; Schmitt, J. Über das Vorkommen von N-Glykolyl-neuraminsäure bei Affen. *Naturwissenschaften* **1965**, *52*, 163.
797. Hedlund, M.; Tangvoranuntakul, P.; Takematsu, H.; Long, J. M.; Housley, G. D.; Kozutsumi, Y.; Suzuki, A.; Wynshaw-Boris, A.; Ryan, A. F.; Gallo, R. L.; Varki, N.; Varki, A. *N*-Glycolylneuraminic Acid Deficiency in Mice: Implications for Human Biology and Evolution. *Mol. Cell. Biol.* **2007**, *27*, 4340–4346.
798. Naito-Matsui, Y.; Davies, L. R. L.; Takematsu, H.; Chou, H.-H.; Tangvoranuntakul, P.; Carlin, A. F.; Verhagen, A.; Heyser, C. J.; Yoo, S.-W.; Choudhury, B.; Paton, J. C.; Paton, A. W.; Varki, N. M.; Schnaar, R. L.; Varki, A. Physiological Exploration of the Long Term Evolutionary Selection Against Expression of *N*-Glycolylneuraminic Acid in the Brain. *J. Biol. Chem.* **2017**, *292*, 2557–2570.
799. Schauer, R.; Stoll, S.; Reuter, G. Differences in the Amount of *N*-Acetyl- and *N*-Glycoloyl-neuraminic Acid, as Well as *O*-Acylated Sialic Acids, of Fetal and Adult Bovine Tissues. *Carbohydr. Res.* **1991**, *213*, 353–359.
800. Ji, S.; Wang, F.; Chen, Y.; Yang, C.; Zhang, P.; Zhang, X.; Troy, F. A., II.; Wang, B. Developmental Changes in the Level of Free and Conjugated Sialic Acids, Neu5Ac, Neu5Gc and KDN in Different Organs of Pig: A LC–MS/MS Quantitative Analyses. *Glycoconjugate J.* **2017**, *34*, 21–30.
801. Faillard, H.; Blohm, M. Synthese der *N*-Glykolyl-neuraminsäure. *Hoppe-Seyler's Z. Physiol. Chem.* **1965**, *341*, 167–171.
802. Schoop, H. J.; Faillard, H. Beitrag zur Biosynthese der Glykolyl-Gruppe der *N*-Glykolyl-neuraminsäure. *Hoppe-Seyler's Z. Physiol. Chem.* **1967**, *348*, 1518–1524.
803. Da Fonseca-Wollheim, F.; Heesen, H.; Holzer, H. The Oxidative Decarboxylation of Hydroxypyruvate to Glycolyl Coenzyme A. *Biochem. Z.* **1964**, *340*, 383–390.
804. Kent, P. W.; Draper, P. Biosynthesis of Intestinal Mucins. Sialic Acids of Sheep Colonic Epithelial Mucin. *Biochem. J.* **1968**, *106*, 293–299.
805. Mårtensson, E.; Raal, A.; Svennerholm, L. Sialic Acids in Blood Serum. *Biochim. Biophys. Acta* **1958**, *30*, 124–129.
806. Von Jagow, R.; Kiese, M.; Lenk, W. Hydroxylation of Acetic Acid to Glycolic Acid in Rabbits. *Biochim. Biophys. Acta* **1968**, *158*, 45–50.
807. Petit, J.-F.; Adam, A.; Wietzerbin-Falszpan, J. Isolation of UDP-*N*-glycolylmuramyl-(Ala, Glu, DAP) From *Mycobacterium phlei*. *FEBS Lett.* **1970**, *6*, 55–57.
808. Schoop, H. J.; Schauer, R. Über den Einbau von molekularem Sauerstoff bei der Biosynthese der *N*-Glykolylneuraminsäure durch eine Neuraminsäure-*N*-acetyl-Hydroxylase. *Hoppe-Seyler's Z. Physiol. Chem.* **1969**, *350*, 13.

809. Schauer, R.; Schoop, H. J.; Faillard, H. Zur Biosynthese der Glykolyl-Gruppe der *N*-Glykolyl-neuraminsäure—Die oxydative Umwandlung der *N*-Acetyl-Gruppe zur Glykolyl-Gruppe. *Hoppe-Seyler's Z. Physiol. Chem.* **1968**, *349*, 645–652.
810. Schoop, H. J.; Schauer, R.; Faillard, H. Zur Biosynthese der *N*-Glykolyl-neuraminsäure—Die oxydative Entstehung von *N*-Glykolyl-neuraminsäure aus *N*-Acetyl-neuraminsäure. *Hoppe-Seyler's Z. Physiol. Chem.* **1969**, *350*, 155–162.
811. Schauer, R. Biosynthese der *N*-Glykoloylneuraminsäure durch eine von Ascorbinsäure bzw. NADPH abhängige *N*-Acetyl-hydroxylierende "*N*-Acetylneuraminate:O_2-Oxidoreduktase" in Homogenaten der Unterkieferspeicheldrüse vom Schwein. *Hoppe-Seyler's Z. Physiol. Chem.* **1970**, *351*, 783–791.
812. Buscher, H.-P.; Casals-Stenzel, J.; Schauer, R.; Mestres-Ventura, P. Biosynthesis of *N*-Glycolylneuraminic Acid in Porcine Submandibular Glands—Subcellular Site of Hydroxylation of *N*-Acetylneuraminic Acid in the Course of Glycoprotein Biosynthesis. *Eur. J. Biochem.* **1977**, *77*, 297–310.
813. Shaw, L.; Schauer, R. The Biosynthesis of *N*-Glycoloylneuraminic Acid Occurs by Hydroxylation of the CMP-Glycoside of *N*-Acetylneuraminic Acid. *Biol. Chem. Hoppe-Seyler* **1988**, *369*, 477–486.
814. Shaw, L.; Schauer, R. Detection of CMP-*N*-Acetylneuraminic Acid Hydroxylase Activity in Fractionated Mouse Liver. *Biochem. J.* **1989**, *263*, 355–363.
815. Schlenzka, W.; Shaw, L.; Schneckenburger, P.; Schauer, R. Purification and Characterization of CMP-*N*-Acetylneuraminic Acid Hydroxylase From Pig Submandibular Glands. *Glycobiology* **1994**, *4*, 675–683.
816. Schneckenburger, P.; Shaw, L.; Schauer, R. Purification, Characterization and Reconstitution of CMP-*N*-Acetylneuraminate Hydroxylase From Mouse Liver. *Glycoconjugate J.* **1994**, *11*, 194–203.
817. Kozutsumi, Y.; Kawano, T.; Yamakawa, T.; Suzuki, A. Participation of Cytochrome b_5 in CMP-*N*-Acetylneuraminic Acid Hydroxylation in Mouse Liver Cytosol. *J. Biochem.* **1990**, *108*, 704–706.
818. Kawano, T.; Kozutsumi, Y.; Takematsu, H.; Kawasaki, T.; Suzuki, A. Regulation of Biosynthesis of *N*-Glycolylneuraminic Acid-Containing Glycoconjugates: Characterization of Factors Required for NADH-Dependent Cytidine 5′Monophosphate-*N*-acetylneuraminic Acid Hydroxylation. *Glycoconjugate J.* **1993**, *10*, 109–115.
819. Shaw, L.; Schneckenburger, P.; Schlenzka, W.; Carlsen, J.; Christiansen, K.; Jürgensen, D.; Schauer, R. CMP-*N*-Acetylneuraminic Acid Hydroxylase From Mouse Liver and Pig Submandibular Glands—Interaction With Membrane-Bound and Soluble Cytochrome b_5-Dependent Electron Transport Chains. *Eur. J. Biochem.* **1994**, *219*, 1001–1011.
820. Shaw, L.; Schneckenburger, P.; Carlsen, J.; Christiansen, K.; Schauer, R. Mouse Liver Cytidine-5′-Monosphosphate-*N*-acetylneuraminic Acid Hydroxylase—Catalytic Function and Regulation. *Eur. J. Biochem.* **1992**, *206*, 269–277.
821. Schlenzka, W.; Shaw, L.; Kelm, S.; Schmidt, C. L.; Bill, E.; Trautwein, A. X.; Lottspeich, F.; Schauer, R. CMP-*N*-Acetylneuraminic Acid Hydroxylase: The First Cytosolic Rieske Iron-Sulphur Protein to Be Described in Eukarya. *FEBS Lett.* **1996**, *385*, 197–200.
822. Kawano, T.; Koyama, S.; Takematsu, H.; Kozutsumi, Y.; Kawasaki, H.; Kawashima, S.; Kawasaki, T.; Suzuki, A. Molecular Cloning of Cytidine Monophospho-*N*-acetylneuraminic Acid Hydroxylase—Regulation of Species- and Tissue-Specific Expression of *N*-Glycolylneuraminic Acid. *J. Biol. Chem.* **1995**, *270*, 16458–16463.
823. Irie, A.; Koyama, S.; Kozutsumi, Y.; Kawasaki, T.; Suzuki, A. The Molecular Basis for the Absence of *N*-Glycolylneuraminic Acid in Humans. *J. Biol. Chem.* **1998**, *273*, 15866–15871.

824. Chou, H.-H.; Takematsu, H.; Diaz, S.; Iber, J.; Nickerson, E.; Wright, K. L.; Muchmore, E. A.; Nelson, D. L.; Warren, S. T.; Varki, A. A Mutation in Human CMP-Sialic Acid Hydroxylase Occurred After the *Homo-Pan* Divergence. *Proc. Natl. Acad. Sci. U.S.A.* **1998**, *95*, 11751–11756.
824a. Hayakawa, T.; Satta, Y.; Gagneux, P.; Varki, A.; Takahata, N. *Alu*-Mediated Inactivation of the Human CMP-*N*-Acetylneuraminic Acid Hydroxylase Gene. *Proc. Natl. Acad. Sci. U. S. A.* **2001**, *98*, 11399–11404.
825. Klein, A.; Roussel, P. *O*-Acetylation of Sialic Acids. *Biochimie* **1998**, *80*, 49–57.
826. Schauer, R.; Srinivasan, G. V.; Wipfler, D.; Kniep, B.; Schwartz-Albiez, R. *O*-Acetylated Sialic Acids and Their Role in Immune Defense. *Adv. Exp. Med. Biol.* **2011**, *705*, 525–548.
827. Sgroi, D.; Varki, A.; Braesch-Andersen, S.; Stamenkovic, I. CD22, a B Cell-Specific Immunoglobulin Superfamily Member, Is a Sialic Acid-Binding Lectin. *J. Biol. Chem.* **1993**, *268*, 7011–7018.
827a. Kelm, S.; Pelz, A.; Schauer, R.; Filbin, M. T.; Tang, S.; de Bellard, M.-E.; Schnaar, R. L.; Mahoney, J. A.; Hartnell, A.; Bradfield, P.; Crocker, P. R. Sialoadhesin, Myelin-Associated Glycoprotein and CD22 Define a New Family of Sialic Acid-Dependent Adhesion Molecules of the Immunoglobulin Superfamily. *Curr. Biol.* **1994**, *4*, 965–972.
828. Kelm, S.; Schauer, R.; Manuguerra, J.-C.; Gross, H.-J.; Crocker, P. R. Modifications of Cell Surface Sialic Acids Modulate Cell Adhesion Mediated by Sialoadhesin and CD22. *Glycoconjugate J.* **1994**, *11*, 576–585.
829. Herrler, G.; Rott, R.; Klenk, H.-D.; Müller, H.-P.; Shukla, A. K.; Schauer, R. The Receptor-Destroying Enzyme of Influenza C Virus Is Neuraminate-*O*-acetylesterase. *EMBO J.* **1985**, *4*, 1503–1506.
830. Klotz, F. W.; Orlandi, P. A.; Reuter, G.; Cohen, S. J.; Haynes, J. D.; Schauer, R.; Howard, R. J.; Palese, P.; Miller, L. H. Binding of *Plasmodium falciparum* 175-Kilodalton Erythrocyte Binding Antigen and Invasion of Murine Erythrocytes Requires *N*-Acetylneuraminic Acid but Not Its *O*-Acetylated Form. *Mol. Biochem. Parasitol.* **1992**, *51*, 49–54.
831. Schauer, R. Biosynthese von *N*-Acetyl-*O*-acetylneuraminsäuren—I. Inkorporation von [^{14}C]Acetat in Schnitte der Unterkieferspeicheldrüse von Rind und Pferd. *Hoppe-Seyler's Z. Physiol. Chem.* **1970**, *351*, 595–602.
832. Schauer, R. Biosynthese von *N*-Acetyl-*O*-acetylneuraminsäuren—II. Untersuchungen über Substrat und intracelluläre Lokalisation der Acetyl-Coenzyme A: *N*-Acetylneuraminat-7- und 8-*O*-Acetyltransferase vom Rind. *Hoppe-Seyler's Z. Physiol. Chem.* **1970**, *351*, 749–758.
833. Tiralongo, J.; Schmid, H.; Thun, R.; Iwersen, M.; Schauer, R. Characterisation of the Enzymatic 4-*O*-Acetylation of Sialic Acids in Microsomes From Equine Submandibular Glands. *Glycoconjugate J.* **2000**, *17*, 849–858.
834. Iwersen, M.; Dora, H.; Kohla, G.; Gasa, S.; Schauer, R. Solubilisation and Properties of the Sialate-4-*O*-acetyltransferase From Guinea Pig Liver. *Biol. Chem.* **2003**, *384*, 1035–1047.
835. Schauer, R.; Wember, M.; Ferreira do Amaral, C. Synthesis of CMP-Glycosides of Radioactive *N*-Acetyl-, *N*-Glycoloyl-, *N*-Acetyl-7-*O*-acetyl- and *N*-Acetyl-8-*O*-acetylneuraminic Acids by CMP-Sialate Synthase From Bovine Submaxillary Glands. *Hoppe-Seyler's Z. Physiol. Chem.* **1972**, *353*, 883–886.
836. Schauer, R.; Wember, M. Studies on the Substrate Specificity of Acylneuraminate Cytidylyltransferase and Sialyltransferase of Submandibular Glands From Cow, Pig and Horse. *Hoppe-Seyler's Z. Physiol. Chem.* **1973**, *354*, 1405–1414.
837. Lrhorfi, L. A.; Srinivasan, G. V.; Schauer, R. Properties and Partial Purification of Sialate-*O*-acetyltransferase From Bovine Submandibular Glands. *Biol. Chem.* **2007**, *388*, 297–306.

838. Vandamme-Feldhaus, V.; Schauer, R. Characterization of the Enzymatic 7-*O*-Acetylation of Sialic Acids and Evidence for Enzymatic *O*-Acetyl Migration From C-7 to C-9 in Bovine Submandibular Glands. *J. Biochem.* **1998**, *124*, 111–121.
839. Diaz, S.; Higa, H. H.; Varki, A. Glycoprotein Sialate 7(9)-*O*-Acetyltransferase From Rat Liver Golgi Vesicles. *Methods Enzymol.* **1989**, *179*, 416–422.
840. Higa, H. H.; Butor, C.; Diaz, S.; Varki, A. *O*-Acetylation and De-*O*-Acetylation of Sialic Acids—*O*-Acetylation of Sialic Acids in the Rat Liver Golgi Apparatus Involves an Acetyl Intermediate and Essential Histidine and Lysine Residues—A Transmembrane Reaction? *J. Biol. Chem.* **1989**, *264*, 19427–19434.
841. Diaz, S.; Higa, H. H.; Hayes, B. K.; Varki, A. *O*-Acetylation and De-*O*-Acetylation of Sialic Acids—7- and 9-*O*-Acetylation of α2,6-Linked Sialic Acids on Endogenous *N*-Linked Glycans in Rat Liver Golgi Vesicles. *J. Biol. Chem.* **1989**, *264*, 19416–19426.
842. Higa, H. H.; Varki, A. Acetyl-Coenzyme A:Polysialic Acid *O*-Acetyltransferase From K1-Positive *Escherichia coli*—The Enzyme Responsible for the *O*-Acetyl plus Phenotype and for *O*-Acetyl Form Variation. *J. Biol. Chem.* **1988**, *263*, 8872–8878.
843. Deszo, E. L.; Steenbergen, S. M.; Freedberg, D. I.; Vimr, E. R. *Escherichia coli* K1 Polysialic Acid *O*-Acetyltransferase Gene, *neuO*, and the Mechanism of Capsule Form Variation Involving a Mobile Contingency Locus. *Proc. Natl. Acad. Sci. U. S. A.* **2005**, *102*, 5564–5569.
844. Bergfeld, A. K.; Claus, H.; Lorenzen, N. K.; Spielmann, F.; Vogel, U.; Mühlenhoff, M. The Polysialic Acid-Specific *O*-Acetyltransferase OatC From *Neisseria meningitidis* Serogroup C Evolved Apart From Other Bacterial Sialate *O*-Acetyltransferases. *J. Biol. Chem.* **2009**, *284*, 6–16.
845. Houliston, R. S.; Endtz, H. P.; Yuki, N.; Li, J.; Jarrell, H. C.; Koga, M.; van Belkum, A.; Karwaski, M.-F.; Wakarchuk, W. W.; Gilbert, M. Identification of a Sialate *O*-Acetyltransferase From *Campylobacter jejuni*—Demonstration of Direct Transfer to the C-9 Position of Terminal α-2,8-Linked Sialic Acid. *J. Biol. Chem.* **2006**, *281*, 11480–11486.
846. Arming, S.; Wipfler, D.; Mayr, J.; Merling, A.; Vilas, U.; Schauer, R.; Schwartz-Albiez, R.; Vlasak, R. The Human Cas1 Protein: A Sialic Acid-Specific *O*-Acetyltransferase? *Glycobiology* **2011**, *21*, 553–564.
847. Baumann, A.-M. T.; Bakkers, M. J. G.; Buettner, F. F. R.; Hartmann, M.; Grove, M.; Langereis, M. A.; de Groot, R. J.; Mühlenhoff, M. 9-*O*-Acetylation of Sialic Acids Is Catalysed by CASD1 *via* a Covalent Acetyl-Enzyme Intermediate. *Nat. Commun.* **2015**, *7673*(1 – 12), 6.
848. Kleineidam, R. G.; Hofmann, O.; Reuter, G.; Schauer, R. Indications for the Enzymatic Synthesis of 9-*O*-Lactoyl-*N*-acetylneuraminic Acid in Equine Liver. *Glycoconjugate J.* **1993**, *10*, 116–119.
849. Kelm, A.; Shaw, L.; Schauer, R.; Reuter, G. The Biosynthesis of 8-*O*-Methylated Sialic Acids in the Starfish *Asterias rubens*—Isolation and Characterisation of *S*-Adenosyl-L-methionine:Sialate-8-*O*-methyltransferase. *Eur. J. Biochem.* **1998**, *251*, 874–884.
850. Yamakawa, N.; Sato, C.; Miyata, S.; Maehashi, E.; Toriyama, M.; Sato, N.; Furuhata, K.; Kitajima, K. Development of Sensitive Chemical and Immunochemical Methods for Detecting Sulfated Sialic Acids and Their Application to Glycoconjugates From Sea Urchin Sperm and Eggs. *Biochimie* **2007**, *89*, 1396–1408.
851. Inoue, S.; Iwasaki, M. Isolation of a Novel Glycoprotein From the Eggs of Rainbow Trout: Occurrence of Disialosyl Groups on All Carbohydrate Chains. *Biochem. Biophys. Res. Commun.* **1978**, *83*, 1018–1023.
852. Wang, F.; Xie, B.; Wang, B.; Troy, F. A., II. LC-MS/MS Glycomic Analyses of Free and Conjugated Forms of the Sialic Acids, Neu5Ac, Neu5Gc and KDN in Human Throat Cancers. *Glycobiology* **2015**, *25*, 1362–1374.

853. Terada, T.; Kitajima, K.; Inoue, S.; Koppert, K.; Brossmer, R.; Inoue, R. Substrate Specificity of Rainbow Trout Testis CMP-3-Deoxy-D-*glycero*-D-*galacto*-nonulosonic Acid (CMP-Kdn) Synthetase—Kinetic Studies of the Reaction of Natural and Synthetic Analogues of Nonulosonic Acid Catalyzed by CMP-Kdn Synthetase. *Eur. J. Biochem.* **1996**, *236*, 852–855.
854. Angata, T.; Kitazume, S.; Terada, T.; Kitajima, K.; Inoue, S.; Troy, F. A., II.; Inoue, Y. Identification, Characterization, and Developmental Expression of a Novel α2→8-KDN-Transferase Which Terminates Elongation of α2→8-Linked Oligo-Polysialic Acid Chain Synthesis in Trout Egg Polysialoglycoproteins. *Glycoconjugate J.* **1994**, *11*, 493–499.
855. Lewis, A. L.; Desa, N.; Hansen, E. E.; Knirel, Y. A.; Gordon, J. I.; Gagneux, P.; Nizet, V.; Varki, A. Innovations in Host and Microbial Sialic Acid Biosynthesis Revealed by Phylogenomic Prediction of Nonulosonic Acid Structure. *Proc. Natl. Acad. Sci. U. S. A.* **2009**, *106*, 13552–13557.
856. Schoenhofen, I. C.; Vinogradov, E.; Whitfield, D. M.; Brisson, J.-R.; Logan, S. M. The CMP-Legionaminic Acid Pathway in *Campylobacter*: Biosynthesis Involving Novel GDP-Linked Precursors. *Glycobiology* **2009**, *19*, 715–725.
857. Schoenhofen, I. C.; McNally, D. J.; Brisson, J.-R.; Logan, S. M. Elucidation of the CMP-Pseudaminic Acid Pathway in *Helicobacter pylori*: Synthesis From UDP-*N*-acetylglucosamine by a Single Enzymatic Reaction. *Glycobiology* **2006**, *16*, 8C–14C.
858. Liu, F.; Aubry, A. J.; Schoenhofen, I. C.; Logan, S. M.; Tanner, M. E. The Engineering of Bacteria Bearing Azido-Pseudaminic Acid-Modified Flagella. *ChemBioChem* **2009**, *10*, 1317–1320.
859. Hinderlich, S.; Tauber, R.; Bertozzi, C. R.; Hackenberger, C. P. R. Werner Reutter: A Visionary Pioneer in Molecular Glycobiology. *ChemBioChem* **2017**, *18*, 1141–1145.
860. Brossmer, R.; Nebelin, E. Synthesis of N-Formyl- and N-Succinyl-D-neuraminic Acid on the Specificity of Neuraminidase. *FEBS Lett.* **1969**, *4*, 335–336.
861. Grünholz, H.-J.; Harms, E.; Opetz, M.; Reutter, W.; Černý, M. Inhibition of *In Vitro* Biosynthesis of *N*-Acetylneuraminic Acid by *N*-Acyl- and *N*-Alkyl-2-amino-2-deoxyhexoses. *Carbohydr. Res.* **1981**, *96*, 259–270.
862. Kayser, H.; Zeitler, R.; Kannicht, C.; Grunow, D.; Nuck, R.; Reutter, W. Biosynthesis of a Nonphysiological Sialic Acid in Different Rat Organs, Using *N*-Propanoyl-D-hexosamines as Precursors. *J. Biol. Chem.* **1992**, *267*, 16934–16938.
863. Kayser, H.; Geilen, C. C.; Paul, C.; Zeitler, R.; Reutter, W. Incorporation of *N*-Acyl-2-amino-2-deoxy-hexoses Into Glycosphingolipids of the Pheochromocytoma Cell Line PC12. *FEBS Lett.* **1992**, *301*, 137–140.
864. Roehlecke, C.; Horstkorte, R.; Reutter, W. Stimulation of Human Peripheral Blood Mononuclear Cells by the Sialic Acid Precursor *N*-Propanoylmannosamine. *Glycoconjugate J.* **2013**, *30*, 813–818.
865. Schmidt, C.; Stehling, P.; Schnitzer, J.; Reutter, W.; Horstkorte, R. Biochemical Engineering of Neural Cell Surfaces by the Synthetic *N*-Propanoyl-Substituted Neuraminic Acid Precursor. *J. Biol. Chem.* **1998**, *273*, 19146–19152.
866. Keppler, O. T.; Stehling, P.; Herrmann, M.; Kayser, H.; Grunow, D.; Reutter, W.; Pawlita, M. Biosynthetic Modulation of Sialic Acid-Dependent Virus-Receptor Interactions of Two Primate Polyoma Viruses. *J. Biol. Chem.* **1995**, *270*, 1308–1314.
867. Wieser, J. R.; Heisner, A.; Stehling, P.; Oesch, F.; Reutter, W. *In Vivo* Modulated *N*-Acyl Side Chain of *N*-Acetylneuraminic Acid Modulates the Cell Contact-Dependent Inhibition of Growth. *FEBS Lett.* **1996**, *395*, 170–173.
868. Gnanapragassam, V. S.; Bork, K.; Galuska, C. E.; Galuska, S. P.; Glanz, D.; Nagasundaram, M.; Bache, M.; Vordermark, D.; Kohla, G.; Kannicht, C.; Schauer, R.; Horstkorte, R. Sialic Acid Metabolic Engineering: A Potential Strategy for the Neuroblastoma Therapy. *PLoS One* **2014**, *e105403*(1–10), 9.

869. Cheng, B.; Xie, R.; Dong, L.; Chen, X. Metabolic Remodeling of Cell-Surface Sialic Acids: Principles, Applications, and Recent Advances. *ChemBioChem* **2016**, *17*, 11–27.
870. Wratil, P. R.; Horstkorte, R.; Reutter, W. Metabolic Glycoengineering With *N*-Acyl Side Chain Modified Mannosamines. *Angew. Chem., Int. Ed.* **2016**, *55*, 9482–9512.
871. Schauer, R.; Veh, R. W.; Sander, M.; Corfield, A. P.; Wiegandt, H. "Neuraminidase-Resistant" Sialic Acid Residues of Gangliosides. *Adv. Exp. Med. Biol.* **1980**, *125*, 283–294.
872. Schauer, R.; Reuter, G.; Stoll, S. Sialate-*O*-acetylesterases: Key Enzymes in Sialic Acid Catabolism. *Biochimie* **1988**, *70*, 1511–1519.
873. Schauer, R.; Shukla, A. K. Isolation and Properties of Two Sialate-*O*-Acetylesterases From Horse Liver With 4- and 9-*O*-Acetyl Specificities. *Glycoconjugate J.* **2008**, *25*, 625–632.
874. Schauer, R.; Reuter, G.; Stoll, S.; Shukla, A. K. Partial Purification and Characterization of Sialate *O*-Acetylesterase From Bovine Brain. *J. Biochem.* **1989**, *106*, 143–150.
875. Smits, S. L.; Gerwig, G. J.; van Vliet, A. L. W.; Lissenberg, A.; Briza, P.; Kamerling, J. P.; Vlasak, R.; de Groot, R. J. Nidovirus Sialate-*O*-acetylesterases—Evolution and Substrate Specificity of Coronaviral and Toroviral Receptor-Destroying Enzymes. *J. Biol. Chem.* **2005**, *280*, 6933–6941.
876. Srinivasan, G. V.; Schauer, R. Assays of Sialate-*O*-acetyltransferases and Sialate-*O*-acetylesterases. *Glycoconjugate J.* **2009**, *26*, 935–944.
877. Hirst, G. K. The Relationship of the Receptors of a New Strain of Virus to Those of the Mumps-NDV-Influenza Group. *J. Exp. Med.* **1950**, *91*, 177–184.
878. Herrler, G.; Rott, R.; Klenk, H.-D. Neuraminic Acid Is Involved in the Binding of Influenza C Virus to Erythrocytes. *Virology* **1985**, *141*, 144–147.
879. Schauer, R.; Reuter, G.; Stoll, S.; Posadas del Rio, F.; Herrler, G.; Klenk, H.-D. Isolation and Characterization of Sialate 9(4)-*O*-Acetylesterase From Influenza C Virus. *Biol. Chem. Hoppe-Seyler* **1988**, *369*, 1121–1130.
880. Herrler, G.; Klenk, H.-D. Structure and Function of the Hef Glycoprotein of Influenza C Virus. *Adv. Virus Res.* **1991**, *40*, 213–234.
881. Garcia-Sastre, A.; Villar, E.; Manuguerra, J. C.; Hannoun, C.; Cabezas, J. A. Activity of Influenza C Virus *O*-Acetylesterase With *O*-Acetyl-Containing Compounds. *Biochem. J.* **1991**, *273*, 435–441.
882. Rosenthal, P. B.; Zhang, X.; Formanowski, F.; Fitz, W.; Wong, C.-H.; Meier-Ewert,- H.; Skehel, J. J.; Wiley, D. C. Structure of the Haemagglutinin-Esterase-Fusion Glycoprotein of Influenza C Virus. *Nature* **1998**, *396*, 92–96.
883. Regl, G.; Kaser, A.; Iwersen, M.; Schmid, H.; Kohla, G.; Strobl, B.; Vilas, U.; Schauer, R.; Vlasak, R. The Hemagglutinin-Esterase of Mouse Hepatitis Virus Strain S Is a Sialate-4-*O*-acetylesterase. *J. Virol.* **1999**, *73*, 4721–4727.
884. Shen, Y.; Tiralongo, J.; Kohla, G.; Schauer, R. Regulation of Sialic Acid *O*-Acetylation in Human Colon Mucosa. *Biol. Chem.* **2004**, *385*, 145–152.
885. Pillai, S. Rethinking Mechanisms of Autoimmune Pathogenesis. *J. Autoimmun.* **2013**, *45*, 97–103.
886. Surolia, I.; Pirnie, S. P.; Chellappa, V.; Taylor, K. N.; Cariappa, A.; Moya, J.; Liu, H.; Bell, D. W.; Driscoll, D. R.; Diederichs, S.; Haider, K.; Netravali, I.; Le, S.; Elia, R.; Dow, E.; Lee, A.; Freudenberg, J.; De Jager, P. L.; Chretien, Y.; Varki, A.; MacDonald, M. E.; Gillis, T.; Behrens, T. W.; Bloch, D.; Collier, D.; Korzenik, J.; Podolsky, D. K.; Hafler, D.; Murali, M.; Sands, B.; Stone, J. H.; Gregersen, P. K.; Pillai, S. Functionally Defective Germline Variants of Sialic Acid Acetylesterase in Autoimmunity. *Nature* **2010**, *466*, 243–247.
887. Yamamoto, M.; Iguchi, G.; Bando, H.; Fukuoka, H.; Suda, K.; Takahashi, M.; Nishizawa, H.; Matsumoto, R.; Tojo, K.; Mokubo, A.; Ogata, T.; Takahashi, Y. A

Missense Single-Nucleotide Polymorphism in the Sialic Acid Acetylesterase (*SIAE*) Gene Is Associated With Anti-PIT-1 Antibody Syndrome. *Endocr. J. (Tokyo)* **2014**, *61*, 641–644.
888. Orizio, F.; Damiati, E.; Giacopuzzi, E.; Benaglia, G.; Pianta, S.; Schauer, R.; Schwartz-Albiez, R.; Borsani, G.; Bresciani, R.; Monti, E. Human Sialic Acid Acetyl Esterase: Towards a Better Understanding of a Puzzling Enzyme. *Glycobiology* **2015**, *25*, 992–1006.
889. Hirst, G. K. Adsorption of Influenza Hemagglutinins and Virus by Red Blood Cells. *J. Exp. Med.* **1942**, *76*, 195–209.
890. McCrea, J. F. Modification of Red-Cell Agglutinability by *Cl. welchii* Toxins. *Aust. J. Exp. Biol. Med. Sci.* **1947**, *25*, 127–136.
891. Burnet, F. M.; McCrea, J. F.; Stone, J. D. Modification of Human Red Cells by Virus Action—I. The Receptor Gradient for Virus Action in Human Red Cells. *Br. J. Exp. Pathol.* **1946**, *27*, 228–236.
892. Gottschalk, A. Neuraminidase: The Specific Enzyme of Influenza Virus and *Vibrio cholerae*. *Biochim. Biophys. Acta* **1957**, *23*, 645–646.
893. IUB Enzyme Nomenclature 1984, Academic Press: New York, NY, USA, 1984.
894. Cabezas, J. A. Some Questions and Suggestions on the Type References of the Official Nomenclature (IUB) for Sialidase(s) and Endosialidase. *Biochem. J.* **1991**, *278*, 311–312.
895. Warren, L.; Spearing, C. W. Mammalian Sialidase (Neuraminidase). *Biochem. Biophys. Res. Commun.* **1960**, *3*, 489–492.
896. Ada, G. L.; Lind, P. E. Neuraminidase in the Chorioallantois of the Chick Embryo. *Nature* **1961**, *190*, 1169–1171.
897. Carubelli, R.; Trucco, R. E.; Caputto, R. Neuraminidase Activity in Mammalian Organs. *Biochim. Biophys. Acta* **1962**, *60*, 196–197.
898. Cassidy, J. T.; Jourdian, G. W.; Roseman, S. The Sialic Acids—VI. Purification and Properties of Sialidase From *Clostridium perfringens*. *J. Biol. Chem.* **1965**, *240*, 3501–3506.
899. Roggentin, T.; Kleineidam, R. G.; Majewski, D. M.; Tirpitz, D.; Roggentin, P.; Schauer, R. An Immunoassay for the Rapid and Specific Detection of Three Sialidase-Producing Clostridia Causing Gas Gangrene. *J. Immunol. Methods* **1993**, *157*, 125–133.
900. Uchida, Y.; Tsukada, Y.; Sugimori, T. Enzymatic Properties of Neuraminidases From *Arthrobacter ureafaciens*. *J. Biochem.* **1979**, *86*, 1573–1585.
901. Ada, G. L.; French, E. L.; Lind, P. E. Purification and Properties of Neuraminidase From *Vibrio cholerae*. *J. Gen. Microbiol.* **1961**, *24*, 409–421.
902. Hoyer, L. L.; Roggentin, P.; Schauer, R.; Vimr, E. R. Purification and Properties of Cloned *Salmonella typhimurium* LT2 Sialidase With Virus-Typical Kinetic Preferences for Sialyl $\alpha 2 \rightarrow 3$ Linkages. *J. Biochem.* **1991**, *110*, 462–467.
903. Drzeniek, R. Substrate Specificity of Neuraminidases. *Histochem. J.* **1973**, *5*, 271–290.
904. Paulson, J. C.; Weinstein, J.; Dorland, L.; van Halbeek, H.; Vliegenthart, J. F. G. Newcastle Disease Virus Contains a Linkage-Specific Glycoprotein Sialidase—Application to the Localization of Sialic Acid Residues in *N*-Linked Oligosaccharides of α_1-Acid Glycoprotein. *J. Biol. Chem.* **1982**, *257*, 12734–12738.
905. Royle, L.; Mattu, T. S.; Hart, E.; Langridge, J. I.; Merry, A. H.; Murphy, N.; Harvey, D. J.; Dwek, R. A.; Rudd, P. M. An Analytical and Structural Database Provides a Strategy for Sequencing O-Glycans From Microgram Quantities of Glycoproteins. *Anal. Biochem.* **2002**, *304*, 70–90.
906. Amano, J.; Nishimura, R.; Mochizuki, M.; Kobata, A. Comparative Study of the Mucin-Type Sugar Chains of Human Chorionic Gonadotropin Present in the Urine of Patients With Trophoblastic Diseases and Healthy Pregnant Women. *J. Biol. Chem.* **1988**, *263*, 1157–1165.

907. Kwiatkowski, B.; Boschek, B.; Thiele, H.; Stirm, S. Endo-*N*-Acetylneuraminidase Associated With Bacteriophage Particles. *J. Virol.* **1982**, *43*, 697–704.
908. Faillard, H. Aminozuckerspaltende Enzyme. In: *Hoppe-Seyler/Thierfelder, Handbuch der physiologisch- und pathologisch-chemischen Analyse. 10. Aufl. Bd. VI/B*; Lang, K. Ed.; Springer: Berlin, Germany, 1966; pp 1234–1261.
909. Drzeniek, R. Viral and Bacterial Neuraminidases. *Curr. Top. Microbiol. Immunol.* **1972**, *59*, 35–74.
910. Corfield, A. P.; Michalski, J.-C.; Schauer, R. The Substrate Specificity of Sialidases From Microorganisms and Mammals. In: *Perspectives in Inherited Metabolic Diseases*; Tettamanti, G., Durand, P., Di Donato, S., Eds.; Vol. 4; Edi Ermes: Milan, Italy, 1981; pp 3–70.
911. Saito, M.; Yu, R. K. Biochemistry and Function of Sialidases. In: *Biology of the Sialic Acids*; Rosenberg, A. Ed.; Plenum Press: New York, USA, 1995; pp 261–313.
912. Achyuthan, K. E.; Achyuthan, A. M. Comparative Enzymology, Biochemistry and Pathophysiology of Human *exo*-α-Sialidases (Neuraminidases). *Comp. Biochem. Physiol. B: Biochem. Mol. Biol.* **2001**, *129B*, 29–64.
913. Monti, E.; Preti, A.; Venerando, B.; Borsani, G. Recent Development in Mammalian Sialidase Molecular Biology. *Neurochem. Res.* **2002**, *27*, 649–663.
914. Giacopuzzi, E.; Bresciani, R.; Schauer, R.; Monti, E.; Borsani, G. New Insights on the Sialidase Protein Family Revealed by a Phylogenetic Analysis in Metazoa. *PLoS One* **2012**, *e44193*(1–17), 7.
915. Miyagi, T.; Yamaguchi, K. Mammalian Sialidases: Physiological and Pathological Roles in Cellular Functions. *Glycobiology* **2012**, *22*, 880–896.
916. Ada, G. L.; French, E. L. Purification of Bacterial Neuraminidase (Receptor-Destroying Enzyme). *Nature* **1959**, *183*, 1740–1741.
917. Kawamura, K. Production and Purification of Receptor Destroying Enzyme (RDE). *Tohoku J. Exp. Med.* **1958**, *67*, 372.
918. Nees, S.; Schauer, R. Induction of Neuraminidase From *Clostridium perfringens* and the Cooperation of This Enzyme With Acetylneuraminate Pyruvate-Lyase. *Behring Inst. Mitt.* **1974**, *55*, 68–78.
919. Thomas, G. H. Sialic Acid Acquisition in Bacteria—One Substrate, Many Transporters. *Biochem. Soc. Trans.* **2016**, *44*, 760–765.
920. Vimr, E. R.; Troy, F. A. Identification of an Inducible Catabolic System for Sialic Acids (*nan*) in *Escherichia coli*. *J. Bacteriol.* **1985**, *164*, 845–853.
921. Corfield, A. P.; Schauer, R. Metabolism of Sialic Acids. In: *Sialic Acids—Chemistry, Metabolism and Function*; Schauer, R. Ed.; Cell Biology Monographs, Vol. 10; Springer-Verlag: Vienna, Austria, 1982; pp 195–261.
922. Kleineidam, R. G.; Furuhata, K.; Ogura, H.; Schauer, R. 4-Methylumbelliferyl-α-Glycosides of Partially *O*-Acetylated *N*-Acetylneuraminic Acids as Substrates of Bacterial and Viral Sialidases. *Biol. Chem. Hoppe Seyler* **1990**, *371*, 715–719.
923. Miyagi, T.; Tsuiki, S. Rat-Liver Lysosomal Sialidase—Solubilization, Substrate Specificity and Comparison With the Cytosolic Sialidase. *Eur. J. Biochem.* **1984**, *141*, 75–81.
924. Miyagi, T.; Konno, K.; Emori, Y.; Kawasaki, H.; Suzuki, K.; Yasui, A.; Tsuiki, S. Molecular Cloning and Expression of cDNA Encoding Rat Skeletal Muscle Cytosolic Sialidase. *J. Biol. Chem.* **1993**, *268*, 26435–26440.
925. Miyagi, T.; Wada, T.; Iwamatsu, A.; Hata, K.; Yoshikawa, Y.; Tokuyama, S.; Sawada, M. Molecular Cloning and Characterization of a Plasma Membrane-Associated Sialidase Specific for Gangliosides. *J. Biol. Chem.* **1999**, *274*, 5004–5011.
926. Monti, E.; Bassi, M. T.; Bresciani, R.; Civini, S.; Croci, G. L.; Papini, N.; Riboni, M.; Zanchetti, G.; Ballabio, A.; Preti, A.; Tettamanti, G.; Venerando, B.; Borsani, G. Molecular Cloning and Characterization of *NEU4*, the Fourth Member of the Human Sialidase Gene Family. *Genomics* **2004**, *83*, 445–453.

927. Miyagi, T.; Takahashi, K.; Shiozaki, K.; Yamaguchi, K.; Hosono, M. Plasma Membrane-Associated Sialidase Confers Cancer Initiation, Promotion and Progression. *Adv. Exp. Med. Biol.* **2015**, *842*, 139–145.
928. Forcella, M.; Mozzi, A.; Stefanini, F. M.; Riva, A.; Epistolio, S.; Molinari, F.; Merlo, E.; Monti, E.; Fusi, P.; Frattini, M. Deregulation of Sialidases in Human Normal and Tumor Tissues. *Cancer Biomarkers* **2018**, *21*, 591–601.
929. Hunter, C. D.; Khanna, N.; Richards, M. R.; Rezaei Darestani, R.; Zou, C.; Klassen, J. S.; Cairo, C. W. Human Neuraminidase Isoenzymes Show Variable Activities for 9-*O*-Acetyl-sialoside Substrates. *ACS Chem. Biol.* **2018**, *13*, 922–932.
930. Roggentin, P.; Schauer, R.; Hoyer, L. L.; Vimr, E. R. The Sialidase Superfamily and Its Spread by Horizontal Gene Transfer. *Mol. Microbiol.* **1993**, *9*, 915–921.
931. Pereira, M. E. A Developmentally Regulated Neuraminidase Activity in *Trypanosoma cruzi*. *Science* **1983**, *219*, 1444–1446.
932. Prioli, R. P.; Mejia, J. S.; Aji, T.; Aikawa, M.; Pereira, M. E. A. *Trypanosoma cruzi:* Localization of Neuraminidase on the Surface of Trypomastigotes. *Trop. Med. Parasitol.* **1991**, *42*, 146–150.
933. Previato, J. O.; Andrade, A. F. B.; Pessolani, M. C. V.; Mendonça-Previato, L. Incorporation of Sialic Acid Into *Trypanosoma cruzi* Macromolecules—A Proposal for a New Metabolic Route. *Mol. Biochem. Parasitol.* **1985**, *16*, 85–96.
934. Schenkman, S.; Jiang, M.-S.; Hart, G. W.; Nussenzweig, V. A Novel Cell Surface trans-Sialidase of *Trypanosoma cruzi* Generates a Stage-Specific Epitope Required for Invasion of Mammalian Cells. *Cell* **1991**, *65*, 1117–1125.
935. Engstler, M.; Reuter, G.; Schauer, R. Purification and Characterization of a Novel Sialidase Found in Procyclic Culture Forms of *Trypanosoma brucei*. *Mol. Biochem. Parasitol.* **1992**, *54*, 21–30.
936. Engstler, M.; Schauer, R.; Brun, R. Distribution of Developmentally Regulated trans-Sialidases in the Kinetoplastida and Characterization of a Shed trans-Sialidase Activity From Procyclic *Trypanosoma congolense*. *Acta Trop.* **1995**, *59*, 117–129.
937. Tiralongo, E.; Schrader, S.; Lange, H.; Lemke, H.; Tiralongo, J.; Schauer, R. Two Trans-Sialidase Forms With Different Sialic Acid Transfer and Sialidase Activities From *Trypanosoma congolense*. *J. Biol. Chem.* **2003**, *278*, 23301–23310.
938. Wilbrink, M. H.; ten Kate, G. A.; Sanders, P.; Gerwig, G. J.; van Leeuwen, S. S.; Sallomons, E.; Klarenbeek, B.; Hage, J. A.; van Vuure, C. A.; Dijkhuizen, L.; Kamerling, J. P. Enzymatic Decoration of Prebiotic Galacto-oligosaccharides (Vivinal GOS) With Sialic Acid Using *Trypanosoma cruzi trans*-Sialidase and Two Bovine Sialoglycoconjugates as Donor Substrates. *J. Agric. Food Chem.* **2015**, *63*, 5976–5984.
939. Luo, Y.; Li, S.-C.; Li, Y.-T.; Luo, M. The 1.8 Å Structures of Leech Intramolecular *trans*-Sialidase Complexes: Evidence of Its Enzymatic Mechanism. *J. Mol. Biol.* **1999**, *285*, 323–332.
940. Gut, H.; King, S. J.; Walsh, M. A. Structural and Functional Studies of *Streptococcus pneumoniae* Neuraminidase B: An Intramolecular *trans*-Sialidase. *FEBS Lett.* **2008**, *582*, 3348–3352.
941. Juge, N.; Tailford, L.; Owen, C. D. Sialidases From Gut Bacteria: A Mini-Review. *Biochem. Soc. Trans.* **2016**, *44*, 166–175.
942. Kitazume, S.; Kitajima, K.; Inoue, S.; Inoue, Y. Detection, Isolation, and Characterization of Oligo/Poly(sialic Acid) and Oligo/Poly(deaminoneuraminic Acid) Units in Glycoconjugates. *Anal. Biochem.* **1992**, *202*, 25–34.
943. Kitajima, K.; Kuroyanagi, H.; Inoue, S.; Ye, J.; Troy, F. A., II.; Inoue, Y. Discovery of a New Type of Sialidase, "KDNase", Which Specifically Hydrolyzes Deaminoneuraminyl (3-Deoxy-D-*glycero*-D-*galacto*-2-nonulosonic Acid) but Not *N*-Acylneuraminyl Linkages. *J. Biol. Chem.* **1994**, *269*, 21415–21419.

944. Li, Y.-T.; Yuziuk, J. A.; Li, S.-C.; Nematalla, A.; Hasegawa, A.; Kimura, M.; Nakagawa, H. A Novel Sialidase Capable of Cleaving 3-Deoxy-D-*glycero*-D-*galacto*-2-nonulosonic Acid (KDN). *Arch. Biochem. Biophys.* **1994**, *310*, 243–246.
945. Angata, T.; Kitajima, K.; Inoue, S.; Chang, J.; Warner, T. G.; Troy, F. A., II.; Inoue, Y. Identification, Developmental Expression and Tissue Distribution of Deaminoneuraminate Hydrolase (KDNase) Activity in Rainbow Trout. *Glycobiology* **1994**, *4*, 517–523.
946. Inoue, S.; Lin, S.-L.; Inoue, Y.; Groves, D. R.; Thomson, R. J.; von Itzstein, M.; Pavlova, N. V.; Li, S.-C.; Li, Y.-T. A Unique Sialidase That Cleaves the Neu5-Gcα2$\rightarrow$5-$O_{glycolyl}$Neu5Gc Linkage: Comparison of Its Specificity With That of Three Microbial Sialidases Toward Four Sialic Acid Dimers. *Biochem. Biophys. Res. Commun.* **2001**, *280*, 104–109.
947. Meindl, P.; Bodo, G.; Palese, P.; Schulman, J.; Tuppy, H. Inhibition of Neuraminidase Activity by Derivatives of 2-Deoxy-2,3-dehydro-*N*-acetylneuraminic Acid. *Virology* **1974**, *58*, 457–463.
948. Schreiner, E.; Zbiral, E.; Kleineidam, R. G.; Schauer, R. Structural Variations on *N*-Acetylneuraminic Acid—20. Synthesis of Some 2,3-Didehydro-2-deoxysialic Acids Structurally Varied at C-4 and Their Behavior Towards Sialidase From *Vibrio cholerae*. *Liebigs Ann. Chem.* **1991**, 129–134.
949. Von Itzstein, M.; Wu, W.-Y.; Kok, G. B.; Pegg, M. S.; Dyason, J. C.; Jin, B.; Van Phan, T.; Smythe, M. L.; White, H. F.; Oliver, S. W.; Colman, P. M.; Varghese, J. N.; Ryan, D. M.; Woods, J. M.; Bethell, R. C.; Hotham, V. J.; Cameron, J. M.; Penn, C. R. Rational Design of Potent Sialidase-Based Inhibitors of Influenza Virus Replication. *Nature* **1993**, *363*, 418–423.
950. Von Itzstein, M.; Wu, W.-Y.; Jin, B. The Synthesis of 2,3-Didehydro-2,4-dideoxy-4-guanidinyl-*N*-acetylneuraminic Acid: A Potent Influenza Virus Sialidase Inhibitor. *Carbohydr. Res.* **1994**, *259*, 301–305.
951. Kim, C. U.; Lew, W.; Williams, M. A.; Liu, H.; Zhang, L.; Swaminathan, S.; Bischofberger, N.; Chen, M. S.; Mendel, D. B.; Tai, C. Y.; Laver, W. G.; Stevens, R. C. Influenza Neuraminidase Inhibitors Possessing a Novel Hydrophobic Interaction in the Enzyme Active Site: Design, Synthesis, and Structural Analysis of Carbocyclic Sialic Acid Analogues With Potent Anti-influenza Activity. *J. Am. Chem. Soc.* **1997**, *119*, 681–690.
952. Rohloff, J. C.; Kent, K. M.; Postich, M. J.; Becker, M. W.; Chapman, H. H.; Kelly, D. E.; Lew, W.; Louie, M. S.; McGee, L. R.; Prisbe, E. J.; Schultze, L. M.; Yu, R. H.; Zhang, L. Practical Total Synthesis of the Anti-influenza Drug GS 4104. *J. Org. Chem.* **1998**, *63*, 4545–4550.
953. Bardsley-Elliot, A.; Noble, S. Oseltamivir. *Drugs* **1999**, *58*, 851–860.
954. McNicholl, I. R.; McNicholl, J. J. Neuraminidase Inhibitors: Zanamivir and Oseltamivir. *Ann. Pharmacother.* **2001**, *35*, 57–70.
955. Gubareva, L. V.; Webster, R. G.; Hayden, F. G. Detection of Influenza Virus Resistance to Neuraminidase Inhibitors by an Enzyme Inhibition Assay. *Antiviral Res.* **2002**, *53*, 47–61.
956. Sriwilaijaroen, N.; Magesh, S.; Imamura, A.; Ando, H.; Ishida, H.; Sakai, M.; Ishitsubo, E.; Hori, T.; Moriya, S.; Ishikawa, T.; Kuwata, K.; Odagiri, T.; Tashiro, M.; Hiramatsu, H.; Tsukamoto, K.; Miyagi, T.; Tokiwa, H.; Kiso, M.; Suzuki, Y. A Novel Potent and Highly Specific Inhibitor Against Influenza Viral N1-N9 Neuraminidases: Insight Into Neuraminidase–Inhibitor Interactions. *J. Med. Chem.* **2016**, *59*, 4563–4577.
957. Taubenberger, J. K. The Virulence of the 1918 Pandemic Influenza Virus: Unraveling the Enigma. *Arch. Virol. Suppl.* **2005**, *19*, 101–115.

958. Klenk, H.-D.; Nagai, Y.; Rott, R.; Nicolau, C. The Structure and Function of Paramyxovirus Glycoproteins. *Med. Microbiol. Immunol.* **1977**, *164*, 35–47.
959. Herrler, G.; Hausmann, J.; Klenk, H.-D. Sialic Acid as Receptor Determinant of Ortho- and Paramyxoviruses. In: *The Biology of Sialic Acids*; Rosenberg, A. Ed.; Plenum Press: New York, NY, USA, 1995; pp 315–336.
960. Cook, G. M. W.; Heard, D. H.; Seaman, G. V. F. A Sialomucopeptide Liberated by Trypsin From the Human Erythrocyte. *Nature* **1960**, *188*, 1011–1012.
961. Cook, G. M. W. Glycobiology of the Cell Surface: Its Debt to Cell Electrophoresis 1940–65. *Electrophoresis* **2016**, *37*, 1399–1406.
962. Bennett, H. S. Morphological Aspects of Extracellular Polysaccharides. *J. Histochem. Cytochem.* **1963**, *11*, 14–23.
963. Reutter, W.; Köttgen, E.; Bauer, C.; Gerok, W. Biological Significance of Sialic Acids. In: *Sialic Acids—Chemistry, Metabolism and Function*; Schauer, R. Ed.; Cell Biology Monographs, Vol. 10; Springer-Verlag: Vienna, Austria, 1982; pp 263–305.
964. Ashwell, G.; Morell, A. G. The Role of Surface Carbohydrates in the Hepatic Recognition and Transport of Circulating Glycoproteins. *Adv. Enzymol. Relat. Areas Mol. Biol.* **1974**, *41*, 99–128.
965. Morell, A. G.; Gregoriadis, G.; Scheinberg, I. H.; Hickman, J.; Ashwell, G. The Role of Sialic Acid in Determining the Survival of Glycoproteins in the Circulation. *J. Biol. Chem.* **1971**, *246*, 1461–1467.
966. Lodish, H. F.; Kong, N.; Wikström, L. Calcium Is Required for Folding of Newly Made Subunits of the Asialoglycoprotein Receptor Within the Endoplasmic Reticulum. *J. Biol. Chem.* **1992**, *267*, 12753–12760.
967. Stewart, W. B.; Petenyi, C. W.; Rose, H. M. The Survival Time of Canine Erythrocytes Modified by Influenza Virus. *Blood* **1955**, *10*, 228–234.
968. Halbhuber, K.-J.; Helmke, U.; Geyer, G. Zur Elimination digerierter Erythrozyten der Ratten. *Folia Haematol. (Leipzig)* **1972**, *97*, 196–203.
969. Jancik, J. M.; Schauer, R.; Andres, K. H.; von Düring, M. Sequestration of Neuraminidase-Treated Erythrocytes—Studies on Its Topographic, Morphologic and Immunologic Aspects. *Cell Tissue Res.* **1978**, *186*, 209–226.
970. Jancik, J.; Schauer, R. Sialic Acid—A Determinant of the Life-Time of Rabbit Erythrocytes. *Hoppe-Seyler's Z. Physiol. Chem.* **1974**, *355*, 395–400.
971. Aminoff, D.; Vorder Bruegge, W. F.; Bell, W. C.; Sarpolis, K.; Williams, R. Role of Sialic Acid in Survival of Erythrocytes in the Circulation: Interaction of Neuraminidase-Treated and Untreated Erythrocytes With Spleen and Liver at the Cellular Level. *Proc. Natl. Acad. Sci. U. S. A.* **1977**, *74*, 1521–1524.
972. Kolb, H.; Kolb-Bachofen, V. A Lectin-Like Receptor on Mammalian Macrophages. *Biochem. Biophys. Res. Commun.* **1978**, *85*, 678–683.
973. Müller, E.; Schröder, C.; Schauer, R.; Sharon, N. Binding and Phagocytosis of Sialidase-Treated Rat Erythrocytes by a Mechanism Independent of Opsonins. *Hoppe-Seyler's Z. Physiol. Chem.* **1983**, *364*, 1419–1429.
974. Schauer, R.; Fischer, C.; Lee, H.; Ruch, B.; Kelm, S. Sialic Acids as Regulators of Molecular and Cellular Interactions. In: *Lectins and Glycoconjugates in Oncology*; Gabius, H.-J., Nagel, G. A., Eds.; Springer-Verlag: Berlin, Germany, 1988; pp 5–23.
975. Schauer, R. Sialic Acids Regulate Cellular and Molecular Recognition. In: *Carbohydrates—Synthetic Methods and Applications in Medicinal Chemistry*; Ogura, H., Hasegawa, A., Suami, T., Eds.; Kodansha Ltd.: Tokyo, Japan, 1992; pp 340–354.
976. Kelm, S.; Schauer, R. The Galactose-Recognizing System of Rat Peritoneal Macrophages; Identification and Characterization of the Receptor Molecule. *Biol. Chem. Hoppe-Seyler* **1988**, *369*, 693–704.
977. Barondes, S. H. Vertebrate Lectins: Properties and Functions. In: *The Lectins: Properties, Functions, and Applications in Biology and Medicine*; Liener, I. E., Sharon, N., Goldstein, I. J., Eds.; Academic Press, Inc.: Orlando, FL, USA, 1986 pp 438–467

978. Gesner, B. M.; Ginsburg, V. Effect of Glycosidases on the Fate of Transfused Lymphocytes. *Proc. Natl. Acad. Sci. U. S. A.* **1964**, *52*, 750–755.
979. Woodruff, J. J.; Gesner, B. M. The Effect of Neuraminidase on the Fate of Transfused Lymphocytes. *J. Exp. Med.* **1969**, *129*, 551–567.
980. Fischer, C.; Kelm, S.; Ruch, B.; Schauer, R. Reversible Binding of Sialidase-Treated Rat Lymphocytes by Homologous Peritoneal Macrophages. *Carbohydr. Res.* **1991**, *213*, 263–273.
981. Jibril, S. E.; von Gaudecker, B.; Kelm, S.; Schauer, R. Interaction of Rat Peritoneal Macrophages With Homologous, Sialidase-Treated Lymphocytes In Vitro. *Biol. Chem. Hoppe-Seyler* **1987**, *368*, 819–829.
982. Roffman, E.; Wilchek, M. Selective Covalent Modification of Newly Synthesized Activation-Induced Sialoglycoconjugates on the T Lymphocyte Membrane. *Biochem. Biophys. Res. Commun.* **1986**, *135*, 473–479.
983. Kreisel, W.; Volk, B. A.; Büchsel, R.; Reutter, W. Different Half-Lives of the Carbohydrate and Protein Moieties of a 110,000-Dalton Glycoprotein Isolated From Plasma Membranes of Rat Liver. *Proc. Natl. Acad. Sci. U. S. A.* **1980**, 77, 1828–1831.
984. Choi, S.-I.; Simone, J. V.; Journey, L. J. Neuraminidase-Induced Thrombocytopenia in Rats. *Br. J. Haematol.* **1972**, *22*, 93–101.
985. Mahajan, V. S.; Pillai, S. Sialic Acids and Autoimmune Disease. *Immunol. Rev.* **2016**, *269*, 145–161.
986. Emmelot, P. Biochemical Properties of Normal and Neoplastic Cell Surfaces. *Eur. J. Cancer* **1973**, *9*, 319–333.
987. Feizi, T. Demonstration by Monoclonal Antibodies That Carbohydrate Structures of Glycoproteins and Glycolipids Are Onco-developmental Antigens. *Nature* **1985**, *314*, 53–57.
988. Alhadeff, J. A. Malignant Cell Glycoproteins and Glycolipids. *Crit. Rev. Oncol. Hematol.* **1989**, *9*, 37–107.
989. Bhavanandan, V. P.; Furukawa, K. Biochemistry and Oncology of Sialoglycoproteins. In: *Biology of the Sialic Acids*; Rosenberg, A. Ed.; Plenum Press: New York, NY, USA, 1995; pp 145–196.
990. Dabelsteen, E. Cell Surface Carbohydrates as Prognostic Markers in Human Carcinomas. *J. Pathol.* **1996**, *179*, 358–369.
991. Dall'Olio, F. Protein Glycosylation in Cancer Biology: An Overview. *J. Clin. Pathol.: Mol. Pathol.* **1996**, *49*, M126–M135.
992. Nishimura, R.; Kobata, A. Cancer Cells and Metastasis—Choriocarcinoma and hCG. In: *Glycoproteins and Disease*; Montreuil, J., Vliegenthart, J. F. G., Schachter, H., Eds.; New Comprehensive Biochemistry; Vol. 30; Elsevier Science B.V: Amsterdam, The Netherlands, 1996; pp 185–195.
993. Yamashita, K.; Kobata, A. Cancer Cells and Metastasis—Hepatocellular Carcinoma. In: *Glycoproteins and Disease*; Montreuil, J., Vliegenthart, J. F. G., Schachter, H., Eds.; New Comprehensive Biochemistry; Vol. 30; Elsevier Science B.V: Amsterdam, The Netherlands, 1996; pp 197–210.
994. Yamashita, K.; Kobata, A. Cancer Cells and Metastasis—Carcinoembryonic Antigen and Related Normal Antigens. In: *Glycoproteins and Disease*; Montreuil, J., Vliegenthart, J. F. G., Schachter, H., Eds.; New Comprehensive Biochemistry, Vol. 30; Elsevier Science B.V: Amsterdam, The Netherlands, 1996; pp 229–239.
995. Kobata, A. Cancer Cells and Metastasis—The Warren-Glick Phenomenon—A Molecular Basis of Tumorigenesis and Metastasis. In: *Glycoproteins and Disease*; Montreuil, J., Vliegenthart, J. F. G., Schachter, H., Eds.; New Comprehensive Biochemistry, Vol. 30; Elsevier Science B.V: Amsterdam, The Netherlands, 1996; pp 211–227.
996. Hanisch, F.-G.; Baldus, S. E. The Thomsen-Friedenreich (TF) Antigen: A Critical Review on the Structural, Biosynthetic and Histochemical Aspects of a Pancarcinoma-Associated Antigen. *Histol. Histopathol.* **1997**, *12*, 263–281.

997. Stephens, A. S.; Day, C. J.; Tiralongo, J. Sialic Acids and Cancer. In: *Sialobiology: Structure, Biosynthesis and Function—Sialic Acid Glycoconjugates in Health and Disease*; Tiralongo, J., Martinez-Duncker, I., Eds.; Bentham Science Publishers: Oak Park, IL, USA, 2013; pp 381–403.
998. Pearce, O. M. T.; Läubli, H. Sialic Acids in Cancer Biology and Immunity. *Glycobiology* **2016**, *26*, 111–128.
999. Vajaria, B. N.; Patel, P. S. Glycosylation: A Hallmark of Cancer? *Glycoconjugate J.* **2017**, *34*, 147–156.
1000. Ambrose, E. J.; James, A. M.; Lowick, J. H. B. Differences Between the Electrical Charge Carried by Normal and Homologous Tumour Cells. *Nature* **1956**, *177*, 576–577.
1001. Purdom, L.; Ambrose, E. J.; Klein, G. A Correlation Between Electrical Surface Charge and Some Biological Characteristics During the Stepwise Progression of a Mouse Sarcoma. *Nature* **1958**, *181*, 1586–1587.
1002. Gasic, G.; Gasic, T. Removal of Sialic Acid From the Cell Coat in Tumor Cells and Vascular Endothellum, and Its Effects on Metastasis. *Proc. Natl. Acad. Sci. U. S. A.* **1962**, *48*, 1172–1177.
1003. Buck, C. A.; Glick, M. C.; Warren, L. Glycopeptides From the Surface of Control and Virus-Transformed Cells. *Science* **1971**, *172*, 169–171.
1004. Warren, L.; Fuhrer, J. P.; Buck, C. A. Surface Glycoproteins of Normal and Transformed Cells: A Difference Determined by Sialic Acid and a Growth-Dependent Sialyl Transferase. *Proc. Natl. Acad. Sci. U. S. A.* **1972**, *69*, 1838–1842.
1005. Bosmann, H. B.; Hall, T. C. Enzyme Activity in Invasive Tumors of Human Breast and Colon. *Proc. Natl. Acad. Sci. U. S. A.* **1974**, *71*, 1833–1837.
1006. Nisselbaum, J. S.; Bernfeld, P. The Properties of Two Glycoproteins Isolated From the Plasma of Normal and Tumor Bearing Mice. *J. Am. Chem. Soc.* **1956**, *78*, 687–689.
1007. Turumi, K.-I.; Dawes, S. M. L. Serum Sialic Acid Levels in Mice With Neoplasms. *Cancer Res.* **1958**, *18*, 575–577.
1008. Kessel, D.; Allen, J. Elevated Plasma Sialyltransferase in the Cancer Patient. *Cancer Res.* **1975**, *35*, 670–672.
1009. Plucinsky, M. C.; Riley, W. M.; Prorok, J. J.; Alhadeff, J. A. Total and Lipid-Associated Serum Sialic Acid Levels in Cancer Patients With Different Primary Sites and Differing Degrees of Metastatic Involvement. *Cancer* **1986**, *58*, 2680–2685.
1010. Müller-Eberhard, H. J.; Kunkel, H. G. The Carbohydrate of γ-Globulin and Myeloma Proteins. *J. Exp. Med.* **1956**, *104*, 253–269.
1011. Miyagi, T.; Goto, T.; Tsuiki, S. Sialidase of Rat Hepatomas: Qualitative and Quantitative Comparison With Rat Liver Sialidase. *Jpn. J. Cancer Res.* **1984**, *75*, 1076–1082.
1012. Miyagi, T.; Wada, T.; Yamaguchi, K.; Shiozaki, K.; Sato, I.; Kakugawa, Y.; Yamanami, H.; Fujiya, T. Human Sialidase as a Cancer Marker. *Proteomics* **2008**, *8*, 3303–3311.
1013. Schengrund, C.-L.; Lausch, R. N.; Rosenberg, A. Sialidase Activity in Transformed Cells. *J. Biol. Chem.* **1973**, *248*, 4424–4428.
1014. Holmes, E. H.; Ostrander, G. K.; Hakomori, S.-I. Biosynthesis of the Sialyl-Le^x Determinant Carried by Type 2 Chain Glycosphingolipids ($IV^3NeuAcIII^3FucnLc_4$, $VI^3NeuAcV^3FucnLc_6$, and $VI^3NeuAcIII^3V^3Fuc_2nLc_6$) in Human Lung Carcinoma PC9 Cells. *J. Biol. Chem.* **1986**, *261*, 3737–3743.
1015. Mann, B.; Klussmann, E.; Vandamme-Feldhaus, V.; Iwersen, M.; Hanski, M.-L.; Riecken, E.-O.; Buhr, H. J.; Schauer, R.; Kim, Y. S.; Hanski, C. Low O-Acetylation of Sialyl-Le^x Contributes to Its Overexpression in Colon Carcinoma Metastases. *Int. J. Cancer* **1997**, *72*, 258–264.

1016. Corfield, A. P.; Myerscough, N.; Warren, B. F.; Durdey, P.; Paraskeva, C.; Schauer, R. Reduction of Sialic Acid O-Acetylation in Human Colonic Mucins in the Adenoma-Carcinoma Sequence. *Glycoconjugate J.* **1999**, *16*, 307–317.
1017. Kohla, G.; Stockfleth, E.; Schauer, R. Gangliosides With *O*-Acetylated Sialic Acids in Tumors of Neuroectodermal Origin. *Neurochem. Res.* **2002**, *27*, 583–592.
1018. Magnani, J. L.; Nilsson, B.; Brockhaus, M.; Zopf, D.; Steplewski, Z.; Koprowski, H.; Ginsburg, V. A Monoclonal Antibody-Defined Antigen Associated With Gastrointestinal Cancer Is a Ganglioside Containing Sialylated Lacto-*N*-fucopentaose II. *J. Biol. Chem.* **1982**, *257*, 14365–14369.
1019. Groux-Degrootte, S.; Rodríguez-Walker, M.; Dewald, J. H.; Daniotti, J. L.; Delannoy, P. Gangliosides in Cancer Cell Signaling. *Prog. Mol. Biol. Transl. Sci.* **2018**, *156*, 197–227.
1020. Higashi, H.; Ikuta, K.; Ueda, S.; Kato, S.; Hirabayashi, Y.; Matsumoto, M.; Naiki, M. Characterization of *N*-Glycolylneuraminic Acid-Containing Glycosphingolipids From a Marek's Disease Lymphoma-Derived Chicken Cell Line, MSB1, as Tumor-Associated Heterophile Hanganutziu-Deicher Antigens. *J. Biochem.* **1984**, *95*, 785–794.
1021. Higashi, H.; Sasabe, T.; Fukui, Y.; Maru, M.; Kato, S. Detection of Gangliosides as *N*-Glycolylneuraminic Acid-Specific Tumor-Associated Hanganutziu-Deicher Antigen in Human Retinoblastoma Cells. *Jpn. J. Cancer Res.* **1988**, *79*, 952–956.
1022. Kawai, T.; Kato, A.; Higashi, H.; Kato, S.; Naiki, M. Quantitative Determination of *N*-Glycolylneuraminic Acid Expression in Human Cancerous Tissues and Avian Lymphoma Cell Lines as a Tumor-Associated Sialic Acid by Gas Chromatography-Mass Spectrometry. *Cancer Res.* **1991**, *51*, 1242–1246.
1023. Nöhle, U.; Beau, J.-M.; Schauer, R. Uptake, Metabolism and Excretion of Orally and Intravenously Administered, Double-Labeled *N*-Glycoloylneuraminic Acid and Single-Labeled 2-Deoxy-2,3-dehydro-*N*-acetylneuraminic Acid in Mouse and Rat. *Eur. J. Biochem.* **1982**, *126*, 543–548.
1024. Okerblom, J.; Varki, A. Biochemical, Cellular, Physiological, and Pathological Consequences of Human Loss of *N*-Glycolylneuraminic Acid. *ChemBioChem* **2017**, *18*, 1155–1171.
1025. Landsteiner, K. Zur Kenntnis der antifermentativen, lytischen und agglutinierenden Wirkung des Blutserums und der Lymphe. *Zbl. Bakt.* **1900**, *27*, 357–362.
1026. Watkins, W. M. Blood-Group Specific Substances. In: *Glycoproteins—Their Composition, Structure and Function*, Gottschalk, A. Ed.; Vol. 5, Part B, 2nd ed.; B.B.A. Library, Elsevier Publishing Company: Amsterdam, The Netherlands, 1972; pp 830–891.
1027. Race, R. R.; Sanger, R. *Blood Groups in Man.* Blackwell Science: Oxford, UK, 1968.
1028. Lisowska, E. Biochemistry of M and N Blood Group Specificities. *Rev. Franç. Transf. Immuno-hématol.* **1981**, *24*, 75–84.
1029. Kewitz, S.; Groß, H. J.; Kosa, R.; Roelcke, D. Anti-Pr Cold Agglutinins Recognize Immunodominant α2,3- or α2,6-Sialyl Groups on Glycophorins. *Glycoconjugate J.* **1995**, *12*, 714–720.
1030. Roelcke, D. Kälteagglutinine: Humane monoklonale Antikörper gegen Glykokonjugat-Antigene von Zelloberflächen. *Funkt. Biol. Med.* **1984**, *3*, 106–127.
1031. Kościelak, J. A Hypothesis on the Biological Role of ABH, Lewis and P Blood Group Determinant Structures in Glycosphingolipids and Glycoproteins. *Glycoconjugate J.* **1986**, *3*, 95–108.
1032. Anstee, D. J. The Blood Group MNSs-Active Sialoglycoproteins. *Semin. Hematol.* **1981**, *18*, 13–31.
1033. Dahr, W.; Kordowicz, M.; Moulds, J.; Gielen, W.; Lebeck, L.; Krüger, J. Characterization of the Ss Sialoglycoprotein and Its Antigens in Rh_{null} Erythrocytes. *Blut* **1987**, *54*, 13–24.

1034. Kannagi, R.; Roelcke, D.; Peterson, K. A.; Okada, Y.; Levery, S. B.; Hakomori, S.-I. Characterization of an Epitope (Determinant) Structure in a Developmentally Regulated Glycolipid Antigen Defined by a Cold Agglutinin F1, Recognition of α-Sialosyl and α-L-Fucosyl Groups in a Branched Structure. *Carbohydr. Res.* **1983**, *120*, 143–157.
1035. Aminoff, D.; Morgan, W. T. J. Hexosamine Components of the Human Blood Group Substances. *Nature* **1948**, *162*, 579–580.
1036. Watkins, W. M.; Morgan, W. T. J. Inhibition by Simple Sugars of Enzymes Which Decompose the Blood-Group Substances. *Nature* **1955**, *175*, 676–677.
1037. Pusztai, A.; Morgan, W. T. J. Studies in Immunochemistry—18. The Isolation and Properties of a Sialomucopolysaccharide Possessing Blood-Group Le^a Specificity and Virus–Receptor Activity. *Biochem. J.* **1961**, *78*, 135–146.
1038. Hewitt, L. F. Chemistry of Antibodies and Serum-Proteins—I. Nitrogen Distribution and Amino-Acids. II. Protein Carbohydrate Groups. *Biochem. J.* **1934**, *28*, 2080–2087.
1039. Clamp, J. R.; Johnson, I. Immunoglobulins. In: *Glycoproteins—Their Composition, Structure and Function*, Gottschalk, A. Ed.; Vol. 5, Part A, 2nd ed.; B.B.A. Library, Elsevier Publishing Company: Amsterdam, The Netherlands, 1972; pp 612–652.
1040. Schultze, H. E. Influence of Bound Sialic Acid on Electrophoretic Mobility of Human Serum Proteins. *Arch. Biochem. Biophys.* **1962**, (Suppl. 1), 290–294.
1041. Sarma, V. R.; Silverton, E. W.; Davies, D. R.; Terry, W. D. The Three-Dimensional Structure at 6 Å Resolution of a Human γG1 Immunoglobulin Molecule. *J. Biol. Chem.* **1971**, *246*, 3753–3759.
1042. Burton, D. R.; Dwek, R. A. Sugar Determines Antibody Activity. *Science* **2006**, *313*, 627–628.
1043. Rademacher, T. W.; Parekh, R. B.; Dwek, R. A. Glycobiology. *Annu. Rev. Biochem.* **1988**, *57*, 785–838.
1044. Arnold, J. N.; Wormald, M. R.; Sim, R. B.; Rudd, P. M.; Dwek, R. A. The Impact of Glycosylation on the Biological Function and Structure of Human Immunoglobulins. *Annu. Rev. Immunol.* **2007**, *25*, 21–50.
1045. Kaneko, Y.; Nimmerjahn, F.; Ravetch, J. V. Anti-inflammatory Activity of Immunoglobulin G Resulting From Fc Sialylation. *Science* **2006**, *313*, 670–673.
1046. Pagan, J. D.; Kitaoka, M.; Anthony, R. M. Engineered Sialylation of Pathogenic Antibodies *In Vivo* Attenuates Autoimmune Disease. *Cell* **2018**, *172*, 564–577.
1047. Hanganutziu, M. Hémagglutinines Hétérogénétiques après Injection de Sérum de Cheval. *C. R. Séances Soc. Biol.* **1924**, *91*, 1457–1459.
1048. Deicher, H. Über die Erzeugung heterospezifischer Hämagglutinine durch Injektion artfremden Serums—I. Mitteilung. *Z. Hyg. Infekt. Krankh.: Med. Mikrobiol. Immunol. Virol.* **1926**, *106*, 561–579.
1049. Higashi, H.; Naiki, M.; Matuo, S.; Okouchi, K. Antigen of "Serum Sickness" Type of Heterophile Antibodies in Human Sera: Identification as Gangliosides With *N*-Glycolylneuraminic Acid. *Biochem. Biophys. Res. Commun.* **1977**, *79*, 388–395.
1050. Merrick, J. M.; Zadarlik, K.; Milgrom, F. Characterization of the Hanganutziu-Deicher (Serum-Sickness) Antigen as Gangliosides Containing *N*-Glycolylneuraminic Acid. *Int. Arch. Allergy Appl. Immun.* **1978**, *57*, 477–480.
1051. Tai, T.; Kawashima, I.; Furukawa, K.; Lloyd, K. O. Monoclonal Antibody R24 Distinguishes Between Different *N*-Acetyl- and *N*-Glycolylneuraminic Acid Derivatives of Ganglioside GD3. *Arch. Biochem. Biophys.* **1988**, *260*, 51–55.
1052. Richman, P. I.; Bodmer, W. F. Monoclonal Antibodies to Human Colorectal Epithelium: Markers for Differentiation and Tumour Characterization. *Int. J. Cancer* **1987**, *39*, 317–328.

1053. Smithson, J. E.; Campbell, A.; Andrews, J. M.; Milton, J. D.; Pigott, R.; Jewell, D. P. Altered Expression of Mucins Throughout the Colon in Ulcerative Colitis. *Gut* **1997**, *40*, 234–240.
1054. Milton, J. D.; Eccleston, D.; Parker, N.; Raouf, A.; Cubbin, C.; Hoffman, J.; Hart, C. A.; Rhodes, J. M. Distribution of O-Acetylated Sialomucin in the Normal and Diseased Gastrointestinal Tract Shown by a New Monoclonal Antibody. *J. Clin. Pathol.* **1993**, *46*, 323–329.
1055. Schwartz-Albiez, R.; Claus, C.; Kniep, B. CD60 Workshop Report. In: *Leucocyte Typing VII*; Mason, D. Ed.; Oxford University Press: Oxford, UK, 2002; pp 185–188.
1056. Kniep, B.; Kniep, E.; Özkucur, N.; Barz, S.; Bachmann, M.; Malisan, F.; Testi, R.; Rieber, E. P. 9-*O*-Acetyl GD3 Protects Tumor Cells From Apoptosis. *Int. J. Cancer* **2006**, *119*, 67–73.
1057. Varki, A.; Kornfeld, S. An Autosomal Dominant Gene Regulates the Extent of 9-*O*-Acetylation of Murine Erythrocyte Sialic Acids—A Probable Explanation for the Variation in Capacity to Activate the Human Alternate Complement Pathway. *J. Exp. Med.* **1980**, *152*, 532–544.
1058. Shi, W.-X.; Chammas, R.; Varki, N. M.; Powell, L.; Varki, A. Sialic Acid 9-*O*-Acetylation on Murine Erythroleukemia Cells Affects Complement Activation, Binding to I-Type Lectins, and Tissue Homing. *J. Biol. Chem.* **1996**, *271*, 31526–31532.
1059. Ourth, D. D.; Bachinski, L. M. Bacterial Sialic Acid Modulates Activation of the Alternative Complement Pathway of Channel Catfish (*Ictalurus punctatus*). *Dev. Comp. Immunol.* **1987**, *11*, 551–564.
1060. Manzi, A. E.; Sjoberg, E. R.; Diaz, S.; Varki, A. Biosynthesis and Turnover of *O*-Acetyl and *N*-Acetyl Groups in the Gangliosides of Human Melanoma Cells. *J. Biol. Chem.* **1990**, *265*, 13091–13103.
1061. Moe, G. R.; Bhandari, T. S.; Flitter, B. A. Vaccines Containing De-*N*-acetyl Sialic Acid Elicit Antibodies Protective Against *Neisseria meningitidis* Groups B and C. *J. Immunol.* **2009**, *182*, 6610–6617.
1062. Devi, S. J. N.; Robbins, J. B.; Schneerson, R. Antibodies to Poly[(2→8)-α-*N*-acetylneuraminic Acid] and Poly[(2→9)-α-*N*-acetylneuraminic Acid] Are Elicited by Immunization of Mice With *Escherichia coli* K92 Conjugates: Potential Vaccines for Groups B and C Meningococci and *E. coli* K1. *Proc. Natl. Acad. Sci. U. S. A.* **1991**, *88*, 7175–7179.
1063. Krug, L. M.; Ragupathi, G.; Ng, K. K.; Hood, C.; Jennings, H. J.; Guo, Z.; Kris, M. G.; Miller, V.; Pizzo, B.; Tyson, L.; Baez, V.; Livingston, P. O. Vaccination of Small Cell Lung Cancer Patients With Polysialic Acid or *N*-Propionylated Polysialic Acid Conjugated to Keyhole Limpet Hemocyanin. *Clin. Cancer Res.* **2004**, *10*, 916–923.
1064. Finne, J.; Bitter-Suermann, D.; Goridis, C.; Finne, U. An IgG Monoclonal Antibody to Group B Meningococci Cross-Reacts With Developmentally Regulated Polysialic Acid Units of Glycoproteins in Neural and Extraneural Tissues. *J. Immunol.* **1987**, *138*, 4402–4407.
1065. Schauer, R. Sialic Acids as Antigenic Determinants of Complex Carbohydrates. *Adv. Exp. Med. Biol.* **1988**, *228*, 47–72.
1066. Aminoff, D.; Morrow, M. P. Effect of Sialidase on Blood Group Specificity of Hog Submaxillary Glycoproteins. *FEBS Lett.* **1970**, *8*, 353–358.
1067. Howie, A. J.; Brown, G. Effect of Neuraminidase on the Expression of the 3-Fucosyl-*N*-acetyllactosamine Antigen in Human Tissues. *J. Clin. Pathol.* **1985**, *38*, 409–416.

1068. Kelley, L. K.; King, B. F.; Johnson, L. W.; Smith, C. H. Protein Composition and Structure of Human Placental Microvillous Membrane. *Exp. Cell Res.* **1979**, *123*, 167–176.
1069. Taylor, P. V.; Hancock, K. W.; Gowland, G. Effect of Neuraminidase on Immunogenicity of Early Mouse Trophoblast. *Transplantation* **1979**, *28*, 256–257.
1070. Blau, E. B.; Haas, J. E. Glomerular Sialic Acid and Proteinuria in Human Renal Disease. *Lab. Invest.* **1973**, *28*, 477–481.
1071. Currie, G. A.; Bagshawe, K. D. The Effect of Neuraminidase on the Immunogenicity of the Landschütz Ascites Tumour: Site and Mode of Action. *Br. J. Cancer* **1968**, *22*, 588–594.
1072. Sedlacek, H. H.; Weise, M.; Lemmer, A.; Seiler, F. R. Immunotherapy of Spontaneous Mammary Tumors in Mongrel Dogs With Autologous Tumor Cells and Neuraminidase. *Cancer Immunol. Immunother.* **1979**, *6*, 47–58.
1073. Bekesi, J. G.; Holland, J. F. Active Immunotherapy in Leukemia With Neuraminidase-Modified Leukemic Cells. *Recent Results Cancer Res.* **1977**, *62*, 78–89.
1074. Kaufmann, S. H. E.; Schauer, R.; Hahn, H. Carbohydrate Surface Constituents of T Cells Mediating Delayed-Type Hypersensitivity That Control Entry Into Sites of Antigen Deposition. *Immunobiology* **1981**, *160*, 184–195.
1075. Bi, S.; Baum, L. G. Sialic Acids in T Cell Development and Function. *Biochim. Biophys. Acta* **2009**, *1790*, 1599–1610.
1076. Wang, B.; Yu, B.; Karim, M.; Hu, H.; Sun, Y.; McGreevy, P.; Petocz, P.; Held, S.; Brand-Miller, J. Dietary Sialic Acid Supplementation Improves Learning and Memory in Piglets. *Am. J. Clin. Nutr.* **2007**, *85*, 561–569.
1077. Wang, B. Molecular Mechanism Underlying Sialic Acid as an Essential Nutrient for Brain Development and Cognition. *Adv. Nutr.* **2012**, *3*, 465S–472S.
1078. Marchalonis, J. J.; Edelman, G. M. Isolation and Characterization of a Hemagglutinin From *Limulus polyphemus*. *J. Mol. Biol.* **1968**, *32*, 453–465.
1079. Roche, A.-C.; Monsigny, M. Purification and Properties of Limulin: A Lectin (Agglutinin) From Hemolymph of *Limulus polyphemus*. *Biochim. Biophys. Acta* **1974**, *371*, 242–254.
1080. Maget-Dana, R.; Veh, R. W.; Sander, M.; Roche, A.-C.; Schauer, R.; Monsigny, M. Specificities of Limulin and Wheat-Germ Agglutinin Towards Some Derivatives of G_{M3} Gangliosides. *Eur. J. Biochem.* **1981**, *114*, 11–16.
1081. Fischer, E.; Khang, N. Q.; Letendre, G.; Brossmer, R. A Lectin From the Asian Horseshoe Crab *Tachypleus tridentatus*: Purification, Specificity and Interaction With Tumour Cells. *Glycoconjugate J.* **1994**, *11*, 51–58.
1082. Crocker, P. R.; Paulson, J. C.; Varki, A. Siglecs and Their Roles in the Immune System. *Nat. Rev. Immunol.* **2007**, *7*, 255–266.
1083. Varki, A.; Schnaar, R. L.; Crocker, P. R. I-Type Lectins. In: *Essentials of Glycobiology*; Varki, A., Cummings, R. D., Esko, J. D., Stanley, P., Hart, G. W., Aebi, M., Darvill, A., Kinoshita, T., Packer, N. H., Prestegard, J. J., Schnaar, R. L., Seeberger, P. H., Eds., 3rd ed.; Cold Spring Harbor Laboratory Press: Cold Spring Harbor, NY, USA, 2017; pp 453–457.
1084. Sharon, N.; Lis, H. *Lectins*, 2nd ed.; Kluwer Academic Publishers: Dordrecht, The Netherlands, 2003.
1085. Sharon, N.; Lis, H. History of Lectins: From Hemagglutinins to Biological Recognition Molecules. *Glycobiology* **2004**, *14*, 53R–62R.
1086. Sharon, N. Lectins: Carbohydrate-Specific Reagents and Biological Recognition Molecules. *J. Biol. Chem.* **2007**, *282*, 2753–2764.
1087. Boyd, W. C.; Shapleigh, E. Specific Precipitating Activity of Plant Agglutinins (Lectins). *Science* **1954**, *119*, 419.
1088. Kilpatrick, D. C.; Green, C. Lectins as Blood Typing Reagents. *Adv. Lectin Res.* **1992**, *5*, 51–94.

1089. Stillmark, H. *Über Ricin, ein giftiges Ferment aus den Samen von* Ricinus comm. *L. und einigen anderen Euphorbiaceen*; Dissertation University of Tartu (Dorpat), Estonia, 1888.
1090. Gabius, H.-J. Eds. *The Sugar Code—Fundamentals of Glycosciences*; Wiley–VCH Verlag GmbH: Weinheim, Germany, 2009.
1091. Flexner, S.; Noguchi, H. Snake Venom in Relation to Haemolysis, Bacteriolysis, and Toxicity. *J. Exp. Med.* **1902**, *6*, 277–301.
1092. Morgan, W. T. J.; Watkins, W. M. Unravelling the Biochemical Basis of Blood Group ABO and Lewis Antigenic Specificity. *Glycoconjugate J.* **2000**, *17*, 501–530.
1093. Rosenberg, A.; Howe, C.; Chargaff, E. Inhibition of Influenza Virus Haemagglutination by a Brain Lipid Fraction. *Nature* **1956**, *177*, 234–235.
1094. Böttcher-Friebertshäuser, E.; Garten, W.; Matrosovich, M.; Klenk, H.-D. The Hemagglutinin: A Determinant of Pathogenicity. *Curr. Top. Microbiol. Immunol.* **2014**, *385*, 3–34.
1095. Matrosovich, M.; Herrler, G.; Klenk, H.-D. Sialic Acid Receptors of Viruses. In: *SialoGlyco Chemistry and Biology II—Tools and Techniques to Identify and Capture Sialoglycans*; Gerardy-Schahn, R., Delannoy, P., von Itzstein, M., Eds.; Topics in Current Chemistry, Vol. 367; Springer: Berlin, Germany, 2015; pp 1–28.
1096. Howe, C.; Lee, L. T.; Rose, H. M. *Collocalia* Mucoid: A Substrate for Myxovirus Neuraminidase. *Arch. Biochem. Biophys.* **1961**, *95*, 512–520.
1097. Kathan, R. H.; Weeks, D. I. Structure Studies of Collocalia Mucoid—I. Carbohydrate and Amino Acid Composition. *Arch. Biochem. Biophys.* **1969**, *134*, 572–576.
1098. Ogura, H.; Shoji, M. *Origin of Perennial Youth and Long Life From* Tsubame no sû *(Edible Bird's Nest)—For Your Beauty and Health—From Sialic Acids*. Yôkihi Publishers: Tokyo, Japan, 2004.
1099. Peumans, W. J.; Van Damme, E. J. M. Lectins as Plant Defense Proteins. *Plant Physiol.* **1995**, *109*, 347–352.
1100. Ram, S.; Sharma, A. K.; Simpson, S. D.; Gulati, S.; McQuillen, D. P.; Pangburn, M. K.; Rice, P. A. A Novel Sialic Acid Binding Site on Factor H Mediates Serum Resistance of Sialylated *Neisseria gonorrhoeae*. *J. Exp. Med.* **1998**, *187*, 743–752.
1101. Kazatchkine, M. D.; Fearon, D. T.; Austen, K. F. Human Alternative Complement Pathway: Membrane-Associated Sialic Acid Regulates the Competition Between B and β1H for Cell-Bound C3b. *J. Immunol.* **1979**, *122*, 75–81.
1102. Lehmann, F.; Tiralongo, E.; Tiralongo, J. Sialic Acid-Specific Lectins: Occurrence, Specificity and Function. *Cell. Mol. Life Sci.* **2006**, *63*, 1331–1354.
1103. Wegener, H.; Mallagaray, A.; Schöne, T.; Peters, T.; Lockhauserbäumer, J.; Yan, H.; Uetrecht, C.; Hansman, G. S.; Taube, S. Human Norovirus GII.4 (MI 001) P Dimer Binds Fucosylated and Sialylated Carbohydrates. *Glycobiology* **2017**, *27*, 1027–1037.
1104. Li, W.; Hulswit, R. J. G.; Widjaja, I.; Raj, V. S.; McBride, R.; Peng, W.; Widagdo, W.; Tortorici, M. A.; van Dieren, B.; Lang, Y.; van Lent, J. W. M.; Paulson, J. C.; de Haan, C. A. M.; de Groot, R. J.; van Kuppeveld, F. J. M.; Haagmans, B. L.; Bosch, B.-J. Identification of Sialic Acid-Binding Function for the Middle East Respiratory Syndrome Coronavirus Spike Glycoprotein. *Proc. Natl. Acad. Sci. U. S. A.* **2017**, *117*, E8505–E8517.
1105. Hellebø, A.; Vilas, U.; Falk, K.; Vlasak, R. Infectious Salmon Anemia Virus Specifically Binds to and Hydrolyzes 4-O-Acetylated Sialic Acids. *J. Virol.* **2004**, *78*, 3055–3062.
1106. Schultze, B.; Krempl, C.; Ballesteros, M. L.; Shaw, L.; Schauer, R.; Enjuanes, L.; Herrler, G. Transmissible Gastroenteritis Coronavirus, but Not the Related Porcine Respiratory Coronavirus, Has a Sialic Acid (*N*-Glycolylneuraminic Acid) Binding Activity. *J. Virol.* **1996**, *70*, 5634–5637.
1107. Tack, J. *Nachweis 9-O-acetylierter Sialoglykoproteine in menschlicher Nasenschleimhaut*; Diploma-Thesis Christian-Albrechts-Universität Kiel, 1994.

1108. Higa, H. H.; Rogers, G. N.; Paulson, J. C. Influenza Virus Hemagglutinins Differentiate Between Receptor Determinants Bearing *N*-Acetyl-, *N*-Glycollyl-, and *N,O*-Diacetylneuraminic Acids. *Virology* **1985**, *144*, 279–282.
1109. Cohnheim, J. *Lectures on General Pathology: A Handbook for Practitioners and Students.* (Translated by McKee, A. B.); The New Sydenham Society: London, UK, 1889.
1110. Sackstein, R. Fulfilling Koch's Postulates in Glycoscience: HCELL, GPS and Translational Glycobiology. *Glycobiology* **2016**, *26*, 560–570.
1111. Hadjamu, J. *Professor Dr. med. Heinrich Schade—Begründer der Molekularpathologie 1876–1935—Leben und Werk* (Düsseldorfer Arbeiten zur Geschichte der Medizin, Heft 39); Michael Triltsch Verlag: Düsseldorf, Germany, 1974.
1112. Gowans, J. L. The Recirculation of Lymphocytes From Blood to Lymph in the Rat. *J. Physiol.* **1959**, *146*, 54–69.
1113. Marchesi, V. T.; Gowans, J. L. The Migration of Lymphocytes Through the Endothelium of Venules in Lymph Nodes: An Electron Microscope Study. *Proc. R. Soc. London, Ser. B* **1964**, *159*, 283–290.
1114. Marchesi, V. T.; Florey, H. W. Electron Micrographic Observations on the Emigration of Leucocytes. *Q. J. Exp. Physiol. Cogn. Med. Sci.* **1960**, *45*, 343–348.
1115. Stoolman, L. M.; Rosen, S. D. Possible Role for Cell-Surface Carbohydrate-Binding Molecules in Lymphocyte Recirculation. *J. Cell Biol.* **1983**, *96*, 722–729.
1116. Rosen, S. D.; Singer, M. S.; Yednock, T. A.; Stoolman, L. M. Involvement of Sialic Acid on Endothelial Cells in Organ-Specific Lymphocyte Recirculation. *Science* **1985**, *228*, 1005–1007.
1117. Rosen, S. D. Ligands for L-Selectin: Homing, Inflammation, and Beyond. *Annu. Rev. Immunol.* **2004**, *22*, 129–156.
1118. Natoni, A.; Macauley, M. S.; O'Dwyer, M. E. Targeting Selectins and Their Ligands in Cancer. *Front. Oncol.* **2016**, *6*, 93.
1119. Bevilacqua, M. P.; Stengelin, S.; Gimbrone, M. A., Jr.; Seed, B. Endothelial Leukocyte Adhesion Molecule 1: An Inducible Receptor for Neutrophils Related to Complement Regulatory Proteins and Lectins. *Science* **1989**, *243*, 1160–1165.
1120. Ley, K. Functions of Selectins. In: *Mammalian Carbohydrate Recognition Systems*; Crocker, P. Ed.; Springer: Berlin, Germany, 2001; pp 177–200.
1121. Etzioni, A.; Frydman, M.; Pollack, S.; Avidor, I.; Phillips, M. L.; Paulson, J. C.; Gershoni-Baruch, R. Brief Report: Recurrent Severe Infections Caused by a Novel Leukocyte Adhesion Deficiency. *N. Engl. J. Med.* **1992**, *327*, 1789–1792.
1122. Varki, A. Selectins and Other Mammalian Sialic Acid-Binding Lectins. *Curr. Opin. Cell Biol.* **1992**, *4*, 257–266.
1123. Schnaar, R. L. Glycobiology Simplified: Diverse Roles of Glycan Recognition in Inflammation. *J. Leukoc. Biol.* **2016**, *99*, 825–838.
1124. Crocker, P. R.; Gordon, S. Properties and Distribution of a Lectin-Like Hemagglutinin Differentially Expressed by Murine Stromal Tissue Macrophages. *J. Exp. Med.* **1986**, *164*, 1862–1875.
1125. Crocker, P. R.; Kelm, S.; Dubois, C.; Martin, B.; McWilliam, A. S.; Shotton, D. M.; Paulson, J. C.; Gordon, S. Purification and Properties of Sialoadhesin, a Sialic Acid-Binding Receptor of Murine Tissue Macrophages. *EMBO J.* **1991**, *10*, 1661–1669.
1126. Crocker, P. R.; Mucklow, S.; Bouckson, V.; McWilliam, A.; Willis, A. C.; Gordon, S.; Milon, G.; Kelm, S.; Bradfield, P. Sialoadhesin, a Macrophage Sialic Acid Binding Receptor for Haemopoietic Cells With 17 Immunoglobulin-Like Domains. *EMBO J.* **1994**, *13*, 4490–4503.
1127. Van den Berg, T. K.; Brevé, J. J. P.; Damoiseaux, J. G. M. C.; Döpp, E. A.; Kelm, S.; Crocker, P. R.; Dijkstra, C. D.; Kraal, G. Sialoadhesin on Macrophages: Its Identification as a Lymphocyte Adhesion Molecule. *J. Exp. Med.* **1992**, *176*, 647–655.
1128. Stamenkovic, I.; Seed, B. The B-Cell Antigen CD22 Mediates Monocyte and Erythrocyte Adhesion. *Nature* **1990**, *345*, 74–77.

1129. Macauley, M. S.; Crocker, P. R.; Paulson, J. C. Siglec-Mediated Regulation of Immune Cell Function in Disease. *Nat. Rev. Immunol.* **2014**, *14*, 653–666.
1130. Trapp, B. D. Myelin-Associated Glycoprotein. Location and Potential Functions. *Ann. N. Y. Acad. Sci.* **1990**, *605*, 29–43.
1131. Strenge, K.; Schauer, R.; Bovin, N.; Hasegawa, A.; Ishida, H.; Kiso, M.; Kelm, S. Glycan Specificity of Myelin-Associated Glycoprotein and Sialoadhesin Deduced From Interactions With Synthetic Oligosaccharides. *Eur. J. Biochem.* **1998**, *258*, 677–685.
1132. Simmons, D.; Seed, B. Isolation of a cDNA Encoding CD33, a Differentiation Antigen of Myeloid Progenitor Cells. *J. Immunol.* **1988**, *141*, 2797–2800.
1133. Sjoberg, E. R.; Powell, L. D.; Klein, A.; Varki, A. Natural Ligands of the B Cell Adhesion Molecule CD22β Can Be Masked by 9-*O*-Acetylation of Sialic Acids. *J. Cell Biol.* **1994**, *126*, 549–562.
1134. Kelm, S.; Schauer, R.; Crocker, P. R. The Sialoadhesins—A Family of Sialic Acid-Dependent Cellular Recognition Molecules Within the Immunoglobulin Superfamily. *Glycoconjugate J.* **1996**, *13*, 913–926.
1135. Cao, H.; Crocker, P. R. Evolution of CD33-Related Siglecs: Regulating Host Immune Functions and Escaping Pathogen Exploitation?*Immunology* **2010**, *132*, 18–26.
1136. Chang, Y.-C.; Nizet, V. The Interplay Between Siglecs and Sialylated Pathogens. *Glycobiology* **2014**, *24*, 818–825.
1137. Crocker, P. R.; Kannagi, R. Introduction to Special Issue: 'Emerging Roles of Siglecs in Health and Disease'. *Glycobiology* **2014**, *24*, 784.
1138. Padler-Karavani, V.; Hurtado-Ziola, N.; Chang, Y.-C.; Sonnenburg, J. L.; Ronaghy, A.; Yu, H.; Verhagen, A.; Nizet, V.; Chen, X.; Varki, N.; Varki, A.; Angata, T. Rapid Evolution of Binding Specificities and Expression Patterns of Inhibitory CD33-Related Siglecs in Primates. *FASEB J.* **2014**, *28*, 1280–1293.
1139. Bochner, B. S. "Siglec"ting the Allergic Response for Therapeutic Targeting. *Glycobiology* **2016**, *26*, 546–552.
1140. Fontaine, G.; Biserte, G.; Montreuil, J.; Dupont, A.; Farriaux, J.-P.; Strecker, G.; Spik, G.; Puvion, E.; Puvion-Dutilleul, F.; Sezille, G.; Picqué, M. T. La Sialurie: Un Trouble Métabolique Original. *Helv. Paediatr. Acta* **1968**, *23*(Suppl. XVII), 1–32.
1141. Strehle, E. M. Sialic Acid Storage Disease and Related Disorders. *Genet. Test.* **2003**, *7*, 113–121.
1142. Gopaul, K. P.; Crook, M. A. The Inborn Errors of Sialic Acid Metabolism and Their Laboratory Investigation. *Clin. Lab.* **2006**, *52*, 155–169.
1143. Thomas, G. H.; Reynolds, L. W.; Miller, C. S. Overproduction of N-Acetylneuraminic Acid (Sialic Acid) by Sialuria Fibroblasts. *Pediatr. Res.* **1985**, *19*, 451–455.
1144. Aula, P.; Raivio, K.; Autio, S.; Thoden, C.-E.; Rapola, J.; Koskela, S.-L.; Yamashina, I. Four Patients With a New Lysosomal Storage Disorder (Salla Disease). *Monogr. Hum. Genet.* **1978**, *10*, 16–22.
1145. Suzuki, K. Sialic Acid in Biochemical Pathology. In: *Biology of the Sialic Acids*; Rosenberg, A. Ed.; Plenum Press: New York, NY, USA, 1995; pp 337–361.
1146. Çoker, M.; Kalkan-Uçar, S.; Kitiş, Ö.; Uçar, H.; Gökşen-Şimşek, D.; Darcan, Ş.; Gökben, S. Salla Disease in Turkish Children: Severe and Conventional Type. *Turk. J. Pediatr.* **2009**, *51*, 605–609.
1147. Hancock, L. W.; Thaler, M. M.; Horwitz, A. L.; Dawson, G. Generalized *N*-Acetylneuraminic Acid Storage Disease: Quantitation and Identification of the Monosaccharide Accumulating in Brain and Other Tissues. *J. Neurochem.* **1982**, *38*, 803–809.
1148. Tietze, F.; Seppala, R.; Renlund, M.; Hopwood, J. J.; Harper, G. S.; Thomas, G. H.; Gahl, W. A. Defective Lysosomal Egress of Free Sialic Acid (*N*-Acetylneuraminic Acid) in Fibroblasts of Patients With Infantile Free Sialic Acid Storage Disease. *J. Biol. Chem.* **1989**, *264*, 15316–15322.

1149. Verheijen, F. W.; Verbeek, E.; Aula, N.; Beerens, C. E. M. T.; Havelaar, A. C.; Joosse, M.; Peltonen, L.; Aula, P.; Galjaard, H.; van der Spek, P. J.; Mancini, G. M. S. A New Gene, Encoding an Anion Transporter, Is Mutated in Sialic Acid Storage Diseases. *Nat. Genet.* **1999**, *23*, 462–465.
1150. Maroteaux, P.; Humbel, R. Les Oligosaccharidoses—Un Nouveau Concept. *Arch. Franç. Pédiat.* **1976**, *33*, 641–643.
1151. Cantz, M. Sialidoses. In: *Sialic Acids—Chemistry, Metabolism and Function*; Schauer, R. Ed.; Cell Biology Monographs, Vol. 10; Springer-Verlag: Vienna, Austria, 1982; pp 307–320.
1152. Cantz, M.; Gehler, J.; Spranger, J. Mucolipidosis I: Increased Sialic Acid Content and Deficiency of an α-*N*-Acetylneuraminidase in Cultured Fibroblasts. *Biochem. Biophys. Res. Commun.* **1977**, *74*, 732–738.
1153. Spranger, J.; Gehler, J.; Cantz, M. Mucolipidosis I—A Sialidosis. *Am. J. Med. Genet.* **1977**, *1*, 21–29.
1154. Durand, P.; Gatti, R.; Cavalieri, S.; Borrone, C.; Tondeur, M.; Michalski, J.-C.; Strecker, G. Sialidosis (Mucolipidosis I). *Helv. Paediatr. Acta* **1977**, *32*, 391–400.
1155. Bonten, E.; van der Spoel, A.; Fornerod, M.; Grosveld, G.; d'Azzo, A. Characterization of Human Lysosomal Neuraminidase Defines the Molecular Basis of the Metabolic Storage Disorder Sialidosis. *Genes Dev.* **1996**, *10*, 3156–3169.
1156. Itoh, K.; Naganawa, Y.; Matsuzawa, F.; Aikawa, S.; Doi, H.; Sasagasako, N.; Yamada, T.; Kira, J.-I.; Kobayashi, T.; Pshezhetsky, A. V.; Sakuraba, H. Novel Missense Mutations in the Human Lysosomal Sialidase Gene in Sialidosis Patients and Prediction of Structural Alterations of Mutant Enzymes. *J. Hum. Genet.* **2002**, *47*, 29–37.
1157. Seyrantepe, V.; Poupetova, H.; Froissart, R.; Zabot, M.-T.; Maire, I.; Pshezhetsky, A. V. Molecular Pathology of NEU1 Gene in Sialidosis. *Hum. Mutat.* **2003**, *22*, 343–352.
1158. D'Azzo, A.; Hoogeveen, A.; Reuser, A. J. J.; Robinson, D.; Galjaard, H. Molecular Defect in Combined β-Galactosidase and Neuraminidase Deficiency in Man. *Proc. Natl. Acad. Sci. U. S. A.* **1982**, *79*, 4535–4539.
1159. Vinogradova, M. V.; Michaud, L.; Mezentsev, A. V.; Lukong, K. E.; El-Alfy, M.; Morales, C. R.; Potier, M.; Pshezhetsky, A. V. Molecular Mechanism of Lysosomal Sialidase Deficiency in Galactosialidosis Involves Its Rapid Degradation. *Biochem. J.* **1998**, *330*, 641–650.
1160. Bonten, E. J.; Wang, D.; Toy, J. N.; Mann, L.; Mignardot, A.; Yogalingam, G.; d'Azzo, A. Targeting Macrophages With Baculovirus-Produced Lysosomal Enzymes: Implications for Enzyme Replacement Therapy of the Glycoprotein Storage Disorder Galactosialidosis. *FASEB J.* **2004**, *18*, 971–973.
1161. D'Azzo, A.; Machado, E.; Annunziata, I. Pathogenesis, Emerging Therapeutic Targets and Treatment in Sialidosis. *Expert Opin. Orphan Drugs* **2015**, *3*, 491–504.
1162. Van Pelt, J.; Kamerling, J. P.; Bakker, H. D.; Vliegenthart, J. F. G. A Comparative Study of Sialyloligosaccharides Isolated From Sialidosis and Galactosialidosis Urine. *J. Inherit. Metab. Dis.* **1991**, *14*, 730–740.
1163. Yardeni, T.; Choekyi, T.; Jacobs, K.; Ciccone, C.; Patzel, K.; Anikster, Y.; Gahl, W. A.; Kurochkina, N.; Huizing, M. Identification, Tissue Distribution, and Molecular Modeling of Novel Human Isoforms of the Key Enzyme in Sialic Acid Synthesis, UDP-GlcNAc 2-Epimerase/ManNAc Kinase. *Biochemistry* **2011**, *50*, 8914–8925.
1164. Broccolini, A.; Gidaro, T.; Morosetti, R.; Sancricca, C.; Mirabella, M. Hereditary Inclusion-Body Myopathy With Sparing of the Quadriceps: The Many Tiles of an Incomplete Puzzle. *Acta Myol.* **2011**, *30*, 91–95.
1165. Argov, Z.; Yarom, R. "Rimmed Vacuole Myopathy" Sparing the Quadriceps—A Unique Disorder in Iranian Jews. *J. Neurol. Sci.* **1984**, *64*, 33–43.

1166. Bennmann, D.; Weidemann, W.; Thate, A.; Kreuzmann, D.; Horstkorte, R. Aberrant *O*-GlcNAcylation Disrupts GNE Enzyme Activity in GNE Myopathy. *FEBS J.* **2016**, *283*, 2285–2294.
1167. Malicdan, M. C. V.; Noguchi, S.; Nishino, I. Autophagy in a Mouse Model of Distal Myopathy With Rimmed Vacuoles or Hereditary Inclusion Body Myopathy. *Autophagy* **2007**, *3*, 396–398.
1168. Cho, A.; Malicdan, M. C. V.; Miyakawa, M.; Nonaka, I.; Nishino, I.; Noguchi, S. Sialic Acid Deficiency Is Associated With Oxidative Stress Leading to Muscle Atrophy and Weakness in GNE Myopathy. *Hum. Mol. Genet.* **2017**, *26*, 3081–3093.
1169. Chan, Y. M.; Lee, P.; Jungles, S.; Morris, G.; Cadaoas, J.; Skrinar, A.; Vellard, M.; Kakkis, E. Substantial Deficiency of Free Sialic Acid in Muscles of Patients With GNE Myopathy and in a Mouse Model. *PLoS One* **2017**, *12,* e0173261 (1–17).
1170. Malicdan, M. C. V.; Noguchi, S.; Tokutomi, T.; Goto, Y.-I.; Nonaka, I.; Hayashi, Y. K.; Nishino, I. Peracetylated *N*-Acetylmannosamine, a Synthetic Sugar Molecule, Efficiently Rescues Muscle Phenotype and Biochemical Defects in Mouse Model of Sialic Acid-Deficient Myopathy. *J. Biol. Chem.* **2012**, *287*, 2689–2705.
1171. Huizing, M.; Krasnewich, D. M. Hereditary Inclusion Body Myopathy: A Decade of Progress. *Biochim. Biophys. Acta* **2009**, *1792*, 881–887.

CHAPTER TWO

Chemical Synthesis of Glycosides of *N*-Acetylneuraminic Acid

Cristina De Meo[1], Bradley T. Jones
Department of Chemistry, Southern Illinois University—Edwardsville, Edwardsville, IL, United States
[1]Corresponding author: e-mail address: cdemes@siue.edu

Contents

1. INTRODUCTION

Sialic acids are a family of 3-deoxynon-2-ulosonic acids (Fig. 1). Their structures are quite different from other common monosaccharides with the presence of a carboxylate group at C-1, a deoxy carbon at C-3, a functionalized C-5 position, and a glycerol chain at C-6. The most abundant member of this family is *N*-acetylneuraminic acid (5-acetamido-3,5-dideoxy-D-*glycero*-D-*galacto*-non-2-ulopyranosonic acid, NANA or Neu5Ac), which bears an acetamido group at the C-5 position. Less widespread, but still common, are *N*-glycolylneuraminic acid (3,5-dideoxy-5-((hydroxyacetyl)-amino)-D-*glycero*-D-*galacto*-non-2-ulopyranosonic acid, Neu5Gc), bearing a (2-hydroxy)acetamido group at C-5, and Kdn (3-deoxy-D-*glycero*-D-*galacto*-non-2-ulopyranosonic acid), which is hydroxylated at the C-5 position

Advances in Carbohydrate Chemistry and Biochemistry, Volume 75
ISSN 0065-2318
https://doi.org/10.1016/bs.accb.2018.09.005

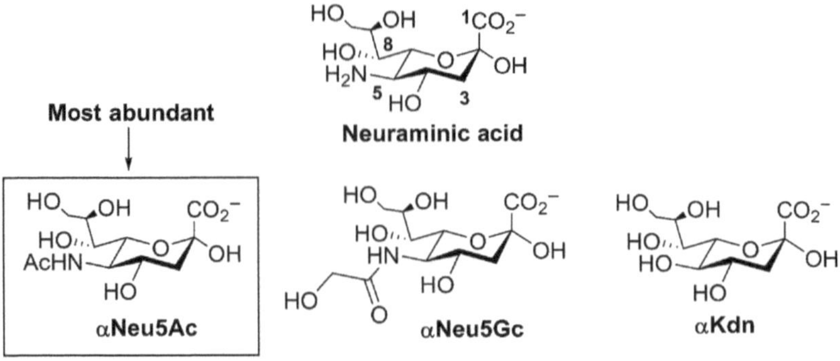

Fig. 1 The most common sialic acids.

(Fig. 1).[1–8] Neu5Gc, while present only in trace amounts in healthy humans, is overexpressed in human cancer cells.[9]

Within the cell, Neu5Ac is found predominantly at the terminal position of glycoconjugates, in α(2→3)- or α(2→6)-linkages with galactosides, and in α(2→6)-linkages with *N*-acetylgalactosaminides. Additional dimeric linkages such as Neu5Acα(2→8)Neu5Ac and Neu5Acα(2→9)Neu5Ac also can be found in glycoconjugates. Polymeric structures (polysialic acids) are found primarily in neural cell-adhesion molecules, as well as in fish eggs, and as capsular polysaccharides in some bacteria. Sialic acids bearing structural modifications such as acetylation, phosphorylation, sulfonation, and glycosylation at position 4 or 5 have also been detected.

The biological properties of sialic acids are numerous, mainly due to their unique structures and terminal positions in glycoconjugate chains. Thus, sialic acid-containing glycoconjugates are involved in cell–cell interaction and recognition, including pathogenic attacks and cellular overexpression in cancerous malignancy. In the study of sialic acids as biological targets, chemical and enzymatic methods are still the major sources of most derivatives. This chapter is dedicated to the description of methodologies in chemical sialylations, with a particular emphasis on *N*-acetylneuraminic acid glycosidic linkages. Several reviews have been written in the past related to this subject.[10–26] Nevertheless, while some overlaps with previous reviews can be found, the most recent recorded advancements are described herein. In addition, due to the connections and intersection of these synthetic approaches, the subdivision of the following subchapters is only a general guideline, and many topics will be described in multiple sections.

2. CHEMICAL SYNTHESIS OF O-SIALOSIDES

The chemical synthesis of α-sialosides is considered still one of the most challenging tasks in the carbohydrate field. Several methodologies have been investigated with the purpose of optimizing the yield of the kinetic product (α-anomer) and minimizing the formation of the thermodynamic product (β-anomer), while suppressing unwanted side reactions (hydrolysis and elimination, Scheme 1).[10,12–14,27–29] In the commonly accepted description of a sialylation reaction, which proceeds via an oxacarbenium ion intermediate, nucleophilic attack of the acceptor can be difficult to control without the use of a participating group at the C-3 position. In addition, the electron-withdrawing carboxylate group at C-1 destabilizes the placement of a positive charge at the anomeric carbon. Additional influence of the acetamido group at C-5 (for *N*-acetylneuraminic acid) also can influence the reactivity of the donor and the stereoselectivity of the reaction.

In the sections following, a review of sialic acid structural modifications, investigated to improve the outcome of chemical sialylations, is described. The classification is based mainly on the position of structural variations (e.g., protecting groups at C-5 and a leaving group at C-2), although some special attention is given to conformationally constrained cyclic protecting groups (e.g., oxazolidinone). A section for mechanistic studies is added as well, in order to summarize the efforts in the field that have been carried out to gain a better understanding of chemical sialylations.

2.1 Modifications at C-1

Structural modifications at the C-1 position have followed two major approaches: (1) introduction of a participating group at C-1 (Approach 1, Fig. 2),

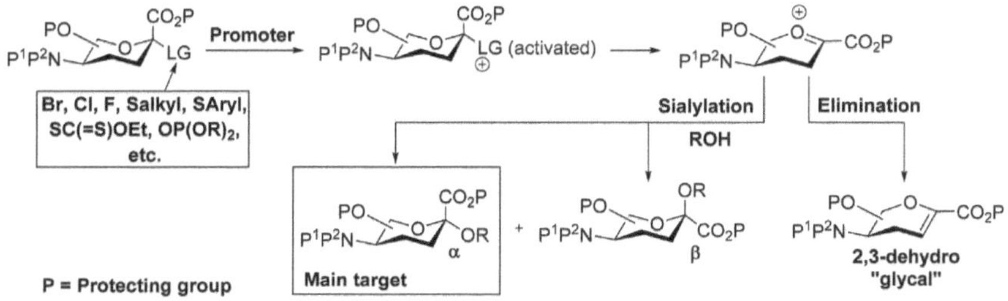

Scheme 1 Overview of a generic sialylation reaction.

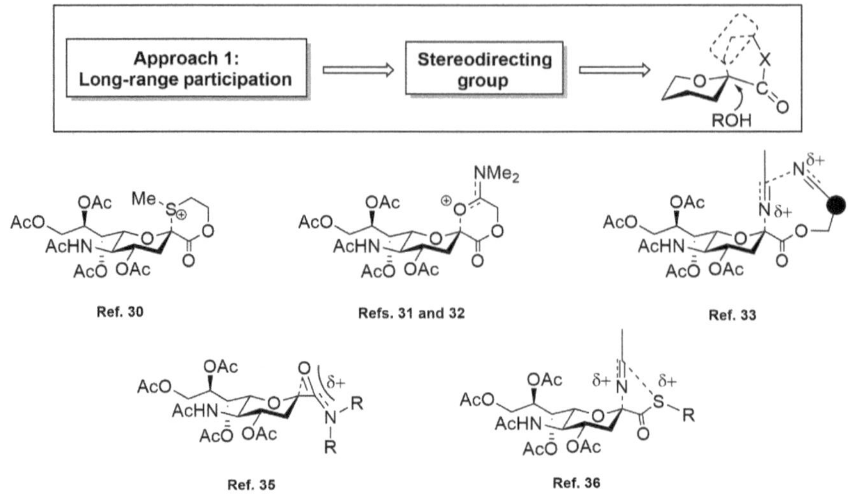

Fig. 2 Stereodirecting groups at C-1.

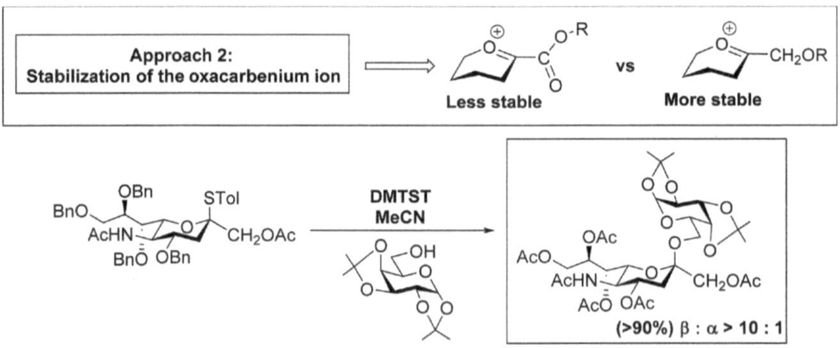

Fig. 3 Stabilization of oxacarbenium by reduction of the carboxylate.

and (2) stabilization of the oxacarbenium ion by reduction of the carboxylate moiety (Approach 2, Fig. 3). Although no remarkable improvement toward the synthesis of α-sialosides has been noted, these methods have helped in gaining a better insight into sialylation reactions.

Strategies involving the first approach, to introduce stereodirecting groups at C-1 in order to direct the nucleophilic attack toward the α-face of the oxacarbenium ion, have been investigated by several groups.

Takahashi and coworkers reported an improvement in the stereoselectivity of the chemical glycosylation outcome when the carboxylic acid group at C-1 was converted into a 2-methylthioethyl ester.[30] When promoted as a glycosyl donor in the presence of a 2-[2-(2-azidoethoxy)ethoxy]ethanol acceptor with *N*-iodosuccinimide (NIS) and trifluoromethanesulfonic (triflic) acid (TfOH) using acetonitrile as a solvent, the standard methyl ester-protected per-O-acetylated 2-methylthio Neu5Ac derivative produced 62% of the desired product with poor stereoselectivity (α/β = 1/1.5). The same acceptor and reaction conditions applied to the equivalent 2-methylthioethyl ester-protected Neu5Ac donor produced only 50% of the targeted product, but with a significant improvement in stereoselectivity (α/β = 4/1, Scheme 2). Interestingly, shifting from the use of acetonitrile as a solvent to the less common 1,2-dimethoxyethane (DME) resulted in a slight reduction of the yield (45%), albeit excellent α-stereoselectivity (α/β = 20/1). When the 2-methylthioethyl ester Neu5Ac donor was coupled with methyl 2,3,4-tri-*O*-methoxybenzyl-β-D-galactopyranosyl acceptor in either acetonitrile or DME using NIS/TfOH, a low-to-moderate yield (39% to 51%) of the desired α(2→6)-linked disaccharide was obtained with good stereoselectivity (α/β = 5/1). When methyl 2,4,6-tri-*O*-methoxybenzyl-β-D-galactopyranosyl acceptor was glycosylated in DME, a slight improvement in stereoselectivity was observed compared to that obtained in acetonitrile (α/β = 5/1 to 10/1), but yields of the target product were lower (21% to 35%, Scheme 2).[30]

An improvement on this strategy was described by Haberman and Gin. In this case, the auxiliary employed at the C-1 position was an *N,N*-dimethylglycolamido (OCH_2CONMe_2) ester functionality (Scheme 3).[31] Using a standard methyl ester-protected per-O-acetylated sialyl chloride donor and methyl 2,3,4-tri-*O*-benzyl-α-D-glucopyranosyl acceptor, activation by silver triflate (AgOTf) gave 60% of the target (2→6)-linked disaccharide as a racemic mixture. In the presence of the *N,N*-dimethylglycolamido ester-protected sialyl chloride donor, sialylation with the same acceptor under the same reaction conditions gave higher stereoselectivity (α/β = 7/1). An even greater improvement was observed with a secondary methyl 2,6-di-*O*-benzyl-α-D-galactopyransyl acceptor. With the latter, the *N,N*-dimethylglycolamido ester-protected sialyl chloride donor afforded 56% of the (2→3)-linked product in high stereoselectivity (α/β = 13/1). The benchmark methyl ester-protected sialyl chloride donor produced only 22% of the corresponding (2→3)-linked disaccharide as a racemic mixture.[31]

Haberman and Gin later employed the same *N,N*-dimethylglycolamido ester auxiliary in a 2-hydroxy-Neu5Ac donor using a *p*-nitrophenyl phenyl sulfoxide [(p-$NO_2C_6H_4$)(Ph)SO] and triflic anhydride (Tf_2O) promoter system to explore dehydrative sialylations with common galactopyranosyl acceptors. When used with a primary acceptor, such as methyl 2,3,4-tri-*O*-benzyl-α-D-galactopyranosyl, this donor yielded 63% of the target

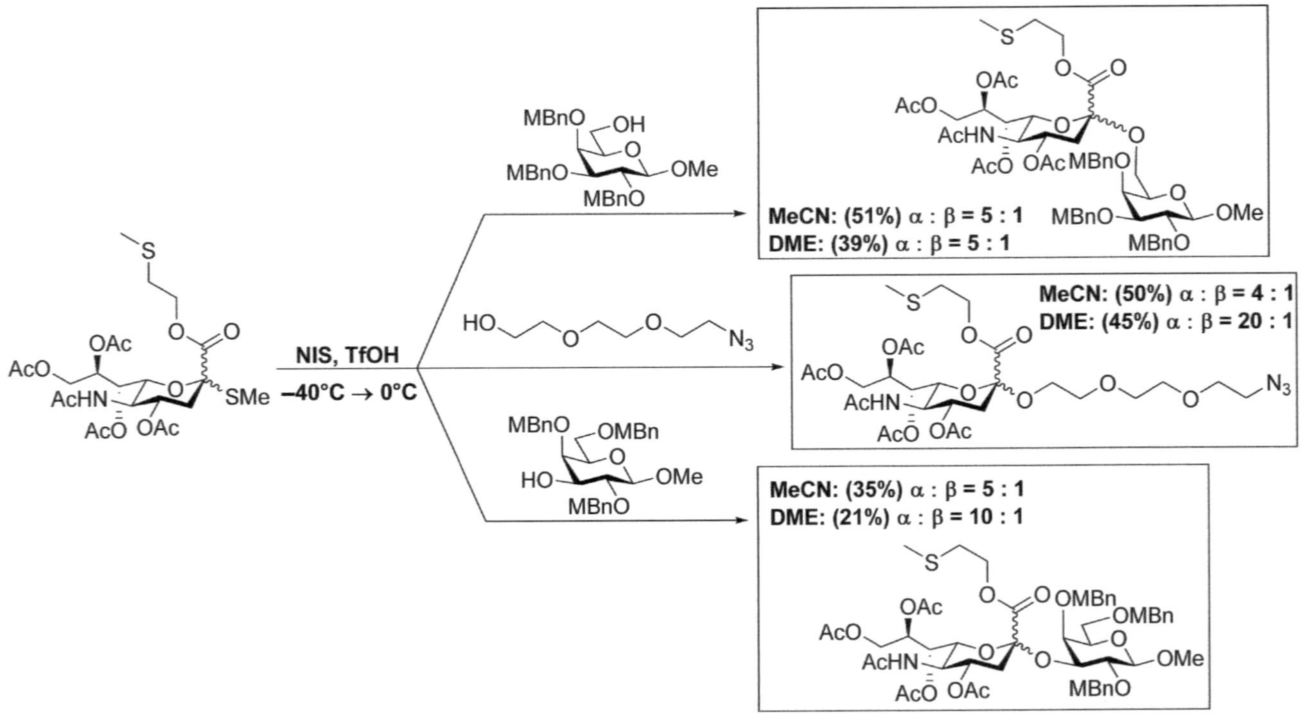

Scheme 2 Effect of a methylthioethyl ester at C-1.

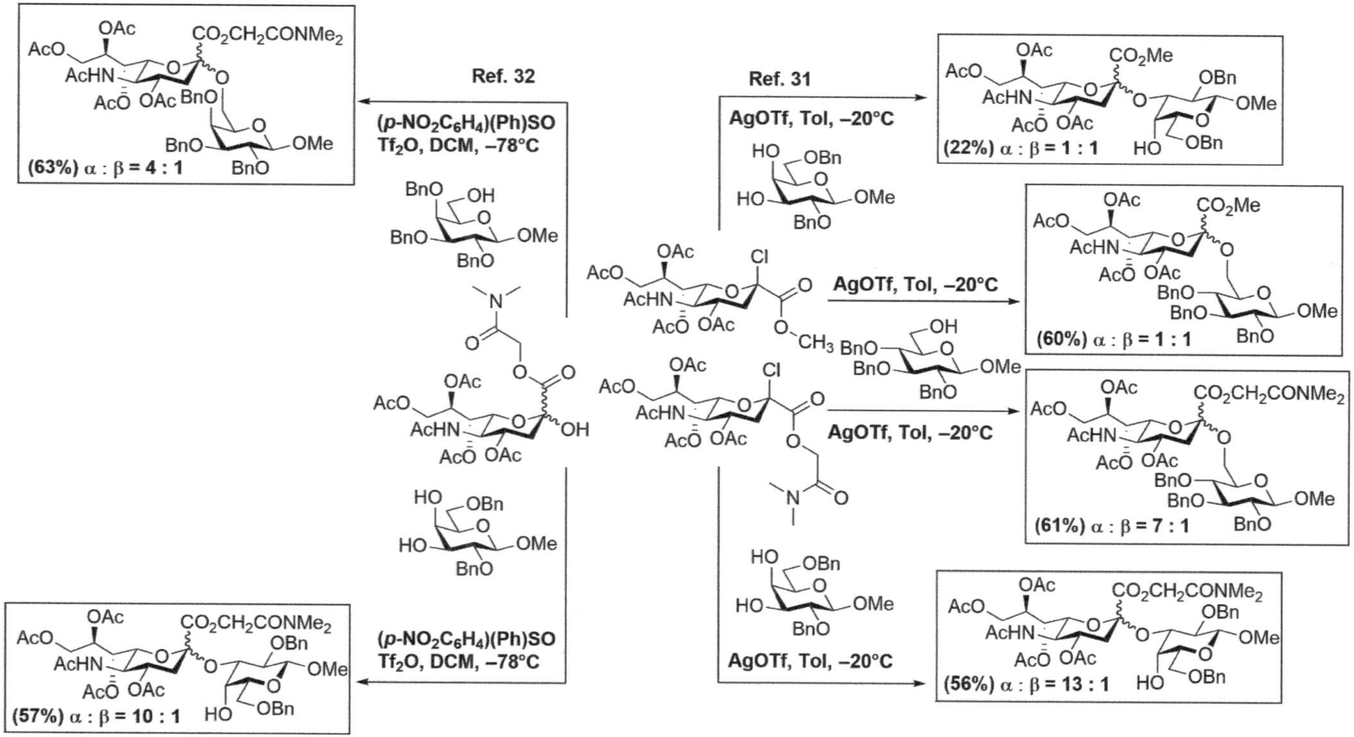

Scheme 3 Effect of an *N,N*-dimethylglycolamido ester at C-1.

(2→6)-linked product (α/β=4/1). The same system applied to a secondary acceptor methyl 2,6-di-*O*-benzyl-α-D-galactopyranosyl produced 57% of the (2→3)-linked disaccharide with higher stereoselectivity (α/β=10/1, Scheme 3).[32]

Ishiwata and Ito expanded this concept by synthesizing a vast array of functionalized esters. A lack of α-stereoselectivity was observed in the presence of dichloromethane (DCM) as solvent, whereas stereoselectivity improved in several donor–acceptor systems when the solvent of choice was acetonitrile.[33] The authors rationalized these findings by a possible electronic stabilization of the generated β-nitrilium ion intermediate performed by some C-1 esters. A more recent investigation of this concept has been reported by Amarasekara and Crich, and it is described in Section 3.[34]

More recently, Kajihara and coworkers converted the common ester functionality at C-1 to an amide in order to decrease the electron-withdrawing property of C-1 and allow for a more stabilized intermediate.[35] The ester–amide conversion can be accomplished by treatment of Neu5Ac with a mixture of ammonium hydroxide and ammonium bicarbonate (for simple acetamide), or methylamine in methanol (for the *N*-methylamide). The *N*,*N*-dimethylamido derivative can be obtained by coupling of the C-1 carboxylic acid with dimethylamine and 1-ethyl-3-(3-dimethylaminopropyl)carbodiimide (EDC) in THF. Preliminary testing of a 2-diethylphosphite donor equipped with the *N*-methylamide modification at C-1 showed higher stereoselectivity when compared to the common methyl ester. This method then was extended to the synthesis of an STn derivative (Scheme 4).

Sato and coworkers synthesized a series of dibenzylphosphite donors bearing different thioester functionalities at the C-1 position. These donors were synthesized by treatment of the C-1 carboxylate in a per-O-acetylated 2-phenylthio precursor with ethyl chloroformate to generate a carboxylic anhydride, which reacts with several thiols to generate various thioesters in moderate-to-good yields.[36] Sialylation reactions promoted by activation of the phosphite leaving group with trimethylsilyl triflate (TMSOTf) demonstrated a slight improvement in stereoselectivity. However, a greater improvement in stereoselectivity was achieved by coupling with a galactal acceptor (Scheme 4).[36]

As in Approach 2, Wong and coworkers explored the reactivity and stereodirecting ability of sialyl donors upon conversion of the electron-withdrawing C-1 carboxylic acid moiety into various functionalized hydroxymethyl groups. A remarkable improvement in reactivity was

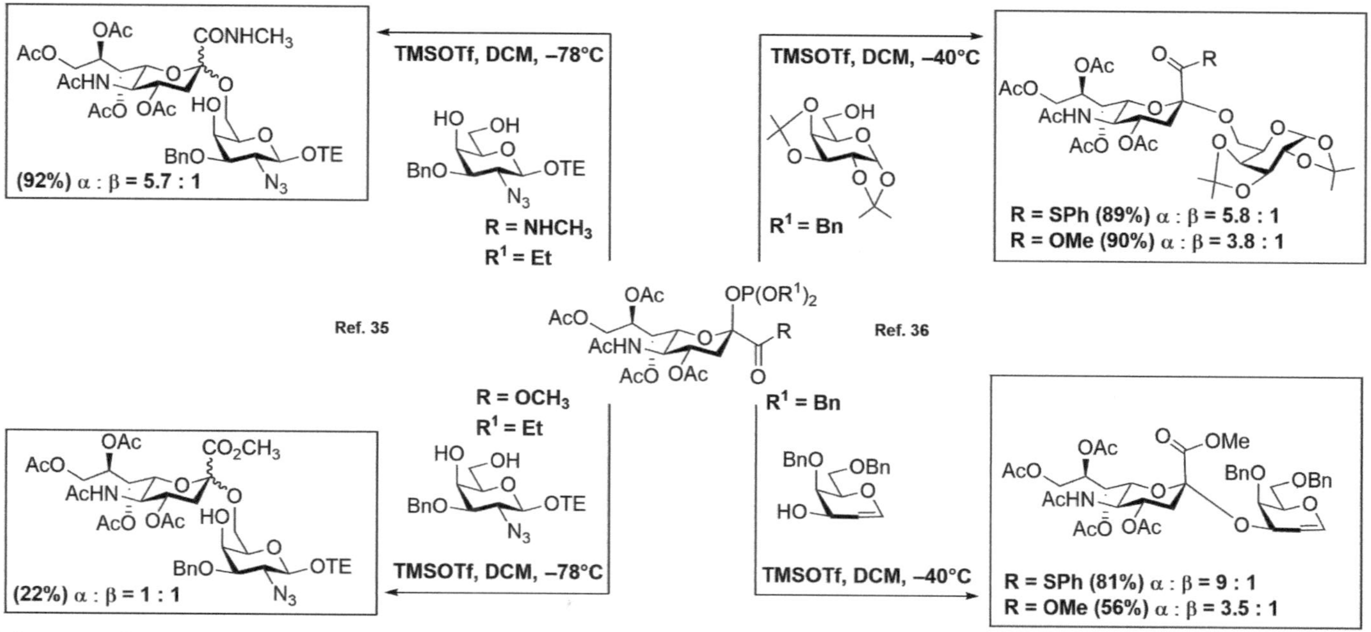

Scheme 4 Effect of *N*-methylformamido groups and thioesters on donors with phosphite leaving groups.

observed, likely due to the improved stability of the oxacarbenium ion. Unfortunately, despite the high yield, a predominance of the β-linked product was observed. The β-stereoselectivity was also dependent upon the reaction solvent and the promoter system used (Fig. 3).[37]

2.2 Modifications at C-2: Leaving Groups

This section, although needed to describe the most frequently used and important leaving group strategies and methodologies, is not intended to be all-inclusive. Many additional examples and applications of some of the leaving groups described below can be found in other sections of this review.

2.2.1 Halides

Sialyl chlorides were the first leaving groups investigated for the synthesis of O-sialosides.[38] 2-Bromo[39] and 2-fluoro[40–42] derivatives also have been synthesized but have seen a narrower application than that of the 2-chloro counterpart. In fact, sialyl chlorides were the most- used donors in the late 1980s and early 1990s.[10,24,43–46] Traditionally, glycosylations using sialyl halide donors are promoted by metal-induced activation. Silver(I) salts were used in the Koenigs–Knorr approach, while mercury(II) salts were used in the Helferich method.[13,47] In general, silver salt promoter systems give faster reaction times and higher stereoselectivities, but also lower yields than mercury-catalyzed sialylations.[13] For a more comprehensive description of the use of sialyl chlorides in the past decades, several reviews and book chapters have been published.[10–13] It is noteworthy that the classic Koenigs–Knorr procedure has been reinvestigated via activation of a sialyl chloride donor in the presence of silver carbonate using galactopyranosyl diols and triols as acceptors for the synthesis of α(2→6)-glycosidic bonds. In general, moderate yields and high stereoselectivities were observed (Scheme 5).[47] Unusual regioselectivity was noticed when an *N*-acetylgalactopyranosyl triol was used as a glycosyl acceptor. In this case, a mixture of α(2→6)-linked disaccharide and α(2→3)-/α(2→6)-linked trisaccharide was observed. The inverted reactivity between O-3 and O-6 was explained by the presence of the acetamido group at C-2 in the acceptor. The use of 2-chloro sialosides in the last decade has been extended to their facile conversion to other leaving groups and intermediates,[48–53] several of which can be produced without any promoter and in high stereoselectivity.[54] The stereoselective outcome is influenced by the solvent of choice: high α-stereoselectivity was observed in dichloromethane or 1,2-dichloroethane (DCE), while lower or complete loss of α-stereoselectivity occurred in acetonitrile.[54]

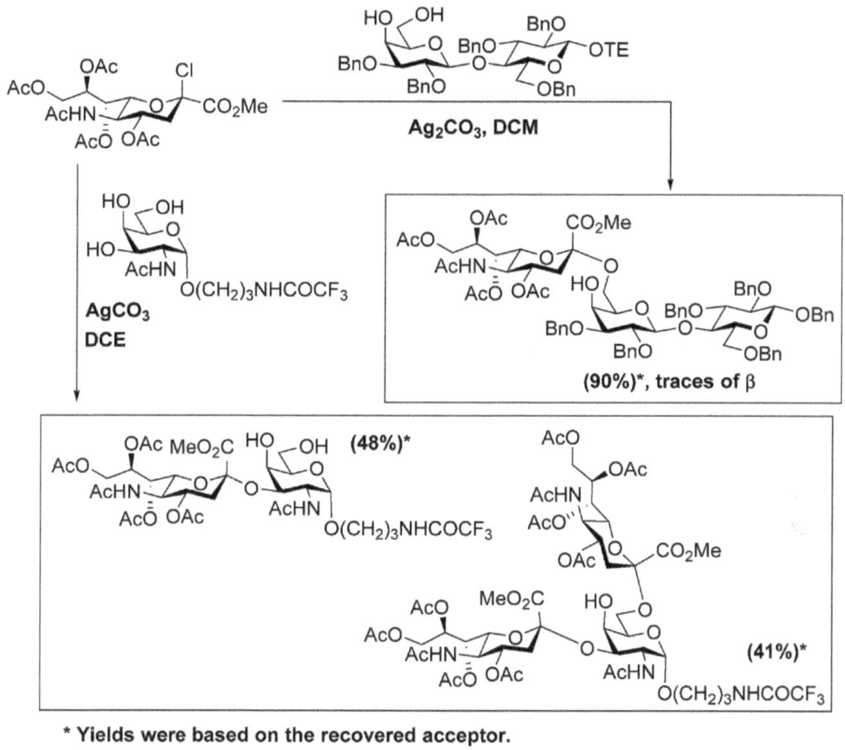

Scheme 5 Examples of sialylations using glycosyl chlorides.

2.2.2 S-Sialosides: Thiosialosides and Other 2-Thio Derivatives

The introduction of 2-alkylthio and 2-arylthio leaving groups by the pioneering work of Hasegawa and coworkers[55–57] and Roy and coworkers[58] is undoubtedly an important milestone in current sialylation methodologies, as elegantly phrased by Linhardt: "[this work] *has ushered in a renaissance with respect of sialic acid glycosylation.*"[13]

Alkyl and aryl thiosialosides are still the most commonly used building blocks for the synthesis of sialosides, due to their stability, long shelf life, and general reliability. Several articles and reviews have described the synthesis and overall performance of methyl-,[59–62] phenyl-,[46,49] and ethyl-[49,58] thioglycosides as donors.[10–14] Their syntheses can be accomplished by reaction of 2-*O*-acetyl derivatives or halides with the corresponding thiols or thiolates. Thioglycosides can be activated by the action of several promoter systems, such as dimethyl(methylthio)sulfonium trifluoromethanesulfonate (DMTST) or NIS/TfOH.[63,64]

Hasegawa and coworkers first used the per-O-acetylated 2-methylthio Neu5Ac donor with both the DMTST and NIS/TfOH promoter systems in the presence of common secondary galactopyranosyl and lactosyl acceptors in the late 1980s. When promoted in the presence of a (2-trimethylsilyl) ethyl 6-O-benzoyl-β-D-galactopyranosyl acceptor using DMTST in acetonitrile at −40°C, the per-O-acetylated 2-methylthio Neu5Ac donor produced the α(2→3)-linked disaccharide in 52% yield, whereas the same donor–acceptor pair promoted by NIS/TfOH gave a slightly higher yield (61%, Scheme 6). Both sialylations were completely α-stereoselective. When the same 2-methylthio donor was reacted with a benzylated lactosyl acceptor in the presence of DMTST in acetonitrile at −40°C, the (2→3)-linked trisaccharide was obtained in a fair yield of 38% as an anomeric mixture with the prevalence of the α-anomer (α/β=4/1). The use of the NIS/TfOH promoter system resulted in a notable increase in yield (69%) and a slight improvement in stereoselectivity (α/β=6/1). The higher outcome obtained with the NIS/TfOH promoter system prompted its preferential use over DMTST in more recent work.[65]

Unfortunately, sialylation of the 8-hydroxy sialyl acceptor in the presence of DMTST produced only trace amounts (5%) of the desired dimeric α(2→8)-linked product.[55] Several advancements have been made in the field for the synthesis of α(2→8) linkages, and these will be described in other sections.

The selection of the alkylthio or arylthio group is, in general, a fine balance between the ease of its stereocontrolled synthesis, foul odor produced during the reaction, and performance in chemical sialylations. Thus, in

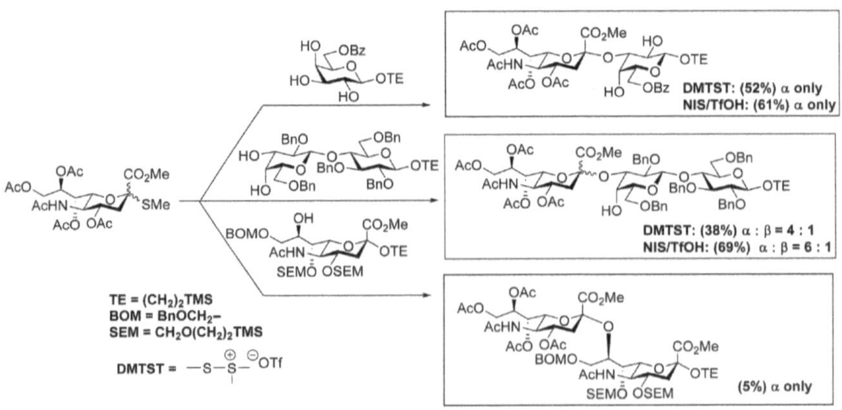

Scheme 6 Early applications of the methylthio sialyl donor.

recent years, arylthio groups such as the phenylthio and tolylthio groups have been used more frequently than their alkylthio counterparts. Many of these applications will be seen in later sections, used in conjunction with other positional modifications. Noteworthy is the recent addition in the thioglycoside list of donors, the adamantylthio[66] group, introduced and tested in sialic acid by Crich and Li.[67] The adamantylthio group can be introduced easily with moderate stereoselectivity (without the unpleasant odor), and it is more reactive than its phenylthio counterpart. When promoted in the presence of an unreactive methyl 2,6-di-*O*-benzyl-β-D-galactopyranosyl acceptor using NIS/TfOH in acetonitrile at −40°C, an oxazolidinone-protected β-phenylthio sialyl donor gave no reaction. The same chemical sialylation performed using the equivalent β-adamantylthio sialyl donor produced 87% of the desired (2→3)-linked disaccharide with modestly high stereoselectivity (α/β = 10/1, Scheme 7).[67]

Other thio leaving groups include anomeric xanthates, introduced by Marra and Sinaÿ. Xanthates are obtained by the reaction of the sialyl chloride with potassium ethoxydithiocarbonate in EtOH.[49] Xanthates can be activated with common promoters for thioglycosides but can also be activated selectively in the presence of thioglycosides,[68] thereby enhancing their potential applications in multistep oligosaccharide sequencing. Good stereoselectivities can be accomplished with DMTST[69] or NIS/TfOH,[59,70,71] as well as with phenylsulfenyl trifluoromethanesulfonate (PhSOTf) and 2,6-di(*tert*-butyl)pyridine (DTBP) at low temperature (−70°C). PhSOTf is generated in situ by reaction between phenylsulfenyl chloride and AgOTf.[72]

When promoted using PhSOTf/DTBP in the presence of a methyl 2,6-di-*O*-benzyl-β-D-galactopyranosyl acceptor, the per-O-acetylated sialyl

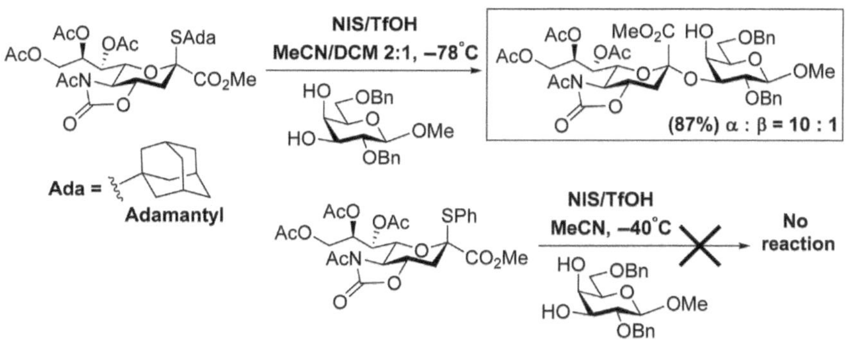

Scheme 7 *S*-Adamantyl performance in sialylation.

xanthate donor produced the desired (2→3)-linked disaccharide in 74% yield and in high stereoselectivity (α/β=19/1). Similar results were reported in coupling with a benzylated lactosyl acceptor, and the desired (2→3)-linked product was obtained with almost complete stereoselectivity (α/β=24/1) in 78% yield.[72] Sialyl xanthates tend to perform better than alkylthio counterparts in several applications. As an example, activation of per-O-acetylated sialyl xanthate donor in the presence of an unreactive threonine-substituted disaccharide acceptor gave 49% of the desired α(2→3)-linked trisaccharide with only traces of the β-anomer. The same acceptor used in the presence of a per-O-acetylated 2-ethylthio donor with NIS/TfOH yielded the desired disaccharide in only 38% yield with a loss of stereoselectivity (α/β=1.5/1, Scheme 8).[68] Due to their enhanced reactivity over comparative alkylthio donors and the ability to activate selectively in the presence of thioglycosides, sialyl xanthates still find wide utility.[73–77]

Sialyl xanthate donors can also be promoted efficiently with methylsulfenyl triflate (MeSOTf), generated in situ by reaction between methylsulfenyl bromide and AgOTf (Scheme 9).[68] When promoted in the presence of an *N*-(2,2,2-trichloroethoxycarbonyl)-protected (*N*-Troc) lactosamine acceptor using MeSOTf, a per-O-acetylated sialyl xanthate donor produced α(2→3)-linked trisaccharide with complete stereoselectivity. Subsequent conversion of the levulinoyl (Lev) protecting group into a standard acetyl group produced the desired trisaccharide in 55% yield over three steps (Scheme 9).[78]

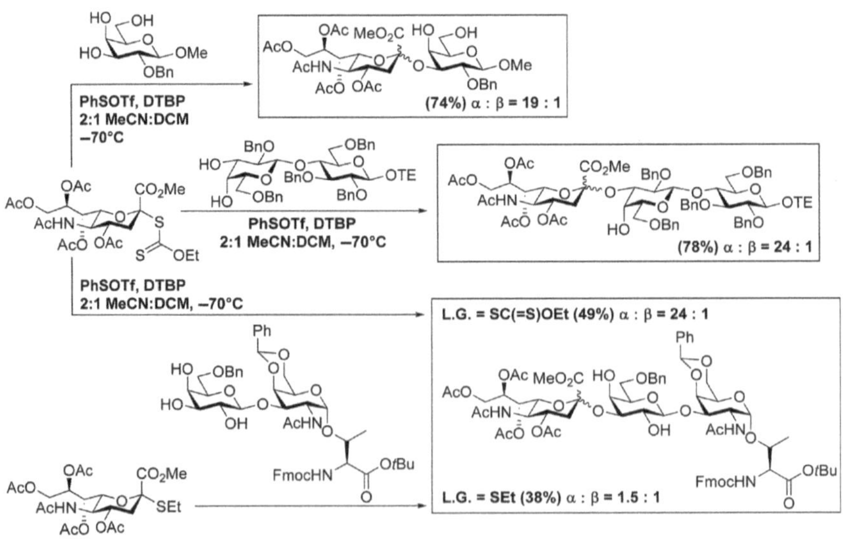

Scheme 8 Example of a xanthate leaving group in sialylation.

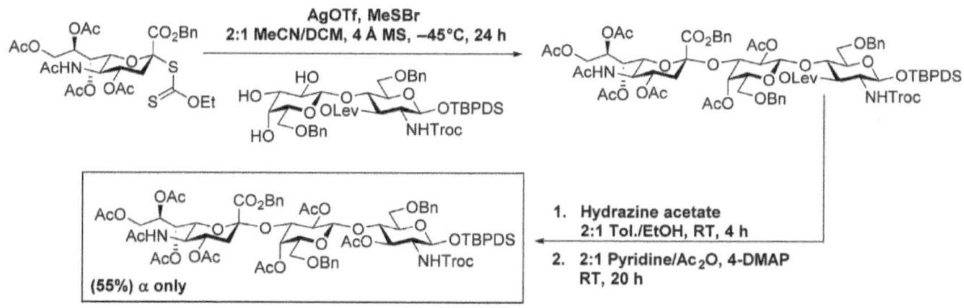

Scheme 9 Activation of xanthates by methylsulfenyl triflate.

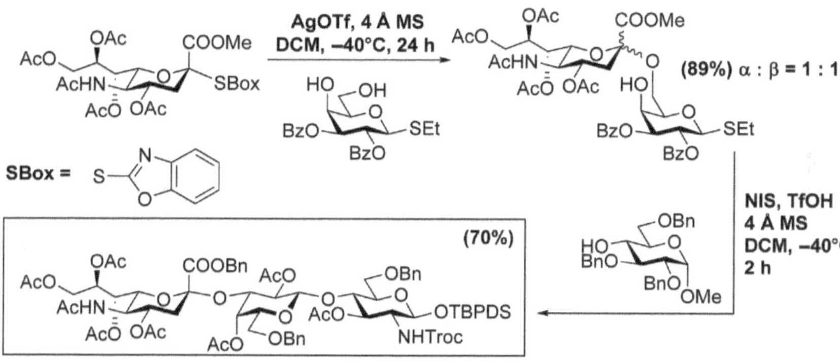

Scheme 10 Selective sialylation using thioimidate as a sialyl donor.

Thioimidates such as the *S*-benzoxazolyl (SBox) group synthesized by De Meo and Parker have been tested as leaving groups in chemical sialylations as well.[79] The major advantage of using thioimidates lies in their selective activation in the presence of thioglycosyl acceptors such as the ethylthio group, as shown in Scheme 10. When promoted in the presence of an ethylthio 2,3-di-*O*-benzoyl-β-D-galactopyranosyl acceptor using AgOTf, the per-O-acetylated 2-*S*-benzoxazolyl Neu5Ac donor produced the (2→6)-linked disaccharide in 89% yield as a racemic mixture. Subsequent addition of a methyl 2,3,6-tri-*O*-benzyl-α-D-glucopyranosyl acceptor and NIS/TfOH furnished only the β(2→4)-linkage and the GM3 analog in 70% yield (62% overall).[79]

Additional uses of the *S*-benzoxazolyl leaving group have been reported by Hung and coworkers, who also noted good yields and stereoselectivities in a one-pot synthesis of several oligosaccharides (Scheme 11).[80]

Improvements on this methodology to provide higher yield and stereoselectivities involved the use of C-5 modifications as well as the use of different promoter systems, such as bismuth(III) triflate ($Bi(OTf)_3$). For example, Takahashi tested several leaving groups in combination of

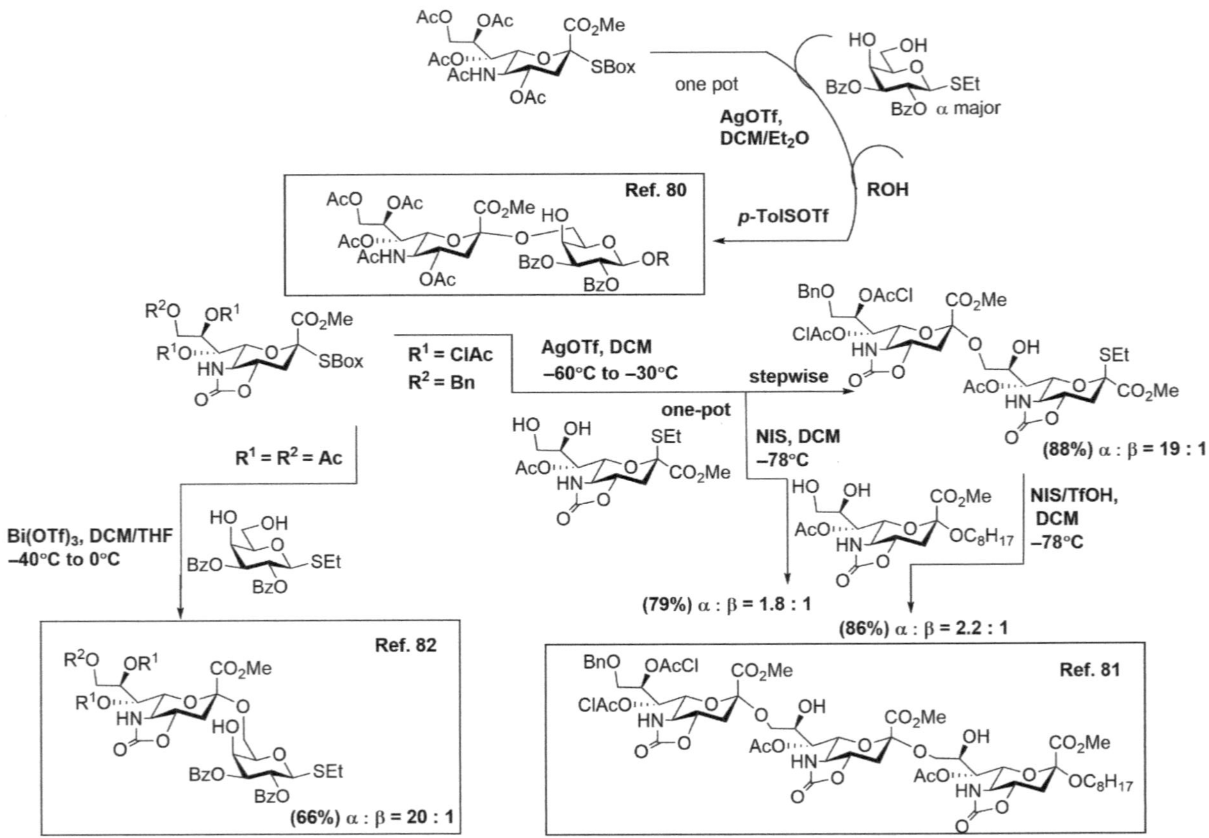

Scheme 11 Additional applications of SBox sialyl donors.

5-*N*,4-*O*-oxazolidinone, including halides and thioimidates (SBox).[81] The synthesis of an α(2→9)-trisialic acid via one-pot sialylation was performed by selectively activating a 5-*N*,4-*O*-carbonyl-protected 2-*S*-benzoxazolyl donor in the presence of an ethylthio sialyl acceptor (Scheme 11).[81] Loss of stereoselectivity observed by activation of the *S*-ethyl leaving group led to an alternative synthetic route that is described in Scheme 46.

5-*N*,4-*O*-Carbonyl-protected SBox sialyl donor was also tested by De Meo and coworkers for the selective synthesis of an α(2→6)-linkage: it was reported to give complete stereoselectivity in the coupling with an ethylthio galactopyranosyl acceptor upon activation with $Bi(OTf)_3$ (Scheme 11).[82]

2.2.3 O-Sialosides: Trifluoroacetimidates, Phosphites, and Phosphates

As an emerging oxygen-based leaving group, *N*-phenyltrifluoroacetimidates[83] were introduced by Cai and Yu in the early 2000s. Treatment of the per-O-acetylated 2-hydroxy-Neu5Ac derivative with 2,2,2-trifluoro-*N*-phenylacetimidoyl chloride and potassium carbonate gave the desired targets except in the case of the *N*-(*p*-methoxyphenyl) derivative.[83] In preliminary tests with this leaving group, a strong solvent effect was noticed with activation by TMSOTf, and complete inversion of stereoselectivity was observed between dichloromethane (β-preferred) and acetonitrile (α-preferred). For example, with a 1,2,3,4-di-*O*-isopropylidene-α-D-galactopyranosyl acceptor, the corresponding disaccharide was obtained in similar yields with an α/β ratio of 1/2.4 in DCM and 1.6/1 in MeCN (Scheme 11). Higher α-stereoselectivity was observed in a 1/1 (v/v) mixture of the two solvents (α/β = 6/1), although the temperature used in this reaction was much lower (−65°C). Under the same conditions, the per-O-acetylated 2-*N*-phenyltrifluoroacetimidate sialyl donor and a methyl 3-*O*-benzoyl-α-D-galactopyranosyl acceptor produced 61% of only the α-anomer (Scheme 12). After this initial and

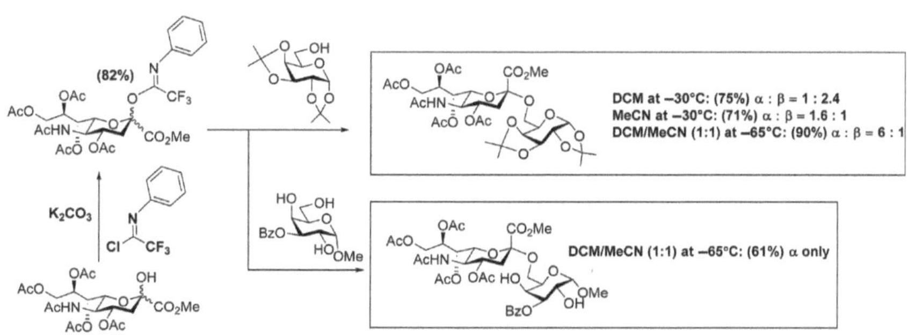

Scheme 12 *N*-Phenyltrifluoroacetimidate synthesis and sialylations.

promising data, and in concomitance with the discovery of the influence of C-5 modifications, trifluoroacetimidate sialyl donors modified at C-5 have been tested[84] (see Section 2.5). In particular, the combination between the *N*-phthalimido group at C-5 and the *N*-phenyltrifluoroacetimidoyl leaving group gave high α-stereoselectivities in several sialylations with primary and secondary acceptors[84] and was used in large-scale synthesis for microfluidic sialylations via a microreactor.[85]

High stereoselectivities and good yields for primary and secondary acceptors were also reported for C-5-trichloroacetyl-modified *N*-phenyltrifluoroacetimidate sialyl donors.[86] Selective sialylation in the presence of tolylthio galactopyranosyl acceptors using TMSOTf in a mixture of dichloromethane and acetonitrile gave the corresponding disaccharides in good yields as only the α-anomers[86] (Scheme 13). A combination of a C-5-trifluoroacetyl modification and an *N*-phenyltrifluoroacetimidate leaving group was used by Ando and Kuketsu in an efficient sialylation with a 1,5-lactam sialyl acceptor (see Section 2.5.3).

Recently, Hung and coworkers reported the use of a 5-phthalimido *N*-phenyltrifluoroacetimidoyl donor for the one-pot synthesis of *N*-acetyl and *N*-glycolylneuraminic acid-capped trisaccharides, as potential influenza A inhibitors, by performing a selective sialylation in the presence of a tolylthio galactopyranosyl acceptor (Scheme 14).[87] Complete stereoselectivity was observed in coupling with a tolylthio galactopyranosyl acceptor in the presence of TMSOTf at −78°C.

In order to reevaluate the performance of the standard 5-acetamido *N*-phenyltrifluoroacetimidate as a sialyl donor, Fukase and coworkers tested it under microfluidic conditions. The authors reported an improvement

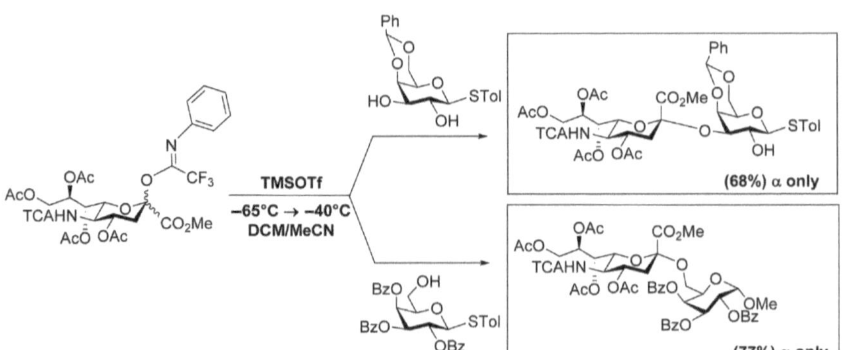

Scheme 13 Selective sialyations with trifluoroacetimidates.

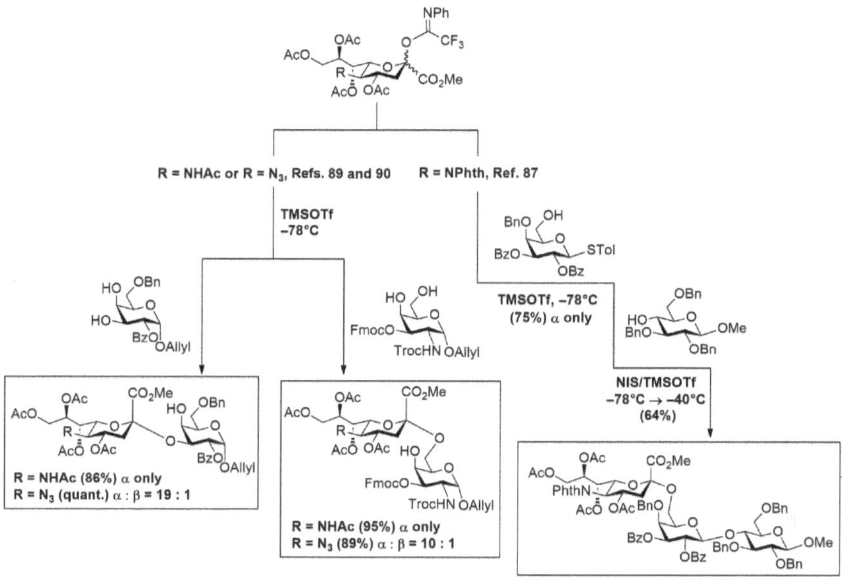

Scheme 14 One-pot synthesis and microfluidic sialylation using *N*-phenyltrifluoroacetimidates.

in yield and stereoselectivity upon steady temperature control at −80°C and with the careful addition of the promoter (TMSOTf). Under these conditions, the microreactor allowed for stabilization of kinetic conditions and therefore higher α-stereoselectivity.[88] More recently, Fukase and coworkers reported the coupling of this Neu5Ac donor, as well as its C-5 azido-modified counterpart for the assembly of a disialylated tetrasaccharide (Scheme 14) and more complex structures.[89,90]

Sialyl phosphites, introduced by Wong and coworkers[91] and by Schmidt and coworkers,[92] have been used in chemical sialylations when promoted with a catalytic amount of TMSOTf.[10] Sialyl phosphites also have been used in the synthesis of glycosyl phosphates, sugar nucleotides, and more complex oligosaccharides.[93,94]

Although many variations such as –O(*n*-Bu), $-OCH_2CH_2Cl$, $-OCH_2CCl_3$, $-O(CH_2)_3O-$, $-OCH_2CH_2Me_2CH_2O-$,[95] 1,2-*O*-cyclopentyl,[96] and dimethyl[97] have been investigated, diethyl and dibenzyl phosphites have been the most frequently used leaving groups in this family. They can be prepared in high yield by reaction of the tetra-acetate hemiketal of Neu5Ac with $ClP(OEt)_2$ in the presence of a hindered base such as *i*-Pr_2NEt,[92,95] or with dibenzyl *N*,*N*-diethylphosphoramidite and 1*H*-tetrazole[91] (Scheme 15). A more stereoselective synthesis of β-sialyl phosphites

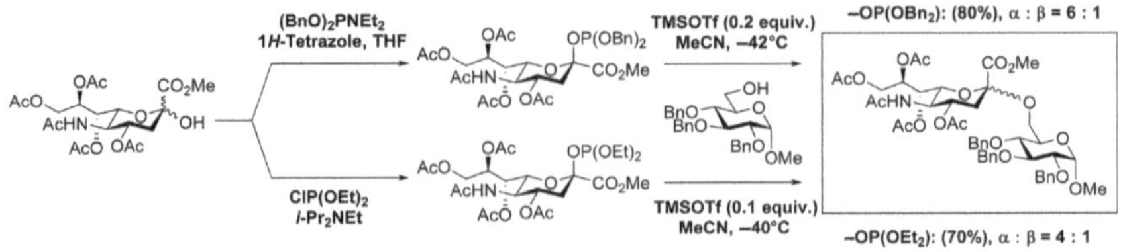

Scheme 15 Syntheses of and glycosylations with sialyl phosphites.

has been described by Bhattacharya and Danishesfky, who used diisopropyl phosphoramidite, obtaining the β-anomer almost quantitatively.[98]

An example of the initial application of diethyl and dibenzyl phosphite sialyl donors is highlighted in Scheme 15. By coupling a methyl 2,3,4-tri-*O*-benzyl-α-D-glucopyranosyl acceptor in the presence of TMSOTf, the two leaving groups gave comparable yields and stereoselectivities. The dibenzyl phosphite derivative produced slightly better results with an 80% yield of the (2→6)-linkage (α/β=6/1).[99] The diethyl phosphite derivative produced 70% of the desired disaccharide with slightly poorer stereoselectivity (α/β=4/1, Scheme 15).[77,85] Unexpectedly, however, a chemical sialylation promoted between a diethyl phosphite donor and an 8-hydroxy sialyl acceptor gave complete β-stereoselectivity.[100,101]

Additional applications of phosphites as leaving groups are depicted in Scheme 16. Moderate yields were obtained when coupling a dibenzyl phosphite sialyl donor in the synthesis of the STn antigen,[102] as well as in the synthesis of a GM1 methyl ganglioside.[98] As it has been observed with other leaving groups, C-5 modifications allow the use of phosphites in a more efficient manner. An *N*-Troc-modified dibenzyl phosphite coupled with a 4,6-di-*O*-benzyl galactal acceptor produced the desired disaccharide in an 80% yield and with an α/β ratio of 11/1.[103] The same donor did not perform as well, however, in the synthesis of α(2→8)-dimeric linkage when compared with *N*-phenyltrifluoroacetimidates.[104]

Lin and coworkers reported complete α-stereoselectivity in the coupling of a C-5-trifluoroacetylated dibenzyl phosphite donor with a 9-hydroxy sialyl acceptor, also bearing a trifluoroacetamido group at C-5 (Scheme 16).[105] Phosphite donors with an azido modification at the C-5 position also have been reported to perform well in couplings with primary acceptors.[106] More recently, a C-5-trichloroacetamide dibenzyl phosphite donor produced high yields and α-anomeric selectivities with a wide variety of glycosyl acceptors.[107]

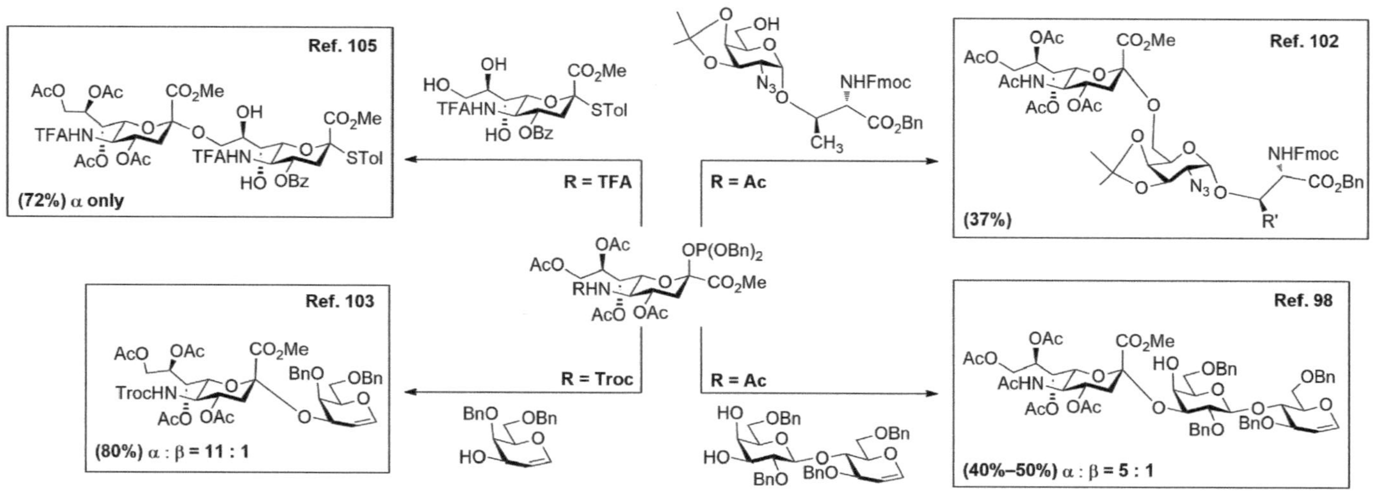

Scheme 16 Additional applications of sialyl phosphite donors.

The synthesis of an *N*-acetyl-5-*N*,4-*O*-carbonyl-protected[108] phosphite donor by conversion of the corresponding thiosialoside precursor was unsuccessful, and it led to the development of phosphates as leaving groups in chemical sialylations.[109] Similar to that of phosphites and *N*-phenyltrifluoroacetimidates, phosphates can be activated selectively in the presence of thioglycosides, allowing for the assembly of complex oligosaccharide sequences in fewer steps. However, due to early reports of low yields and poor stereoselectivities,[91,94] it was not until recently that Wong and coworkers have demonstrated the utility of an *N*-acetyl-5-*N*,4-*O*-carbonyl-protected dibutyl phosphate sialyl donor.[109] The introduction of a phosphate at the anomeric position can be accomplished from the corresponding *N*-acetyl-5-*N*,4-*O*-carbonyl-protected thiosialyl donor using dibutyl phosphate and NIS/TfOH as a promoter system (Scheme 17). The activation of a phosphate leaving group is dependent on its anomeric configuration. The α-phosphate is more reactive and therefore can be

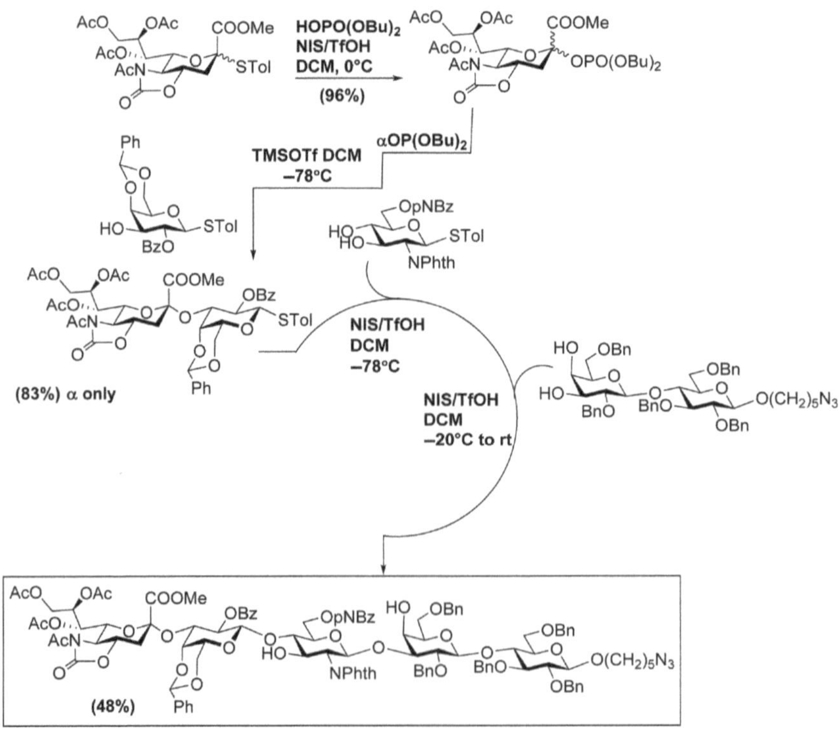

Scheme 17 Synthesis of an *N*-acetyl-5-*N*,4-*O*-carbonyl phosphate donor and its application in a one-pot sialylation.

activated at −78°C, whereas a higher temperature (−50°C) is needed to activate the β-phosphate counterpart, compromising the α-stereoselectivity of the sialylation reaction.[109] Coupling of an α-dibutyl phosphate sialyl donor with a wide variety of acceptors, both primary and secondary, including thiogalactosides and -sialosides, led to complete α-stereoselectivity and high yields.[109] This finding, combined with the measurement of the relative reactivity values (RRVs) of strategic building blocks, allowed for the programmable one-pot synthesis of several oligomers (Scheme 17). The strategy then was extended to the synthesis of an α(2→9)-oligosialic acid dodecamer.[110] In addition, Crich and coworkers reported excellent yields and stereoselectivities using a dibutyl phosphate sialyl donor equipped with 5-*N*-acetyl-5-*N*,4-*O*-oxazolidinone for the synthesis of C- and S-linked sialosides.[111,112]

The combination of 5-*N*-acetyl-5-*N*,4-*O*-oxazolidinone protection and a dibutyl phosphate leaving group in a sialyl donor has been extensively studied and applied to several synthetic targets in recent years,[113–119] due to the advantage of oxazolidinone ring stabilization (minimizing elimination) and the reactive phosphate which can be activated over thioglycosides. Additional examples will be discussed in the following subsections.

2.3 Modifications at C-3

In the past, introducing modifications at the C-3 position of Neu5Ac glycosyl donors was a popular strategy for controlling anomeric selectivity in glycosylation reactions. The participating auxiliary group at C-3 is capable of assisting in the formation of 2,3-*trans*-glycosides via neighboring group participation.[10] When installed in the C-3-equatorial position, the neighboring group assists in the formation of α-glycosides, while in the C-3-axial position promotes the formation of β-glycosides.

An auxiliary group in the axial position may also inhibit elimination side product formation that occurs frequently in glycosylation reactions involving a Neu5Ac donor.[120] The C-3 auxiliary groups typically are installed by transformation of the 2,3-dehydro derivative of Neu5Ac, and they must be removed following glycosylation (Scheme 18).[121] As a consequence, since the 1990s, research efforts largely have shifted away from C-3 auxiliary strategies in favor of other, more direct, positional modifications.

2.3.1 3-Bromo Auxiliaries

The 3-bromo auxiliary is one of the earliest examples of C-3 neighboring group participation in stereoselective chemical glycosylations with Neu5Ac

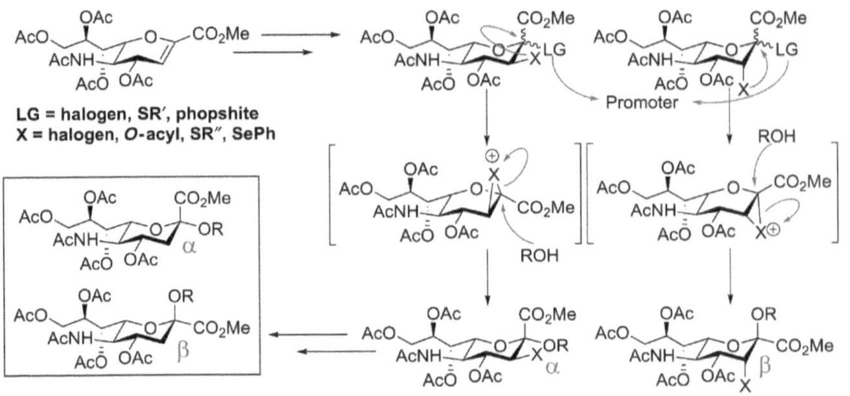

Scheme 18 Generic C-3 auxiliary-assisted sialylation reaction.

donors.[122] The diaxially substituted 2,3-dibromo derivative of Neu5Ac can be synthesized in high yield (93%) by direct addition of bromine to the double bond of the 2,3-dehydro derivative (Scheme 19).[123] The resulting diaxial 2,3-dibromo glycosyl donor resulted in the formation of β-glycosides, including a modest yield (55%) of a β(2→8)-Neu5Ac dimer.[122] The 3*eq*-bromo-hemiketal derivative of Neu5Ac can be prepared in high yield (84%) by reaction of the 2,3-dehydro derivative with *N*-bromosuccinimide (NBS) at −20°C in aqueous dimethyl sulfoxide (DMSO).[123]

Conversion into β-diethyl phosphite as a leaving group was performed in excellent yield (93%) by reaction with diethyl chlorophosphite and *i*-Pr_2NEt.[124] Glycosylation of the resulting 3*eq*-bromo diethyl phosphite sialyl donor with a lactosyl acceptor resulted in a good yield (61%) of the α-glycoside, only. Unfortunately, attempts at (2→8)-Neu5Ac dimerization using the 3*eq*-bromo diethyl phosphite sialyl derivative resulted in only a modest yield (58%) of the β-glycoside. Following 3-bromo-assisted glycosylations, the bromo auxiliary can be removed nearly quantitatively via reduction with tributyltin hydride (Bu_3SnH) and azobisisobutyronitrile (AIBN, Scheme 19).[124]

2.3.2 3-O-Acyl Auxiliaries

By introduction of 3-*O*-acyl groups in the equatorial position in Neu5Ac glycosyl donors, the subsequent chemical glycosylations largely produced α-glycosides. The equatorial 3-*O*-thionocarbonyl substituent can participate as a neighboring group, forming a five-membered cyclic intermediate which only allows nucleophilic attack from the α-face of the donor.[10] The 3*eq*-*O*-thionocarbonyl substituent has been tested in chemical glycosylation with both chloride and diethyl phosphite as leaving groups.[124]

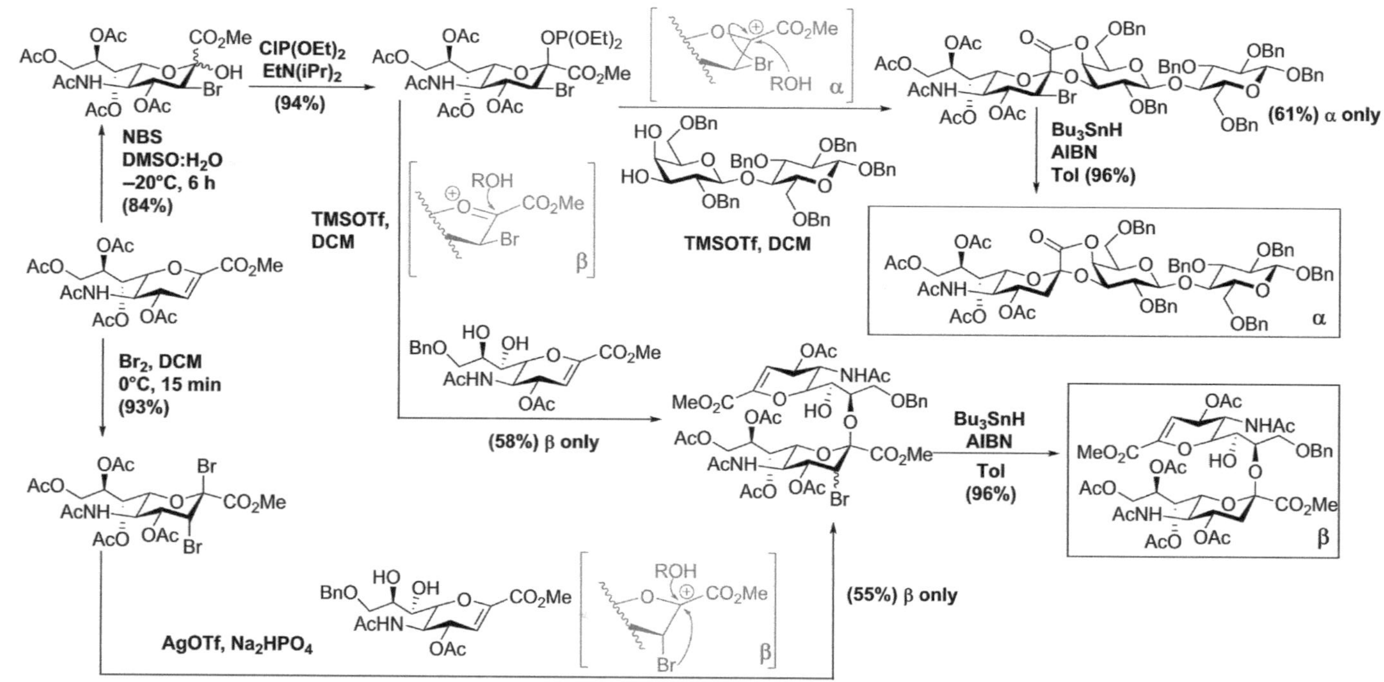

Scheme 19 Syntheses of and glycosylations with a 3-bromo sialyl donor.

To prepare the 2-chloro-3-*O*-thionocarbonyl derivative of Neu5Ac, the 3*ax*-bromo-β-hemiketal derivative of Neu5Ac first can be synthesized in modest yield (73%) by reaction of the 2,3-dehydro derivative with NBS at 80°C (Scheme 20). The 3*ax*-bromo-β-hemiketal derivative can be then converted in high yield (92%) by reaction with diazabicycloundecene (DBU) in acetonitrile to a 2,3-epoxy derivative, which is opened subsequently in the presence of titanium tetrachloride ($TiCl_4$) to afford the 2-chloro-3*eq*-hydroxy derivative in an excellent yield (95%).[123] The equatorial 3-hydroxy substituent can be then converted in high yield (89%) to a thionocarbonyl functionality by reaction with chlorophenylthionocarbonate and 4-dimethylaminopyridine (DMAP) in acetonitrile.[101]

When used as a glycosyl donor with a benzylated lactosyl acceptor, the 2-chloro-3*eq*-*O*-thionocarbonyl derivative resulted in an excellent yield (89%) of the α-glycoside (Scheme 20). This donor was successful even in the formation of a difficult α(2→8) glycosidic linkage between two Neu5Ac molecules, with a modest yield of 68% and the possibility of additional iterative oligosaccharide synthesis.[124]

Conversely, the 2-diethoxyphosphino-3*eq*-*O*-thionocarbonyl derivative can be prepared first by reaction of the 2,3-dehydro derivative of Neu5Ac with osmium tetroxide (OsO_4) and *N*-methylmorpholine *N*-oxide (NMO) in acetone/water/*tert*-butanol to give the equatorial 2,3-dihydroxy derivative in excellent yield (95%). The 3-hydroxy substituent of the 2,3-dihydroxy derivative is protected selectively with the thionocarbonyl functionality by reaction with *N*,*N*-dimethyl-α-chlorobenzimidium chloride, treatment with pyridine, and a subsequent reaction with hydrogen sulfide (H_2S) to give an overall yield of 92%. The anomeric hydroxyl group then can be converted to a diethyl phosphite leaving group in high yield (93%) by reaction with $ClP(OEt)_2$ and *i*-Pr_2NEt.[124]

When used as a glycosyl donor, with a benzylated lactosyl acceptor, the 2-diethoxyphosphino-3*eq*-thionocarbonyl derivative produced a yield of 88% of only the α-glycoside (Scheme 20). Glycosylation for the synthesis of α(2→8) dimers was also successful, obtaining the desired target in 83% yield and complete α-stereoselectivity.[124] Following glycosylation reactions, the auxiliary group can be reduced easily and nearly quantitatively with Bu_3SnH/AIBN.[101,124]

2.3.3 3-S-Aryl (and 3-Se-Aryl) Auxiliaries

Of the various C-3 auxiliary substituents studied in the past, the arylthio functionality has proven to be one of the most reliable and extensively

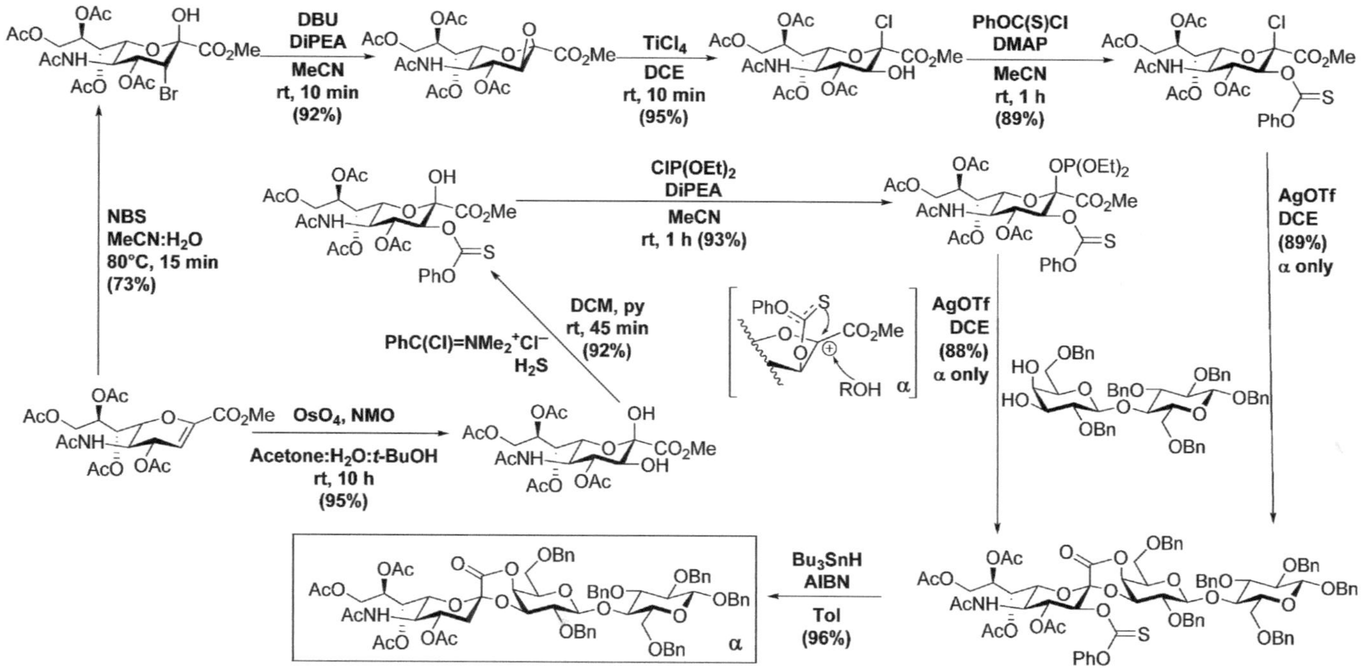

Scheme 20 Synthesis of and sialylation with a 3-*O*-thionocarbonyl sialyl donor and subsequent reduction to the 3-deoxy compound.

investigated. When used as glycosyl donors, Neu5Ac derivatives substituted with 3*eq*-arylthio or 3*eq*-arylseleno auxiliaries tend to promote the formation of α-glycosides via a pathway believed to proceed through an episulfonium (or episelenium) intermediate.[10] The α-directing effects of the 3*eq*-phenylthio and 3*eq*-phenylseleno donor modifications were reported first by Ito and Ogawa.[125–127] The introduction of a phenylseleno moiety was accomplished by reaction of the 2,3-dehydro derivative of Neu5Ac with phenylselenyl acetate (PhSeOAc). The latter is generated in situ via reaction of phenylselenyl chloride (PhSeCl) and silver acetate (AgOAc), in the presence of catalytic TMSOTf. Although this synthesis afforded a 3*ax*–3*eq* 11/1 diastereomeric mixture, subsequent epimerization of the C-3 auxiliary to favor the formation of the equatorial product (2/1) was accomplished by treatment with sodium methoxide (NaOMe). The 3*ax*-phenylseleno derivative can be then recycled via epimerization to improve the overall yield (83%). The hemiketal formed was then converted to several halide sialyl donors (Scheme 21).[125]

Similarly, the less labile 3-phenylthio derivative was introduced by treatment of the 2,3-dehydro derivative with NBS to obtain a mixture of bromohydrins in excellent yield (97%) with the 3*ax*-bromo-2-hydroxy derivative favored over the 3*eq*-bromo-2-hydroxy derivative (4/1). Displacement of the 3-bromo with a phenylthio group could be accomplished by treatment with thiophenol (HSPh) and potassium *tert*-butoxide with retention of configuration at C-3. The axial derivative then could be epimerized with catalytic DBU in toluene at 20°C to give an 83% yield of the 2-hydroxy-3*eq*-phenylthio derivative overall. Also in this case, the displacement of the hemiketal (2-hydroxyl group) allowed the synthesis of several sialyl halide donors in high yield (Scheme 21).[128]

An overview of the performance of this class of sialyl donors is in Table 1. Overall, the 2-fluoro-3*eq*-phenylthio derivative outperformed the 2-fluoro-3*eq*-phenylseleno derivative when promoted with silver triflate (AgOTf)/tin chloride ($SnCl_2$) in the presence of common glycosyl acceptors at room temperature in carbon tetrachloride. For example, glycosylation of the 2-fluoro-3*eq*-phenylseleno derivative with a common 1,2:3,4-di-*O*-isopropylidene galactopyranosyl acceptor yielded a modest 46% of only the α(2→6) glycoside (Entry 1).[125] When promoted under the same conditions, the 2-fluoro-3*eq*-phenylthio derivative produced 72% of mostly the α(2→6) glycoside (Entry 2).[128] Similarly, when promoted in the presence of a common per-O-benzylated lactosyl acceptor, the 2-fluoro-3*eq*-phenylseleno derivative yielded only 25% of the (2→3)-linkage with

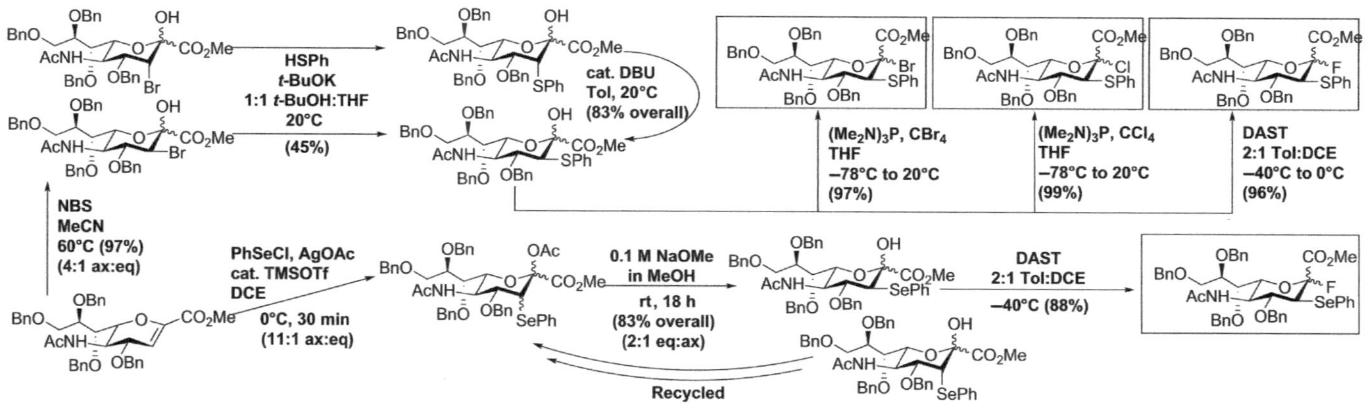

Scheme 21 Synthesis of 3-*S*- and 3-*Se*-sialyl donors.

Table 1 Sialylation of 3-*S*- and 3-*Se*-Sialyl Donors

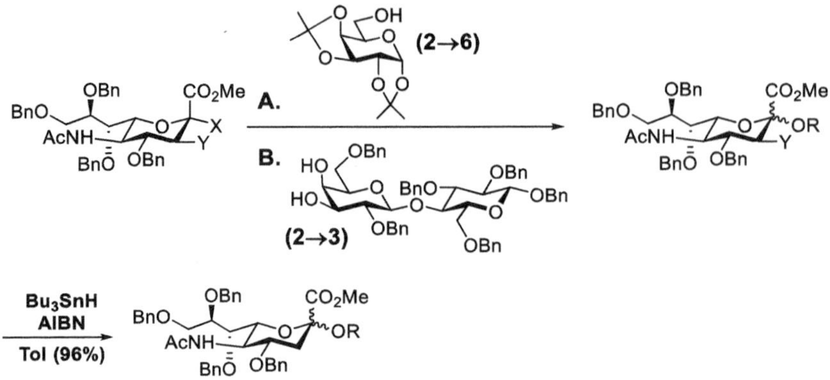

Entry	X	Y	Promoter	Solvent	Temperature (°C)	Time (h)	Acceptor	Yield (%)	α : β Ratio
1	F	SePh	AgOTf $SnCl_2$	CCl_4	20	18	A	46	α only
2	F	SPh	AgOTf $SnCl_2$	CCl_4	20	18	A	72	20 : 1
3	F	SePh	AgOTf $SnCl_2$	CCl_4	20	18	B	25	4 : 1
4	F	SPh	AgOTf $SnCl_2$	CCl_4	20	18	B	45	3.5 : 1
5	Cl	SPh	$Hg(CN)_2$ $HgBr_2$	CCl_4	30	72	A	46	α only
6	Br	SPh	$Hg(CN)_2$ $HgBr_2$	CCl_4	20	18	A	72	α only
7	Br	SPh	$Hg(CN)_2$ $HgBr_2$	CCl_4	20	18	B	78	α only
8	Cl	SPh	$Hg(CN)_2$ $HgBr_2$	CCl_4	40	40	B	64	30 : 1

the α-glycoside predominating over the β-glycoside (α/β = 4/1, Entry 3). Under the same conditions, the 2-fluoro-3*eq*-phenylthio derivative produced a 45% yield, again favoring the α-glycoside (α/β = 3.5/1, Entry 4).[126]

Sialylations of the 2-chloro and 2-bromo derivatives by activation with mercury(II) cyanide ($Hg(CN)_2$)/mercury(II) bromide ($HgBr_2$) promoter system are listed in Entries 5–8. When promoted in the presence of the galactopyranosyl acceptor, the 2-chloro-3*eq*-phenylthio derivative gave a 46% yield of only the α(2→6) glycoside, although the reaction temperature was raised to 30°C (Entry 5). Using the same promoter/acceptor system at room temperature, the 2-bromo-3*eq*-phenylthio derivative gave 72% of only the α(2→6) glycoside (Entry 6). Coupling with a lactosyl acceptor resulted in 78% of only the α(2→3) glycoside in the case 2-bromo-3*eq*-phenylthio derivative (Entry 7), while the 2-chloro-3*eq*-phenylthio derivative promoted at 40°C yielded a 64% yield (α/β=30/1, Entry 8).[126]

The 2-bromo-3*eq*-phenylthio derivative has been applied successfully to a number of difficult chemical glycosylations to great effect. For example, coupling with a 4,7,9-tri-*O*-benzyl sialyl glycal acceptor at room temperature in carbon tetrachloride overnight afforded 64% of only the α(2→8) glycoside (Scheme 22). The product was transformed into an additional iteration of the 2-bromo-3*eq*-phenylthio donor, and subsequent glycosylation with the same conditions and a lactosyl acceptor afforded the final tetrasaccharide in 48% yield with complete α-stereoselectivity (60:1).[127] Interestingly, at room temperature, carbon tetrachloride seemed to outperform acetonitrile. Thus, coupling 2-bromo-3*eq*-phenylthio with a lactosyl acceptor produced 89% of only the α(2→3) glycoside in carbon tetrachloride at room temperature, whereas the same reaction conducted in acetonitrile produced the desired target in only 49% yield. In both cases, the only other major product isolated was the 2,3-dehydro derivative (Scheme 22).[127]

It also was found that certain derivatives of arylsulfenyl chloride (ArSCl) can be added directly to the double bond of the 2,3-dehydro derivative with relatively high stereoselectivity (Scheme 23).[129] Thus, phenylsulfenyl chloride (PhSCl) was added to the 2,3-dehydro derivative in dichloromethane at 30°C, affording a diastereomeric mixture of the 2-chloro-3*eq*-phenylthio and the 2-chloro-3*ax*-phenylthio derivative in 77% and 15% yield, respectively.[129] 2,4-Dimethylbenzenesulfenyl chloride could be added to the 2,3-dehydro derivative in dichloromethane at 0°C to produce 85% of the 2-chloro-3*eq*-arylthio derivative and 4% of the 2-chloro-3*ax*-phenylthio derivative. Upon recrystallization, the 2-chloro-3*ax*-phenylthio derivative is retained in the mother liquor.[50]

As a sialyl donor, the 2-chloro-3*eq*-phenylthio derivative was not as favorable as the 2-bromo equivalent,[130] but it has proven useful as a precursor to some 2-alkylthio-3*eq*-phenylthio derivatives.[131,132] From the 2*ax*-chloro-3*eq*-phenylthio derivative, the 2-*O*-acetyl-3*eq*-phenylthio derivative was be synthesized in excellent yield (96%) by reaction with mercury(II) acetate ($Hg(OAc)_2$) in a 10:1 mixture of acetic acid and

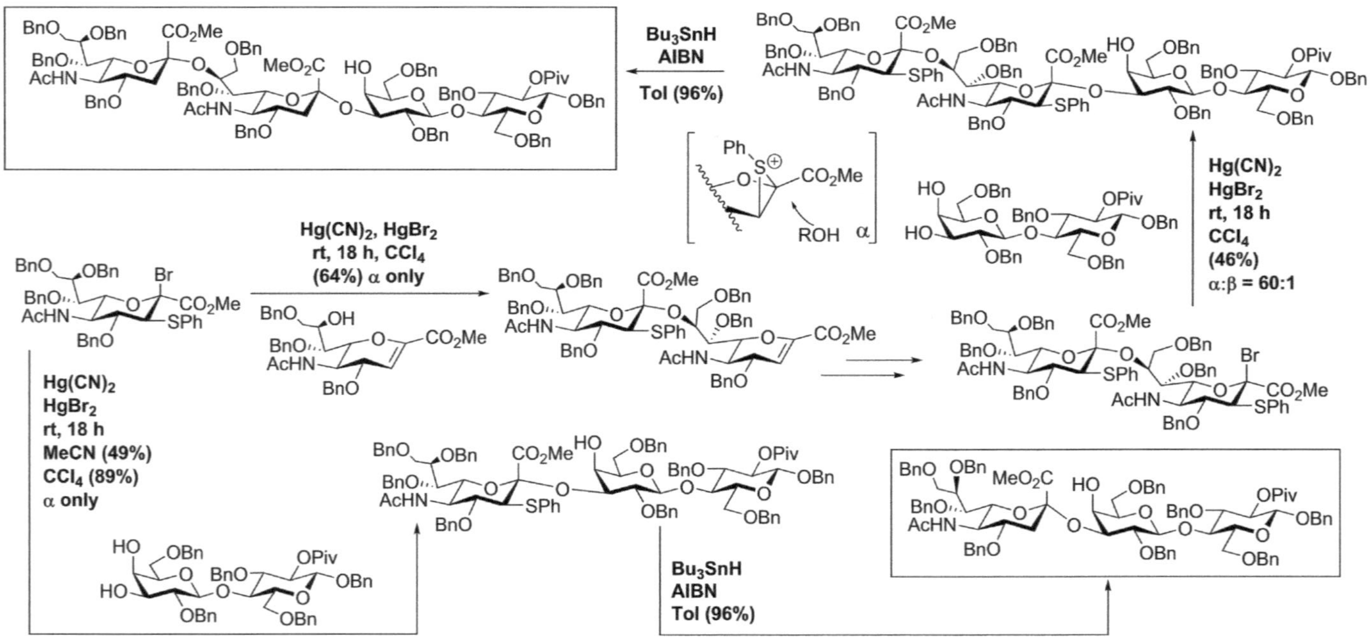

Scheme 22 Examples of sialylations of a 2-bromo-3*eq*-phenylthio donor.

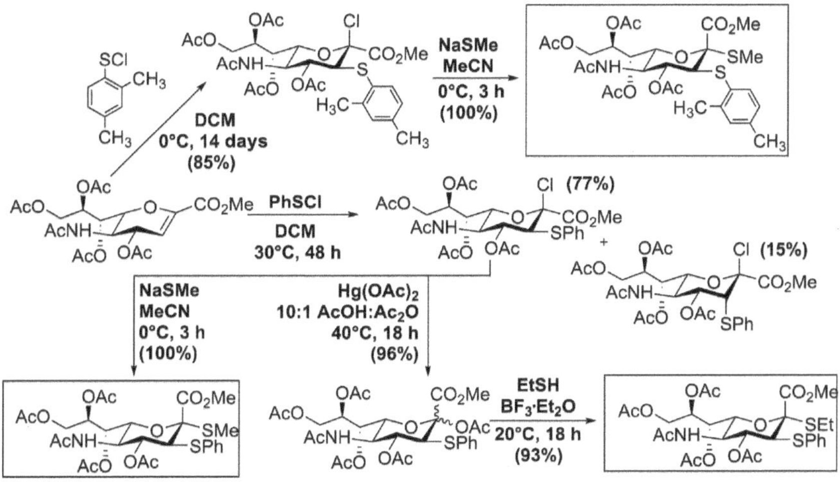

Scheme 23 Synthesis of a 2-alkylthio-3*eq*-phenylthio donor.

acetic anhydride overnight at 40°C. An anomeric mixture favoring the diequatorial derivative (5:1) was then produced, which could be reacted with ethanethiol (EtSH) and boron trifluoride etherate ($BF_3 \cdot Et_2O$) to afford only the 2*eq*-ethylthio-3*eq*-phenylthio derivative in high yield (93%).[131] The 2-chloro-3*eq*-phenylthio derivative was then reacted with sodium thiomethoxide (NaSMe) to afford the 2*eq*-methylthio-3*eq*-phenylthio derivative in quantitative yield.[132] Similarly, the 2*eq*-methylthio-3*eq*-(2,4-dimethylphenylthio) derivative was synthesized quantitatively by reaction of the 2-chloro-3*eq*-(2,4-dimethylphenylthio) derivative with NaSMe (Scheme 23).[50]

In stark contrast to the sialylation of 2-halo-3*eq*-arylthio derivatives being conducted in carbon tetrachloride as a solvent and at room temperature, sialylations of 2-alkylthio-3*eq*-arylthio derivatives were performed at low temperatures using MeCN or an MeCN/DCM mixture as the solvent (Scheme 24). Activation of the O-benzylated 2*eq*-methylthio-3*eq*-phenylthio sialyl donor with NIS/TfOH in the presence of a common galactopyranosyl acceptor at −40°C in acetonitrile produced the α(2→3) disaccharide in 85% yield.[133] When promoted with phenylsulfenyl chloride (PhSCl), silver triflate (AgOTf), and 2,4-di-*tert*-butylpyridine (DTBP) in the presence of a common lactosyl acceptor at −40°C in acetonitrile, the acetylated 2*eq*-methylthio-3*eq*-phenylthio derivative yielded the α(2→3) glycoside in 83% yield.[132] Under the same glycosylation conditions with the same lactosyl acceptor, the 2*eq*-methylthio-3*eq*-(2,4-dimethylphenylthio)

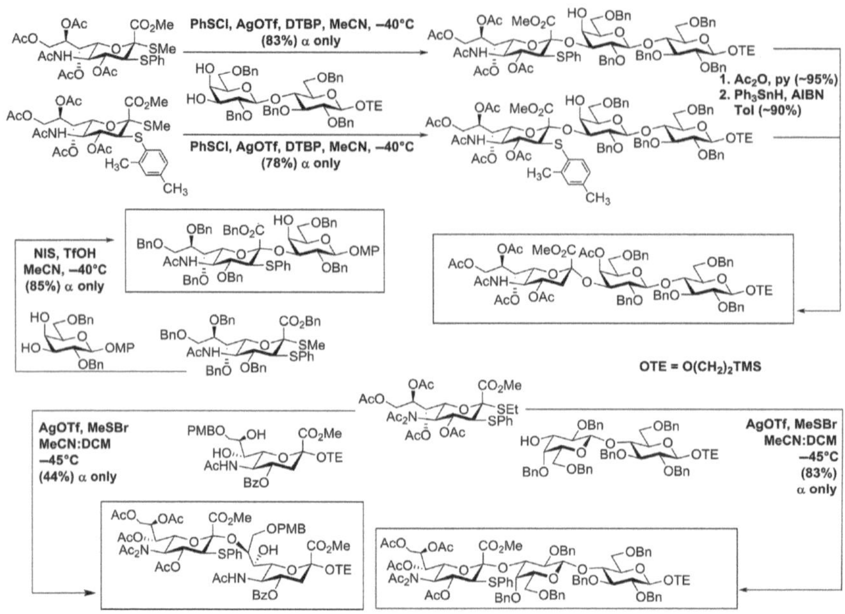

Scheme 24 Examples of sialylations with a 2-alkylthio-3*eq*-phenylthio donor.

yielded only the α(2→3) glycoside in 78% yield.[50] The acetylated 2*eq*-ethylthio-3*eq*-phenylthio derivative performed well as a glycosyl donor when promoted with AgOTf and methylsulfenyl bromide (MeSBr) in the presence of a lactosyl acceptor at −45°C in a mixture of acetonitrile and dichloromethane as solvent. An 83% yield of only the α(2→3) glycoside was obtained. The same donor promoted under the same conditions in the presence of a Neu5Ac acceptor afforded the α(2→8) glycoside in 44% yield (Scheme 24).[134]

2.4 Modifications at C-4

Exploration of the influence of modifications at the C-4 position of Neu5Ac has been a fairly recent endeavor compared with modifications at the C-2, C-3, and C-5 positions.

2.4.1 4-O-tert-Butyldimethylsilyl Ether Derivative

Investigating the effect of varying the protecting group at only the C-4 position on chemical sialylations is quite new in the field. One of the first studies involved the introduction of a *tert*-butyldimethylsilyl ether

(TBDMS) group, which was installed at the C-4 position starting from the per-*O*-hydroxy derivative of Neu5Ac as described in Scheme 25.[135]

For comparative studies, De Meo and coworkers activated both the benchmark per-O-acetylated Neu5Ac derivative and the acetylated 4-*O*-TBDMS donor with NIS and a catalytic amount of triflic acid in the presence of common primary acceptors. Both Neu5Ac donors were synthesized with a β-tolylthio leaving group. Both the 1,2:3,4-di-*O*-isopropylidene-α-D-galactopyranosyl (**I**, Table 2) and methyl 2,3,4-tri-*O*-benzyl-α-D-galactopyranosyl acceptors (**II**, Table 2) were used, and chemical glycosylations were promoted either in acetonitrile at −40°C or in dichloromethane at −72°C.[135]

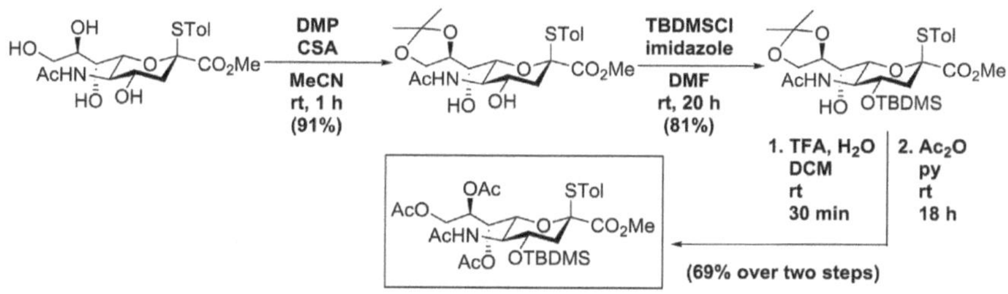

Scheme 25 Synthesis of a 4-*O*-*tert*-butyldimethylsilyl donor.

Table 2 Comparison of Sialylation Data Between 4-*O*-TBDMS and 4-*O*-Ac Donors

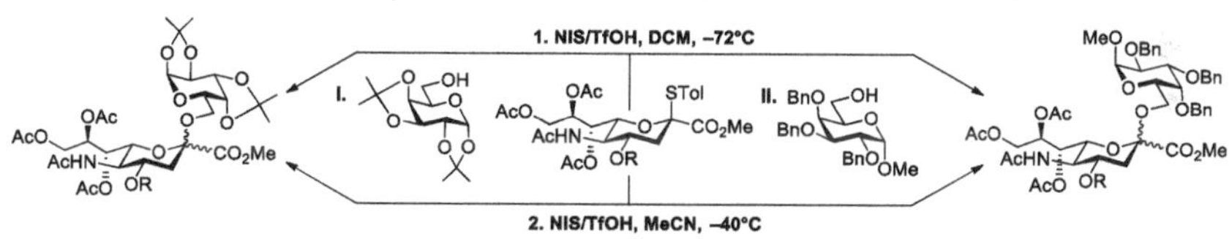

Entry	Acceptor	Conditions	R	Yield (%)
1	I	1	Ac	96
2	I	1	TBDMS	94
3	I	2	Ac	84
4	I	2	TBDMS	79
5	II	1	Ac	99
6	II	1	TBDMS	N.R.
7	II	2	TBDMS	84

In general, changing the acetyl functionality at the C-4 position to a TBDMS group seemed to have a significant effect on glycosylation stereoselectivity in favor of the α-anomeric product (Table 2). In DCM, glycosylation of the acetylated 4-*O*-TBDMS donor vs the benchmark per-O-acetylated donor with the 6OH isopropylidene-protected acceptor resulted in comparably excellent yields (96% to 94%) with a slight improvement in α-stereoselectivity (α/β=1.5/1 to 2.8/1, Entries 1 and 2). The same coupling in acetonitrile as a solvent resulted in comparable yields (84% to 79%), however with a stark shift to complete α-stereoselectivity (α/β=4.4/1 to α only, Entries 3 and 4). Interestingly, for the coupling with a benzylated galactosyl acceptor, no activation was observed in DCM, while complete α-stereoselectivity was observed in acetonitrile (Entries 5–7).

2.4.2 4-O-Picoloyl (and 4-O-Benzoyl) Derivatives

Hydrogen-bond-mediated aglycone delivery (HAD), developed by Demchenko and coworkers,[136–139] has been used to great effect with glucosyl, galactosyl, and mannosyl donors. In this method, a picoloyl substituent installed in the donor is believed to facilitate hydrogen bonding with the acceptor. The resulting chemical glycosylation produces glycosides with high facial selectivity, *syn* relative to the picoloyl substituent. Inspired by this success, the effect of the picoloyl group installed at the C-4 position of Neu5Ac was also investigated by De Meo and coworkers.[140]

Both the 4-O-picoloylated and 4-O-benzoylated derivatives of Neu5Ac have been synthesized starting from a per-*O*-hydroxyl Neu5Ac derivative with an α-phenylthio leaving group (Scheme 26). Introduction of an isopropylidene protecting group to the glycerol chain allows selective introduction at C-4 of either the benzoyl group via benzoic anhydride (Bz_2O) or the picoloyl group via picolinic acid/1-ethyl-3-(3-dimethylaminopropyl)

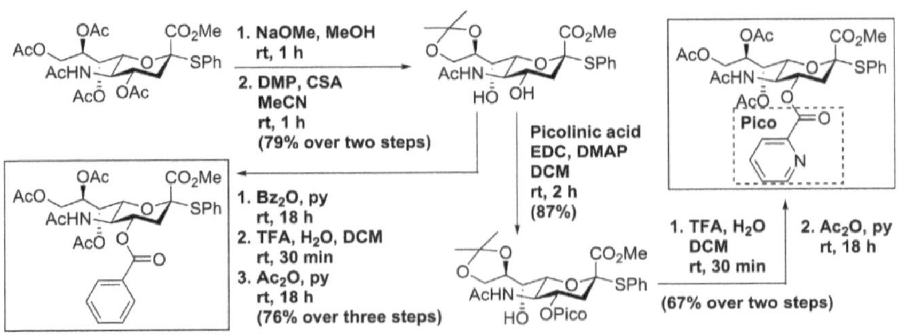

Scheme 26 Synthesis of 4-*O*-picoloyl and 4-*O*-benzoyl donors.

carbodiimide (EDC). Again, deprotection and acetylation of the glycerol chain furnish the final acetylated 4-*O*-Bz or 4-*O*-Pico thioglycoside in relatively high yield (Scheme 26).

Additionally, a series of Neu5Ac and $Neu5Ac_2$ derivatives bearing either a picoloyl functionality or a standard acetyl group at the C-4 position were synthesized to serve as glycosyl donors with a common 1,2:3,4-di-*O*-isopropylidene-α-D-galactopyranosyl acceptor. Donors were synthesized with both α-phenylthio and α-ethylthio leaving groups to expand the scope of the comparison, and glycosylation reactions were promoted at −45°C in dichloromethane using NIS and a catalytic amount of triflic acid (Table 3).[140]

A benchmark glycosylation reaction with a per-O-acetylated Neu5Ac donor bearing an α-phenylthio leaving group promoted with NIS and catalytic triflic acid in the presence of a 1,2:3,4-di-*O*-isopropylidene-α-D-galactopyranosyl acceptor at −45°C in dichloromethane resulted in 68% of the desired glycoside over the course of 1 h, with the α-anomer very slightly favored over the β-anomer (α/β = 1.1/1, Entry 1). The analogous 4-*O*-Pico derivative produced the expected glycoside in 70% yield over

Table 3 Comparison of Sialylation Data Between 4-*O*-Picoloyl and 4-*O*-Ac Donors

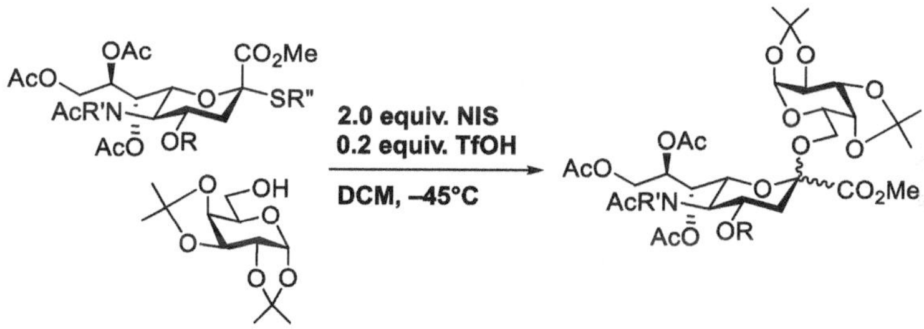

Entry	R	R′	R″	Time (min)	Yield (%)	α : β Ratio
1	Ac	H	Ph	60	68	1.1 : 1
2	Pico	H	Ph	300	70	3.4 : 1
3	Ac	H	Et	10	94	1.1 : 1
4	Pico	H	Et	45	75	4.8 : 1
5	Ac	Ac	Ph	5	73	1 : 2.2
6	Pico	Ac	Ph	240	94	3.8 : 1
7	Pico	Ac	Et	15	60	4.0 : 1

the course of 5 h when promoted under the same conditions, with a notable shift in stereoselectivity favoring the α-anomer (α/β=3.4/1, Entry 2). By using an ethylthio leaving group, the results obtained were very similar. The use of a 4-O-Pico substituent in lieu of the standard acetyl group resulted in a slight loss of yield, but enhanced α-stereoselectivity (Entries 3 and 4). The per-O-acetylated Neu5Ac$_2$ donor coupled with the isopropylidene-protected acceptor in good yield (73%), but stereoselectivity was poor (α/β=1/2.2, Entry 5). With either a phenylthio or an ethylthio leaving group, the 4-O-Pico-protected Neu5Ac$_2$ derivative demonstrated enhanced stereoselectivity (Entries 6 and 7).

Using the 4-O-Pico α-phenylthio Neu5Ac donor and the same 1,2:3,4-di-O-isopropylidene-α-D-galactopyranosyl acceptor, De Meo and coworkers proceeded to explore the effects of varying individual reaction parameters. It was found that an excess of triflic acid relative to the donor sugar had a tremendous impact on both glycosylation reactivity and stereochemical outcome (Table 4).[140]

Table 4 Effect of Triflic Acid on Sialylation With a 4-O-Pico Donor

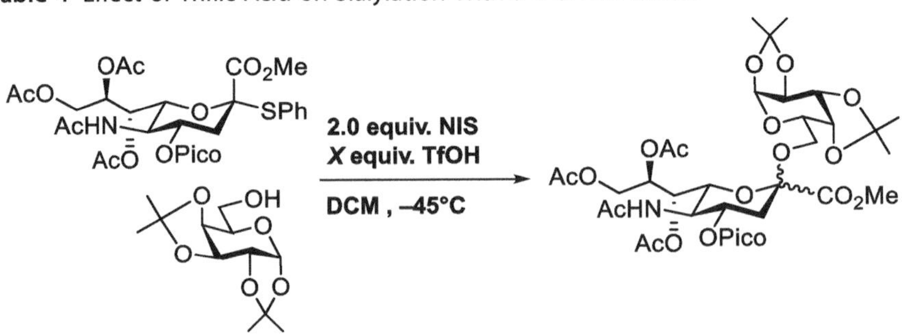

Entry	X	Time (h)	Yield (%)	α : β Ratio
1	0.2	16	64	3.0 : 1
2	0.7	5	70	3.0 : 1
3	1.2	2	N/A[a]	3.0 : 1
4	2.0	0.75	93	14.2 : 1
5	4.0	0.67	N/A[a]	9.0 : 1
6	5.0	0.67	N/A[a]	8.0 : 1

[a]Anomeric ratios were determined from integration of analogous ^{1}H NMR signals in the crude reaction mixture.

When promoted in the presence of the 1,2:3,4-di-*O*-isopropylidene-α-D-galactopyranosyl acceptor using 2.0 equiv. of NIS and 0.2 equiv. of TfOH per donor, in dichloromethane at −45°C, the 4-*O*-Pico α-phenylthio Neu5Ac derivative produced the desired (2→6) glycoside in 64% yield, with the α-anomer slightly favored (α/β = 3/1, Entry 1). The same glycosylation using 0.7 equiv. of TfOH yielded the expected product in the similar yield (70%), faster rate, and no change in stereochemical outcome (Entry 2). Similarly, the use of 1.2 equiv. of TfOH resulted in a shorter reaction time, but stereoselectivity of the reaction remained unchanged (Entry 3). A dramatic shift occurred when an excess of TfOH was used (2 equiv., Entry 4). The glycosylation reaction yielded the desired sialoside in 93% yield, with a dramatic improvement in α-stereoselectivity (α/β = 14.2/1, Entry 4). The use of 4 equiv. of triflic acid resulted in a reaction time of only 40 min, but the stereochemical outcome of the reaction shifted unfavorably (α/β = 9.0/1, Entry 5). Five equivalents of triflic acid produced no change in reaction time (40 min), but a slight loss of stereoselectivity was observed (Entry 6).

Inspired by these observations, De Meo and coworkers expanded the comparative analysis of triflic acid concentration by promoting analogous glycosylations in the presence of a methyl 2,3,4-tri-*O*-benzyl-α-D-galactopyranosyl or a methyl 2,3,4-tri-*O*-benzoyl-α-D-galactopyranosyl acceptor in an excess of TfOH (Scheme 27).

Coupling with methyl 2,3,4-tri-*O*-benzyl-α-D-galactopyranosyl acceptor in the presence of 2 equiv. of TfOH gave the desired α(2→6)-linkage in faster reaction time, with a higher α-stereoselectivity (α/β = 1.2/1 to 6.5/1) but slightly lower yield (93% to 70%). Similar outcome was recorded for the coupling with the methyl 2,3,4-tri-*O*-benzoyl-α-D-galactopyranosyl acceptor, obtaining the corresponding α(2→6) disaccharide in faster reaction time and higher stereoselectivity (α/β = 3.2/1 to 8.5/1).

As shown in Table 3, comparison data of the effect of TfOH on the stereoselectivity of sialylation were done by comparing 4-O-picoloylated and 4-O-benzoylated donors, bearing at C-5 the common acetamido group as well as its N-acetylated modification. All of these glycosyl donors were promoted in dichloromethane at −45°C in the presence of a common isopropylidene-protected galactopyranosyl acceptor, using 2 equiv. of NIS and either 0.7 or 2 equiv. of TfOH per equivalent of donor sugar (Table 5).

With the benchmark standard per-O-acetylated Neu5Ac derivative, increasing TfOH concentration in the glycosylation reaction did not have as profound of an effect as was observed with the 4-O-Pico Neu5Ac

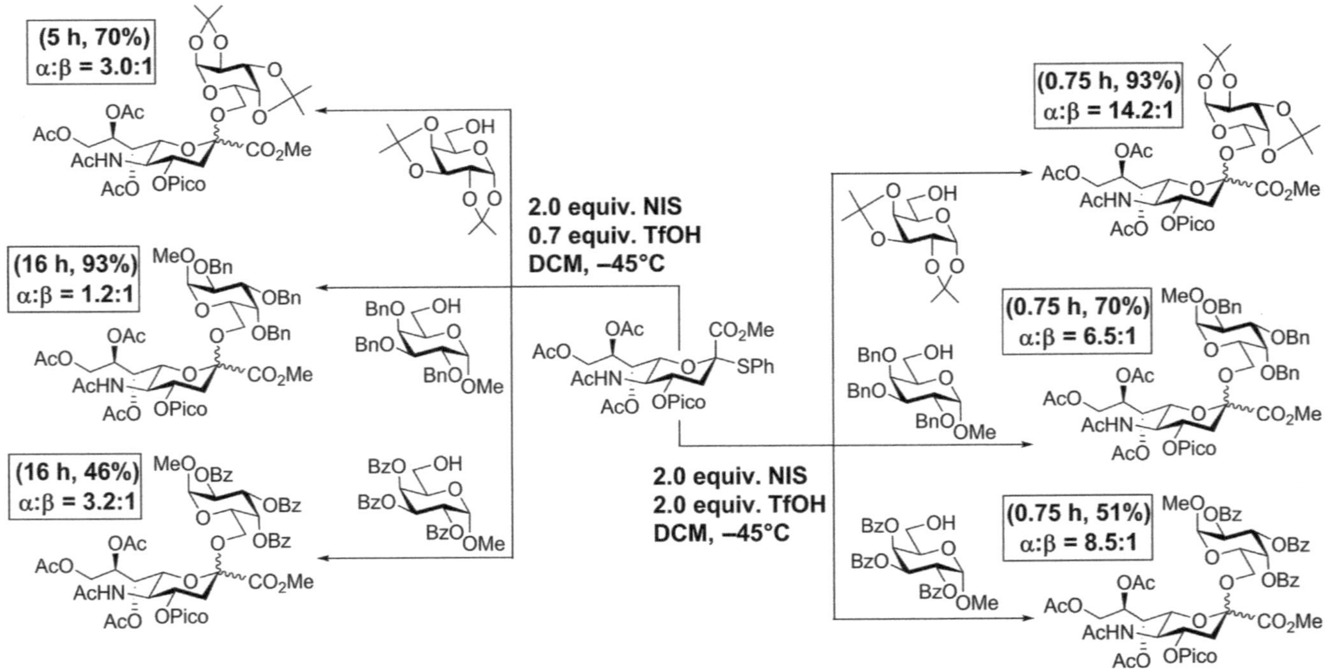

Scheme 27 Comparative analysis of the triflic acid effect for sialylation with primary –OH acceptors.

Table 5 Probing the Effect of TfOH and the 4-*O*-Pico Group for the C-5 *N*-Acetylacetacetamido Donor

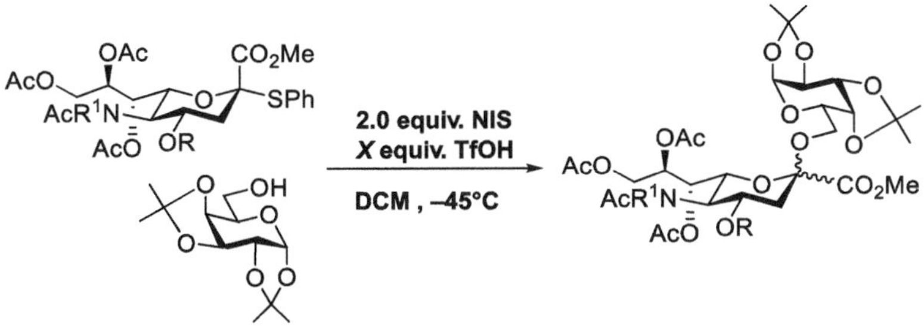

Entry	*X*	R	R[1]	Time (h)	Yield (%)	α : β Ratio
1	0.7	Ac	H	1	68	1.1 : 1
2	2.0	Ac	H	0.33	63	1.5 : 1
3	0.7	Pico	H	5	70	3.0 : 1
4	2.0	Pico	H	0.75	93	14.2 : 1
5	2.0	Bz	H	0.25	44	1.0 : 1
6	0.7	Pico	Ac	4	94	3.8 : 1
7	2.0	Pico	Ac	0.4	63	4.8 : 1

derivative. With the per-O-acetylated derivative, increasing the TfOH concentration from 0.7 equiv. to 2 equiv. per equivalent of glycosyl donor resulted in a shorter reaction time (1 h to 20 min), but also in a slightly lower yield (68% to 63%), with only a minor improvement in stereoselectivity (α/β = 1.1/1 to 1.5/1, Entries 1 and 2). These results contrast strikingly with the 4-*O*-Pico Neu5Ac donor. The same increase in TfOH concentration for the 4-*O*-Pico Neu5Ac derivative resulted in a much shorter reaction time (5 h to 45 min), a significantly higher product yield (70% to 93%), and a dramatic shift in stereoselectivity favoring the α-anomer (α/β = 3.0/1 to 14.2/1, Entries 3 and 4).

When the 4-*O*-Pico substituent was changed to a 4-*O*-Bz substituent, activation of the subsequent Neu5Ac donor with 2 equiv. of TfOH resulted in a shorter reaction time (15 min), a much lower product yield (44%), and a remarkable shift toward a racemic mixture (α/β = 1.0/1, Entry 5). In sialylations promoted using the 4-*O*-Pico Neu5Ac$_2$ derivative, an increase in TfOH concentration resulted in a shorter reaction time (4 h to 25 min), a lower yield of disaccharide (94% to 63%), and a slight improvement in anomeric stereoselectivity ($\alpha/\beta = 3.8/1$ to $4.8/1$).[140] These data suggest a synergistic effect between the picoloyl group and TfOH that leads to higher α-stereoselectivity. This effect is directly correlated to the presence of the pyridine ring, as it was not observed for the analogous benzoyl group. Additional investigation of picoloylated donors is described in Section 2.7.4.

2.5 Modifications at C-5

The discovery by Demchenko and Boons in 1998 that protecting groups at C-5 can dramatically influence the reactivity of chemical sialylations triggered a paradigm shift among carbohydrate chemists.[141] Success with C-5 positional modifications in accomplishing stereospecific glycosidic linkages without the need for a participating auxiliary group has made these strategies popular. In recent years, a significant amount of research has been conducted to better understand the influence of C-5 positional modifications in chemical sialylations. Currently, the assembly of complex sialic acid-containing oligosaccharides is typically performed with C-5-modified sialyl donors; however, in some cases, the unmodified C-5-acetamido sialyl donor might give better stereoselectivity.[142]

2.5.1 N-Acetylacetamido Derivatives

In their initial account of the effects of the *N*-acetylacetamido group at the C-5 position, Demchenko and Boons noted a significant increase in the reactivity of sialyl donors bearing an additional acetyl functionality at the C-5 nitrogen group ($Neu5Ac_2$).[141] The $Neu5Ac_2$ thioglycoside can be prepared easily from the standard per-O-acetylated Neu5Ac thioglycoside by treatment with isopropenylacetate in the presence of *p*-toluenesulfonic acid at 65°C. Coupling of a 2-(trimethylsilyl)ethyl 6-*O*-benzoyl-β-D-galactopyranosyl acceptor with a standard C-5-acetamido methylthio Neu5Ac donor at −40°C in acetonitrile using NIS and TfOH resulted in 61% of the desired α(2→3) disaccharide after 2–6 h.[56] The same secondary acceptor, coupled with the C-5-acetylacetamido-modified sialyl donor at −40°C in acetonitrile, gave a 72% yield of the expected α(2→3) disaccharide in less than 5 min. Subsequent removal of the additional *N*-acetyl group can be accomplished by treatment with sodium methoxide in methanol (Scheme 28).[141]

The exact cause behind the increased reactivity in $Neu5Ac_2$ donors is not fully understood, although Demchenko and Boons suggested that it may have been related to the diminished nucleophilicity of the *N*-acetylacetamido substituent relative to the standard acetamido group.[141] More recently, work

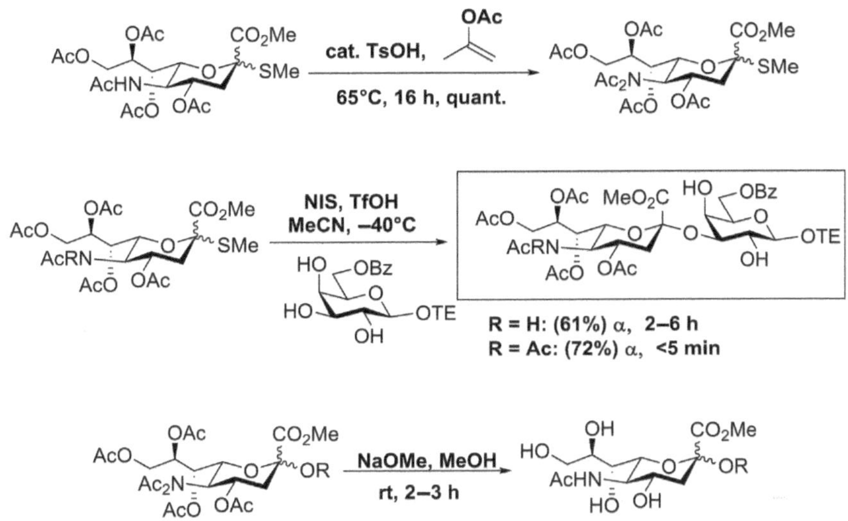

Scheme 28 Synthesis of and sialylations with a C-5-acetylacetamido sialyl donor.

reported by Kononov et al. suggests that the increase in reactivity and influence over stereoselectivity may also be due to a disruption in the intermolecular hydrogen-bond networks.[143]

The per-O-acetylated methylthio Neu5Ac$_2$ derivative performed relatively well in the synthesis of (2→9)- and (2→8)-linked dimers. It was observed by Demchenko and Boons that both the yield and stereoselectivity improved for (2→8) and (2→9) sialylations when the *N*-acetylacetamido functionality was employed in both donor and acceptor. For example, coupling between the per-O-acetylated methylthio Neu5Ac$_2$ donor and 8-OH of a Neu5Ac acceptor with NIS and TfOH in acetonitrile at −40°C resulted in 16% of the desired disaccharide as a racemic α/β mixture. The same donor, coupled to 8-OH of the equivalent Neu5Ac$_2$ acceptor, gave a 50% yield of the target with a diastereomeric ratio slightly favoring the α-anomer (α/β = 1.7/1). Similarly, the same chemical glycosylation promoted between the Neu5Ac$_2$ donor and 9-OH of a Neu5Ac acceptor produced 67% of the preferred sialoside in an anomeric mixture again favoring the α-anomer (α/β = 2/1). Conversion of the acceptor into a C-5-acetylacetamido moiety improved the yield to 98% with a slight improvement to stereoselectivity as well (α/β = 2.5/1).[144]

By using an equatorial phenylthio substituent as a C-3 auxiliary in a per-O-acetylated α-ethylthio Neu5Ac$_2$ donor, Hossain and Magnusson reported

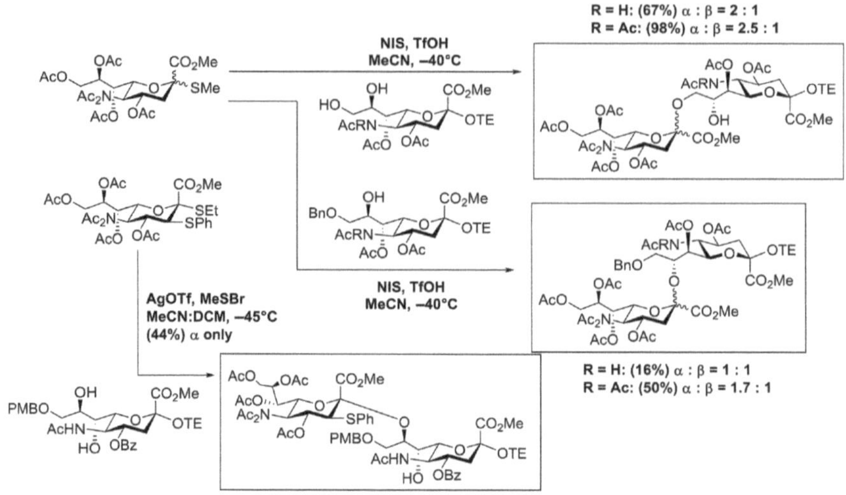

Scheme 29 Synthesis of α(2→8) and α(2→9) dimers by 5-acetylacetamido donors.

a 44% yield of only the α(2→8) dimer when coupled with a Neu5Ac acceptor using methylsulfenyl bromide (MeSBr) and silver triflate (AgOTf) in a mixture of acetonitrile/dichloromethane at −45°C (Scheme 29).[134]

Crich's group reported efficient glycosylations between a per-O-acetylated β-phenylthio Neu5Ac$_2$ donor and a wide range of primary, secondary, and tertiary acceptors, promoting with excess diphenyl sulfoxide (Ph_2SO), 2,4,6-tri-*tert*-butylpyrimidine (TTBP), and triflic anhydride (Tf_2O) in dichloromethane at −78°C. It was noted that the presence of excess Ph_2SO suppresses the formation of the elimination side product. Some examples of sialylations by this methodology are listed in Table 6.[145]

Kononov and coworkers reported the use of a Neu5Ac$_2$ (as well as Neu5Ac) glycosyl chloride precursor along with various anionic nucleophiles to synthesize a myriad of compounds with high stereoselectivity and in the absence of a promoter system.[54,146] Yields ranged from 65% to 82%, and products synthesized included α-sialyl phosphate, β-sialyl dibenzyl phosphate, α-sialyl azide, α-sialyl xanthate, and the α-phenylthio Neu5Ac$_2$ derivative.[54] The methodology used to synthesize the α-sialyl phosphate was then employed to synthesize a novel polyprenyl sialyl phosphate (Scheme 30).[146]

Recently, Fukase and coworkers reported the successful synthesis of a complex N-glycan with a dodecasaccharide containing two sialic acid residues. It was noted that reactivity improved in each of the key sialylation steps when a Neu5Ac$_2$ residue was employed rather than a Neu5Ac residue.[90]

Table 6 Glycosidation of a Phenyl Thiosialoside Donor With Diphenyl Sulfoxide and Triflic Anhydride

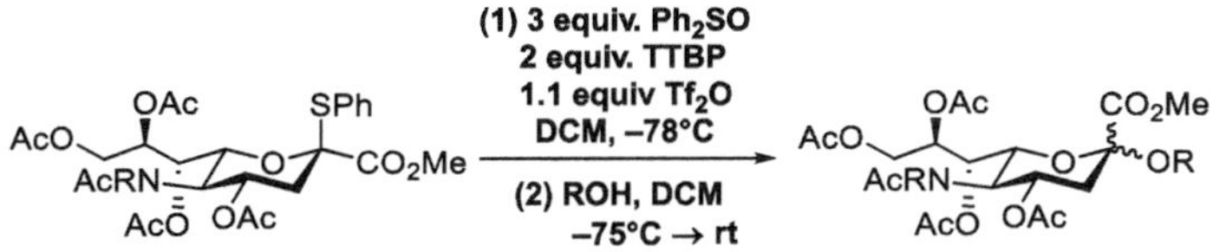

R	Yield (%)	α : β Ratio	R	Yield (%)	α : β Ratio
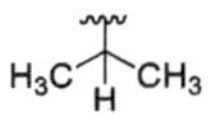	97	2.3 : 1	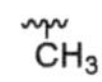	94	1.5 : 1
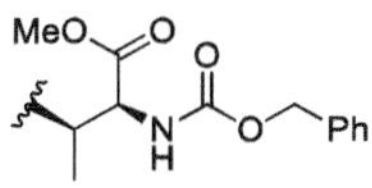	91	1 : 1		89	3.8 : 1
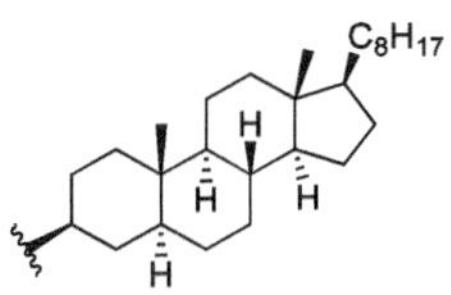	92	2 : 1	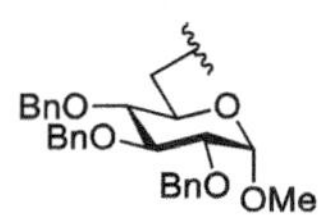	97	1.2 : 1
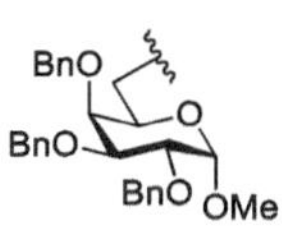	92	6 : 1	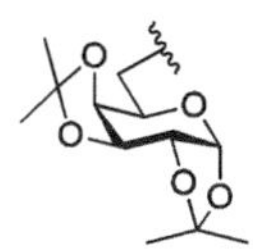	94	10 : 1
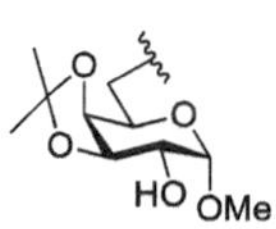	92	α only	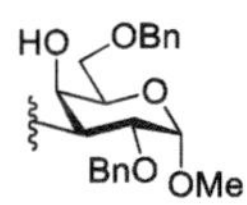	76	1 : 1.2
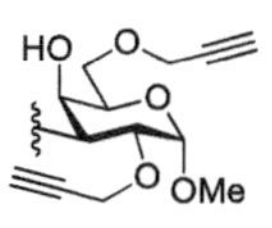	81	1 : 1.7	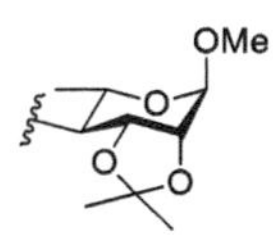	82	1 : 7

2.5.2 Azido Derivatives

An *N*-acetylneuraminic acid derivative bearing an azido (N_3) functionality at the C-5 position, in lieu of the standard acetamido group, had been synthesized as early as 1990 in the laboratory of Zbiral.[147] Over a decade later, Schmidt and coworkers[148] and Wong and coworkers[106] described a facile

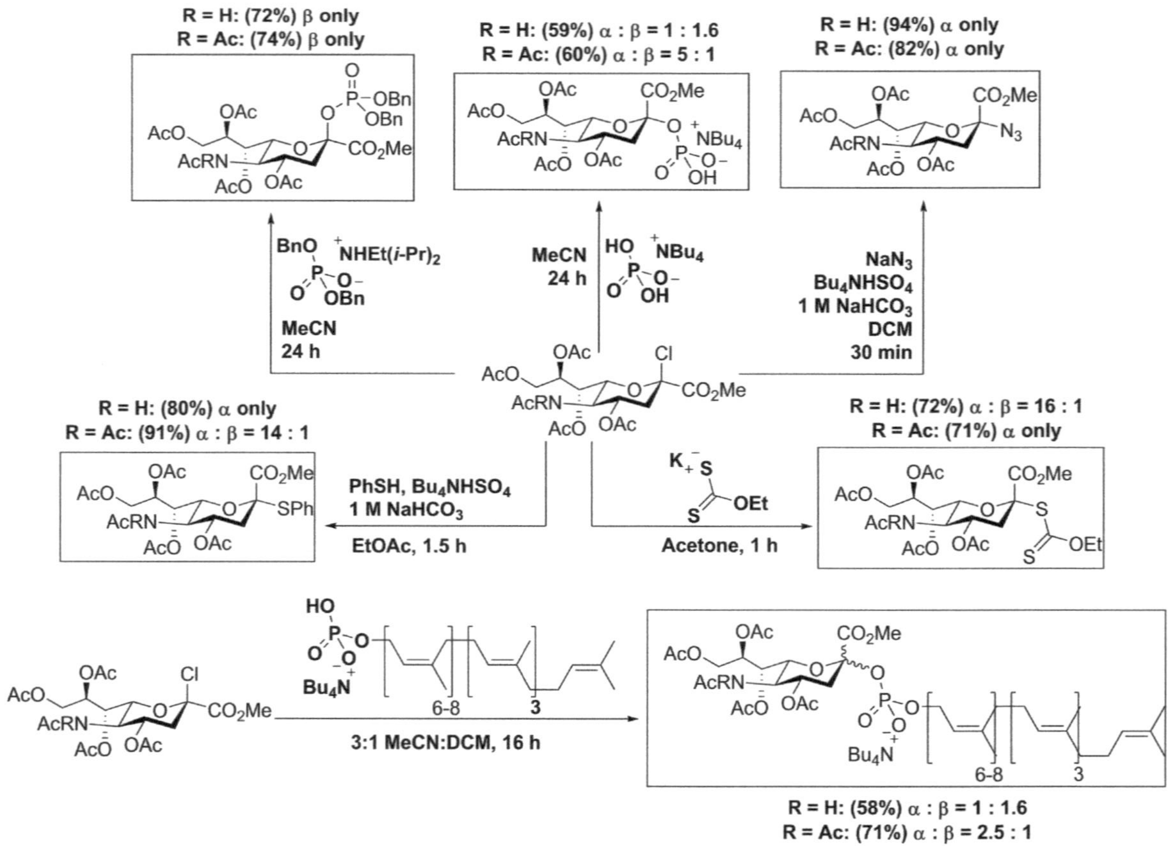

Scheme 30 Conversion of anomeric sialyl chlorides and the synthesis of polyprenyl sialyl phosphate.

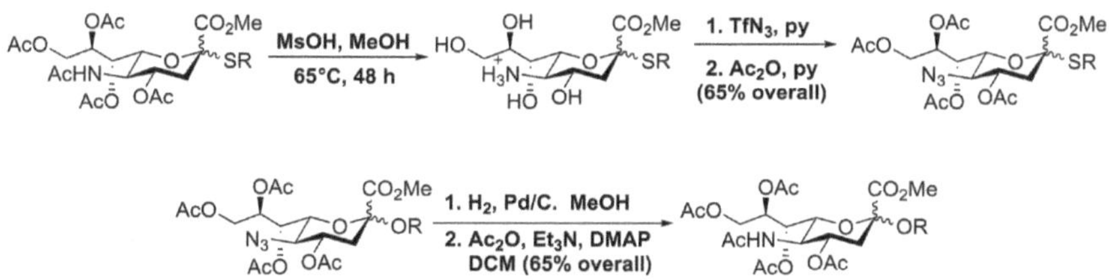

Scheme 31 Synthesis and removal of a C-5 azido group.

method of preparing the 5-azido-5-deoxy thioglycoside and glycosyl phosphite by complete deacetylation of the Neu5Ac derivative to give the free amino group, which could be then converted into an azido group by reaction with trifluoromethanesulfonyl azide (TfN_3) followed by in situ acetylation to furnish the acetylated 5-azido-5-deoxy sialoside in ~65% overall yield. The 5-azido group can be reduced by hydrogenation and subsequently acetylated to yield the original acetamido group (Scheme 31).[149]

The 5-azido functionality has been investigated in chemical sialylations as a means of reducing hydrogen bonds which inhibit reactivity. Lin and Wong carried out chemical glycosylations using a 5-azido-5-deoxy-9-hydroxyl sialyl acceptor and α-tolylthio sialyl donors bearing, respectively, a 5-azido group, a 5-*N*-acetylacetamido group, or standard 5-acetamido group. When promoted using NIS and TfOH at −40°C in acetonitrile, the standard Neu5Ac donor and 5-azido-5-deoxy sialyl acceptor produced 75% of the desired product with an α/β ratio of 3/1. Surprisingly, the $Neu5Ac_2$ donor produced inferior results with a 30% yield of mixed stereoisomers, slightly favoring the β-anomer (α/β = 4/5). Coupling the 5-azido-5-deoxy sialyl donor along with the 5-azido-5-deoxy sialyl acceptor gave, in 65% yield, a complete α-stereoselectivity of the desired target.[106]

Mukaiyama and coworkers reported successful glycosylation results using 5-azido-5-deoxy sialyl fluoride donors promoted with tin(IV) chloride ($SnCl_4$) and silver(I) tetrakis(pentafluorophenyl)borate ($AgB(C_5F_6)_4$) in propionitrile at −10°C, while the equivalent Neu5Ac and $Neu5Ac_2$ derivatives yielded no disaccharide.[150] The generally higher α-stereoselectivity observed with the 5-azido-5-deoxy derivatives has been attributed to a greater electron-withdrawing capability relative to the standard acetamide, as well as to the reduced steric hindrance observed in the α-face of the sialyl donor with a linear azido functionality (Scheme 32).[106]

Sialyl donors bearing a 5-azido modification have not performed as well with more sterically hindered secondary acceptors (such as the 8-hydroxy

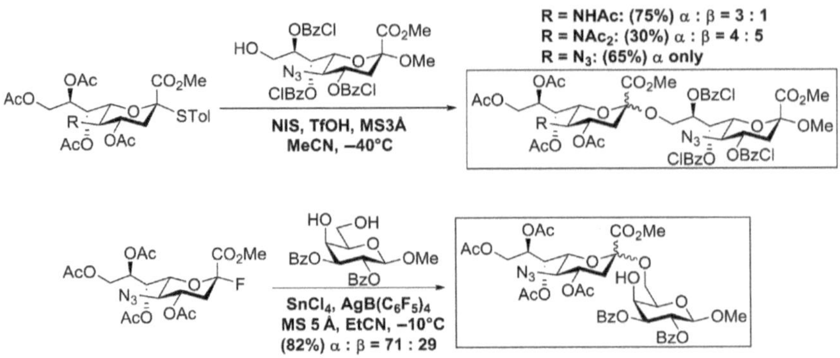

Scheme 32 Sialylation reactions of 5-azido-5-deoxy-sialyl donors.

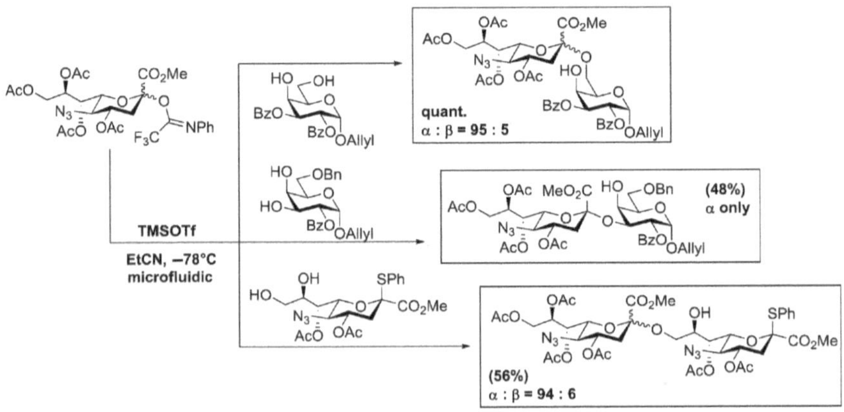

Scheme 33 Sialylation of a 5-azido-5-deoxy-sialyl *N*-phenyltrifluoroacetimidate donor under microfluidic conditions.

sialyl acceptor).[151] More recently, Fukase and coworkers reported the use of a microfluidic technique in conjunction with a 5-azido-5-deoxy-sialyl *N*-phenyltrifluoroacetimidate donor to obtain modest-to-quantitative yields with a range of acceptors. When promoted with trimethysilyl triflate (TMSOTf) in propionitrile at −78°C in the presence of an allyl 2,3-di-*O*-benzoyl-α-D-galactopyranosyl acceptor under microfluidic conditions, a quantitative yield of the (2→6) disaccharide was produced with an overwhelming majority being the α-anomer (α/β=95/5). The same donor and conditions applied to a 5-azido-5-deoxy-9-hydroxy sialyl acceptor resulted in 56% of the (2→9)-linked glycoside with a similar anomeric selectivity (α/β=94/6). When coupled with an allyl 2-*O*-benzoyl-6-*O*-benzyl-α-D-galactopyranosyl acceptor, the same donor yielded 48% of the α(2→3)-linked product with complete stereoselectivity (Scheme 33).[152]

2.5.3 Trifluoroacetamido Derivatives

The diminished nucleophilicity and increased donor reactivity observed with the *N*-acetylacetamido group relative to the standard acetamido group led Boons and coworkers to the development of a Neu5TFA derivative.[153] The 5-trifluoroacetamido-5-deoxy thioglycoside can be prepared from a standard Neu5Ac thioglycoside by complete deacetylation to afford the 5-amino-5-deoxy derivative, followed by subsequent treatment with methyl trifluoroacetate and triethylamine in methanol.[39] Upon glycosylation, deprotection of the 5-trifluoroacetamido group to afford the original acetamide can be accomplished by deacylation with 1 M sodium hydroxide in methanol, followed by selective N-acetylation using acetic anhydride in methanol (Scheme 34).[153]

5-Trifluoroacetamido donors have proven to be far more reactive than their respective Neu5Ac and $Neu5Ac_2$ derivatives, especially toward more sterically hindered secondary acceptors. When promoted with NIS and TfOH in acetonitrile at −35°C in the presence of a benzylated lactosyl acceptor, the per-O-acetylated 5-trifluoroacetamido-5-deoxy methylthio donor produced the α(2→3) trisaccharide (84%, α only). Similar to dimerization reactions using $Neu5Ac_2$ derivatives, dimerization reactivity using Neu5TFA derivatives improves upon protection of both the donor and the acceptor with trifluoroacetamido groups. Glycosylation of the per-O-acetylated 5-trifluoroacetamido donor with a partially protected 5-trifluoroacetamido-5-deoxy-8-hydroxyl sialyl acceptor using NIS and TfOH in acetonitrile at −35°C resulted in 55% yield of only the α (2→8)-linked dimer. The same reaction carried out using an equivalent Neu5Ac sialyl acceptor produced only 34% of the α(2→8) product.[153]

Similarly, Meijer and Ellervik coupled a per-O-acetylated 5-trifluoroacetamido-5-deoxy β-phenylthio donor to a 9-azido-protected 5-trifluoroacetamido sialyl acceptor using iodine monobromide (IBr) and silver triflate (AgOTf) in a mixture of dichloromethane/acetonitrile at −72°C to produce 59% of only the α (2→8)-linked disaccharide. The disaccharide subsequently was transformed into a lactam (86%) by the treatment with triphenylphosphine (PPh_3) in wet tetrahydrofuran at 60°C (Scheme 35).[154]

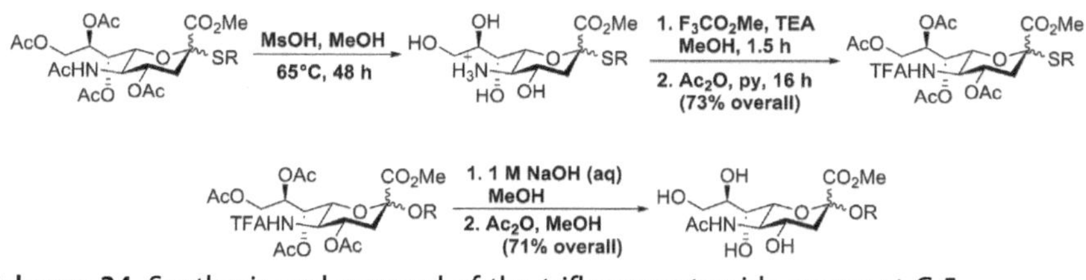

Scheme 34 Synthesis and removal of the trifluoroacetamido group at C-5.

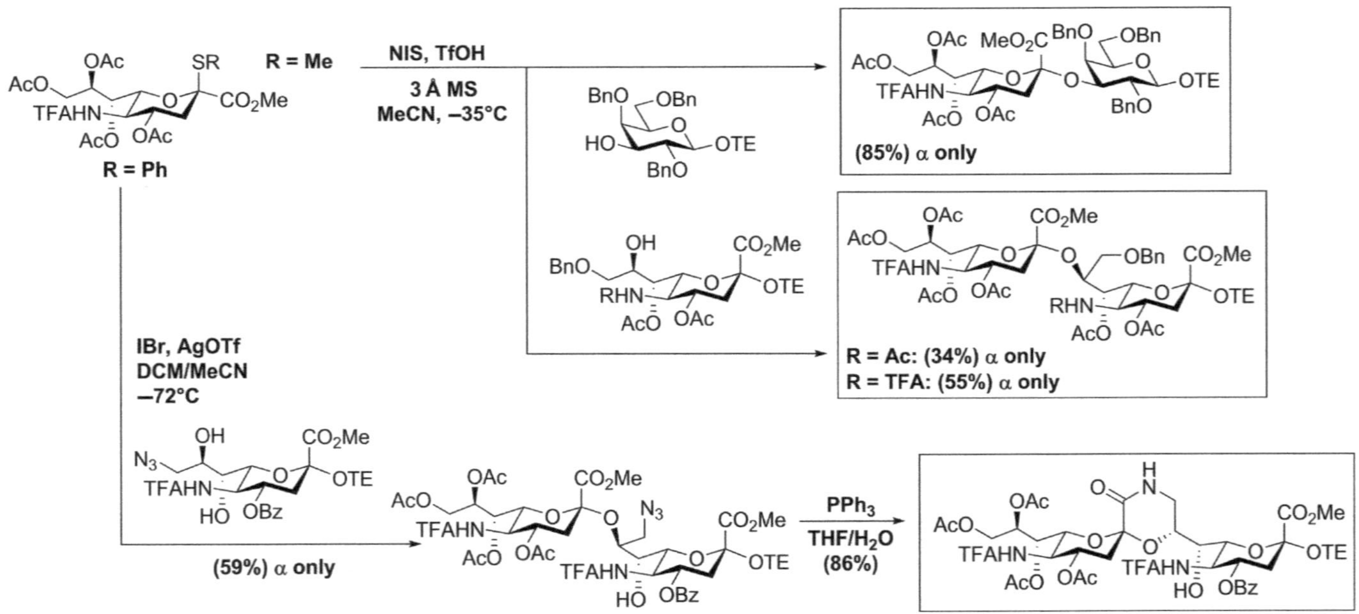

Scheme 35 Effect of the 5-trifluoroacetamido group in sialylations of phenylthio and methylthio donors.

As a consequence of its high reactivity, coupling between a 5-trifluoroacetamido donor and less sterically hindered primary acceptors was in general less α-stereoselective.[155] For example, when promoted in the presence of a methyl 2,3,4-tri-*O*-benzyl-β-D-galactopyranosyl acceptor using NIS and TfOH in acetonitrile at −40°C, the per-O-acetylated 5-trifluoroacetamido-5-deoxy methylthio donor produced the desired (2→6)-linked disaccharide in 96% yield, but as an anomeric mixture (α/β=3/1). In contrast, the same glycosylation carried out using a methyl 2,4,6-tri-*O*-benzyl-β-D-galactopyranosyl acceptor produced the corresponding disaccharide in 84% yield as only the α-anomer.[155] Nifantiev and coworkers reported the use of a per-O-chloroacetylated 5-trifluoroacetamido-5-deoxy 2-ethylthio sialyl donor with an acetylated lactosyl acceptor to obtain 48% of only the α(2→3)-linked trisaccharide.[156]

When equipped with a phosphite leaving group, however, the per-O-acetylated 5-trifluoroacetamido donor group and a partially protected 5-trifluoroacetamido thiosialyl primary acceptor gave highly α-stereoselective results. An initial α(2→9)-linkage was produced in 77% yield when coupled with NIS and TMSOTf in acetonitrile at −40°C, and conversion of the thioglycoside to a new glycosyl phosphite allowed for the iterative synthesis of (2→9) oligosialosides.[157] It was noted by Lin and coworkers that stereoselectivity decreased for each iterative glycosylation (Scheme 36).[157]

Ando and Koketsu described a highly stereoselective strategy for the iterative synthesis of α(2→8)-linked oligosialosides using alternating

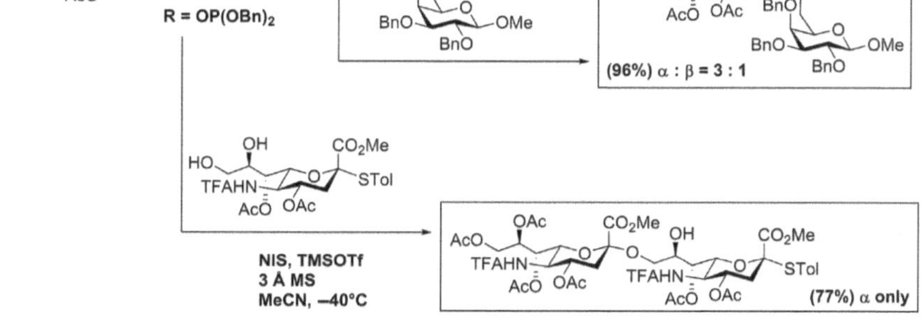

Scheme 36 Sialylations with 5-trifluoroacetamido donors.

1,5-lactamization and glycosylation steps.[158,159] A 4,7,9-tri-*O*-benzyl-8-*O*-chloroacetyl-5-deoxy-5-trifluoroacetamido donor was prepared bearing an *N*-phenyltrifluoroacetimidate as leaving group. This donor could be converted into a 8-hydroxy 1,5-lactamized acceptor in high yield (84%) by treatment with 1 M aqueous sodium hydroxide, tetrahydrofuran, and methanol in a 1 : 2 : 2 ratio, respectively, followed by reaction with HBTU and *i*-Pr_2Net, and an additional treatment with the 1 : 2 : 2 1 M aq NaOH/THF/MeOH solution. Upon activation by TMSOTf in propionitrile at −80°C, the (2→8)-linked disaccharide was obtained in 86% yield with excellent stereoselectivity (α/β = 16/1). An additional lactamization step (83% yield), followed by a glycosylation, resulted in 66% of only the α-linked trisaccharide. The 2,3-dehydro derivative side product formed from the preceding glycosylation reaction cannot undergo lactamization, allowing for a facile purification method. A third iterative step resulted in the formation of only the α(2→8)-linked tetrasaccharide with a 44% yield (Scheme 37).[158]

2.5.4 N-*Trichloroethoxycarbonyl Derivatives*

Derivatives of *N*-acetylneuraminic acid have been synthesized with several different carbamate protecting groups at the C-5 position, including *N*-*tert*-butoxycarbonyl (Boc),[160] *N*-allyloxycarbonyl (Alloc), *N*-fluorenylmethoxycarbonyl (Fmoc),[161] and *N*-carboxybenzyl (Cbz),[162] although none have been studied as extensively or have proven to be as successful as the *N*-2,2,2-trichloroethoxycarbonyl (Troc) group. The Troc protecting group was introduced to the field of sialic acid chemistry by the laboratories of Wu[163] and Kiso.[164] The Troc group can be introduced to the 5-amino sialyl derivative following N,O-deacetylation by using either *N*-(2,2,2-trichloroethoxycarbonyloxy)succinimide (TrocSu) in an aqueous solution of sodium bicarbonate[163] or 2,2,2-trichloroethyl chloroformate (TrocCl) in a solution of triethylamine/methanol. The Troc group can be removed easily and selectively in the presence of most *O*-acyl groups by treatment with activated zinc in glacial acetic acid to afford the free 5-amino derivative (Scheme 38).[164]

Kiso and coworkers reported the coupling between the per-O-acetylated 5-*N*-Troc α-phenylthio sialyl donor with a 4-methoxyphenyl 2,6-di-*O*-benzyl-β-D-galactopyranosyl acceptor using NIS and TfOH in acetonitrile at −50°C, which produced 54% of the α(2→3)-linked disaccharide. The same reaction carried out at −25°C and using a standard per-O-acetylated Neu5Ac α-phenylthio donor yielded 23% of only the α-anomer after 13 h (Scheme 39).[164]

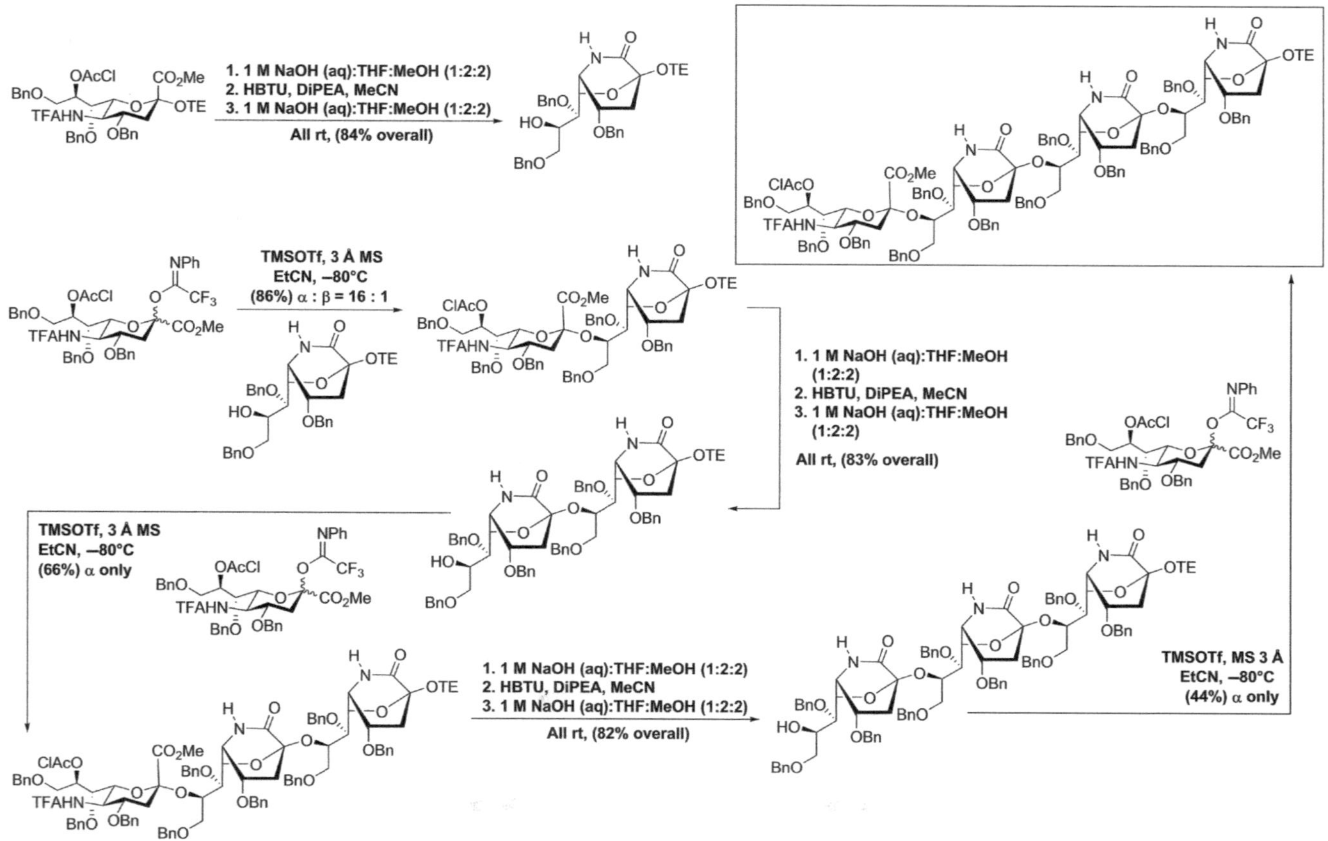

Scheme 37 Iterative synthesis of α(2→8)-linked oligosialosides via 1,5-lactamization.

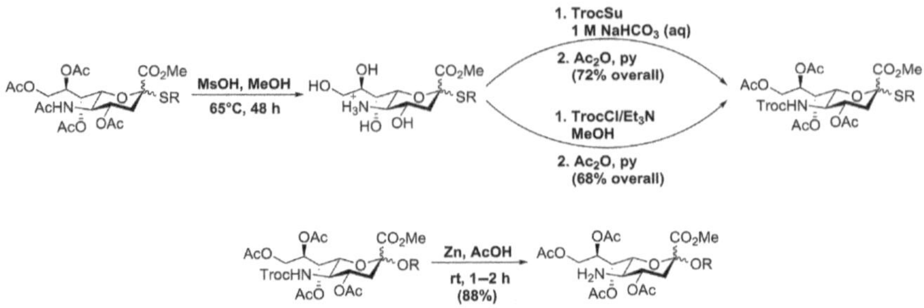

Scheme 38 Introduction and removal of *N*-Troc functionality at C-5.

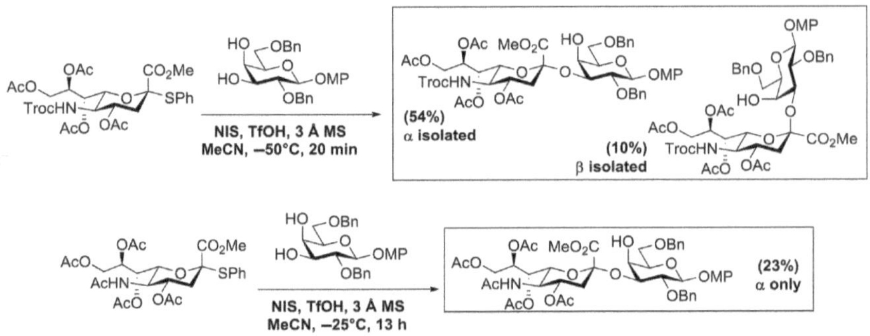

Scheme 39 Sialylation of a 5-*N*-Troc β-phenylthio donor.

Takahashi and coworkers demonstrated the use of the per-O-acetylated 5-*N*-Troc β-phenylthio donor in an efficient one-pot trisaccharide synthesis.[161] The coupling between the 5-*N*-Troc sialyl donor and a 2-azido-4-*O*-benzyl-2-deoxy-α-D-galactopyranosyl acceptor, using NIS and TfOH in acetonitrile at −35°C, gave a (2→6)-linkage with an α/β ratio of ~4/1. Subsequent dilution with dichloromethane and addition of 2-*O*-benzoyl-3,4,6-tri-*O*-benzyl-β-galactopyranosyl fluoride donor at −35°C furnished the Galβ(2→3)GalN-linkage in a 77% overall yield (Scheme 40).[161]

Similarly, Seeberger and coworkers reported a highly efficient (2→3)-linkage synthesis between a per-O-acetylated 5-*N*-Troc β-sialyl phosphite donor and a benzylated 3-hydroxy-D-galactal acceptor using TMSOTf in propionitrile at −78°C, with 86% of the α(2→3)-linked disaccharide isolable and an overall α/β ratio of 11/1.[103]

More recently, Kiso and coworkers utilized a 4,7,9-tri-*O*-acetyl-8-*O*-methoxy 5-*N*-Troc β-phenylthio donor with a diol disaccharide GalN-Gal acceptor and NIS/TfOH in propionitrile at −60°C to furnish the disialylated tetrasaccharide in 59% yield. Forty-one percent of the isolated tetrasaccharide contained the desired α-linkage between both sialyl subunits (Scheme 40).[165]

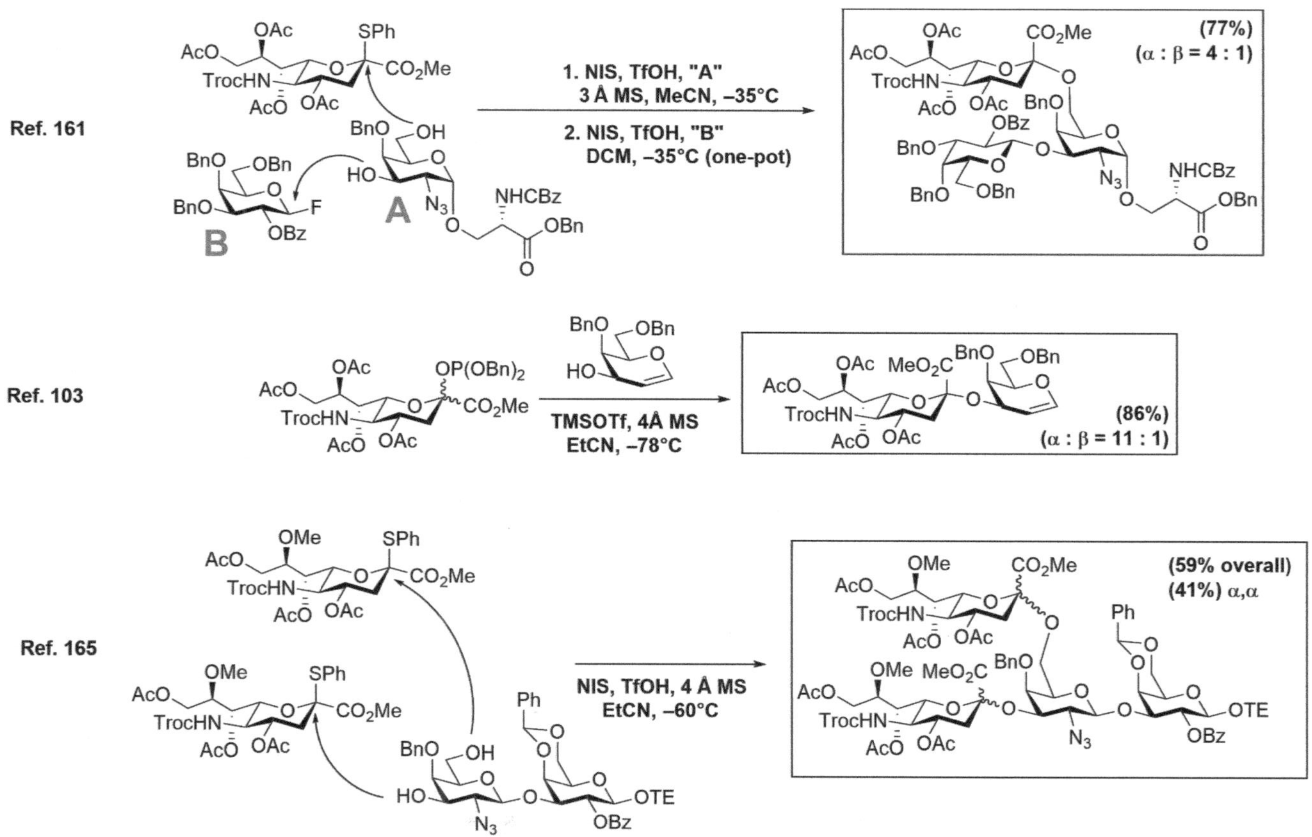

Scheme 40 Application of *N*-Troc-modified sialyl donors.

2.5.5 N-Phthalimido Derivatives

The *N*-phthalimido group was first reported to be used in chemical sialylations in 1994 by Sugimoto and coworkers, who introduced the group by N,O-deacetylation of the corresponding Neu5Ac thioglycosides, followed by treatment with phthalic anhydride and triethylamine in methanol.[166] The high reactivity and α-stereoselectivity observed with the functional group were then explored by Fukase and coworkers.[84] These workers reported that when the donor was equipped with an *N*-phenyltrifluoroacetimidate leaving group, the per-O-acetylated *N*-phthalimido donor efficiently coupled to an allyl 2,3-di-*O*-benzoyl-α-D-galactopyranosyl acceptor in an α(2→6)-linkage when using TMSOTf as the promoter in propionitrile at −78°C. The same donor and reaction conditions applied to an allyl 2-*O*-benzoyl-6-*O*-benzyl-α-D-galactopyranosyl acceptor yielded 77% of the target α(2→3)-linked disaccharide with only trace amounts of the β-anomer formed. Computational studies conducted by Fukase and coworkers suggested that a fixed dipole induced by the *N*-phthaloyl moiety favors the formation of a β-nitrilium intermediate upon donor activation, and an S_N2-like attack produces the final α-glycosidic product. Upon glycosylation, the *N*-phthalimido group can be converted back to the standard acetamido group by treatment with methylhydrazine acetate ($MeNHNH_2$/AcOH) in ethanol at 80°C for 24h, followed by acetylation with acetic anhydride in pyridine (Scheme 41).[84]

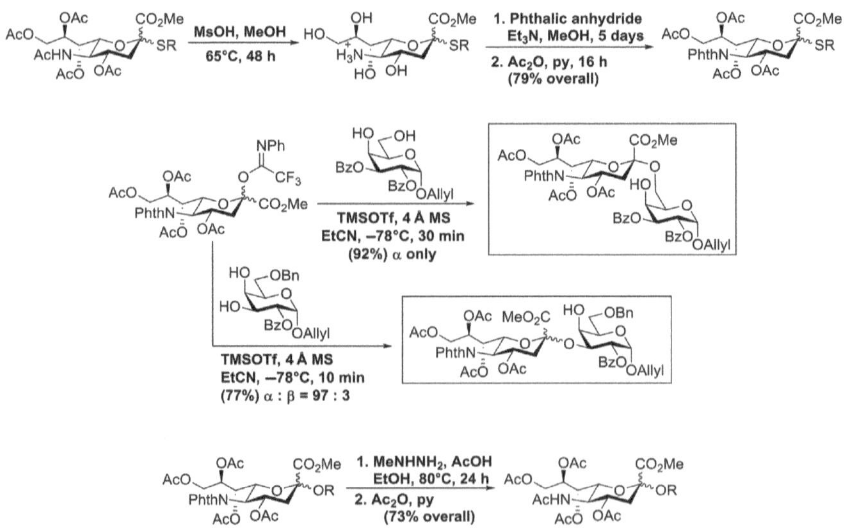

Scheme 41 Synthesis of and sialylation with an *N*-phthalimido-modified sialyl donor.

2.5.6 Isothiocyanate Derivatives

Most recently, Crich and coworkers have explored the use of isothiocyanate (NCS) as a C-5 modification.[167] The isothiocyanate function can be installed from the polyol *N*-Boc-protected thioglycoside by carbamate removal via treatment with HCl in diethyl ether, followed by installation of the NCS group using phenyl chlorothionoformate (PhOC(=S)Cl) and aqueous sodium bicarbonate. The isothiocyanate group offers two major advantages when compared to other C-5 modifications: (1) its α-directing propriety, including reactions with less reactive acceptors, and (2) the opportunity for its conversion into a large variety of C-5-modified moieties at the sialoside level. For example, the NIS/TfOH-promoted coupling of a per-O-acetylated 5-isothiocyanato β-adamantylthio donor with a methyl 2,3,4-tri-*O*-benzyl-β-D-galactopyranosyl acceptor gave 80% of only the α(2→6)-linked disaccharide (Scheme 42).[167] More impressive it is the successful sialylation with an unreactive acceptor such as a methyl 2,4,6-tri-*O*-benzyl-β-D-galactopyranosyl acceptor, yielding 87% of only the α(2→3)-disaccharide. In Scheme 42, a series of postglycosylation derivatizations of the isothiocyanate group are shown as examples of its versatility in the diversification of the C-5 position, including the replacement of the C-5–N-5 bond with C–C or C–H. Furthermore, an isothiocyanate group can also be converted by reactions with thioacids into amides, and by reactions with amines into thioureas. The latter can be then converted into guanidine compounds (Scheme 42).[167]

2.6 Cyclic Protecting Groups

2.6.1 Oxazolidinone trans-*Fused Ring*

The synthesis of a 5-*N*,4-*O*-oxazolidinone-protected sialyl donor, as well as its N-acylated version, is one of the most promising innovations in the field in the last decade. The oxazolidinone group was originally introduced by Takahashi and coworkers for the synthesis of α(2→8) glycosidic bonds.[168] Farris and De Meo then reported the application of oxazolidinone-protected sialyl donors to the synthesis of α(2→6) and α(2→3) glycosidic bonds.[169] Concomitantly, Crich and Li synthesized and tested the N-acetylated oxazolidinone variation.[67,108] The latter, while it produces similar overall yields and similar or better stereoselectivities than the non-N-acetylated counterpart, has the advantage of allowing for the direct conversion into the natural C-5-acetamido group by mild reaction conditions.[108]

The introduction of an oxazolidinone *trans*-fused ring is accomplished by initial N-deprotection in acidic medium (methanesulfonic acid), followed

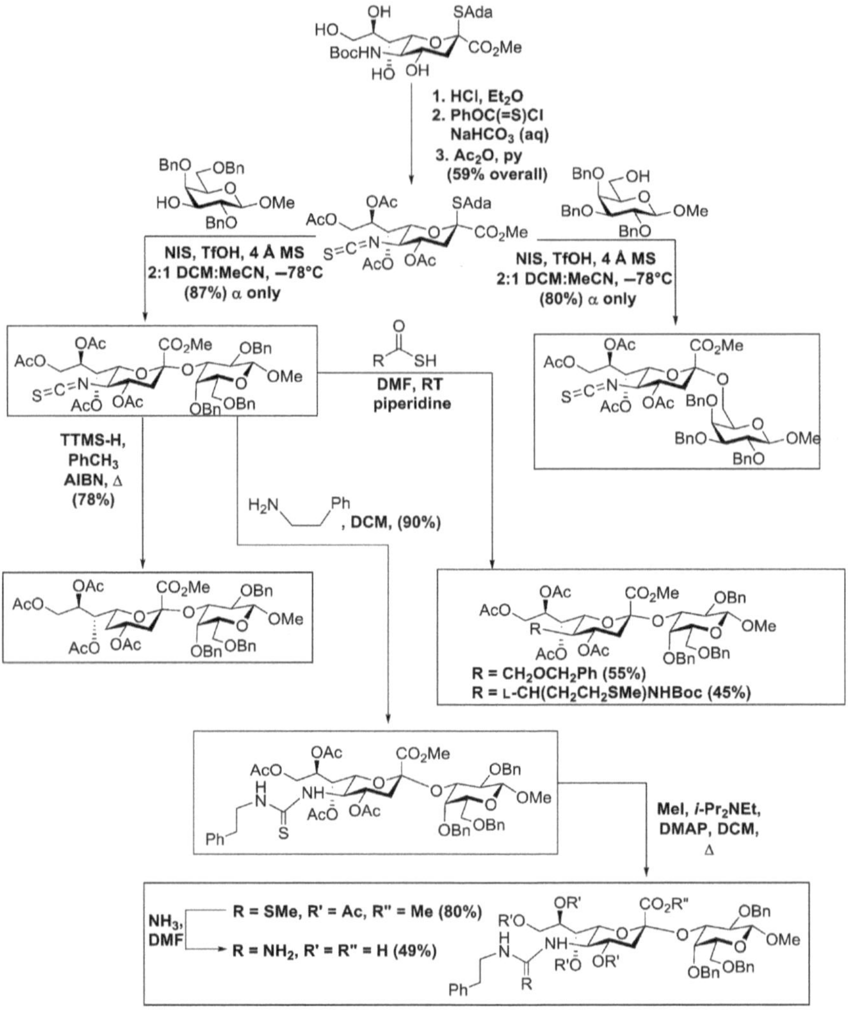

Scheme 42 Syntheses, sialylations, and C-5 diversifications of isothiocyanate sialyl donors.

by N-functionalization by treatment with *p*-nitrophenyl chloroformate (NPC, Schemes 43 and 44). The N-acetylated sialyl donor can then be obtained using acetyl chloride and *i*-Pr_2NEt.

Takahashi and coworkers reported the synthesis of a 9-*O*-benzyl-5-*N*,4-*O*-carbonyl-7,8-dichloroacetyl β-phenylthio donor and its coupling with a 9-*O*-benzyl-5-*N*,4-*O*-carbonyl-protected acceptor for the synthesis of α(2→8) dimers (Scheme 43).[168] Thus, the desired disaccharide was obtained in an excellent yield as the α-anomer, only. Subsequent removal of the

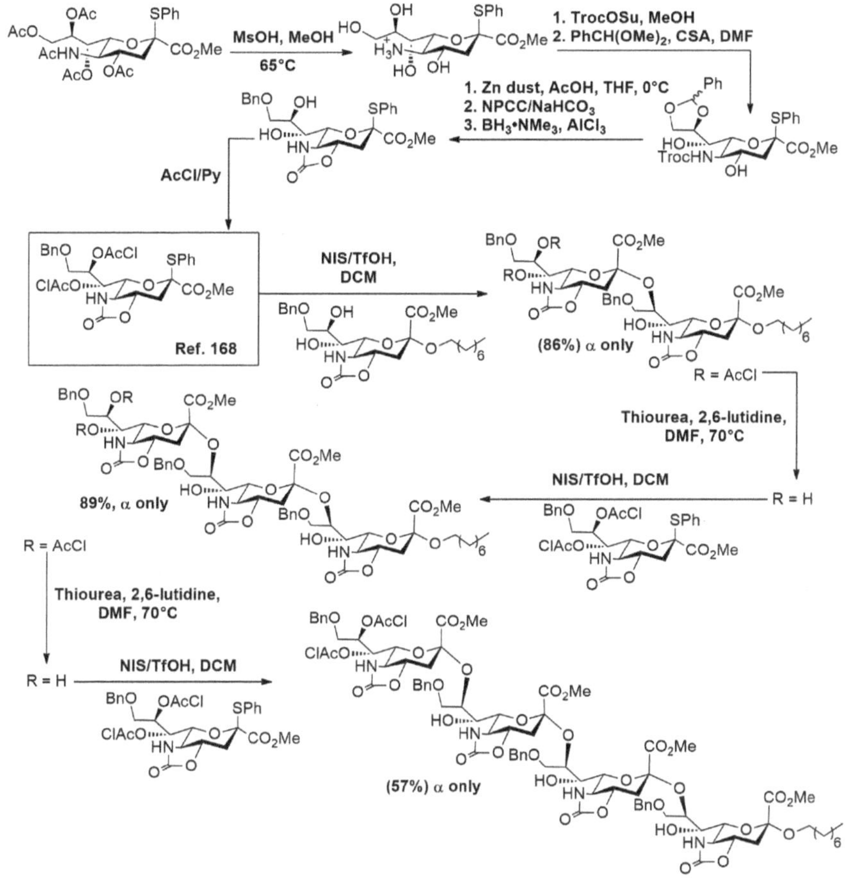

Scheme 43 Synthesis of 5-*N*,4-*O*-oxazolidinone thioglycosyl donors and an example of their application to the iterative synthesis of a tetrasaccharide.

chloroacetyl group allowed the additional coupling for the synthesis of a trisaccharide, in which the product was formed in high yield and with complete α-stereoselectivity. This iterative process was carried on to yield an α(2→8)-linked tetrasaccharide.[168] Similarly, the Takahashi group reported the iterative synthesis of an α(2→9) tetrasaccharide that was accomplished in high yields and stereoselectivities with a minimal number of protecting group manipulations.[170] In this process, a much higher stereoselectivity was obtained when a 8,9-diol acceptor was used instead of the 8-acetylated counterpart.[170]

For the synthesis of α(2→6) glycosidic linkages, Farris and De Meo reported that good yields and stereoselectivities can be achieved with a

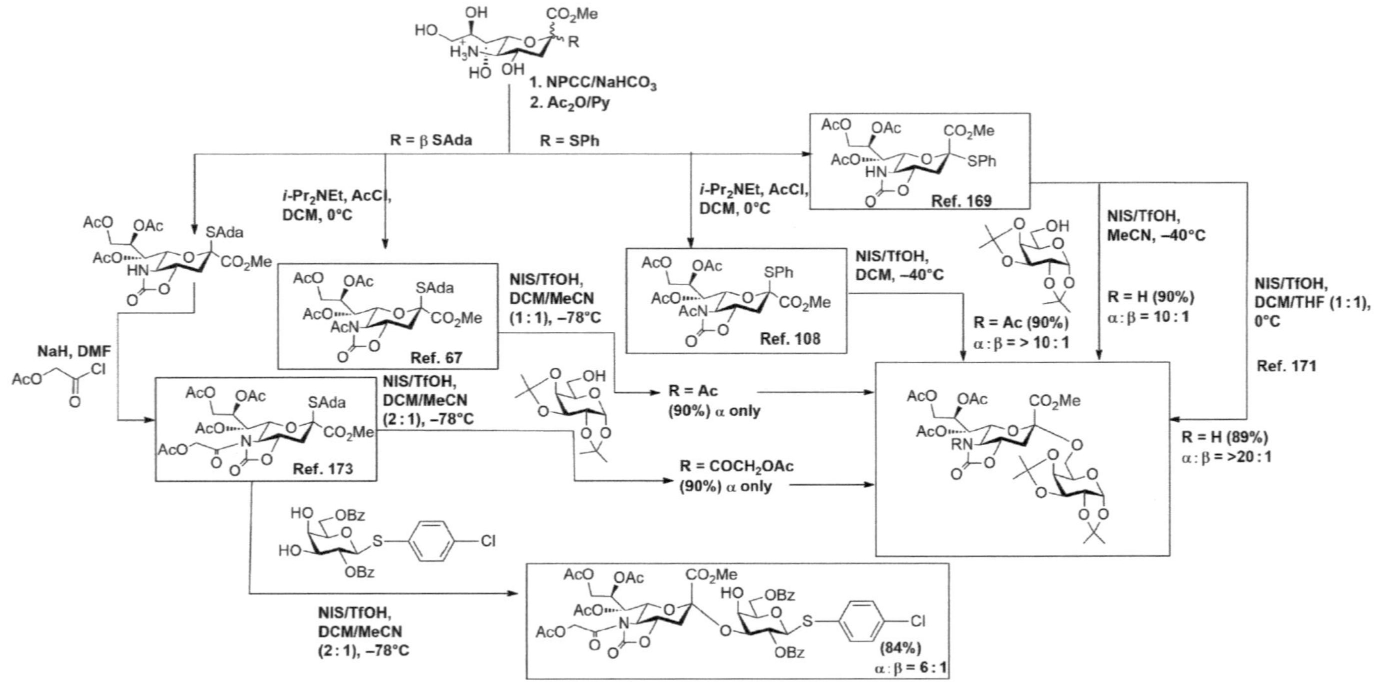

Scheme 44 Synthesis and application of 5-*N*,4-*O*-oxazolidinone and *N*-acetyl-5-*N*,4-*O*-oxazolidinone thioglycosyl donors,

5-*N*,4-*O*-oxazolidinone α-phenylthio donor when promoted with the NIS/TfOH system in acetonitrile (Scheme 44).[169] However, higher stereoselectivity was observed in a tetrahydrofuran/dichloromethane 1/1 mixture, allowing the coupling at an unusual temperature of 0°C.[171] Noteworthy, successful sialylations were also performed in neat tetrahydrofuran at room temperature.[171]

The *N*-acetyl-5-*N*,4-*O*-oxazolidinone β-phenylthio donor was synthesized and studied by Crich and Li.[108] Initial testing was done by activation with diphenyl sulfoxide and triflic anhydride in the presence of a hindered base such as 2,4,6-tri-*tert*-butylpyrimidine.[108,145] Interestingly, sialylations favored the β-anomer formation with this promoter system. This stereoselectivity was linked to the combination of the *N*-acetyloxazolidinone ring and the chosen promoter, since the same effect was not observed for the *N*-acetylacetamido counterpart.[145] However, with the NIS/TfOH promoter system, the *N*-acetyl-5-*N*,4-*O*-oxazolidinone β-phenylthio donor yielded the desired targets in high yields and stereoselectivities (Scheme 44). Unfortunately, no activation was observed at temperature of −40°C in MeCN for the coupling with secondary acceptors. The ability of the similar 5-*N*,4-*O*-oxazolidinone α-phenylthio donor to be instead activated in these conditions[169] shows how small modifications in the structure of a donor can have a big impact in the overall outcome of sialylations. As more reactive donor, an *N*-acetyl-5-*N*,4-*O*-oxazolidinone β-adamantylthio sialoside was then synthesized and tested.[67] Complete stereoselectivities and excellent yields were observed when these conditions were applied to the coupling with primary acceptors (Scheme 43). In order to improve the stereoselectivity in the coupling with secondary acceptors, an investigation of the solvent effect revealed a nitrile α-directing effect, which was found optimal with a 2/1 ratio of MeCN/DCM.[67] More recently, Crich and coworkers reported the activation of an *N*-acetyl-5-*N*,4-*O*-oxazolidinone β-phenylthio donor at −78°C using *p*-nitrobenzenesulfenyl chloride as the promoter.[172]

An *N*-acetoxyacetimide derivative was also synthesized and applied to the synthesis of *N*-glycolyl α-sialosides.[173] Also in this case, the presence of a 5-*N*,4-*O*-oxazolidinone group allowed the synthesis of several targets with good yields and high stereoselectivities (Scheme 43). In particular, the combination of an *S*-adamantyl and the oxazolidinone ring allowed for a successful selective sialylation in the presence of a galactopyranosyl acceptor equipped with the β-4-chlorophenylthio anomeric group (Scheme 44). This strategy was then applied to the one-pot synthesis of several trisaccharides.[173]

Based on the promising data from sialylations with the oxazolidinone *trans*-fused ring, many variations of this theme were investigated. For example, Xing and coworkers reported the synthesis and application of an *N*-acetyl-5-*N*,4-*O*-oxazolidinone α-tolylthio donor,[174] which demonstrated a higher reactivity when compared to phenylthio counterpart, activation at lower temperatures, successful sialylations with secondary acceptors, and moderate-to-good stereoselectivities in DCM/MeCN (2/1). This donor was then applied to the synthesis of tumor-associated antigen GM3.[175] Recently, Xing and coworkers reported a systematic investigation of sialylations by preactivating an *N*-acetyl 5-*N*,4-*O*-oxazolidinone α-tolylthio sialyl donor with diphenyl sulfoxide (Ph_2SO), triflic anhydride (Tf_2O) by itself or in the presence of a hindered base such as 2,4,6-tri-*tert*-butylpyrimidine (TTBPm) or 2,4,6-tri-*tert*-butylpyridine (TTBPy).[176] It was noted that sialylation outcomes were strongly dependent on the equivalents of promoter, preactivation, and reaction times, as well as the equivalents of TTBPm or TTBPy used. In general, a decrease of stereoselectivity was observed for all the acceptors tested upon an increase in diphenyl sulfoxide concentration, while good yields and stereoselectivity were recorded with an optimal concentration of hindered base between 0 and 1 equiv. (Scheme 45). Thus, when coupled with 2,3,4-tri-O-benzylated glucopyranosyl acceptor, the yield and stereoselectivity ranged from poor yield (41%), albeit with high α-stereoselectivity to quantitative, but with lower stereoselectivity (Scheme 45). More recently, the same group reported the first synthesis of an *N*-acetyl-5-*N*,4-*O*-oxazolidinone-protected sialyl sulfoxide (Scheme 45).[177] Its synthesis can be accomplished from the corresponding 5-*N*,4-*O*-oxazolidinone phenylthio precursor by the treatment with *m*-chloroperoxybenzoic acid (*m*-CPBA). Interestingly, this reaction was unsuccessful when applied to the common C-5-acetamido phenylthio donor. Nifantiev and coworkers reported a detailed investigation on this reaction which, based on their findings, generates a sulfone from a glycal precursor.[178] Thus, the successful synthesis of a sulfoxide from an oxazolidinone precursor might then be related to its notorious stability toward elimination. In order to test the novel sulfoxide sialyl donor, a series of sialyations were performed. Also, in this case, a lower stereoselectivity was observed in the system by increasing the equivalents of Ph_2SO. However, when different aromatic sulfoxides were tested, di-tolyl sulfoxide gave high yield and stereoselectivity. This, the coupling with a 2,3,4-tri-*O*-benzyl-glucosyl acceptor yielded the desired target in 73% yield ($\alpha/\beta = 16.7/1$). Even more stereoselective and

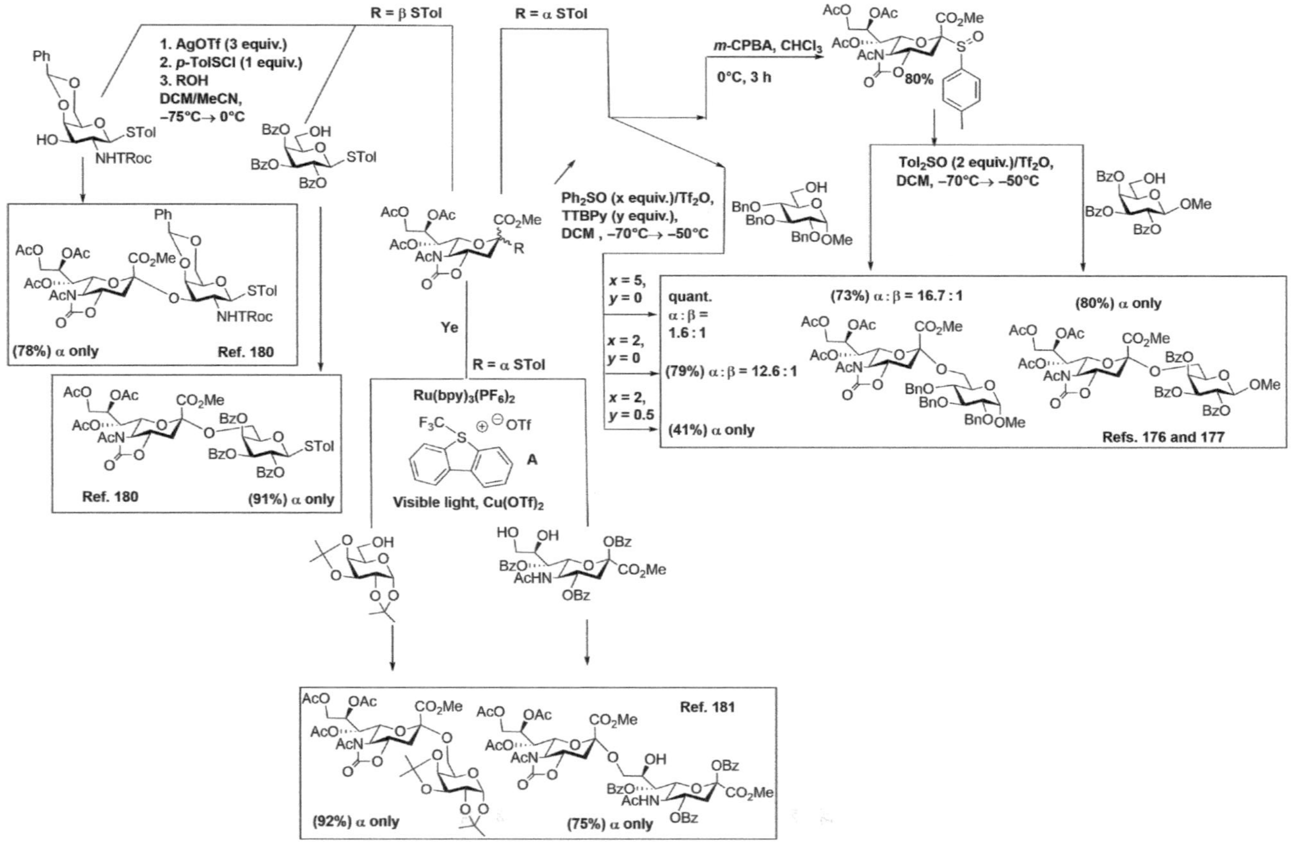

Scheme 45 Application of 5-*N*,4-*O*-oxazolidinone-phenylthio and sulfoxide sialyl donors by the preactivation method.

higher yielding was the coupling with a 2,3,4-tri-*O*-benzoyl-galactopyranosyl acceptor (80%, α only, Scheme 45). This methodology has been also expanded to α-C-sialylations.[179]

As an example of selective sialylations based on the preactivation of the donor, Sun and Jiang reported sialylation reactions between an *N*-acetyl-5-*N*,4-*O*-oxazolidinone β-tolylthio sialyl donor with several *S*-tolyl acceptors by preactivation with AgOTf and *p*-toluenesulfonyl chloride.[180] Excellent yields and complete stereoselectivities were obtained for almost all the acceptors tested, both primary and secondary. For example, sialylation of the STol donor with a benzylidene-protected thiogalactopyranosyl acceptor gave the desired α-sialylation product in 81% yield. Similarly, sialylation with a 2,3,4-tri-*O*-benzoyl *S*-tolyl donor gave the α-sialylation product in 51% yield with exclusive α-stereoselectivity (Scheme 45).[180]

An alternative promotion system for STol donor bearing an *N*-acetyl 5-*N*,4-*O*-oxazolidinone has been developed by Ye and coworkers.[181] Activation occurred by irradiation with blue light in the presence of $Ru(bpy)_3(PF_6)_2$ (catalyst), Umemoto's reagent (source of CF_3 radical, **A**, Scheme 45), and $Cu(OTf)_2$ (additive). Complete stereoselectivities and high yields were observed for some couplings as highlighted in Scheme 45. Interestingly, no activation of the corresponding *S*-adamantyl donor was noted, probably due to steric hindrance of the adamantyl group, creating a potentially useful group for future selective activation protocols.

Functionalization of C-5 as the 5-*N*,4-*O*-oxazolidinone or its acetylated counterpart has been heavily applied to the synthesis of α(2→9)-oligosialic acids, and many examples are described in several sections throughout this review. Takahashi and coworkers described an efficient method for the synthesis of a tetrasialic acid by coupling a 5-*N*,4-*O*-oxazolidinone *S*-phenyl donor with a sialyl 8,9-diol as acceptor (Scheme 46). Every synthetic step proceeded with high yields and sialylations with very high stereoselectivities.[170] A similar synthesis was reported later on by Lin and coworkers, using in this case an *S*-tolyl sialyl donor. Also, in this case excellent yields and stereoselectivities were obtained (Scheme 46).[182] Wong and coworkers' application of *N*-acetyl 5-*N*,4-*O*-oxazolidinone in phosphate sialyl donors for the programmable one-pot synthesis of several oligomers, including a α(2→9)-oligosialic acid dodecamer, has been described in Section 2.7 (see Scheme 55).[109,110]

Recently, Corcilius and Payne described the synthesis of sialyl Tn and 2,6-sialyl T cancer-associated antigens by the application of the Crich group's *N*-acetyl-5-*N*,4-*O*-oxazolidinone group and *S*-phenyl sialyl donor to obtain in this case complete stereoselectivity and excellent yields (Scheme 47). It should be noted that the conversion of the oxazolidinone

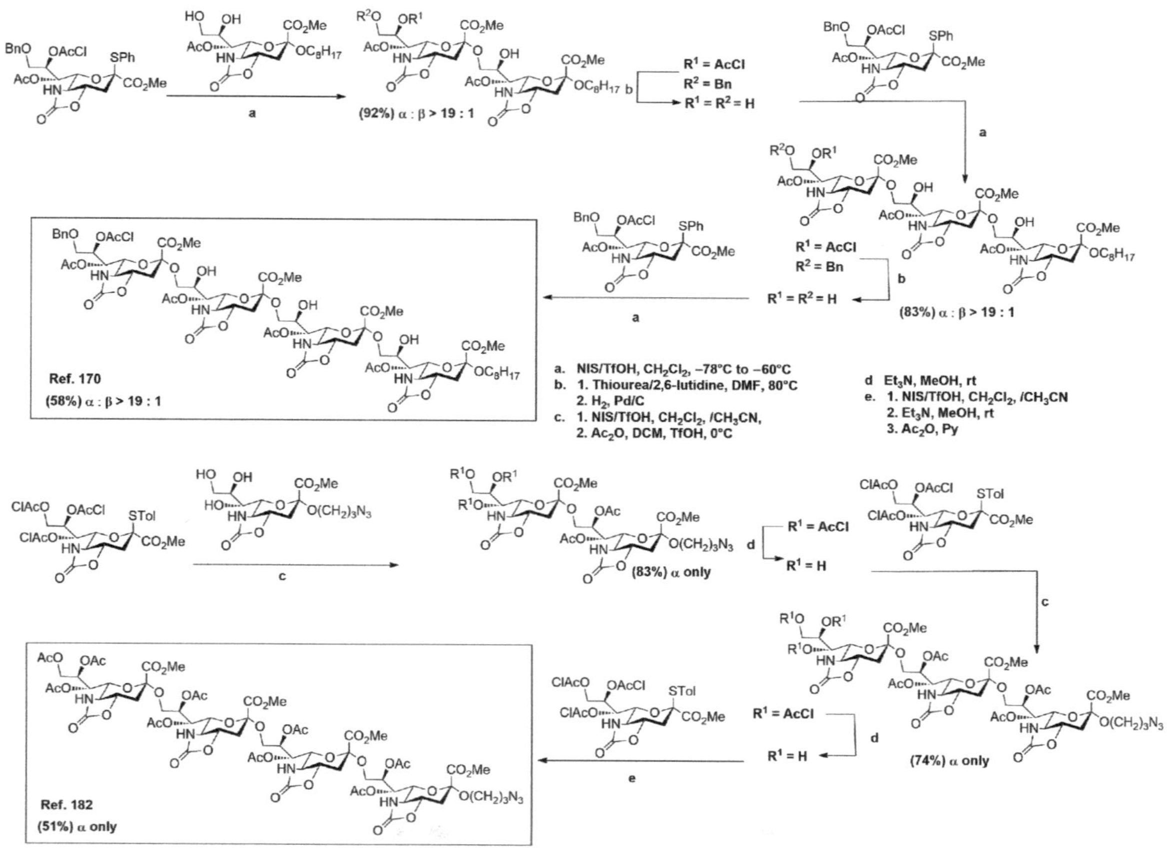

Scheme 46 Application of 5-*N*,4-*O*-oxazolidinone for the synthesis of an α(2→9)-tetramer.

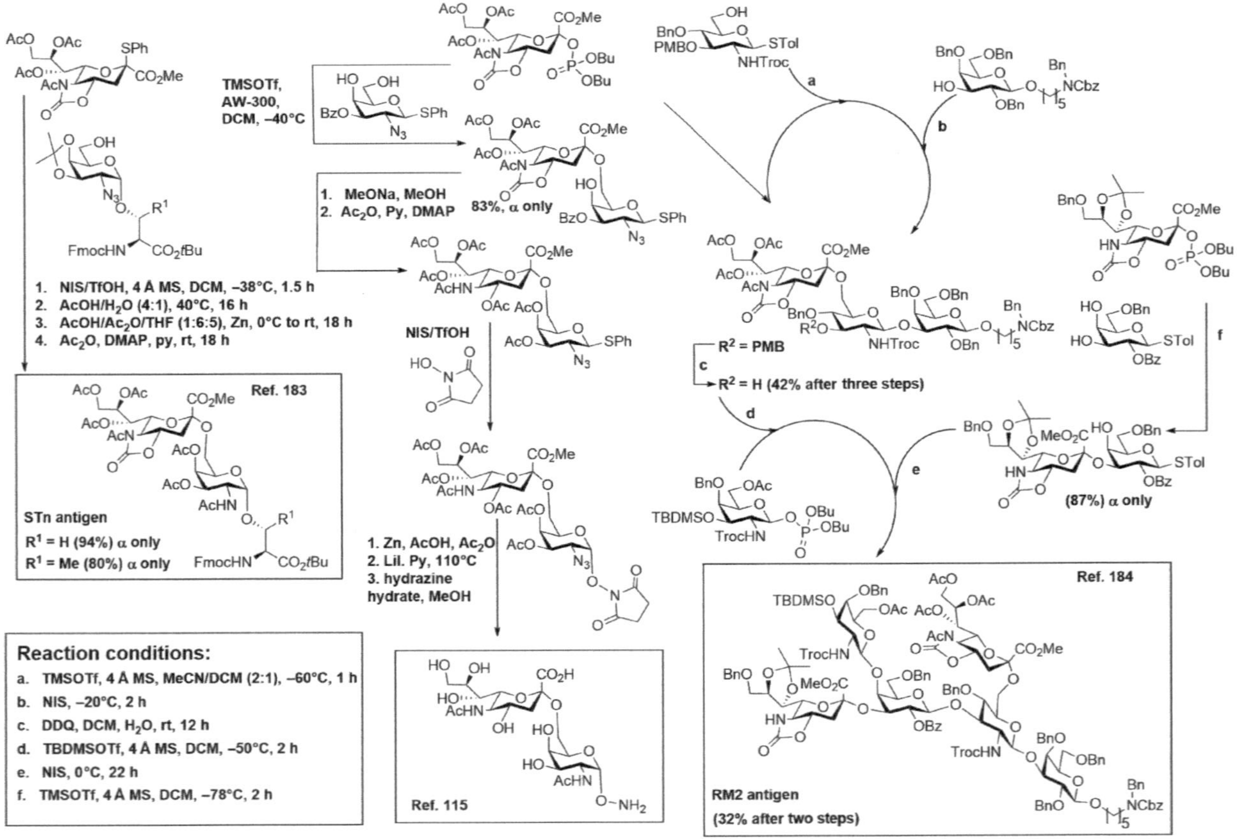

Scheme 47 5-*N*,4-*O*-Oxazolidinone-protected donors for the synthesis of STn and RM2 antigens.

ring to the natural acetamido group was performed following the solid-phase assembly of the glycopeptides, demonstrating the versatility of this C-5 modification.[183]

More recently, Andreana and coworkers reported the use of an *N*-acetyl-5-*N*,4-*O*-oxazolidinone-protected sialyl phosphate donor in the synthesis of α-aminoxy STn antigen (Scheme 47).[115] The coupling with a 6-OH galactosyl acceptor led to a good yield and complete α-stereoselectivity of the disaccharide. Conversion of the oxazolidinone group into the acetamido function was followed by coupling with *N*-hydroxysuccinimide to introduce the α-aminoxy (O–NH_2) arm precursor. The α-aminoxy STn antigen was obtained by sequential conversion of the azide into the acetamido group, methyl ester removal, and removal of *N*-hydroxysuccinimide with concomitant global deprotection (Scheme 47).

5-*N*,4-*O*-Oxazolidinone and its N-acetylated version have been applied by Wong in the first synthesis of tumor-associated antigen RM2.[184] To perform an efficient coupling with both primary and secondary acceptors, the combination of oxazolidinone at C-5 and dibutyl phosphate as the leaving group was chosen. The synthetic strategy involved a [1+2+3] one-pot sequential methodology as previously applied by the same research group.[109] Thus, one-pot sialylation between *N*-acetyl-5-*N*,4-*O*-oxazolidinone-protected dibutyl phosphate donor and glucosamine acceptor was performed by activation with TMSOTf at −60°C. Addition of NIS and rise of the temperature to −20°C allowed for the additional installment of the galactopyranosyl moiety. Removal of the PMB group gave a trisaccharide in 42% yield (Scheme 47). For the synthesis of the α(2→3)-linkage with galactopyranosyl acceptor, the authors applied a 8,9-*O*-isopropylidene-5-*N*,4-*O*-oxazolidinone-protected dibutyl phosphate donor.[185] This sialylation proved to be also very successful in terms of both yields and stereoselectivity, affording the desired target in 87% yield as only α-anomer. The subsequent glycosylation with a galactopyranosyl donor in one pot allowed the RM2 hexasaccharide in high overall stereoselectivity and in a minimal number of steps.

Noteworthy is the application of 5-*N*,4-*O*-oxazolidinone-protected dibutyl phosphate for the successful synthesis of the trisaccharide moiety of ganglioside HLG-2, which is characterized by an α(2→4) dimeric linkage. Also in that case, complete α-stereoselectivity was reported.[113]

A modification of the oxazolidinone moiety as *N*-*tert*-butoxycarbonyl (Boc) 5-*N*,4-*O*-oxazolidinone has been reported by Boltje and coworkers in order to allow a straightforward conversion of the oxazolidinone ring into different N-5-functionalized derivatives.[186] The introduction of a Boc group can be accomplished with the corresponding anhydride in high yield. Sialylation with an *N*-Boc-modified oxazolidinone ring proceeded in high yield and stereoselectivity, typical for all oxazolidinone modifications (Scheme 48). Upon sialyation, the removal of the oxazolidinone ring to afford a deprotected amine occurred with no cleavage of the glycosidic bond or the isopropylidene group. The crude amine was then functionalized as either the *N*-acetyl or the *N*-glycolyl derivative. This strategy was applied to the synthesis of a fucose-bearing tetrasaccharide without compromising the fucosyl linkage as shown in Scheme 48.

Buda and Crich reported a successful and extremely versatile, stereospecific, and chemoselective deamination procedure that allows the conversion of the oxazolidinone and acetamido moieties into differentially functionalized sialic acids at C-5.[187] An oxidative deamination of peracetylated *N*-acetylneuraminic acid was also described by Schreiner and Zbiral in the early 1990s.[147] The synthetic breakthrough of this strategy is the ability to obtain derivatives of sialic acid that are not currently accessible in the field, such as mercapto analogs at C-5. As described in Scheme 49, upon the conversion of 5-*N*-acetyl and 5-*N*,4-*O*-oxazolidinone into the standard 5-acetamido derivative, the addition of nitrosyl tetrafluoroborate, trifluoroethanol (TFE), and sodium 2-propoxide in 2-propanol yields the corresponding 5-diazo derivative. Nucleophilic displacement can then be performed with several nucleophiles (Scheme 49). Reaction can be carried out on either monosaccharides or disaccharides without compromising the glycosidic linkage. Thus, this strategy provides an indirect route to the synthesis of Kdn derivatives (Scheme 49).

2.6.2 Miscellaneous Cyclic Substituents

In order to explore alternative cyclic groups, Crich and Wu synthesized and tested a β-phenylthio donor bearing a 5-*N*,7-*O*-oxazinanone ring. The synthesis was accomplished as depicted in Scheme 50. Selective protection of the amine at C-5 with Fmoc was followed by the introduction of the 8,9-*O*-isopropylidene group and the 4-*O*-TBDMS group. Removal of the *N*-Fmoc group by piperidine and subsequent reaction with 1,1′-carbonyldiimidazole in THF gave the desired target in 66% yield. Unexpectedly, the activation of this donor failed to occur with common promoters such as NIS/TfOH and BPS/TTBP/Tf_2O.[188] Instead it required a more powerful promoter such as Ph_2SO/Tf_2O.[32,145] When coupled with various acceptors, the 5-*N*,7-*O*-oxazinanone-protected sialyl donor yielded

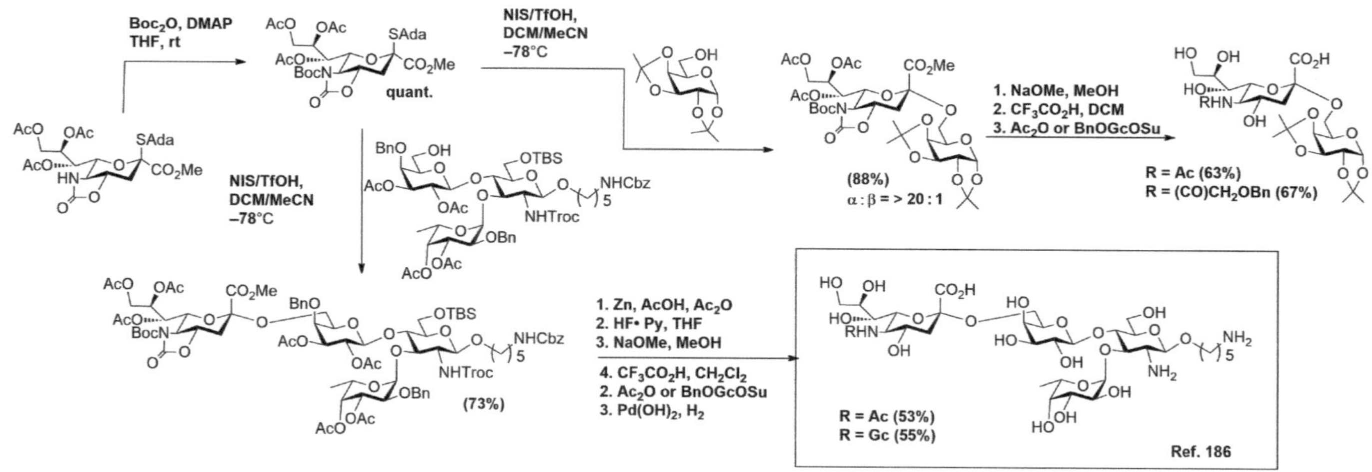

Scheme 48 The synthesis of *N*-Boc-5-*N*,4-*O*-oxazolidinone-protected donors, their application, and deprotection.

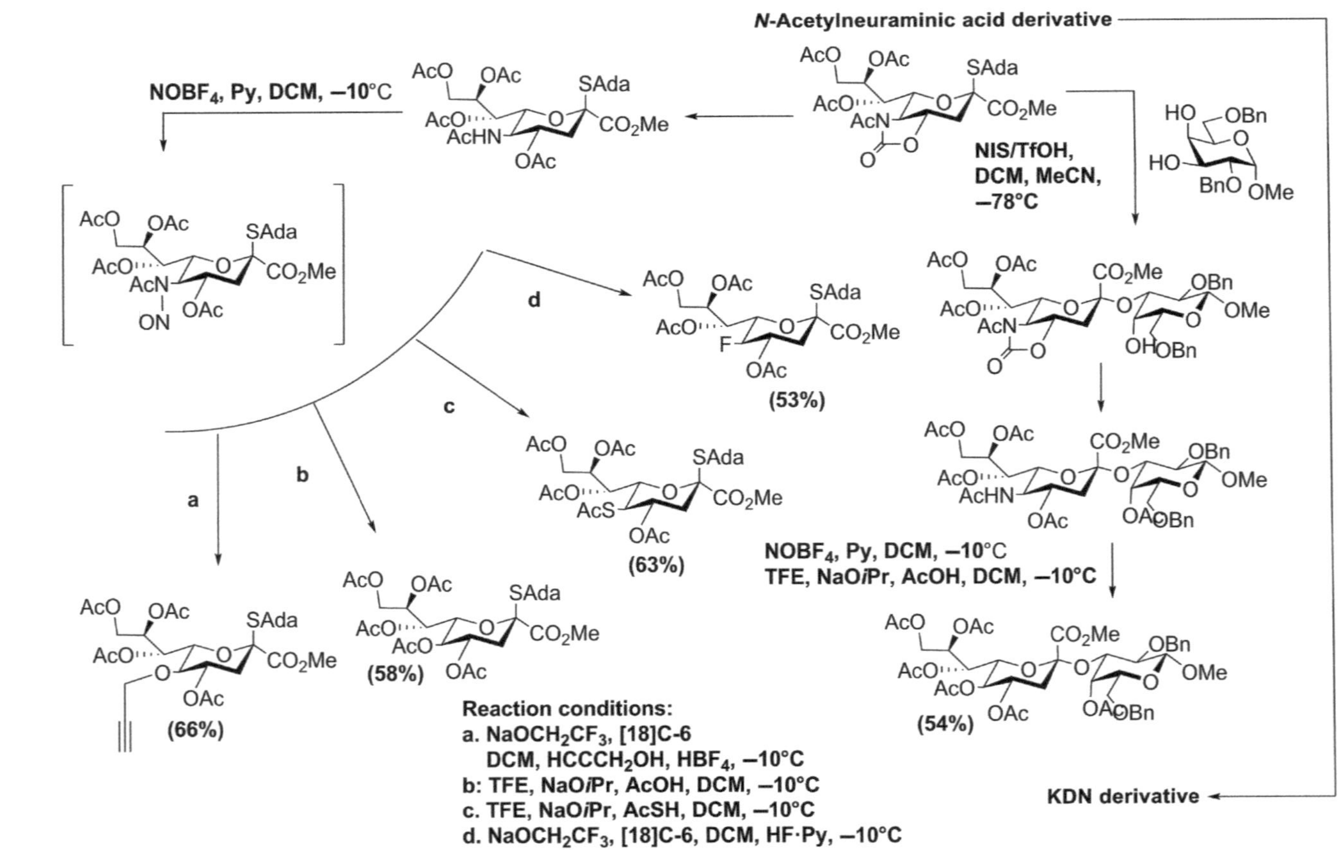

Scheme 49 Oxidative deamination: a straightforward functionalization of the C-5 position.

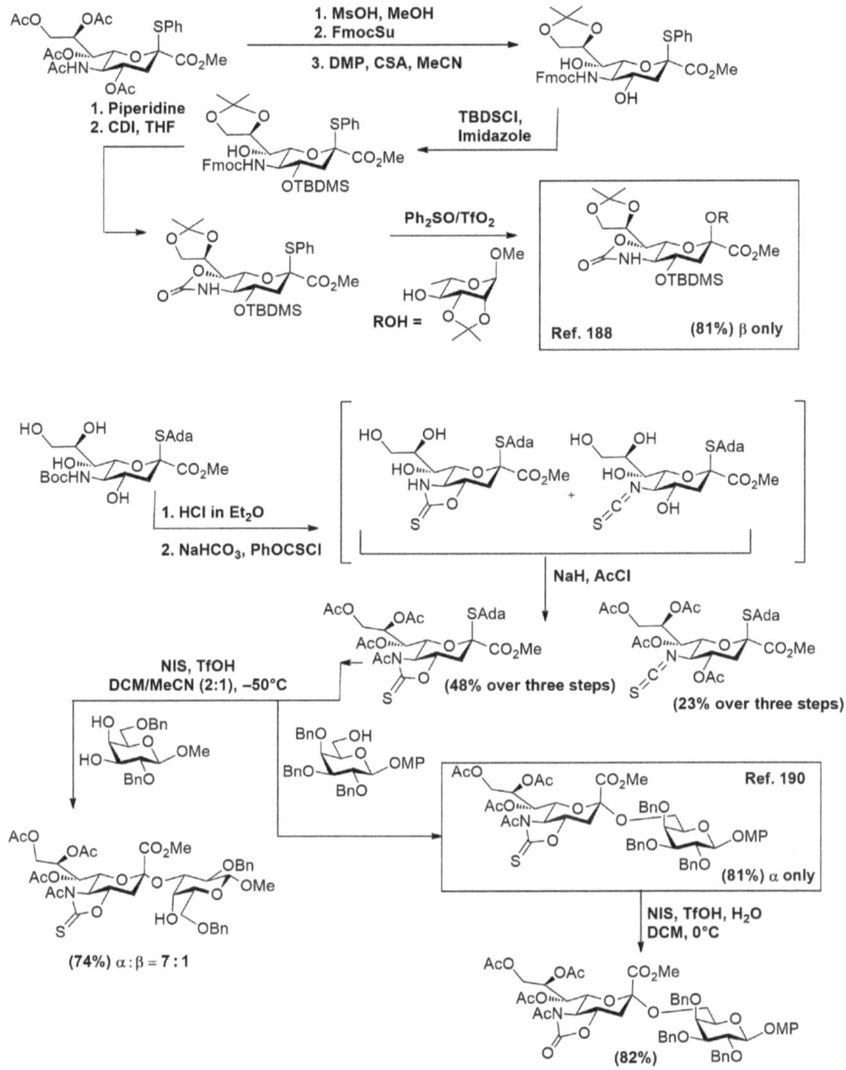

Scheme 50 Studies on oxazinanone and oxazolidinthione effects in sialylations.

the corresponding products as the β-anomers, only (Scheme 50). The preference of this system for the unnatural β-anomer might be due to the 5,7-oxazinanone ring and/or due to the promoter system used. Regardless, a powerful disarming effect of this cyclic system is clear, which can be due to the electron-withdrawing propriety of the ring, as well as the imposition of a *trans–gauche* conformation of the C-6–C-7 bond, as proposed by Crich.

The disarming effect of the 5-*N*,7-*O*-oxazinanone modification was also confirmed by Koketsu and coworkers by the synthesis and testing of a per-acetylated 5-*N*,7-*O*-oxazinanone α-phenylthio donor.[189]

The synthesis of a 5-*N*,7-*O*-oxazolidinthione was also reported by Rajender and Crich (Scheme 50).[190] Removal of the Boc group at the N-5 position, followed by treatment with phenyl thionochloroformate and acetylation of the residue, gave a mixture of the desired oxazolidinthione with the isothiocyanate as by-product. Attempts to activate the novel donor under the standard reaction conditions failed, showing a disarming effect of this *trans*-fused ring. Ultimately, activation was achieved by NIS/TfOH at −50°C. Thus, good yields and stereoselectivities were obtained, which were lower in the same cases than those from the corresponding oxazolidinone group, probably due to the need of working at higher temperatures. On the other hand, the oxazolidinthione modification proved to be stable in the presence of MeONa/MeOH and could be easily converted into the oxazolidinone group by the treatment with NIS/TfOH in the presence of water.

Hanashima et al. reported the synthesis of silylene–oxazolidinone-protected "double-locked" phenylthio donors (Scheme 51). From a 5-*N*,4-*O*-oxazolidinone precursor it was possible to introduce a di-*tert*-butylsiloxanylidene (DTBS) at positions N-5/O-7 and O-8/O-9 by treatment with di-*tert*-butylsilyl bis(trifluoromethanesulfonate) ($t\mathrm{Bu_2Si(OTf)_2}$) in one step (**A**, Scheme 51). On the other hand, the reaction with 1,3-dichloro-1,1,3,3-tetraisopropyldisiloxane ($\mathrm{TIPDSiCl_2}$) allowed the regioselective introduction of a TIPDS group at positions O-8/O-9. Addition of (a) acetic anhydride for the synthesis of **B** or (b) $\mathrm{TBu_2Si(OTf)_2}$ for the synthesis of **C** gave additional donors that were tested in subsequent experiments. For the coupling with primary acceptors, the double-locked donors gave higher yields and stereoselectivities (Scheme 51), which were similar to those reported from simple oxazolidinone systems. Coupling with secondary acceptors for the synthesis of α(2→3) linkages proceeded with stereoselectivities ranging from complete α- to mainly β-, based on the nature of the acceptor (Scheme 51). This strategy was applied by the same group to synthesize starfish ganglioside AG2 pentasaccharide.[191]

2.7 Modifications at the Glycerol Chain

2.7.1 O-Benzoyl (O-Bz) Derivatives

While investigating the effects of various C-5 modifications on chemical glycosylations, Wang and Ye reported a strong β-directing effect of

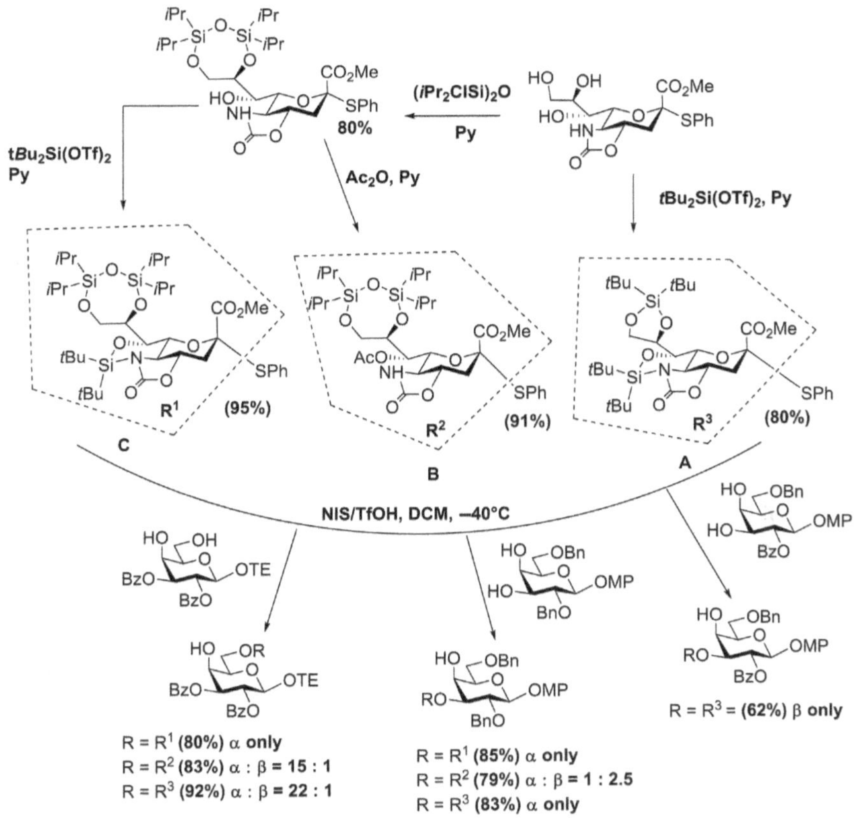

Scheme 51 Studies of the silylene–oxazolidinone effects in sialylations.

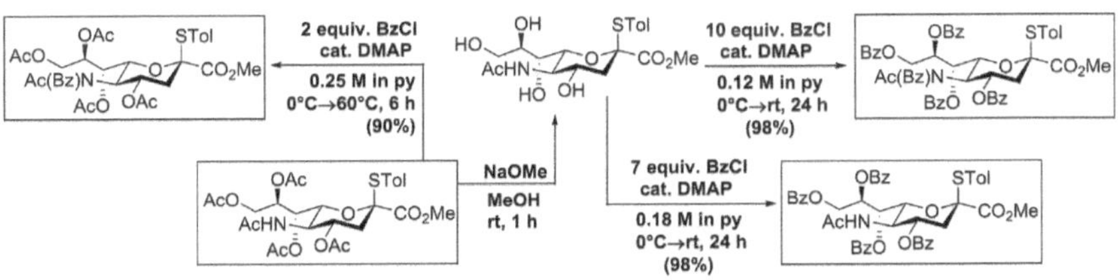

Scheme 52 Synthesis of various O-benzoylated sialyl donors.

per-O-benzoylated-*N*-acetylbenzamido derivative of Neu5Ac.[135,192] In an effort to distinguish the influence of the 5-*N*-acetylbenzamido modification from the influence of per-O-benzoylation, De Meo and coworkers synthesized a series of complementary sialyl donors (Scheme 52).[135] The original per-O-benzoylated *N*-acetylbenzamido derivative was compared with the per-O-acetylated-*N*-acetylbenzamido derivative, as well as the per-O-acetylated and per-O-benzoylated derivatives of Neu5Ac, bearing the standard 5-acetamido group. The per-O-acetylated *N*-acetylbenzamido derivative can be synthesized in high yield (90%) from a common per-O-acetylated Neu5Ac thioglycoside by reaction with benzoyl chloride and 4-DMAP.[135] The fully unprotected 5-acetamido derivative can be used to obtain both the per-O-benzoylated acetamido and per-O-benzoylated *N*-acetylbenzamido derivatives when reacted with benzoyl chloride in different concentrations.[135]

Both the per-O-benzoylated and per-O-acetylated variations of the 5-acetamido and *N*-acetylbenzamido derivatives were promoted as glycosyl donors with NIS/TfOH in the presence of the common 1,2:3,4-di-*O*-isopropylidene-α-D-galactopyranose and methyl 2,3,4-tri-*O*-benzyl-α-D-galactopyranoside primary acceptors. As can be seen in Table 7, when O-protecting groups were switched from acetyl to benzoyl groups in the standard 5-acetamido donors, the subsequent glycosylations with both acceptors resulted in a loss of yield and a shift toward β-stereoselectivity. Interestingly, the same change from *O*-acetyl to *O*-benzoyl protecting groups in the *N*-acetylbenzamido donors resulted in a slightly higher glycosylation yield, but again stereoselectivity has suffered (Table 7).[135]

The results of this comparative study suggest that the initial β-directing effect observed with the per-O-benzoylated *N*-acetylbenzamido derivative is likely to be the result of combined effects from both the *N*-benzamido

Table 7 Studies on the Effect of *O*-Benzoyl Groups in Sialylations

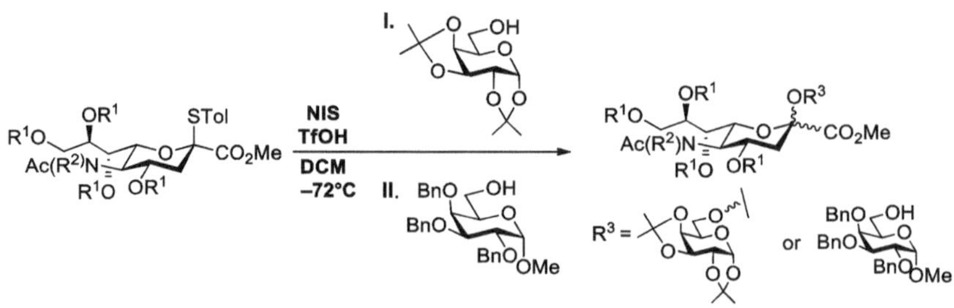

Acceptor	R¹	R²	Yield	α : β Ratio	Acceptor	R¹	R²	Yield (%)	α : β Ratio
I	Ac	H	96%	1.5 : 1	II	Ac	H	99	1 : 4
I	Bz	H	63%	1 : 1.4	II	Bz	H	64	1 : 6
I	Bz	Bz	86%	1 : 6.7	II	Bz	Bz	90	β only
I	Ac	Bz	62%	2 : 1	II	Ac	Bz	86	1.1 : 1

substitution and the per-*O*-benzoyl protection. It also should be noted that steric strain associated with both the per-O-benzoylated and per-O-acetylated *N*-acetylbenzamido derivatives results in hindered rotation about the C-5–N bond. Thus, two distinct sets of NMR signals can be observed for the *N*-acetylbenzamido derivatives at a temperature of −40°C, corresponding to the two distinct rotamers.[135]

2.7.2 O-Chloroacetyl (O-CA or O-ClAc) Derivatives

The chloroacetyl functionality has seen a surge in popularity as an O-protecting group in recent years, in particular for being selectively removed in the presence of other protecting groups such as acetyl and benzyl. The *O*-chloroacetyl protection can be accomplished by the coupling of chloroacetic acid in the presence of *N,N′*-dicyclohexylcarbodiimide (DCC). The cloroacetyl group can be cleaved by treatment with thiourea and 2,6-lutidine in DMF (Scheme 53).[158,185]

In combination with certain C-5 modifications, *O*-chloroacetyl sialyl donors have been used to produce some excellent glycosylation results, as seen in many sections of this review. In combination with the cyclic oxazolidinone functional group, several complex sialic acid-containing oligosaccharides have been synthesized. For example, Takahashi reported the application of 5-*N*,4-*O*-oxazolidinone with C-9-*O*-benzyl and C-7,8-*O*-chloroacetyl protecting groups in the glycerol chain to furnish (2→8)-linked sialic acid dimers and trimers prior to the assembly of a complex GP1c ganglioside epitope.[185] However, stereoselectivity did suffer in certain glycosylation steps.

Guo and coworkers used an oxazolidinone-protected 7,8,9-O-chloroacetylated sialyl phosphate donor to obtain an α(2→3)-linkage in 64% yield with a selectively protected galactopyranosyl acceptor.[114] Guo and coworkers also described an efficient method of synthesizing α(2→9)-linked

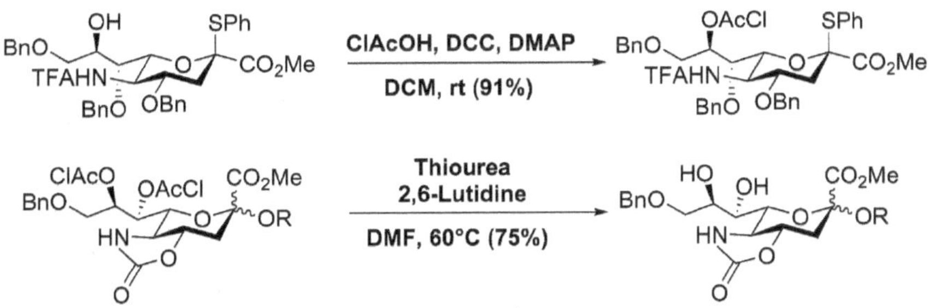

Scheme 53 Synthesis and deprotection of O-chloroacetylated sialyl donors.

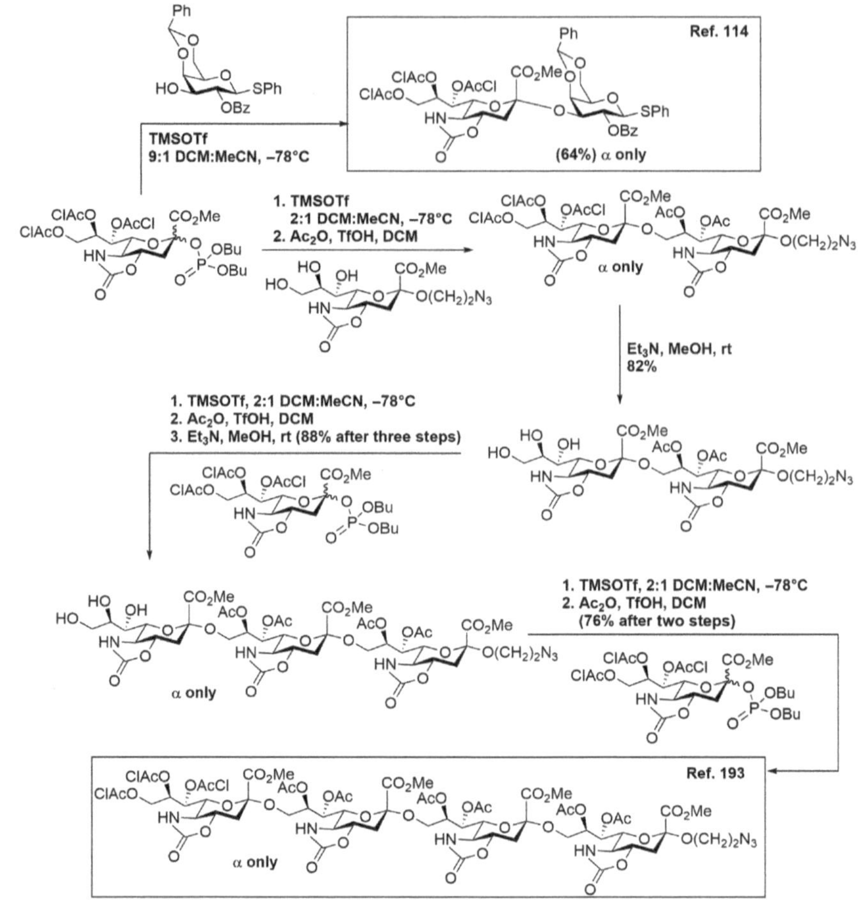

Scheme 54 Application of *O*-chloroacetyl groups in sialylations.

oligosialic acids using the same sialyl phosphate donor. By activation with TMSOTf in a mixture of acetonitrile/dichloromethane at low temperature, and with subsequent transformation of the glycosylation product into an acceptor, a (2→9)-tetrasialic acid was produced having only α-glycosidic linkages (Scheme 54).[193]

Perhaps one of the best examples of the application of the chloroacetyl group for the synthesis of oligosaccharides is the synthesis of the α(2→9) dodecamer reported by Wong and coworkers (the synthesis of precursor octamer is outlined in Scheme 55).[110] In an iterative sequence of selective sialylations, selective removal of the chloroacetyl group, acetylations, and conversion of thioglycosides into phosphate leaving groups, it was possible to synthesize the desired target in high yield and with high stereoselectivities.

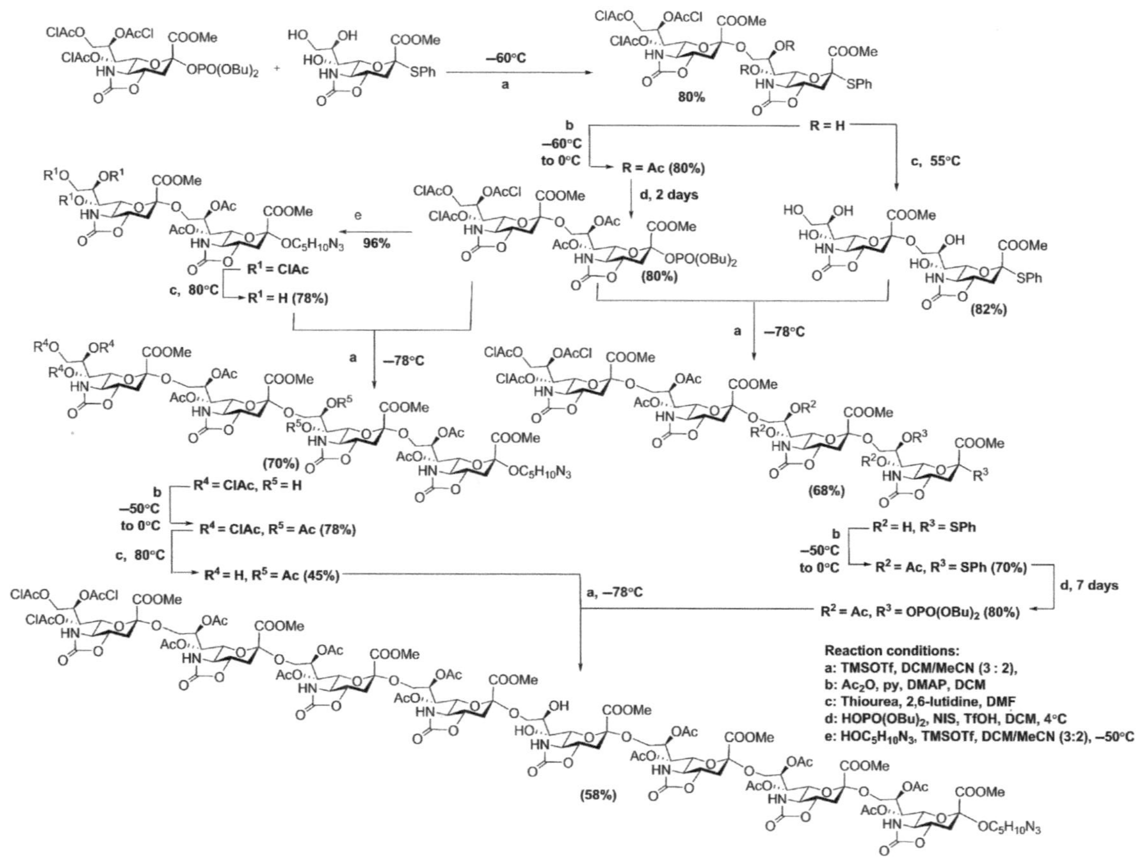

Scheme 55 Wong and coworkers' synthesis of an $\alpha(2\rightarrow9)$ octamer as the precursor of a dodecamer.

2.7.3 O-Trifluoroacetyl (O-TFA) Derivatives

Recently, Kononov and coworkers investigated the effects on reactivity and stereoselectivity of several O-protecting groups in C-5-trifluoroacetamido donors. Acetyl, chloroacetyl, trifluoroacetyl, and benzoyl O-protecting groups were used in a comparative study by activation with NIS/TfOH in acetonitrile at −40°C in the presence of a common 1,2:3,4-di-*O*-isopropylidene-α-D-galactopyranosyl acceptor.[194] The introduction of *O*-trifluoroacetyl groups was accomplished by the treatment of the sialic acid 4,7,8,9-tetraol precursor with trifluoroacetic anhydride and catalytic sodium trifluoroacetate.[195] Sialylation outcomes are shown in Table 8. Thus, both the yield and stereochemical outcome of the sialylation reactions changed dramatically as a consequence of the O-protecting group variation. The highest yield (86%) was obtained from the per-O-chloroacetylated sialyl donor (Entry 3), whereas the lowest yield (38%) was obtained from the per-O-trifluoroacetylated donor (Entry 4). α-Stereoselectivity was maximized (α/β = 15.5/1) with the use of a chloroacetyl group at the C-4 position and protection of the glycerol chain with trifluoroacetyl groups (Entry 7), suggesting an important synergistic effect that can be applied to

Table 8 Effects in Varying O-Protecting Groups on Sialylation Results

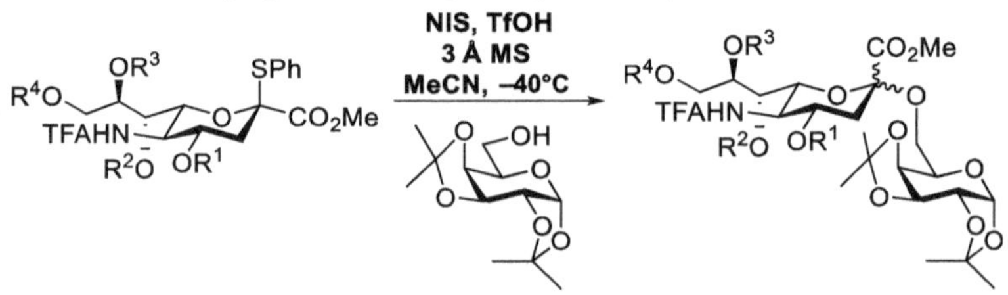

Entry	R^1 (C-4)	R^2 (C-7)	R^3 (C-8)	R^4 (C-9)	Yield (%)	α : β Ratio
1	Ac	Ac	Ac	Ac	82	2 : 1
2	Bz	Bz	Bz	Bz	61	4.6 : 1
3	ClAc	ClAc	ClAc	ClAc	86	10.9 : 1
4	TFA	TFA	TFA	TFA	38	9.8 : 1
5	Ac	TFA	TFA	TFA	59	7.2 : 1
6	Bz	TFA	TFA	TFA	49	6.3 : 1
7	ClAc	TFA	TFA	TFA	55	15.5 : 1

optimizing building block strategies in sialylation (Table 8).[194] Additional discussion of this work can be found in Section 3.

2.7.4 O-Picoloyl (O-Pico) Derivatives

Use of the picoloyl substituent has been extended beyond the C-4 position[140] to the glycerol chain as well.[196] Wu and Tsai synthesized a series of 2-phenylthio Neu5Ac donors with a mixed *O*-benzoyl and *O*-picoloyl protecting group pattern at the C-4, -7, -8, and -9 positions. These donors first were coupled to methyl 2,3,4-tri-*O*-benzyl-α-D-glucopyranosyl acceptor using NIS and an excess of TfOH in dichloromethane, and the results were compared with those obtained using a per-O-benzoylated 2-phenylthio Neu5Ac donor. Yields and stereoselectivities of these sialylation reactions varied greatly, but 70% of the α(2→6) disaccharide was obtained as the exclusive product using the 4,9-*O*-benzoyl/7,8-*O*-picoloyl-protected donor derivative (Table 9).

Wu and Tsai also tested the efficacy of the 4,9-*O*-benzoyl/7,8-*O*-picoloyl-protected donor in this glycosylation system with several other glycosyl acceptors. Most primary acceptors afforded moderate to excellent yields of the α-linked glycosylation products with exclusive stereoselectivity. However, an attempted glycosylation with a more sterically

Table 9 Effects of the *O*-Picoloyl Substituent on Chemical Sialylations at Various Positions

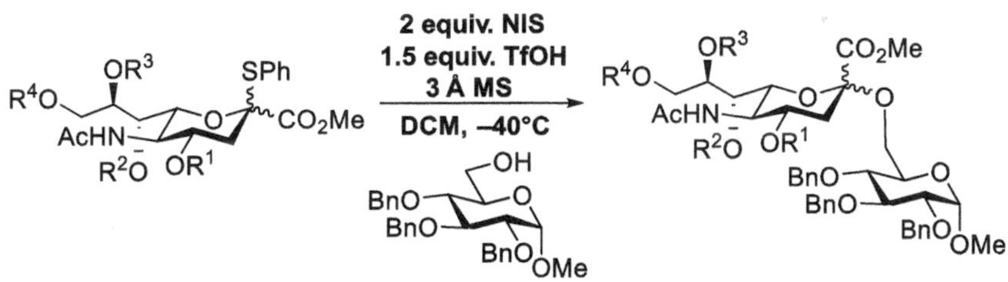

Entry	R^1 (C-4)	R^2 (C-7)	R^3 (C-8)	R^4 (C-9)	Yield (%)	α : β Ratio
1	Bz	Bz	Bz	Bz	90	1 : 3.3
2	Bz	Pico	Bz	Bz	57	1 : 1.2
3	Bz	Bz	Pico	Bz	62	8.0 : 1
4	Pico	Bz	Bz	Pico	86	1.3 : 1
5	Bz	Pico	Pico	Bz	70	α only

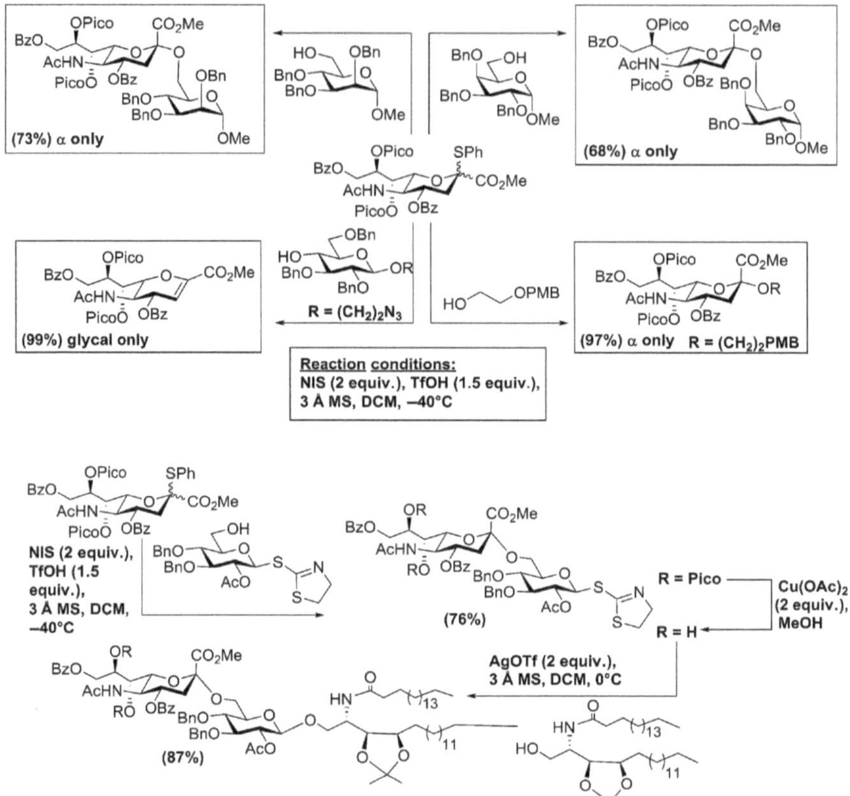

Scheme 56 Synthesis and testing of a 7,8-*O*-picoloyl sialyl donor and its application in the synthesis of ganglioside Hp-s1.

hindered secondary acceptor produced the 2,3-dehydro derivative instead (Scheme 56). This methodology was then successfully applied to the synthesis of ganglioside Hp-s1 (Scheme 56).[196] Selective sialylation of the galactopyranosyl acceptor equipped with the anomeric *S*-thiazolinyl (STaz) group yielded the corresponding disaccharide in 76% yield with complete α-selectivity. Interestingly, the subsequent glycosylation of a protected phytoceramide acceptor occurred in high yield only upon removal of the picoloyl group, affording the protected target in 86% yield (Scheme 56). More recently, De Meo and coworkers reported the synthesis of a series of Neu5Ac donors bearing mixed *O*-acetyl and *O*-picoloyl in several positions[197]; however, glycosylation results acquired with these donors have not yet been reported.

3. MECHANISTIC STUDIES OF O-SIALYLATIONS

Within the area of understanding the mechanism of chemical sialylation reactions, and especially pertaining to the influence of 5-N-protecting groups thereupon, a significant amount of work has been undertaken. In 1998, during the synthesis of a Neu5Acα(2→3)Galβ(1→3)GlcN trisaccharide from a Neu5Acα(2→3)Gal methylthio donor and a glucosaminyl acceptor in the presence of methyl triflate, the desired linkage was accomplished, but the nucleophilic 5-acetamido moiety became methylated. The modified Neu5-Ac_2α(2→3)Gal methylthio donor bearing the *N*-acetylacetamido moiety at C-5′ produced the desired linkage in higher yield and with no methylated product. Demchenko and Boons postulated that the higher reactivity of the Neu5Ac_2 donor may be associated with the reduced nucleophilicity (and the enhanced electron-withdrawing nature) of the *N*-acetylacetamido group relative to the standard acetamido group (Scheme 57).[141]

Additional evidence supporting the enhanced nucleophilicity of the glycosyl acetamido group has been produced by Auzanneau and coworkers. Activation of glycosyl trichloroacetimidate donors in the presence of glucosaminyl acceptors resulted in the capture of the activated donor in the form of glycosyl imidates.[198,199] When attempting the (1→4) glycosylation of a Galβ(1→3)GlcNAc acceptor with a mannosyl trichloroacetimidate donor using triethylsilyl triflate, the glycosyl imidate product was formed in 86% yield. Conversion of the glucosaminyl acetamido group to an *N*-acetylacetamido group produced the desired (1→4)-glycosidic linkage in high yield (95%, Scheme 58).

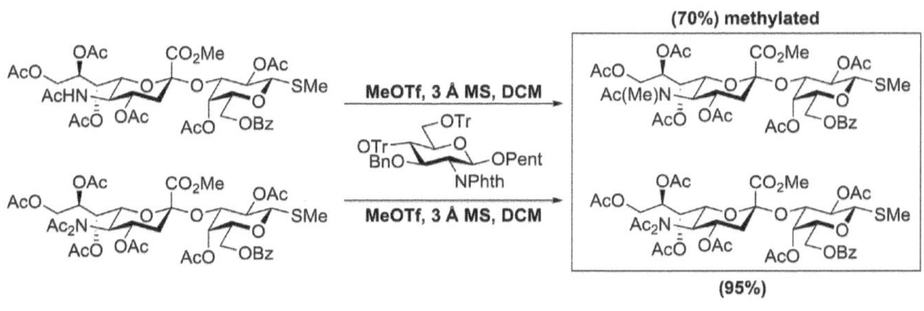

Scheme 57 Initial explanation of the C-5 effect by reduced nucleophilicity at the 5-N position.

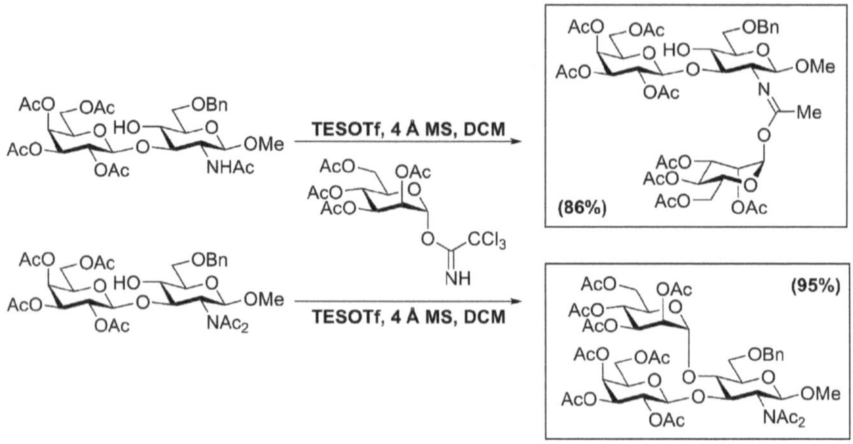

Scheme 58 Synthesis of a glycosyl imidate.

Kononov and coworkers have suggested that the increased reactivity of sialyl donors equipped with highly electron-withdrawing 5-N-protecting groups is attributed to their influence on the intermolecular hydrogen-bond network formed between sugars in solution. When promoted with NIS/TfOH in acetonitrile at −40°C in the presence of a galactosyl diol acceptor, the standard per-O-acetylated β-phenylthio Neu5Ac donor produced a mixture of the (2→3)- and (2→4)-linked products with a maximum yield of about 40% over the course of three hours. The same acceptor and glycosylation conditions applied to the equivalent Neu5Ac$_2$ donor produced the (2→3)- and (2→4)-linked products in about 50% yield after only 5 min. Interestingly, however, when equal parts of Neu5Ac and Neu5Ac$_2$ donors were promoted in the presence of the diol acceptor, the yields of both sets of (2→3)- and (2→4)-linked products were roughly the same, but both donors were almost completely consumed in only 5 min. It was suggested that since the standard 5-acetamido group can act as both a hydrogen-bond donor and an acceptor, supramolecular aggregates of carbohydrates can form. As a result, aggregated Neu5Ac glycosyl donor will show a lower reactivity profile relative to the unbound Neu5Ac. The *N*-acetylacetamido group has two hydrogen-bond acceptor sites, but no hydrogen-bond donor sites. Thus, when both Neu5Ac and Neu5Ac$_2$ donors are promoted in the same glycosylation system, smaller or weaker supramolecular aggregates are formed from the disruption of the hydrogen-bond network. The result of this is a much more efficient activation of the Neu5Ac donor in the presence of Neu5Ac$_2$ than the activation of the Neu5Ac donor alone. These findings were also corroborated with variable temperature infrared spectroscopy experiments (Scheme 59).[200]

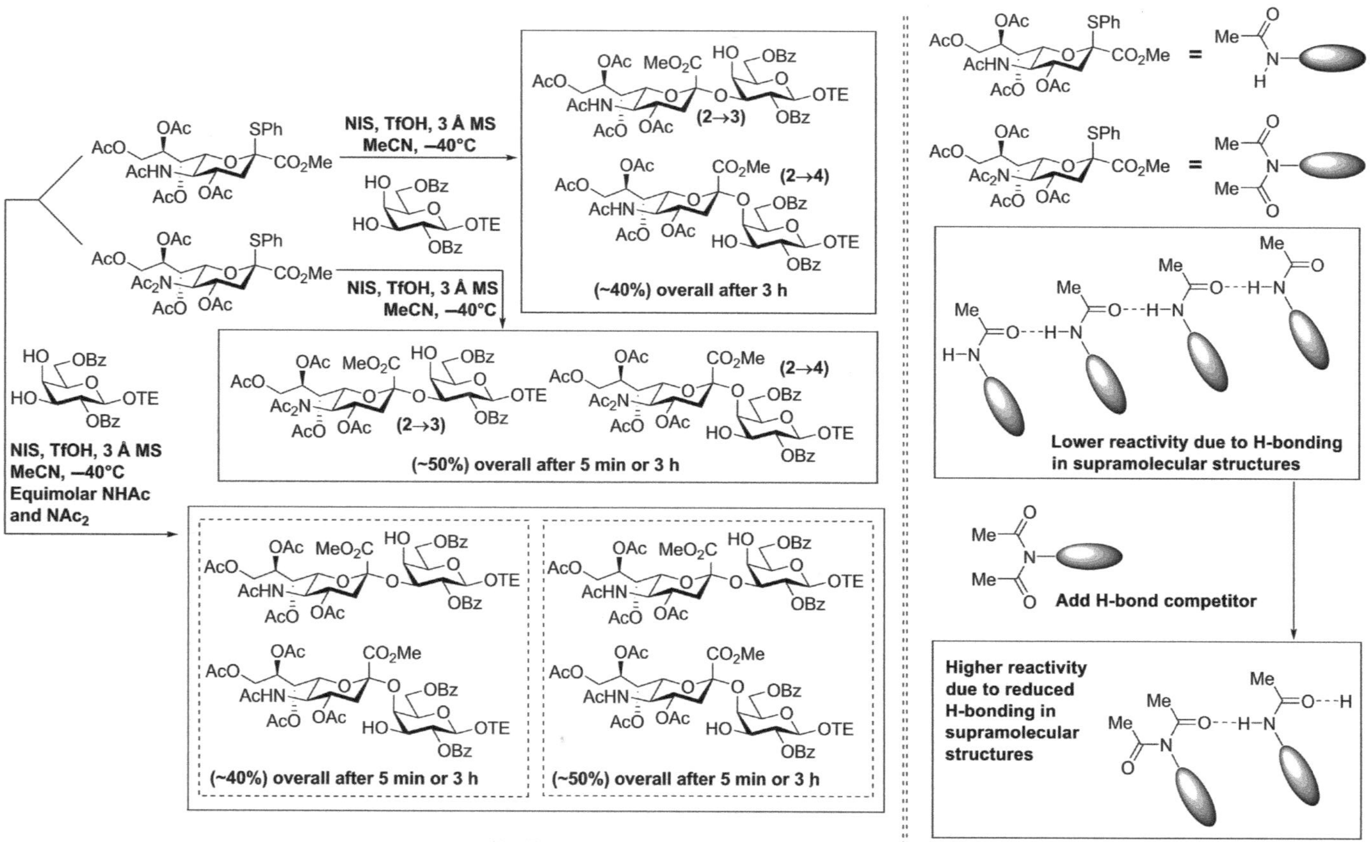

Scheme 59 Supramolecular approach to chemical sialylations.

Kononov and coworkers went on to explore the effects of external amides used as additives in the presence of the standard per-O-acetylated β-phenylthio Neu5Ac donor.[143] Numerous additives were tested at various concentrations in the glycosylation of 1,2:3,4-di-*O*-isopropylidene-α-D-galactopyranose. The additives varied in number and orientation of hydrogen-bond acceptors (carbonyl groups), and their effects on the reactivity of the same donor promoted under the same conditions were profound. An abbreviated insight into these effects is summarized in Table 10, where the influences of *N*,*N*-dimethylformamide and *N*,*N*-dimethylacetamide are compared. It should be noted that the authors ruled out the possibility of these amides forming anomeric imidates as reactive intermediates, as well as the influence of TfOH concentration on the glycosylation outcome.

When promoted in the presence of the isopropylidene-protected acceptor using NIS/TfOH in acetonitrile at −40°C, the target disaccharide was produced in 69% yield. When the same glycosylation system was promoted with 1 equiv. of *N*,*N*-dimethylformamide or *N*,*N*-dimethylacetamide per equivalent of the glycosyl donor, very little effect on the donor reactivity (as determined from the glycosylation product yield) was observed. The addition of 3 equiv. of *N*,*N*-dimethylformamide or *N*,*N*-dimethylacetamide, however, caused a noteworthy loss in donor reactivity. Kononov and coworkers attributed this to the disruption of the hydrogen-bond network formed by sugars in solution in favor of more tightly held and less reactive supramolecular aggregates in the presence of these additives.[143]

Infrared spectroscopic measurements of the Neu5Ac ester and amide carbonyl moieties corroborated these findings. In the absence of any additive, the ester carbonyl resonated at a wavenumber of 1732 cm^{-1}, and the amide carbonyl resonated at 1676 cm^{-1}. Addition of 1 equiv. of *N*,*N*-dimethylformamide or *N*,*N*-dimethylacetamide to a solution of Neu5Ac caused an increase in the resonant frequency of the ester carbonyl but did not affect that of the amide carbonyl. When 3 equiv. of *N*,*N*-dimethylformamide or *N*,*N*-dimethylacetamide were added to a solution of Neu5Ac, increases in the resonant frequencies of both the ester and amide carbonyls were observed. These IR stretching frequency increases were rationalized in the scope of decreased hydrogen bonding between Neu5Ac donors strengthening the C=O carbonyl bond (Table 10).[143]

In order to gain a better understanding of the modus operandi of the 5-*N*,4-*O*-oxazolidinone group in sialylations, studies involving the persistent radical effect (PRE) were done by Crich and coworkers, who synthesized several 2,2,6,6-tetramethylpiperidine *N*-oxide (TEMPO) sialosides with different C-5-protecting groups (Fig. 4). While the α-stereoselectivities

Table 10 Effects of Hydrogen-Bonding Additives on Glycosylation Reactivity and Infrared Carbonyl Stretching Frequencies

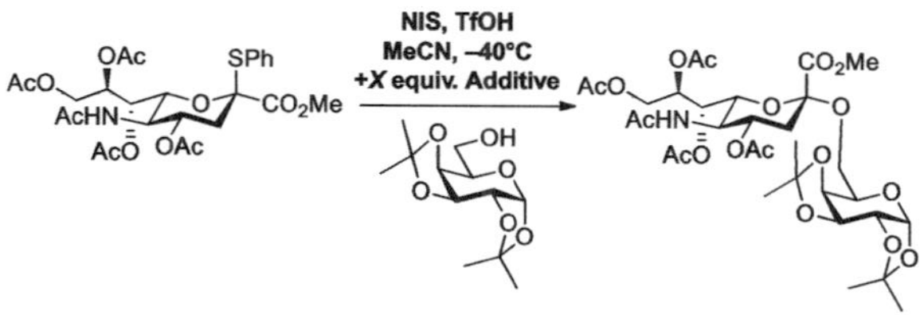

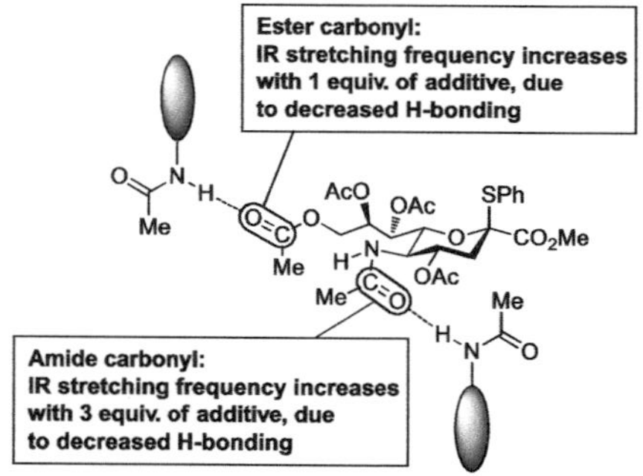

Additive	Equivalents	Yield Disaccharide (%)	Ester Carbonyl IR (cm^{-1})	Amide Carbonyl (cm^{-1})
None	N/A	69	1732	1676
DMF	1	67	1738 (+6)	1676 (+0)
DMF	3	57	1746 (+14)	1686 (+10)
DMA	1	72	1738 (+6)	1676 (+0)
DMA	3	63	1749 (+17)	1686 (+10)

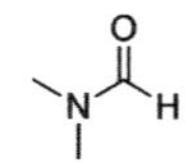

DMF = *N,N*-Dimethylformamide

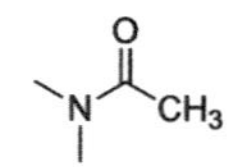

DMA = *N,N*-Dimethylacetamide

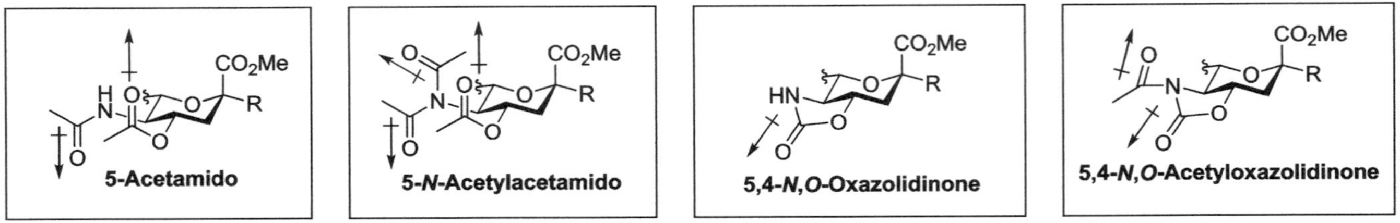

R	C-5 Group	α : β Ratio in C_6D_6	α : β Ratio in $C_2D_6Cl_2$	α : β Ratio in CD_3CN
TEMPO	5-Acetamido	2.6 : 1	2.9 : 1	2.9 : 1
TMIO	5-Acetamido	2.5 : 1	2.7 : 1	3.6 : 1
TEMPO	5,4-*N*,*O*-Oxazolidinone	6.9 : 1	7.0 : 1	7.0 : 1
TMIO	5,4-*N*,*O*-Oxazolidinone	3.5 : 1	6.0 : 1	6.2 : 1
TEMPO	5,4-*N*,*O*-Acetyloxazolidinone	5.9 : 1	6.2 : 1	6.7 : 1
TMIO	5,4-*N*,*O*-Acetyloxazolidinone	3.2 : 1	3.3 : 1	5.5 : 1

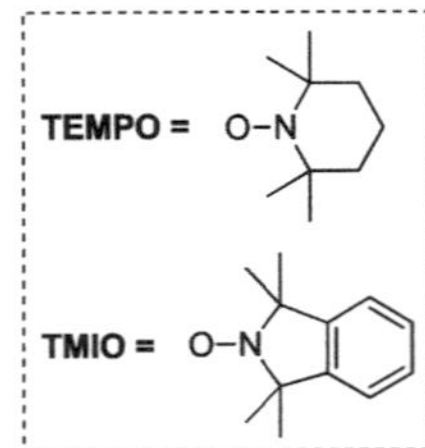

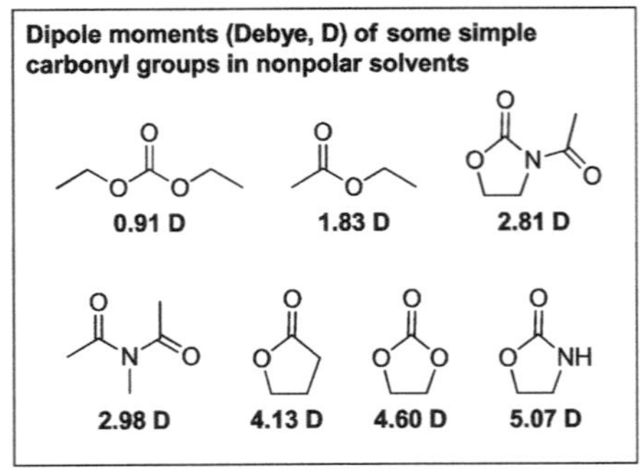

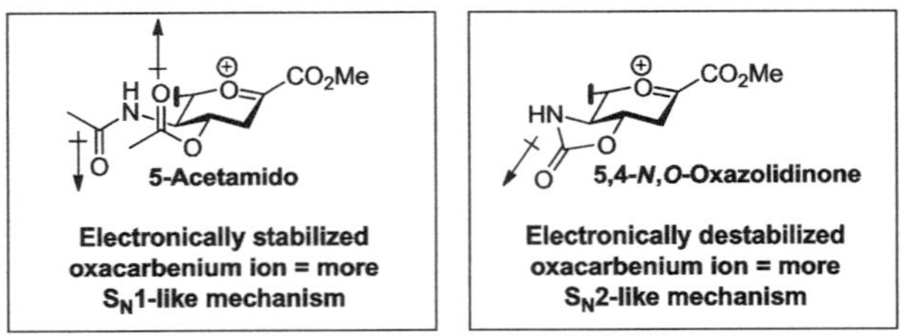

Fig. 4 Anomeric equilibrium ratios of TEMPO/TMIO sialosides, examples of dipole moments, and stabilities of oxacarbenium ions.

observed could be explained by the steric factors associated with the TEMPO group, the differences in the α/β ratio of products between acyclic and cyclic groups at C-5 were proposed to be correlated with an electronic effect.[201] Furthermore, Crich and coworkers helped distinguish the steric contributions of the 2-*O*-hydroxylamine to the anomeric equilibriums from the electronic contributions of the C-5 substituents by carrying out similar experiments on PREs using both TEMPO and the larger 1,1,3,3-tetramethylisoindoline *N*-oxide (TMIO). The larger TMIO group generally resulted in poorer stereoselectivity in the anomeric equilibrium relative to the TEMPO group, but the electron-withdrawing influence of C-5-protecting groups was apparent regardless (Fig. 4).[202]

Thus, the simple 5-*N*,4-*O*-oxazolidinone group, having the largest dipole, is characterized by an overall dipole moment that aligns parallel to the C-4–O-4 and C-5–N-5 bonds, enhancing its electron-withdrawing nature. The N-acetylated 5-*N*,4-*O*-oxazolidinone has the dipole from the *N*-acetyl group assuming a position antiparallel to the oxazolidinone ring, reducing, therefore, its electron-withdrawing ability. In acyclic systems such as that of the simple acetamide, the carbonyl dipole assumes an orthogonal position toward the C-5–N-5 bond. The dipole moments of some simple cyclic and acyclic carbonyl groups can be found in Fig. 4. Additional considerations and testing led to the assumption that: (a) the oxazolidinone ring does not influence the pyranose ring toward reaction at one side vs the other; (b) the oxacarbenium ion is in a locked $^4H_5$ half-chair due to the presence of the fused ring, while for the acyclic case, a conformational equilibrium is possible.[201,202]

Kancharla and Crich also investigated the influence of the glycerol side-chain conformation on the reactivity and stereochemical outcome of sialylations promoted with both the standard per-O-acetylated β-adamantylthio Neu5Ac donor and an unnatural (7*R*)-isomer. The most favorable conformations for the glycerol side chains were determined from the NMR coupling constants of each isomer, especially the $J_{6,7}$ coupling constant. The natural (7*S*)-derivative was found to be more reactive and more α-selective than the equivalent unnatural (7*R*)-derivative. Both the diminished reactivity and poorer stereoselectivity in the (7*R*)-isomer were attributed to the side-chain conformation. It was postulated that the closer proximity of the natural O-7 substituent to the anomeric center relative to the unnatural O-7 led to stabilization of the positive charge generated upon activation of the donor, thereby resulting in higher reactivity for the natural isomer. The poorer α-selectivity in the unnatural (7*R*)-derivative was attributed to the alignment of the glycerol chain perpendicularly below the plane of the sugar in the α-face. Steric occupation of the α-face by the glycerol chain in the unnatural isomer led to higher β-glycosidic predisposition than the natural isomer (Scheme 60).[203]

An additional investigation on the sialylation outcome of a series of xylylene glycol monoesters of *S*-adamantyl glycosides revealed changes in the anomeric selectivity depending on the configuration of the leaving group and the nature of the C-5 group (Scheme 61).[34] Thus, when the C-5 *N*-acetamido was present, a higher degree of β-stereoselectivity was observed from the α-*S*-adamantyl donor (α/β=1/5), which supports an S_N2-like mechanism occurring with inversion of configuration, by intramolecular anti-attack of the nucleophile (Scheme 61). However, the lack of a full inversion of configuration suggests concomitant alternative pathways. When comparing the 5-*N*-acetamido group with the N-acetylated 5-*N*,4-*O*-oxazolidinone and 5-isothiocyanate, the higher degree of α-stereoselectivity was in accordance with the stabilization of S_N2 transition states.

Finally, an investigation was done in the coupling of the α-*S*-adamantyl donor in the presence of acetonitrile, a solvent that has been widely used in sialic acid glycosylations by its generally being more α-stereodirecting. This stereocontrol is probably mediated by the formation of a β-nitrilium ion, although its existence in sialylations has never been confirmed. An HPLC/ESIMS-based monitoring of this reaction, as well as the isolation as side product of a β-acetamide at the anomeric carbon, strongly supports the participation of acetonitrile in sialylations.

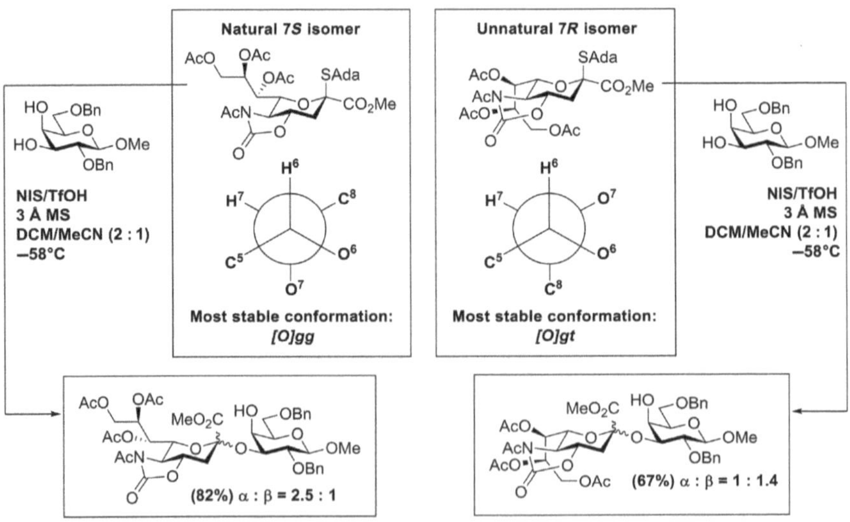

Scheme 60 Effect of the side-chain conformation on donor reactivity and sialylation stereoselectivity.

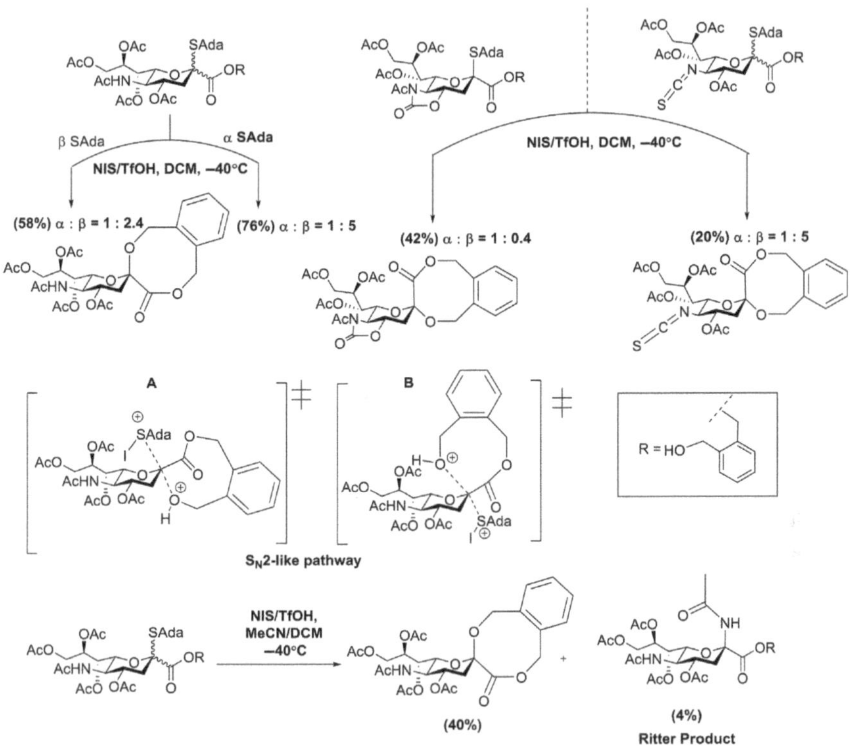

Scheme 61 Effects of C-1 xylylene monoesters on chemical sialylations.

4. CONCLUSIONS

The myriad of methodologies investigated over the past few decades has led to huge advances in chemical sialylations that have enhanced our ability to synthesize sialic acid-containing glycoconjugates. New developments and discoveries allow for the assembly of previously inaccessible complex oligosaccharide structures in high yields and in good/high stereoselectivities. With this review the authors want to acknowledge the extraordinary effort and pioneering accomplishments of many scientists in the field. A particular emphasis has been made in describing the discoveries and applications of C-5 modifications and the use of thioglycosides and phosphates as leaving groups.

Can we conclude that we have solved all the problems in sialylations? Can we now associate sialylations with common glycosylations in terms of feasibility and overall performance? Sadly, not yet.

The general complexity of glycosidic linkages still requires careful selection of building blocks to be used for the assembly of such structures. All of the following have to be taken into consideration: (a) the type of glycosidic linkage to be formed (e.g., 2→3 vs 2→6 or dimers), (b) the nature of the acceptor (reactivity/stability), and (c) the overall number of steps in the synthesis of the building blocks as well as in further modifications of the products. Clearly, there is no "universal donor" and every synthetic strategy must be finely tuned in order to maximize the yield and anomeric configuration of the target.

An effective strategy for the glycosylation of primary acceptors is the use of a glycosyl donor bearing oxazolidinone protection at C-5 with either a phosphate or a thioglycoside as the leaving group. In addition, the opportunity of performing selective sialylations between phosphates/thioglycosides or between armed/disarmed thioglycosides allows for the expeditious synthesis of many targets. Oxazolidinone systems are also efficient in the synthesis of α(2→8) and α(2→9) dimeric linkages. For these targets, selectively removable *O*-chloroacetyl protecting groups allow one to shorten the number of steps required for the multistep oligosaccharide synthesis. For the synthesis of α(2→3)-linked sialosides, the jury is still out, but it is already known that highly reactive donors (e.g., C-5 azido and C-5 trifluoroacetamido) are beneficial for these reactions.

Thus, while some building blocks can indeed be considered "ideal" (efficient, short sequences with good scalability), methodologies in sialylations are still limited in applicability for a wider range of acceptors as well as variations in temperature and solvent systems. In order to expand and strengthen current and future methodologies, investigation of the mechanism by which some sialylations proceed is bringing light to some important synthetic milestones. As the knowledge in the mechanisms advances, so will a more targeted design of building blocks, which will undoubtedly open additional doors and allow sialylations to become more routine for every synthetic chemist.

ACKNOWLEDGMENTS

The authors thank the National Science Foundation (NSF, CHE-RUI#1465003) and the National Institutes of Health (NIH, GM120673) for the generous support.

REFERENCES

1. Varki, A.; Gagneux, P. Multifarious Roles of Sialic Acids in Immunity. *Ann. N. Y. Acad. Sci.* **2012**, *1253*, 16–36.
2. Schauer, R. Ed.; *Cell Biology Monographs Series*, Vol. 10; *Sialic Acids: Chemistry, Metabolism and Function;* Springer-Verlag: New York, 1982.

3. Kelm, S.; Schauer, R. Sialic Acids in Molecular and Cellular Interactions. *Int. Rev. Cytol.* **1997**, *175*, 137–240.
4. Schauer, R. Achievements and Challenges of Sialic Acid Research. *Glycoconjugate J.* **2000**, *17*, 495–499.
5. Varki, A.; Schnaar, R. L.; Schauer, R. Sialic Acids and Other Nonulosonic Acids. In *Essentials of Glycobiology*; Varki, A., Cummings, R. D., Esko, J. D., Stanley, P., Hart, G. W., Aebi, M., Darvill, A. G., Kinoshita, T., Packer, N. H., Prestegard, J. H., Schnaar, R. L., Seeberger, P. H., Eds. 3rd ed.; Cold Spring Harbor Laboratory Press: Cold Spring Harbor, NY, 2015.
6. Angata, T.; Varki, A. Chemical Diversity in the Sialic Acids and Related α-Keto Acids: An Evolutionary Perspective. *Chem. Rev.* **2002**, *102*, 439–470.
7. Chen, X.; Varki, A. Advances in the Biology and Chemistry of Sialic Acids. *ACS Chem. Biol.* **2010**, *5*, 163–176.
8. Inoue, S.; Kitajima, K. KDN (Deaminated Neuraminic Acid): Dreamful Past and Exciting Future of the Newest Member of the Sialic Acid Family. *Glycoconjugate J.* **2006**, *23*, 277–290.
9. Padler-Karavani, V.; Yu, H.; Cao, H.; Chokhawala, H.; Karp, F.; Varki, N.; Chen, X.; Varki, A. Diversity in Specificity, Abundance, and Composition of Anti-Neu5Gc Antibodies in Normal Humans: Potential Implications for Disease. *Glycobiology* **2008**, *18*, 818–830.
10. Boons, G. J.; Demchenko, A. V. Recent Advances in O-Sialylation. *Chem. Rev.* **2000**, *100*, 4539–4565.
11. Boons, G.-J.; Demchenko, A. V. The Chemistry of Sialic Acid. In: *Carbohydrate-Based Drug Discovery*; Wong, C.-H. Ed.; Wiley: New York, 2003; pp 55–102. Chapter 3.
12. De Meo, C.; Boons, G. J.; Demchenko, A. V. Synthesis of Glycosides of Sialic Acid: Chemical Aspects (Scope and Limitations). In: *Comprehensive Glycoscience—From Chemistry to Systems Biology*; Kamerling, J. P. Ed.; Elsevier: Amsterdam, The Netherlands, 2007.
13. Ress, D. K.; Linhardt, R. J. Sialic Acid Donors: Chemical Synthesis and Glycosylation. *Curr. Org. Synth.* **2004**, *1*, 31–46.
14. Navaluri, C.; Crich, D. Stereocontrolled Synthesis of Sialosides. In *Glycochemical Synthesis: Strategies and Applications*; Hung, S.-C., Zulueta, M. M. L., Eds.; John Wiley & Son, Inc: Hoboken, New Jersey, 2016.
15. Lih, Y.-H.; Wu, C.-Y. Chemical Synthesis of Sialosides. In *Selective Glycosylations: Synthetic Methods and Catalysts*; Bennett, C. S. Ed.; Wiley-VCH Verlag GmbH & Co. KGaA: Weinheim, 2017; pp 353–370.
16. Sun, B. Recent Developments of Highly Selective Sialylation. *Curr. Org. Chem.* **2016**, *20*, 1465–1476.
17. Ando, H. Synthetic Approach Toward Complexity of Sialic Acid-Containing Glycans. *Biosci. Biotechnol. Biochem.* **2015**, *79*, 343–346.
18. Tanaka, H. Chemical Synthesis of Oligosialic Acid by Using 5-*N*,4-*O*-Carbonyl-Protected Sialyl Donors. In *Glycoscience: Biology and Medicine*; Endo, T., Seeberger, P. H., Hart, G. W., Wong, C.-H., Taniguchi, N., Eds.; Springer Japan: Tokyo, 2014; pp 1–5.
19. Adak, A. K.; Yu, C.-C.; Liang, C.-F.; Lin, C.-C. Synthesis of Sialic Acid-Containing Saccharides. *Curr. Opin. Chem. Biol.* **2013**, *17*, 1030–1038.
20. Cao, H.; Chen, X. General Consideration on Sialic Acid Chemistry. *Methods Mol. Biol.* **2012**, *808*, 31–56.
21. Hanashima, S. Recent Strategies for Stereoselective Sialylation and Their Application to the Synthesis of Oligosialosides. *Trends Glycosci. Glycotechnol.* **2011**, *23*, 111–121.
22. Liang, F. F.; Chen, L.; Xing, G. W. Recent Advances in Sialylation. *Chin. J. Org. Chem.* **2009**, *29*, 1317–1324.

23. Kiso, M. Selective alpha-Sialylation. In *Glycoscience: Chemistry and Chemical Biology*; Ando, H., Fraser-Reid, B., Tatsuta, K., Thiem, J., Eds.; Springer-Verlag: Heidelberg, 2008; pp 1313–1359.
24. Okamoto, K.; Goto, T. Glycosidation of Sialic Acid. *Tetrahedron* **1990**, *46*, 5835–5857.
25. Halcomb, R.; Chappell, M. Recent Developments in Technology for Glycosylation With Sialic Acid. *J. Carbohydr. Chem.* **2002**, *21*, 723–768.
26. De Meo, C.; Priyadarshani, U. C-5 Modifications in *N*-Acetyl-neuraminic Acid: Scope and Limitations. *Carbohydr. Res.* **2008**, *343*, 1540–1552.
27. Magnusson, G.; Nilsson, U. J. In *Glycoscience: Chemistry and Chemical Biology*; Fraser-Reid, B., Kuniaki, T., Thiem, J., Eds.; Vol. 2; Springer: Berlin, 2001; pp 1543–1587.
28. Kiefel, M.; von Itzstein, M. Recent Advances in the Synthesis of Sialic Acid Derivatives and Sialylmimetics as Biological Probes. *Chem. Rev.* **2002**, *102*, 471–490.
29. Ando, T.; Ando, H.; Kiso, M. Sialic Acid and Glycobiology: A Chemical Approach. *Trends Glycosci. Glycotechnol.* **2001**, *13*, 573–586.
30. Takahashi, T.; Tsukamoto, H.; Yamada, H. A New Method for the Formation of the α-Glycoside Bond of Sialyl Conjugates Based on Long-Range Participation. *Tetrahedron Lett.* **1997**, *38*, 8223–8226.
31. Haberman, J. M.; Gin, D. Y. A New C(1)-Auxiliary for Anomeric Stereocontrol in the Synthesis of α-Sialyl Glycosides. *Org. Lett.* **2001**, *3*, 1665–1668.
32. Haberman, J. M.; Gin, D. Y. Dehydrative Sialylation With C-2-Hemiketal Sialyl Donors. *Org. Lett.* **2003**, *5*, 2539–2541.
33. Ishiwata, A.; Ito, Y. Preparation of Sialyl Donors Carrying Functionalized Ester Substituents: Effects on the Selectivity of Glycosylation. *Synlett* **2003**, 1339–1343.
34. Amarasekara, H.; Crich, D. Synthesis and Intramolecular Glycosylation of Sialyl Mono-esters of *o*-Xylylene Glycol. The Importance of Donor Configuration and Nitrogen Protecting Groups on Cyclization Yield and Selectivity; Isolation and Characterization of a *N*-Sialyl Acetamide Indicative of Participation by Acetonitrile. *Carbohydr. Res.* **2016**, *435*, 113–120.
35. Okamoto, R.; Souma, S.; Kajihara, Y. Efficient Synthesis of MUC4 Sialylglycopeptide Through the New Sialylation Using 5-Acetamido-neuraminamide Donors. *J. Org. Chem.* **2008**, *73*, 3460–3466.
36. Hanashima, S.; Akai, S.; Sato, K.-I. Thioester-Assisted α-Sialylation Reaction. *Tetrahedron Lett.* **2008**, *49*, 5111–5114.
37. Ye, X.-S.; Huang, X.; Wong, C. H. Conversion of the Carboxy Group of Sialic Acid Donors to a Protected Hydroxymethyl Group Yields an Efficient Reagent for the Synthesis of the Unnatural β-Linkage. *Chem. Commun. (Cambridge)* **2001**, 974–975.
38. Kuhn, R.; Lutz, P.; MacDonald, D. L. Synthese Anomerer Sialinsäure-Methylketoside. *Chem. Ber.* **1966**, *99*, 611–617.
39. Byramova, N. E.; Tuzikov, A. B.; Bovin, N. V. A Simple Procedure for the Synthesis of the Methyl and Benzyl Glycosides of Neu5Ac and 4-*epi*-Neu5Ac. Conversion of the Benzyl and Methyl Glycosides of Neu5Ac Into *N*-Trifluoroacetylneuraminic Acid Benzyl Glycosides. *Carbohydr. Res.* **1992**, *237*, 161–175.
40. Sharma, M. N.; Eby, R. Synthesis and Conformational Studies of 2-β-Chloro, 2-α-Fluoro, and 2-β-Fluoro Derivatives of 2-Deoxy-*N*-acetyl-neuraminic Acid. *Carbohydr. Res.* **1984**, *127*, 201–210.
41. Kunz, H.; Waldmann, H. Directed Stereoselective Synthesis of α- and β-*N*-Acetyl Neuraminic Acid–Galactose Disaccharides Using 2-Chloro and 2-Fluoro Derivatives of Neuraminic Acid Allyl Ester. *J. Chem. Soc., Chem. Commun* **1985**, 638–640.

42. Mugunthan, G.; Kartha, K. P. R. A Facile and Convenient Synthesis of Disarmed Glycosyl Fluorides Using In Situ-Generated Iodine Monofluoride (See Refs. 1, 38, 39, 40, 41, 42, 43 therein). *Tetrahedron Lett.* **2012**, *53*, 5631–5634.
43. Deninno, M. P. The Synthesis and Glycosidation of *N*-Acetylneuraminic Acid. *Synthesis* **1991**, 583–593.
44. Paulsen, H.; Tietz, H. Synthesis of Trisaccharide Moieties From *N*-Acetylneuraminic Acid and *N*-Acetyllactosamine. *Angew. Chem., Int. Ed. Engl.* **1982**, *21*, 927–928.
45. Marra, A.; Sinaÿ, P. Acetylation of *N*-Acetylneuraminic Acid and Its Methyl Ester. *Carbohydr. Res.* **1989**, *190*, 317–322.
46. Kirchner, E.; Thiem, F.; Dernick, R.; Heukeshoven, J.; Thiem, J. Studies on the Glycosylation of *N*-Acetylneuraminic Acid. *J. Carbohydr. Chem.* **1988**, *7*, 453–486.
47. Pazynina, G.; Tuzikov, A.; Chinarev, A.; Obukhova, P.; Bovin, N. Simple Stereoselctive Synthesis of $\alpha(2\rightarrow6)$-Sialooligosaccharides. *Tetrahedron Lett.* **2002**, *43*, 8011–8013.
48. Tropper, F. D.; Andersson, F. O.; Cao, S.; Roy, R. Synthesis of *S*-Glycosyl Xanthates by Phase Transfer Catalyzed Substitution of Glycosyl Halides. *J. Carbohydr. Chem.* **1992**, *11*, 741–750.
49. Marra, A.; Sinaÿ, P. Stereoselective Synthesis of 2-Thioglycosides of *N*-Acetylneuraminic Acid. *Carbohydr. Res.* **1989**, *187*, 35–42.
50. Martichonok, V.; Whitesides, G. M. Stereoselective α-Sialylation With Sialyl Xanthate and Phenylsulfenyl Triflate as a Promotor. *J. Org. Chem.* **1996**, *61*, 1702–1706.
51. Gregar, T. Q.; Gervay-Hague, J. Synthesis of Oligomers Derived From Amide-Linked Neuraminic Acid Analogues. *J. Org. Chem.* **2004**, *69*, 1001–1009.
52. Sun, X.-L.; Kanie, Y.; Guo, C.-T.; Kanie, O.; Suzuki, Y.; Wong, C.-H. Syntheses of C-3-Modified Sialylglycosides as Selective Inhibitors of Influenza Hemagglutinin and Neuraminidase. *Eur. J. Org. Chem.* **2000**, 2643–2653.
53. Shpirt, A. M.; Kononov, L. O.; Torgov, V. I.; Shibaev, V. N. Conversion of *N*-Acetylneuraminic Acid Glycosyl Chloride Into Dibenzyl Glycosyl Phosphate: O-Glycosylation in the Absence of a Promoter. *Russ. Chem. Bull.* **2004**, *53*, 717–719.
54. Orlova, A. V.; Shpirt, A. M.; Kulikova, N. Y.; Kononov, L. O. *N,N*-Diacetylsialyl Chloride—A Novel Readily Accessible Sialyl Donor in Reactions With Neutral and Charged Nucleophiles in the Absence of a Promoter. *Carbohydr. Res.* **2010**, *345*, 721–730.
55. Kanie, O.; Kiso, M.; Hasegawa, A. Glycosylation Using Methylthioglycosides of *N*-Acetylneuraminic Acid and Dimethyl (Methylthio) Sulfonium Triflate. *J. Carbohydr. Chem.* **1988**, *7*, 501–506.
56. Hasegawa, A.; Ohki, H.; Nagahama, T.; Ishida, H.; Kiso, M. A Facile, Large-Scale Preparation of the Methyl 2-Thioglycoside of *N*-Acetylneuraminic Acid, and Its Usefulness for the α-Stereoselective Synthesis of Sialoglycosides. *Carbohydr. Res.* **1991**, *212*, 277–281.
57. Hasegawa, A.; Nagahama, T.; Ohki, H.; Hotta, K.; Ishida, H.; Kiso, M. Synthetic Studies on Sialogylcoconjugates 25: Reactivity of Glycosyl Promoters in α-Glycosylation of *N*-Acetyl-Neuraminic Acid With the Primary and Secondary Hydroxyl Groups in the Suitably Protected Galactose and Lactose Derivatives. *J. Carbohydr. Chem.* **1991**, *10*, 493–498.
58. Cao, S.; Meunier, S. J.; Andersson, F. O.; Letellier, M.; Roy, R. Mild Stereoselective Syntheses of Thioglycosides Under PTC Conditions and Their Use as Selective and Latent Glycosyl Donors. *Tetrahedron: Asymmetry* **1994**, *5*, 2303–2312.
59. Nifantiev, N. E.; Tsvetkov, Y. E.; Shashkov, A. S.; Kononov, L. O.; Menshov, V. M.; Tuzikov, A. B.; Bovin, N. V. Selectin Receptors 4: Synthesis of Tetrasaccharides Sialyl Lewis^{A} and Sialyl Lewis^{X} Containing a Spacer Group. *J. Carbohydr. Chem.* **1996**, *15*, 939–953.

60. Tsvetkov, Y. E.; Schmidt, R. R. Model Investigations for Intramolecular Glycoside Bond Formation With Neuraminic Acid Derivatives—Generation of Crown-Ether Type Structures. *Carbohydr. Lett.* **1996**, *2*, 149–156.
61. Simeoni, L. A.; Byramova, N. E.; Bovin, N. V. The Synthesis of Trisaccharides Neu5Ac(α2→6)[Gal(α1→3)]GalNAc α and Neu5AC (α2→6)[Gal(β1→3)] GalNAc α as Protected Trifluoroacetamidopropyl Glycosides. *Bioorg. Khim.* **1999**, *25*, 62–69.
62. Cheshev, P. E.; Khatuntsvesa, E. A.; Gerbst, A. G.; Tsvetkov, Y. E.; Shashkov, A. S.; Nifant'ev, N. E. Synthesis of Aminoethyl Glycosides of the Carbohydrate Chains of Glycolipids Gb3, Gb4, and Gb5. *Russ. J. Bioorg. Chem.* **2003**, *29*, 372–381.
63. Oscarson, S. Thioglycosides. In *Carbohydrates in Chemistry and Biology*; Ernst, B., Hart, G. W., Sinaÿ, P., Eds.; Vol. 1; Wiley-VCH: Weinheim, New York, 2000; pp 93–116.
64. Oscarson, S. General Synthetic Methods: Reactions at the Anomeric Center: S-Glycosylation. In *Glycoscience: Chemistry and Chemical Biology*; Fraser-Reid, B., Tatsuta, K., Thiem, J., Eds.; Vol. 1; Springer: New York, 2001; pp 643–671.
65. Hasegawa, A.; Nagahama, T.; Ohki, H.; Hotta, K.; Ishida, H.; Kiso, M. Communication: Synthetic Studies on Sialoglycoconjugates 25: Reactivity of Glycosyl Promoters in α-Glycosylation of *N*-Acetyl-Neuraminic Acid With the Primary and Secondary Hydroxyl Groups in the Suitably Protected Galactose and Lactose Derivatives. *J. Carbohydr. Chem.* **1991**, *10*, 493–498.
66. Li, Z.; Gildersleeve, J. C. Mechanistic Studies and Methods to Prevent Aglycon Transfer of Thioglycosides. *J. Am. Chem. Soc.* **2006**, *128*, 11612–11619.
67. Crich, D.; Li, W. Selective Sialylations at −78 °C in Nitrile Solvents With a 1-Adamantanyl Thiosialoside. *J. Org. Chem.* **2007**, *72*, 7794–7797.
68. Dziadek, S.; Brocke, C.; Kunz, H. Biomimetic Synthesis of the Tumor-Associated (2,3)-Sialyl-T Antigen and Its Incorporation Into Glycopeptide Antigens From the Mucins MUC1 and MUC4. *Chem. Eur. J.* **2004**, *10*, 4150–4162.
69. Marra, A.; Sinaÿ, P. A Novel Stereoselective Synthesis of *N*-Acetyl-α-neuraminosyl-galactose Disaccharide Derivatives, Using Anomeric *S*-Glycosyl Xanthates. *Carbohydr. Res.* **1990**, *195*, 303–308.
70. Schmidt, R. R. Chemical Synthesis of Sialylated Glycoconjugates. In *Synthetic Oligosaccharides: Indispensable Probes for the Life Sciences*; Kovac, P. Ed.; ACS Symposium Series; Vol. 560; American Chemical Society: Washington, DC, 1994; pp 276–296.
71. Greilich, U.; Brescello, R.; Jung, K. H.; Schmidt, R. R. Synthesis of Ganglioside GM_1 via a GM_3 Intermediate. *Liebigs Ann.* **1996**, 663–672.
72. Martichonok, V.; Whitesides, G. M. Stereoselective α-Sialylation With Sialyl Xanthate and Phenylsulfenyl Triflate as a Promoter. *J. Org. Chem.* **1996**, *61*, 1702–1706.
73. Zheng, C.; Qu, H.; Liao, W.; Bavaro, T.; Terreni, M.; Sollogoub, M.; Ding, K.; Zhang, Y. Chemoenzymatically Synthesized GM3 Analogues as Potential Therapeutic Agents to Recover Nervous Functionality After Injury by Inducing Neurite Outgrowth. *Eur. J. Med. Chem.* **2018**, *146*, 613–620.
74. Xiao, A.; Zheng, X.-J.; Song, C.; Gui, Y.; Huo, C.-X.; Ye, X.-S. Synthesis and Immunological Evaluation of MUC1 Glycopeptide Conjugates Bearing *N*-Acetyl Modified STn Derivatives as Anticancer Vaccines. *Org. Biomol. Chem.* **2016**, *14*, 7226–7237.
75. Filser, C.; Kowalczyk, D.; Jones, C.; Wild, M. K.; Ipe, U.; Vestweber, D.; Kunz, H. Synthetic Glycopeptides From the E-Selectin Ligand 1 With Varied Sialyl Lewis^X Structure as Cell-Adhesion Inhibitors of E-Selectin. *Angew. Chem., Int. Ed.* **2007**, *46*, 2108–2111.

76. Kuhn, A.; Kunz, H. Saccharide-Induced Peptide Conformation in Glycopeptides of the Recognition Region of LI-Cadherin. *Angew. Chem., Int. Ed.* **2007**, *46*, 454–458.
77. Gaidzik, N.; Kaiser, A.; Kowalczyk, D.; Westerlind, U.; Gerlitzki, B.; Sinn, H. P.; Schmitt, E.; Kunz, H. Synthetic Antitumor Vaccines Containing MUC1 Glycopeptides With Two Immunodominant Domains—Induction of a Strong Immune Response Against Breast Tumor Tissues. *Angew. Chem., Int. Ed.* **2011**, *50*, 9977–9981.
78. Daum, M.; Broszeit, F.; Hoffmann-Röder, A. Synthesis of a Fluorinated Sialophorin Hexasaccharide–Threonine Conjugate for Fmoc Solid-Phase Glycopeptide Synthesis. *Eur. J. Org. Chem.* **2016**, 3709–3720.
79. De Meo, C.; Parker, O. Thioimidoyl Approach to the Synthesis of α-Sialosides. *Tetrahedron: Asymmetry* **2005**, *16*, 303–307.
80. Wang, C.-C.; Lee, J.-C.; Luo, S.-Y.; Kulkarni, S. S.; Huang, Y.-W.; Lee, C.-C.; Chang, K.-L.; Hung, S.-C. Regioselective One-Pot Protection of Carbohydrates. *Nature* **2007**, *446*, 896–899.
81. Tanaka, H.; Tateno, Y.; Nishiura, Y.; Takahashi, T. Efficient Synthesis of an α(2,9) Trisialic Acid by One-Pot Glycosylation and Polymer-Assisted Deprotection. *Org. Lett.* **2008**, *10*, 5597–5600.
82. Harris, B. N.; Patel, P. P.; Gobble, C. P.; Stark, M. J.; De Meo, C. C-5 Modified *S*-Benzoxazolyl Sialyl Donors: Towards More Efficient Selective Sialylations. *Eur. J. Org. Chem.* **2011**, 4023–4027.
83. Cai, S.; Yu, B. Efficient Sialylation With Phenyltrifluoroacetimidates as Leaving Groups. *Org. Lett.* **2003**, *5*, 3827–3830.
84. Tanaka, K.; Goi, T.; Fukase, K. Highly Efficient Sialylation Towards α(2-3)- and α(2-6)-Neu5Ac-Gal Synthesis: Significant 'Fixed Dipole Effect' of *N*-Phthalyl Group on α-Selectivity. *Synlett* **2005**, 2958–2962.
85. Tanaka, S.-I.; Goi, T.; Tanaka, K.; Fukase, K. Highly Efficient α-Sialylation by Virtue of Fixed Dipole Effects of *N*-Phthalyl Group: Application to Continuous Flow Synthesis of α(2-3)-and α(2-6)-Neu5Ac-Gal Motifs by Microreactor. *J. Carbohydr. Chem.* **2007**, *26*, 369–394.
86. Sun, B.; Srinivasan, B.; Huang, X. Pre-activation-Based One-Pot Synthesis of an α-(2,3)-Sialylated Core-Fucosylated Complex Type Bi-antennary N-Glycan Dodecasaccharide. *Chem. Eur. J.* **2008**, *14*, 7072–7081.
87. Hsu, Y.; Ma, H.-H.; Lico, L. S.; Jan, J.-T.; Fukase, K.; Uchinashi, Y.; Zulueta, M. M. L.; Hung, S.-C. One-Pot Synthesis of *N*-Acetyl- and *N*-Glycolylneuraminic Acid Capped Trisaccharides and Evaluation of Their Influenza A(H1N1) Inhibition. *Angew. Chem., Int. Ed.* **2014**, *53*, 2413–2416.
88. Uchinashi, Y.; Nagasaki, M.; Zhou, J.; Tanaka, K.; Fukase, K. Reinvestigation of the C5-Acetamide Sialic Acid Donor for α-Selective Sialylation: Practical Procedure Under Microfluidic Conditions. *Org. Biomol. Chem.* **2011**, *9*, 7243–7248.
89. Zhou, J.; Manabe, Y.; Tanaka, K.; Fukase, K. Efficient Synthesis of the Disialylated Tetrasaccharide Motif in N-Glycans Through an Amide-Protection Strategy. *Chem. Asian J.* **2016**, *11*, 1436–1440.
90. Nagasaki, M.; Manabe, Y.; Minamoto, N.; Tanaka, K.; Silipo, A.; Molinaro, A.; Fukase, K. Chemical Synthesis of a Complex-Type N-Glycan Containing a Core Fucose. *J. Org. Chem.* **2016**, *81*, 10600–10616.
91. Kondo, H.; Ichikawa, Y.; Wong, C. H. β-Sialyl Phosphite and Phosphoramidite: Synthesis and Application to the Chemoenzymic Synthesis of CMP-Sialic Acid and Sialyl Oligosaccharides. *J. Am. Chem. Soc.* **1992**, *114*, 8748–8750.
92. Martin, T. J.; Schmidt, R. R. Efficient Sialylation With Phosphite as Leaving Group. *Tetrahedron Lett.* **1992**, *33*, 6123–6126.

93. Sim, M. M.; Kondo, H.; Wong, C. H. Synthesis of Dibenzyl Glycosyl Phosphites Using Dibenzyl *N,N*-Diethylphosphoramidite as Phosphitylating Reagent: An Effective Route to Glycosyl Phosphates, Nucleotides, and Glycosides. *J. Am. Chem. Soc.* **1993**, *115*, 2260–2267.
94. Nagorny, P.; Fasching, B.; Li, X.; Chen, G.; Aussedat, B.; Danishefsky, S. J. Toward Fully Synthetic Homogeneous β-Human Follicle-Stimulating Hormone (β-hFSH) With a Biantennary N-Linked Dodecasaccharide. Synthesis of β-hFSH With Chitose Units at the Natural Linkage Sites. *J. Am. Chem. Soc.* **2009**, *131*, 5792–5799.
95. Martin, T. J.; Brescello, R.; Toepfer, A.; Schmidt, R. R. Synthesis of Phosphites and Phosphates of Neuraminic Acid and Their Glycosyl Donor Properties—Convenient Synthesis of GM3. *Glycoconjugate J.* **1993**, *10*, 16–25.
96. Veeneman, G. H.; van der Hulst, R. G. A.; van Boeckel, C. A. A.; Philipsen, R. L. A.; Ruigt, G. S. F.; Tonnaer, J. A. D. M.; van Delft, T. M. L.; Konings, P. N. M. Synthesis of Sialic Acid–Lipid Conjugates and Their Neuritogenic Effects on N1E.115 Neuroblastoma Cells. *Bioorg. Med. Chem. Lett.* **1995**, *5*, 9–14.
97. Watanabe, Y.; Nakamoto, C.; Yamamoto, T.; Ozaki, S. Glycosylation Using Glycosyl Phosphite as a Glycosyl Donor. *Tetrahedron* **1994**, *50*, 6523–6536.
98. Bhattacharya, S. K.; Danishefsky, S. J. A Total Synthesis of the Methyl Glycoside of Ganglioside GM1. *J. Org. Chem.* **2000**, *65*, 144–151.
99. Kondo, H.; Aoki, S.; Ichikawa, Y.; Halcomb, R. L.; Ritzen, H.; Wong, C. H. Glycosyl Phosphites as Glycosylation Reagents: Scope and Mechanism. *J. Org. Chem.* **1994**, *59*, 864–877.
100. Tsvetkov, Y. E.; Schmidt, R. R. 8-O-Sialylation of Derivatives of Neuraminic Acid 1,7-Lactone Unusual Stereoselectivity. *Tetrahedron Lett.* **1994**, *35*, 8583–8586.
101. Castro-Palomino, J. C.; Tsvetkov, Y. E.; Schneider, R.; Schmidt, R. R. 8-O-Sialylation of Neuraminic Acid Acceptor Reactivity and Anomeric Stereocontrol. *Tetrahedron Lett.* **1997**, *38*, 6837–6840.
102. Schwartz, J. B.; Kuduk, S. D.; Chen, X. T.; Sames, D.; Glunz, P. W.; Danishefsky, S. J. A Broadly Applicable Method for the Efficient Synthesis of α-O-Linked Glycopeptides and Clustered Sialic Acid Residues. *J. Am. Chem. Soc.* **1999**, *121*, 2662–2673.
103. Hanashima, S.; Castagner, B.; Esposito, D.; Nokami, T.; Seeberger, P. H. Synthesis of a Sialic Acid α(2 − 3) Galactose Building Block and Its Use in a Linear Synthesis of Sialyl Lewis^{X}. *Org. Lett.* **2007**, *9*, 1777–1779.
104. Tanaka, H.; Nishiura, Y.; Adachi, M.; Takahashi, T. Synthetic Study of α(2-8) Oligosialoside Using *N*-Troc Sialyl *N*-Phenyltrifluoroimidate. *Heterocycles* **2006**, *67*, 107–112.
105. Lin, C.-C.; Adak, A. K.; Horng, J.-C.; Lin, C.-C. Phosphite-Based Sialic Acid Donors in the Synthesis of α(2→9) Oligosialic Acids. *Tetrahedron* **2009**, *65*, 4714–4725.
106. Yu, C.-S.; Niikura, K.; Lin, C.-C.; Wong, C. H. The Thioglycoside and Glycosyl Phosphite of 5-Azido Sialic Acid: Excellent Donors for the α-Glycosylation of Primary Hydroxy Groups. *Angew. Chem., Int. Ed.* **2001**, *40*, 2900–2903.
107. Sun, B.; Jiang, H. An Efficient Approach for Total Synthesis of Aminopropyl Functionalized Ganglioside GM1b. *Tetrahedron Lett.* **2012**, *53*, 5711–5715.
108. Crich, D.; Li, W. O-Sialylation With *N*-Acetyl-5-*N*,4-*O*-carbonyl-Protected Thiosialoside Donors in Dichloromethane: Facile and Selective Cleavage of the Oxazolidinone Ring. *J. Org. Chem.* **2007**, *72*, 2387–2391.
109. Hsu, C.-H.; Chu, K.-C.; Lin, Y.-S.; Han, J.-L.; Peng, Y.-S.; Ren, C.-T.; Wu, C.-Y.; Wong, C.-H. Highly α-Selective Sialyl Phosphate Donors for Efficient Preparation of Natural Sialosides. *Chem. Eur. J.* **2010**, *16*, 1754–1760.

110. Chu, K.-C.; Ren, C.-T.; Lu, C.-P.; Hsu, C.-H.; Sun, T.-H.; Han, J.-L.; Pal, B.; Chao, T.-A.; Lin, Y.-F.; Wu, S.-H.; Wong, C.-H.; Wu, C.-Y. Efficient and Stereoselective Synthesis of α(2→9) Oligosialic Acids: From Monomers to Dodecamers. *Angew. Chem., Int. Ed.* **2011**, *50*, 9391–9395.
111. Noel, A.; Delpech, B.; Crich, D. Stereoselective, Electrophilic α-C-Sialidation. *Org. Lett.* **2012**, *14*, 1342–1345.
112. Noel, A.; Delpech, B.; Crich, D. Highly Stereoselective Synthesis of Primary, Secondary, and Tertiary α-S-Sialosides Under Lewis Acidic Conditions. *Org. Lett.* **2012**, *14*, 4138–4141.
113. Xu, F.-F.; Wang, Y.; Xiong, D.-C.; Ye, X.-S. Stereoselective Synthesis of the Trisaccharide Moiety of Ganglioside HLG-2. *J. Org. Chem.* **2014**, *79*, 797–802.
114. Mondal, P. K.; Liao, G.; Mondal, M. A.; Guo, Z. Chemical Synthesis of the Repeating Unit of Type Ia Group B Streptococcus Capsular Polysaccharide. *Org. Lett.* **2015**, *17*, 1102–1105.
115. Shi, M.; Kleski, K. A.; Trabbic, K. R.; Bourgault, J.-P.; Andreana, P. R. Sialyl-Tn Polysaccharide A1 as an Entirely Carbohydrate Immunogen: Synthesis and Immunological Evaluation. *J. Am. Chem. Soc.* **2016**, *138*, 14264–14272.
116. Gao, J.; Guo, Z. Chemical Synthesis of the Repeating Unit of Type V Group B Streptococcus Capsular Polysaccharide. *Org. Lett.* **2016**, *18*, 5552–5555.
117. Sardar, M. Y. R.; Mandhapati, A. R.; Park, S.; Wever, W. J.; Cummings, R. D.; Chaikof, E. L. Convergent Synthesis of Sialyl Lewis^{X}-O-Core-1 Threonine. *J. Org. Chem.* **2018**, *83*, 4963–4972.
118. Kurimoto, K.; Yamamura, H.; Miyagawa, A. Sialylation With Systematically Protected Allyl Galactoside Acceptors Using Sialyl Phosphate Donors. *Tetrahedron Lett.* **2016**, *57*, 1442–1445.
119. Shao, L.; Zhang, H.; Li, Y.; Gu, G.; Cai, F.; Guo, Z.; Gao, J. Chemical Synthesis of the Repeating Unit of Type II Group B Streptococcus Capsular Polysaccharide. *J. Org. Chem.* **2018**, *83*, 5920–5930.
120. De Meo, C.; Wallace, C. E.; Geringer, S. A. Synthesis and Elimination of C-3-Labeled Thiosialosides. *Org. Lett.* **2014**, *16*, 2676–2679.
121. Okamoto, K.; Kondo, T.; Goto, T. An Effective Synthesis of α-Glycosides of *N*-Acetyl Neuraminic Acid by Use of 2β-Halo-3β-hydroxy-4,7,8,9,-tetra-*O*-acetyl-*N*-acetylneuraminic Acid Methyl Ester. *Tetrahedron Lett.* **1987**, *43*, 5919–5928.
122. Okamoto, K.; Kondo, H.; Goto, T. A Stereospecific Synthesis of β-Glycosides of *N*-Acetylneuraminic Acid and Secondary Alcohols. *Chem. Lett.* **1986**, *15*, 1449–1452.
123. Okamoto, K.; Kondo, T.; Goto, T. Glycosylation of 4,7,8,9-Tetra-*O*-acetyl-2-deoxy-2β,3β-epoxy-*N*-acetylneuraminic Acid Methyl Ester. *Bull. Chem. Soc. Jpn.* **1987**, *60*, 637–643.
124. Castro-Palomino, J. C.; Tsvetkov, Y. E.; Schmidt, R. R. 8-O-Sialylation of Neuraminic Acid. *J. Am. Chem. Soc.* **1998**, *120*, 5434–5440.
125. Ito, Y.; Ogawa, T. An Efficient Approach to Stereoselective Glycosylation of *N*-Acetylneuraminic Acid: Use of Phenylselenyl Group as a Stereocontrolling Auxillary. *Tetrahedron Lett.* **1987**, *28*, 6221–6224.
126. Ito, Y.; Ogawa, T. Benzeneselenenyl Triflate as a Promoter of Thioglycosides: A New Method for O-Glycosylation Using Thioglycosides. *Tetrahedron Lett.* **1988**, *29*, 1061–1064.
127. Numata, M.; Sugimoto, M.; Ito, Y.; Ogawa, T. An Efficient Synthesis of Ganglioside GM3: Highly Stereocontrolled Glycosylations by Use of Auxiliaries. *Carbohydr. Res.* **1990**, *203*, 205–217.
128. Ito, Y.; Ogawa, T. Highly Stereoselective Glycosylation of *N*-Acetylneuraminic Acid Aided by a Phenylthio Substituent as a Stereocontrolling Auxilliary. *Tetrahedron Lett.* **1988**, *29*, 3987–3990.

129. Tadao, K.; Hiroyuki, A.; Toshio, G. Efficient Synthesis of 2α-Glycoside of *N*-Acetylneuraminic Acid via Phenylsulfenyl Chloride Adduct of 2,3-Dehydro-2-deoxy-*N*-acetylneuraminic Acid Methyl Ester Tetra-*O*-acetate. *Chem. Lett.* **1988**, *17*, 1657–1660.
130. Ito, Y.; Ogawa, T. Highly Stereoselective Glycosylation of Sialic Acid Aided by Stereocontrolling Auxiliaries. *Tetrahedron* **1990**, *46*, 89–102.
131. Ercegovic, T.; Magnusson, G. A New *N*-Acetylneuraminic Acid Donor for Highly Stereoselective α-Sialylation. *J. Chem. Soc., Chem. Commun.* **1994**, 831–832.
132. Martichonok, V.; Whitesides, G. M. Studies on α-Sialylation Using Sialyl Donors With an Auxiliary 3-Thiophenyl Group. *Carbohydr. Res.* **1997**, *302*, 123–129.
133. Singh, L.; Nakahara, Y.; Ito, Y.; Nakahara, Y. The First Total Synthesis of the Core Class II Disialylated Hexasaccharide as a Building Block for Glycopeptide Synthesis. *Tetrahedron Lett.* **1999**, *40*, 3769–3772.
134. Hossain, N.; Magnusson, G. A Novel Donor for Stereoselective α-Sialylation; Efficient Synthesis of an α(2-8)-Linked Bis-sialic Acid Unit. *Tetrahedron Lett.* **1999**, *40*, 2217–2220.
135. Premathilake, H. D.; Gobble, C. P.; Pornsuriyasak, P.; Hardimon, T.; Demchenko, A. V.; De Meo, C. How O-Substitution of Sialyl Donors Affects Their Stereoselectivity. *Org. Lett.* **2012**, *14*, 1126–1129.
136. Yasomanee, J. P.; Demchenko, A. V. Hydrogen Bond-Mediated Aglycone Delivery (HAD): A Highly Stereoselective Synthesis of 1,2-cis α-D-Glucosides From Common Glycosyl Donors in the Presence of Bromine. *Chem. Eur. J.* **2015**, *21*, 6572–6581.
137. Yasomanee, J. P.; Demchenko, A. V. Hydrogen Bond-Mediated Aglycone Delivery: The Synthesis of Linear and Branched α-Glucans. *Angew. Chem., Int. Ed.* **2014**, *53*, 10453–10456.
138. Pistorio, S. G.; Yasomanee, J. P.; Demchenko, A. V. Hydrogen Bond-Mediated Aglycone Delivery: Focus on β-Mannosylation. *Org. Lett.* **2014**, *16*, 716–719.
139. Yasomanee, J. P.; Demchenko, A. V. The Effect of Remote Picolinyl and Picoloyl Substituents on the Stereoselectivity of Chemical Glycosylation. *J. Am. Chem. Soc.* **2012**, *134*, 20097–20102.
140. Escopy, S.; Geringer, S.; De Meo, C. Combined Effect of the Picoloyl Protecting Group and Triflic Acid in Sialylation. *Org. Lett.* **2017**, *19*, 2638–2641.
141. Demchenko, A. V.; Boons, G. J. A Novel and Versatile Glycosyl Donor for the Preparation of Glycosides of *N*-Acetylneuraminic Acid. *Tetrahedron Lett.* **1998**, *39*, 3065–3068.
142. Gong, J.; Liu, H.; Nicholls, J. M.; Li, X. Studies on the Sialylation of Galactoses With Different C-5 Modified Sialyl Donors. *Carbohydr. Res.* **2012**, *361*, 91–99.
143. Kononov, L. O.; Malysheva, N.; Kononova, E.; Orlova, A. V. Intermolecular Hydrogen-Bonding Pattern of a Glycosyl Donor: The Key to Understanding the Outcome of Sialylation. *Eur. J. Org. Chem.* **2008**, 3251–3255.
144. Demchenko, A. V.; Boons, G. J. A Novel Direct Glycosylation Approach for the Synthesis of Dimers of *N*-Acetylneuraminic Acid. *Chem. Eur. J.* **1999**, *5*, 1278–1283.
145. Crich, D.; Wenju, L. Efficient Glycosidation of a Phenyl Thiosialoside Donor With a Diphenyl Sulfoxide and Triflic Anhydride in Dichloromethane. *Org. Lett.* **2006**, *8*, 959–962.
146. Shpirt, A. M.; Kononov, L. O.; Maltsev, S. D.; Shibaev, V. N. Chemical Synthesis of Polyprenyl Sialyl Phosphate, a Probable Biosynthetic Intermediate of Bacterial Polysialic Acid. *Carbohydr. Res.* **2011**, *346*, 2849–2854.
147. Schreiner, E.; Zbiral, E. Structural Variations on *N*-Acetylneuraminic Acid, 16. A Convenient Approach to 3-Deoxy-D-*glycero*-D-*galacto*-nonulosonic Acid (KDN), 5-Azido-5-deoxy-KDN and 5-Deoxy-KDN, and Their 4-Methylumbelliferyl 2α-Glycosides. *Liebigs Ann. Chem.* **1990**, 581–586.

148. Schneider, R.; Freyhardt, C. C.; Schmidt, R. R. 5-Azido Derivatives of Neuraminic Acid—Synthesis and Structure. *Eur. J. Org. Chem.* **2001**, 1655–1661.
149. Voight, E. A.; Rein, C.; Burke, S. D. Synthesis of Sialic Acids via Desymmetrization by Ring-Closing Metathesis. *J. Org. Chem.* **2002**, *67*, 8489–8499.
150. Mukaiyama, T.; Mandai, H.; Jona, H. A Catalytic and α-Selective Sialylation Using Novel 5-Azido Sialyl Fluoride. *Chem. Lett.* **2002**, (12), 1182–1183.
151. Lu, K.-C.; Tseng, S.-Y.; Lin, C.-C. 5-Azido Neuraminic Acid Thioglycoside as Sialylation Donor. *Carbohydr. Res.* **2002**, *337*, 755–760.
152. Uchinashi, Y.; Tanaka, K.; Manabe, Y.; Fujimoto, Y.; Fukase, K. Practical and Efficient Method for α-Sialylation With an Azide Sialyl Donor Using a Microreactor. *J. Carbohydr. Chem.* **2014**, *33*, 55–67.
153. De Meo, C.; Demchenko, A. V.; Boons, G. J. A Stereoselective Approach for the Synthesis of α-Sialosides. *J. Org. Chem.* **2001**, *66*, 5490–5497.
154. Meijer, A.; Ellervik, U. Interhalogens (ICl/IBr) and AgOTf in Thioglycoside Activation Synthesis of Bislactam Analogues of Ganglioside GD3. *J. Org. Chem.* **2004**, *69*, 6249–6256.
155. De Meo, C.; Demchenko, A. V.; Boons, G. J. Trifluoroacetamido Substituted Sialyl Donors for the Preparation of Sialyl Galactosides. *Aust. J. Chem.* **2002**, *55*, 131–134.
156. Khatuntseva, E. A.; Tsvetkov, Y. E.; Stepanenko, R. N.; Vlasenko, R. Y.; Petrov, R. V.; Nifantiev, N. E. Synthesis of 2-Aminoethyl 3′-(*N*-Glycolylneuraminyl)-β-lactoside and Neoglycoconjugates Based Thereon—A Prototype of a Cancer Vaccine and Artificial Antigens for the Control of Immune Response. *Russ. Chem. Bull.* **2010**, *58*, 450–456.
157. Lin, C.-C.; Huang, K. T.; Lin, C.-C. *N*-Trifluoroacetyl Sialyl Phosphite Donors for the Synthesis of α(2-9)-Oligosialic Acid. *Org. Lett.* **2005**, 7, 4169–4172.
158. Tanaka, H.; Ando, H.; Ishida, H.; Kiso, M.; Ishihara, H.; Koketsu, M. Synthetic Study on α(2→8)-Linked Oligosialic Acid Employing 1,5-Lactamization as a Key Step. *Tetrahedron Lett.* **2009**, *50*, 4478–4481.
159. Ando, H.; Koike, Y.; Koizumi, S.; Ishida, H.; Kiso, M. 1,5-Lactamized Sialyl Acceptor for Various Disialoside Syntheses: Novel Method for the Synthesis of Glycan Portions of Hp-s6 and HLG-2 Ganglosides. *Angew. Chem., Int. Ed.* **2005**, *44*, 6759–6763.
160. Fujita, S.; Numata, M.; Sugimoto, M.; Tomita, K.; Ogawa, T. Total Synthesis of the Modified Gangloside de-*N*-Acetyl-GM3 and Some Analogs. *Carbohydr. Res.* **1992**, *228*, 347–370.
161. Tanaka, H.; Adachi, M.; Takahashi, T. One-Pot Synthesis of Sialo-Containing Glycosyl Amino Acids by Use of an *N*-Trichloroethoxycarbonyl-β-thiophenyl Sialoside. *Chem. Eur. J.* **2005**, *11*, 849–862.
162. Sparks, M. A.; Williams, K. W.; Lukacs, C.; Schrell, A.; Priebe, G.; Spaltenstein, A.; Whitesides, G. M. Synthesis of Potential Inhibitors of Hemagglutination by Influenza-Virus-Chemoenzymic Preparation of N-5 Analogs of *N*-Acetylneuraminic Acid. *Tetrahedron* **1993**, *49*, 1–12.
163. Ren, C. T.; Chen, C. S.; Wu, S. H. Synthesis of a Sialic Acid Dimer Derivative, 2'α-*O*-Benzyl Neu5Ac-α-(2→5)Neu5Gc. *J. Org. Chem.* **2002**, *67*, 1376–1379.
164. Ando, H.; Koike, Y.; Ishida, H.; Kiso, M. Extending the Possibility of an *N*-Troc-Protected Sialic Acid Donor Toward Variant Sialo-glycoside Synthesis. *Tetrahedron Lett.* **2003**, *44*, 6883–6886.
165. Tamai, H.; Ando, H.; Ishida, H.; Kiso, M. First Synthesis of a Pentasaccharide Moiety of Ganglioside GAA-7 Containing Unusually Modified Sialic Acids Through the Use of *N*-Troc-Sialic Acid Derivative as a Key Unit. *Org. Lett.* **2012**, *14*, 6342–6345.

166. Fujita, S.; Numata, M.; Sugimoto, M.; Tomita, K.; Ogawa, T. Total Synthesis of a Modified Ganglioside, De-*N*-acetyl GM_2. *Carbohydr. Res.* **1994**, *263*, 181–196.
167. Mandhapati, A. R.; Rajender, S.; Shaw, J.; Crich, D. The Isothiocyanato Moiety: An Ideal Protecting Group for the Stereoselective Synthesis of Sialic Acid Glycosides and Subsequent Diversification. *Angew. Chem., Int. Ed.* **2015**, *54*, 1275–1278.
168. Tanaka, H.; Nishiura, Y.; Takahashi, T. Stereoselective Synthesis of Oligo-α-(2-8)-sialic Acids. *J. Am. Chem. Soc.* **2006**, *128*, 7124–7125.
169. Farris, M. D.; De Meo, C. Application of 4,5-*O*,*N*-Oxazolidinone Protected Thiophenyl Sialosyl Donor to the Synthesis of α-Sialosides. *Tetrahedron Lett.* **2007**, *48*, 1225–1227.
170. Tanaka, H.; Nishiura, Y.; Takahashi, T. Stereoselective Synthesis of α(2,9) Di- to Tetrasialic Acids, Using a 5,4-*N*,*O*-Carbonyl Protected Thiosialoside. *J. Org. Chem.* **2009**, *74*, 4383–4386.
171. De Meo, C.; Farris, M.; Ginder, N.; Gulley, B.; Priyadarshani, U.; Woods, M. Solvent Effect in the Synthesis of Sialosyl α(2-6) Galactosides: Is Acetonitrile the Only Choice? *Eur. J. Org. Chem.* **2008**, 3673–3677.
172. Crich, D.; Cai, F.; Yang, F. A Stable, Commercially Available Sulfenyl Chloride for the Activation of Thioglycosides in Conjunction With Silver Trifluoromethanesulfonate. *Carbohydr. Res.* **2008**, *343*, 1858–1862.
173. Crich, D.; Wu, B. Stereoselective Iterative One-Pot Synthesis of *N*-Glycolylneuraminic Acid-Containing Oligosaccharides. *Org. Lett.* **2008**, *10*, 4033–4035.
174. Liang, F.; Chen, L.; Xing, G.-W. α-Selective Sialylation With *N*-Acetyl-5-*N*,4-*O*-oxazolidinone-Protected *p*-Toluenethio-sialoside. *Synlett* **2009**, 425–428.
175. Xing, G.-W.; Chen, L.; Liang, F.-f. Facile Synthesis of Tumor-Associated Carbohydrate Antigen Ganglioside GM3 From Sialic Acid, Lactose, and Serine. *Eur. J. Org. Chem.* **2009**, 5963–5970.
176. Wang, Y.-J.; Jia, J.; Gu, Z.-Y.; Liang, F.-F.; Li, R.-C.; Huang, M.-H.; Xu, C.-S.; Zhang, J.-X.; Men, Y.; Xing, G.-W. Tunable Stereoselectivity During Sialylation Using an *N*-Acetyl-5-*N*,4-*O*-oxazolidinone-Protected *p*-Toluene 2-Thio-sialoside Donor With Tf_2O/Ph_2SO/TTBPy. *Carbohydr. Res.* **2011**, *346*, 1271–1276.
177. Gu, Z.-y.; Zhang, J.-x.; Xing, G.-w. *N*-Acetyl-5-*N*,4-*O*-oxazolidinone-Protected Sialyl Sulfoxide: An α-Selective Sialyl Donor With $Tf_2O/(Tol)_2SO$ in Dichloromethane. *Chem. Asian J.* **2012**, 7, 1524–1528.
178. Kononov, L. O.; Komarova, B. S.; Nifantiev, N. E. Selective Preparation of Glycosyl Sulfone or Glycal by Treatment of Phenyl Thioglycoside of *N*-Acetylneuraminic Acid With *m*-Chloroperbenzoic Acid. *Russ. Chem. Bull.* **2002**, *51*, 698–702.
179. Gu, Z.-y.; Zhang, X.-t.; Zhang, J.-x.; Xing, G.-w. Highly Efficient α-C-Sialylation Promoted by $(p\text{-Tol})_2SO/Tf_2O$ With *N*-Acetyl-5-*N*,4-*O*-oxazolidione-Protected Thiosialoside as Donor. *Org. Biomol. Chem.* **2013**, *11*, 5017–5022.
180. Sun, B.; Jiang, H. Pre-Activation Based, Highly α-Selective O-Sialylation With *N*-Acetyl-5-*N*,4-*O*-carbonyl-Protected *p*-Tolyl Thiosialoside Donor. *Tetrahedron Lett.* **2011**, *52*, 6035–6038.
181. Yu, Y.; Xiong, D.-C.; Mao, R.-Z.; Ye, X.-S. Visible Light Photoredox-Catalyzed O-Sialylation Using Thiosialoside Donors. *J. Org. Chem.* **2016**, *81*, 7134–7138.
182. Lin, C.-C.; Lin, N.-P.; Sahabuddin, L. S.; Reddy, V. R.; Huang, L.-D.; Hwang, K. C.; Lin, C.-C. 5-*N*,4-*O*-Carbonyl-7,8,9-tri-*O*-chloroacetyl-Protected Sialyl Donor for the Stereoselective Synthesis of α-(2→9)-Tetra-sialic Acid. *J. Org. Chem.* **2010**, *75*, 4921–4928.
183. Corcilius, L.; Payne, R. J. Stereoselective Synthesis of Sialylated Tumor-Associated Glycosylamino Acids. *Org. Lett.* **2013**, *15*, 5794–5797.

184. Chuang, H.-Y.; Ren, C.-T.; Chao, C.-A.; Wu, C.-Y.; Shivatare, S. S.; Cheng, T.-J. R.; Wong, C.-H. Synthesis and Vaccine Evaluation of the Tumor-Associated Carbohydrate Antigen RM2 From Prostate Cancer. *J. Am. Chem. Soc.* **2013**, *135*, 11140–11150.
185. Tanaka, H.; Nishiura, Y.; Takahashi, T. An Efficient Convergent Synthesis of GP1c Ganglioside Epitope. *J. Am. Chem. Soc.* **2008**, *130*, 17244–17245.
186. Boltje, T. J.; Heise, T.; Rutjes, F. P. J. T.; van Delft, F. L. A Divergent Method to Prepare 5-Amino-, 5-*N*-Acetamido-, and 5-*N*-Glycolylsialosides. *Eur. J. Org. Chem.* **2013**, 5257–5261.
187. Buda, S.; Crich, D. Oxidative Deamination of *N*-Acetyl Neuraminic Acid: Substituent Effects and Mechanism. *J. Am. Chem. Soc.* **2016**, *138*, 1084–1092.
188. Crich, D.; Wu, B. Imposing the Trans/Gauche Conformation on a Sialic Acid Donor With a 5-*N*,7-*O*-Oxazinanone Group: Effect on Glycosylation Stereoselectivity. *Tetrahedron* **2008**, *64*, 2042–2047.
189. Tanaka, H.; Ando, H.; Ishihara, H.; Koketsu, M. Sialylation Reactions With 5-*N*,7-*O*-Carbonyl-Protected Sialyl Donors: Unusual Stereoselectivity With Nitrile Solvent Assistance. *Carbohydr. Res.* **2008**, *343*, 1585–1593.
190. Rajender, S.; Crich, D. Exploration of the Oxazolidinthione Protecting System for the Synthesis of Sialic Acid Glycosides. *J. Carbohydr. Chem.* **2013**, *32*, 324–335.
191. Hanashima, S.; Yamaguchi, Y.; Ito, Y.; Sato, K.-i. Synthesis of the Starfish Ganglioside AG2 Pentasaccharide. *Tetrahedron Lett.* **2009**, *50*, 6150–6153.
192. Wang, Y.; Ye, X.-S. Sialylation Reactions With *N*,*N*-Acetyl, Benzoyl *O*-Perbenzoyl-Protected Sialyl Donor. *Tetrahedron Lett.* **2009**, *50*, 3823–3826.
193. Liao, G.; Zhou, Z.; Guo, Z. Synthesis and Immunological Study of α-2,9-Oligosialic Acid Conjugates as Anti-group C Meningitis Vaccines. *Chem. Commun. (Cambridge)* **2015**, *51*, 9647–9650.
194. Podvalnyy, N. M.; Malysheva, N. N.; Panova, M. V.; Zinin, A. I.; Chizhov, A. O.; Orlova, A. V.; Kononov, L. O. Stereoselective Sialylation With O-Trifluoroacetylated Thiosialosides: Hydrogen Bonding Involved? *Carbohydr. Res.* **2017**, *451*, 12–28.
195. Podvalnyy, N. M.; Zinin, A. I.; Malysheva, N. N.; Kunetskiy, R. A.; Kononov, L. O. Convenient Synthesis and Purification of N-Trifluoroacetylated Neuraminic Acid Tetraol: A Key Precursor for the Synthesis of New Sialyl Donors Containing Labile *O*-Acyl Protecting Groups. *J. Carbohydr. Chem.* **2014**, *33*, 213–224.
196. Wu, Y.-F.; Tsai, Y.-F. Assistance of the C-7,8-Picoloyl Moiety for Directing the Glycosyl Acceptors Into the α-Orientation for the Glycosylation of Sialyl Donors. *Org. Lett.* **2017**, *19*, 4171–4174.
197. Shadrick, M.; Yu, C.; Geringer, S.; Ritter, S.; Behm, A.; Cox, A.; Lohman, M.; De Meo, C. Facile and Robust Methods for the Regioselective Acylation of *N*-Acetylneuraminic Acid. *New J. Chem.* **2018**, *42*, 14138–14141.
198. Liao, L.; Auzanneau, F. I. The Amide Group in *N*-Acetylglucosamine Glycosyl Acceptors Affects Glycosylation Outcome. *J. Org. Chem.* **2005**, *70*, 6265–6273.
199. Cheng, A.; Hendel, J. L.; Colangelo, K.; Bonin, M.; Auzanneau, F.-I. Convenient Temporary Methyl Imidate Protection of *N*-Acetylglucosamine and Glycosylation at O-4. *J. Org. Chem.* **2008**, *73*, 7574–7579.
200. Kononov, L.; Malysheva, N.; Kononova, E.; Garkusha, O. The First Example of Synergism in Glycosylation. Possible Reasons and Consequences. *Russ. Chem. Bull.* **2006**, *55*, 1311–1313.
201. Kancharla, P. K.; Navuluri, C.; Crich, D. Dissecting the Influence of Oxazolidinones and Cyclic Carbonates in Sialic Acid Chemistry. *Angew. Chem., Int. Ed.* **2012**, *124*, 11105–11109.

202. Kancharla, P. K.; Kato, T.; Crich, D. Probing the Influence of Protecting Groups on the Anomeric Equilibrium in Sialic Acid Glycosides With the Persistent Radical Effect. *J. Am. Chem. Soc.* **2014**, *136*, 5472–5480.
203. Kancharla, P. K.; Crich, D. Influence of Side Chain Conformation and Configuration on Glycosyl Donor Reactivity and Selectivity as Illustrated by Sialic Acid Donors Epimeric at the 7-Position. *J. Am. Chem. Soc.* **2013**, *135*, 18999–19007.

AUTHOR INDEX

Note: Page numbers followed by "*f*" indicate figures, "*t*" indicate tables, and "*s*" indicate schemes.

A

B

C

D

E

F

G

H

K

L

Q

R

S

U

V

W

X

Y

Z

SUBJECT INDEX

Note: Page numbers followed by "*f*" indicate figures, "*t*" indicate tables, and "*s*" indicate schemes.

P

R

S

Kolibri 360 GmbH
Ferdinand-Jühlke-Straße 7
99095 Erfurt, Deutschland
produktsicherheit@kolibri360.de